Soil Emission of Nitrous Oxide and its Mitigation

David Ussiri • Rattan Lal

Soil Emission of Nitrous Oxide and its Mitigation

Springer

David Ussiri
Illinois State Geological Survey
Prairie Research Institute
University of Illinois
at Urbana-Champaign
Champaign, Illinois, USA

Rattan Lal
Carbon Management and Sequestration Center
School of Environment and Natural
Resources
The Ohio State University
Columbus, Ohio, USA

ISBN 978-94-007-5363-1 ISBN 978-94-007-5364-8 (eBook)
DOI 10.1007/978-94-007-5364-8
Springer Dordrecht Heidelberg New York London

Library of Congress Control Number: 2012950990

Printed on acid-free paper

Springer is part of Springer Science+Business Media (www.springer.com)

Foreword

Humans are now the prime drivers of global change, which includes climate change. Atmospheric levels of heat-trapping greenhouse gases (GHGs) have increased, and the effects of human-induced climate change are now apparent. Global average temperature is rising, glaciers are melting worldwide, rainstorms are intensifying in many areas, and summer heat waves have become more intense while winters are less severe. Whether the governments and decision makers will act sufficiently fast to stem the global warming is uncertain because global warming presents unique challenges and the crucial research information is not readily available.

Nitrogen is essential for all life forms on Earth. Nature and its biodiversity can only exist because of the availability of reactive nitrogen in the system. Mankind has developed the technology to produce synthetic reactive nitrogen, which is used as a fertilizer to increase food production. Food production for about half of the current global human population of seven billion is being achieved because of the availability of synthetic nitrogenous fertilizers. Moreover, mankind also creates reactive nitrogen indirectly as a by-product from the transportation and energy sectors. Once reactive nitrogen molecule has been created, it is highly mobile and remains in the environment for a considerable time while contributing to several undesirable effects. Yet, anthropogenic activities have more than doubled the availability of reactive nitrogen in the biosphere, primarily through agricultural activities. Increasing nitrogen availability is leading to several unintended environmental consequences, including enhanced emissions of nitrous oxide (N_2O). It is a long-lived radiatively active greenhouse gas (GHG) with an atmospheric lifetime of approximately 120 years, and heat trapping effects about 310 times more powerful than that of carbon dioxide (CO_2) on molecular basis. Despite being a potent GHG, it also plays a significant role in the depletion of stratospheric ozone. The volume entitled *Soil Emission of Nitrous Oxide and Its Mitigation* by David Ussiri and Rattan Lal comprises a clear and concise analysis of the global budget of N_2O and the factors controlling its emission. It also describes the anthropogenic sources of N_2O with major emphasis on agricultural activities. The volume includes an extensive synthesis of mitigation techniques to reduce emissions from agricultural soils and diverse sources of nitrogenous fertilizers. Although significant research information about N_2O emissions exist, this information has remained compartmentalized across a wide range of disciplines. This book fulfills the need for comprehensive multidisciplinary synthesis of the up-to-date scientific knowledge on N_2O and its pathways through nature. The volume is of interest to graduate students pursuing their career in soils and environmental sciences, academicians, policy makers and land managers. The comprehensive synthesis is timely and of relevance to an issue of global significance.

School of Environment and Natural Resources
The Ohio State University
210 Kottman Hall
2021 Coffey Rd
Columbus OH 43210–8522

Ronald L Hendrick
Ronald Hendrick, Director

Foreword

The human population reached seven billion in 2011, and continues to grow unabated. The burden of feeding an ever-increasing population requires greater food production from a decreasing per capita land area, and dwindling water and nutrient resources. Meeting this challenge has resulted in greater applications of nitrogen-containing fertilizers to increase total food production. Yet, there are the consequences of this unparalleled release of anthropogenic nitrogen into the environment. One consequence is an increase in atmospheric concentration of nitrous oxide (N_2O). Agriculture is the major source of N_2O emissions while the combustion of fossil fuels and industrial activities also have contributed to the increase. Carbon dioxide (CO_2) is the major greenhouse gas (GHG) resulting in global climate change. Despite having a larger capacity to absorb long-wave radiation, N_2O – one of the potent GHG – has received far less attention than CO_2.

Nitrogen has been the subject of more intense research than most elements in the periodic table because of its ever increasing role in agriculture. The abiotic and biotic chemical processes that influence the efficacy of nitrogen-containing fertilizers, and the environmental impacts of nitrogen-transformation by-products have been studied by soil chemists, geochemists, agronomists, agricultural engineers, oceanographers, limnologists and climate researchers. A wealth of information exits on N_2O and its formation pathways in nature. Unfortunately, this information is scattered over a wide range of disciplines with different perspectives and priorities. Thus, there is a strong need to collate and synthesize this information and make it readily available in one reference for researchers, academicians and land users. Comprehensive and up-to-date, *Soil Emission of Nitrous Oxide and Its Mitigation* by David Ussiri and Rattan Lal fills this gap admirably. The information presented in this book is crucial to any comprehensive plans for the mitigation of greenhouse gases, and developing any strategy of reducing anthropogenic global warming and global climate change.

Illinois State Geological Survey
Prairie Research Institute
University of Illinois at Urbana-Champaign
615 East Peabody Drive
Champaign, Illinois 61820

William R. Roy

Preface

Since the start of industrial age around 1750, significant increases have occurred in the atmospheric concentrations of several trace gases which have environmental impact at global and regional scales. Three of these gases namely carbon dioxide (CO_2), methane (CH_4), and nitrous oxide (N_2O) – the so-called greenhouse gases (GHGs) – contribute to global warming. The N_2O also causes the depletion of stratospheric ozone layer. Emission of GHGs to the atmosphere is of primary concern worldwide because of their impact on radiative forcing and the climate change. In addition to three primary GHGs listed above, some halocarbons and aerosols present in trace concentrations also impact the radiative forcing. Atmospheric concentrations of these gases have been increasing rapidly since 1950s as a result of human activities. Molecules of these gases trap outgoing long wave radiation energy in the lower atmosphere, which raises the Earth's surface temperature. In its fourth assessment report, the Intergovernmental Panel on Climate Change (IPCC 2007) reported a linear trend in global surface temperature between 1906 and 2005 with a warming of 0.74 °C (range 0.56–0.92 °C), with a more rapid warming trend over the past 50 years. This warming is primarily attributed to human-induced emission of these heat-trapping gases into the atmosphere.

Despite its low concentration in the atmosphere, N_2O is the fourth largest contributor to the greenhouse effect. Its contribution accounts for 7% of the current atmospheric radiative forcing. Therefore, this volume is focuses on emission of N_2O and its mitigation strategies. It begins with an overview of the global N cycle, sources of N input to soil, the general soil–atmosphere exchange of N_2O, as well as key soil processes moderating N_2O fluxes. The volume integrates information from different disciplines and presents a holistic approach to N_2O formation in terrestrial and aquatic ecosystems and emission to the atmosphere, its effects on radiative forcing, and possible sinks and mitigation options. The volume is specifically prepared to provide academic and research information for undergraduate and graduate students, scientists, researchers, land managers and policy makers interested in understanding the environmental impacts of N_2O emission, sources, sinks of N_2O, role of different land use and land management on N_2O fluxes and mitigation options. It places particular emphasis on the roles of anthropogenic activities on both direct and indirect increases in N_2O emissions.

The N_2O is produced in soils and sediments during N transformation by microbial processes of nitrification and denitrification as well as fossil fuel combustion and some industrial processes. Nitrification is the predominant process producing N_2O under warmer conditions. In contrast, denitrification is predominantly a process under wet and anaerobic conditions when ammonium (NH_4^+) and nitrate (NO_3^-) source is present in the soil and sediments. Historical data obtained from ice cores extending back 2,000 years indicate relatively stable N_2O concentration around 270 $\pm$ 7 ppbv in pre-industrial revolution, followed by steady rise during the past 200 years to 319 ppbv in 2005. In the past 20 years, the atmospheric levels of N_2O have been increasing almost linearly at 0.26% year^{-1}. This increase is generally attributed to the anthropogenic emissions. Human activities in the past century have accelerated release and removal of reactive N to the global atmosphere by as much as three to fivefold. Human activities that emit N_2O are those related to the transformation of fertilizer nitrogen into N_2O and its emission from agricultural soils, biomass and fossil fuel burning, animal husbandry, and industrial processes such as nylon manufacture and nitric acid production. In addition, N_2O is produced naturally from a wide range of biological sources in soils, water and sediments. Agricultural soils contribute about 65–70% of the total N_2O produced by the terrestrial ecosystems.

The N_2O is a long-lived radiatively active trace gas with atmospheric lifetime of 118 years. The only known main sink for atmospheric N_2O is stratospheric photo-reactions. Atmospheric N_2O is also one of the major sources of ozone depleting reactions in the stratosphere.

The forth assessment report of the intergovernmental Panel on Climate Change (IPCC 2007) estimated N_2O emissions from both natural (i.e., oceans, tropical soils, wet forests, dry savannas, temperate soils-both forest and grasslands) and anthropogenic sources (i.e., fertilized agricultural soils, biomass burning, industrial processes, animal feeds and manures, sewage processing, landfills, atmospheric inorganic N deposition) at 8.5–27.7 Tg N_2O-N year^{-1} (Tg $=$ teragram $=$ 10^{12} g $=$ million metric ton). Agriculture is the single largest anthropogenic source of N_2O fluxes contributing 1.7–4.8 Tg N_2O-N year^{-1}. This source is likely to be increased with the projected increase in use of nitrogenous fertilizers to meet the ever increasing demand for food of the growing and affluent world population. Natural sources of N_2O emission are estimated at 11.0 Tg N_2O-N year^{-1}. In addition to supply of N, temperature and moisture regimes are the principal controls of N_2O emission.

Collation, synthesis, and critical review of the available literature show that there are large uncertainties associated with the measured data on N_2O fluxes with the emphasis on spatial and temporal variability of measured fluxes. This uncertainty underscores the research needs to increase database by monitoring fluxes for long-term studies. In addition, there is a need for reliable flux data to improve model predictions, and for development of equipment for monitoring of fluxes for long-term plots under a wide range of land use and management systems. Also, there is a strong need for robust models incorporating larger emission scenarios and more emission factors to improve the quality of N_2O data. Yet, there is a strong need for development of nitrogenous fertilizers and the mode and method of their applications which enhance the nitrogen use efficiency while decreasing the losses of reactive nitrogen into the environment. While the Haber–Bosch process of manufacturing nitrogenous fertilizers have drastically enhanced crop yields and saved billions from starvation, the use efficiency of fertilizers must be improved while feeding the globe and mitigating global warming.

Illinois State Geological Survey David Ussiri
Prairie Research Institute
University of Illinois at Urbana-Champaign
Champaign, Illinois, USA

Carbon Management and Sequestration Center Rattan Lal
School of Environment and Natural Resources
The Ohio State University
Columbus, Ohio, USA

Contents

Chapter 1
The Role of Nitrous Oxide on Climate Change

Contents

Abstract Nitrous oxide is the third most important anthropogenic greenhouse gas in the atmosphere, contributing about 6% to the radiative forcing by long-lived greenhouse gases. In addition to its role as greenhouse gas, N_2O also play significant role in ozone (O_3) depletion. Its atmospheric abundance prior to industrialization was 270 parts per billion by volume (ppbv). Its current abundance is ~324 ppbv. Over past 10 years, the mean annual growth rate is ~0.8 ppbv year^{-1}. It is emitted into the atmosphere from natural and anthropogenic sources. The anthropogenic activities have increased the concentration of long-lived greenhouse gases dramatically since industrialization, and altered the composition of global atmosphere, with great implications for current and future climate. Once emitted, the N_2O is very stable in the stratosphere; the only known sink is the photo-oxidation tractions in the stratosphere. These reactions release oxides of N that contributes to ozone depletion. The magnitude of its current and projected future emissions leads to concerns about the timing of the recovery of ozone.

Keywords Greenhouse gases • Climate change • Radiative forcing • Global warming potential • Ozone depletion • Ozone depletion potential

David Ussiri and R. Lal, *Soil Emission of Nitrous Oxide and its Mitigation*,
DOI 10.1007/978-94-007-5364-8_1, © Springer Science+Business Media Dordrecht 2013

Abbreviations

UV	Ultraviolet GHGs greenhouse gases
IR	Infrared radiation
PFCs	Perfluorochlorocarbons
HFCs	Hydrofluorocarbons
GWP	Global warming potential
ODP	Ozone depleting potential
ODS	Ozone depleting substance
DU	Dobson unit
RF	Radiative forcing
IPCC	Intergovernmental Panel on Climate Change
FAR	First Assessment Report
RF	Radiative forcing
BP	Before present

1.1 Introduction

Nitrous oxide (N_2O) is a colorless gas of slightly sweet odor and taste under ambient conditions, which was discovered by Joseph Priestly in 1772 from the reduction of nitric oxide (NO) with iron or iron sulfur mixtures. Its presence in the atmosphere has been known since 1939 (Adel 1939). However, its importance to global environment was only realized in the early 1970s when atmospheric scientists hypothesized that N_2O released into the atmosphere through denitrification of nitrates in soils and waters triggers reactions in the stratosphere that may lead to destruction of the ozone layer (Crutzen 1970, 1972, 1974). Ozone layer protects the earth from biologically harmful ultraviolet (UV) radiations from the Sun. Later on, the investigations of its radiative properties led to classification as an important greenhouse gas (GHG) which could modify the radiation energy balance of the earth—atmosphere system (Wang et al. 1976; Ramanathan et al. 1985). These two characteristics, in combination with observation of its increasing concentration in the atmosphere (Prinn et al. 1990; MacFarling Meure et al. 2006; Forster et al. 2007) and long atmospheric lifetime (114 $\pm$ 10 years; Forster et al. 2007; Wuebbles 2009; Prather and Hsu 2010) makes N_2O an important factor in global climate system and atmospheric chemistry of the ozone. In view of its potential to influence the global climate, it is necessary to gain knowledge about its global budget, source, sinks, and emission mitigations.

Measurements of entrapped N_2O in polar ice indicate that the global concentration of N_2O is higher now than at any time during the past 45,000 years (Leuenberger and Siegenthaler 1992; Schilt et al. 2010). After the last ice age, N_2O concentration increased and then remained at ~270 parts per billion volume (ppbv = nmol mol^{-1}) for about 10,000 years until the nineteenth century. Globally, the average mixing ratio has increased from ~270 ppbv since pre-industrial times in 1750s to current concentrations of 324 ppbv and continues to grow, currently at a rate of 0.8 ± 0.2 nmol mol^{-1} $year^{-1}$ or 0.25 ± 0.05% $year^{-1}$ and by 0.25% during 2000–2010 (WMO 2011).

1.2 Greenhouse Effect

The radiation from the sun is the source of Earth's energy that drives global climate. The climate system can be described as weather generating heat engine driven by the solar radiation energy input and thermal radiation output (Peixoto and Oort 1992). About 99.9% of the electromagnetic radiation emitted by the sun is in wavelengths of 0.15–4.0 μm (i.e., 1 μm = 10^{-6} m). The balance between the solar energy that Earth receives from the Sun and that which it radiates out to the space is a major driver of the Earth's climate. Quantification of the amount of energy flow in and out of the Earth system and identification of the factors determining the balance between the incoming and outgoing energy helps in understanding the climate change. The Sun emits radiation over a spectrum of energies that exist in the form of waves; the radiation wavelength is the inverse of energy. On the high energy side of solar spectrum is UV and on the low energy side is infrared (IR) radiation. Nearly half of solar radiation is in the visible spectrum (i.e., ~0.4–0.7 μm, short wave electromagnetic spectrum) and the other part is mostly in near infrared and UV electromagnetic spectrum. The amount of energy reaching the top of Earth's atmosphere during daytime is estimated at 1,370 Watts (W) s^{-1} m^{-2}. Averaged over the entire planet, the Earth's atmosphere receives 342 W m^{-2}, predominantly as short wave electromagnetic radiation (Kiehl and Trenberth 1997; Le Treut et al. 2007; Fig. 1.1). About 31% (i.e., 107 W m^{-2}) of this solar energy is reflected back to space by clouds, aerosols, atmospheric gases, and the Earth's surface. The fraction of total energy reflected back to space can be referred to as *albedo*, and is influenced by surface conditions such as snow, ice, land versus sea, vegetation type, line of sight to the sun and other surface factors. It is also affected by atmospheric conditions, including cloud cover and wavelength of solar energy received (Pinty et al. 2008). Therefore, total albedo is highly variable from one place to another. The remaining energy (235 W m^{-2}) is mostly absorbed by the Earth's surface—land and ocean (168 W m^{-2}) and to a lesser extent by atmosphere (67 W m^{-2}). The energy absorbed by the vegetation layer drives the plant processes such as evapotranspiration, photosynthesis, and C assimilation, while the remaining fraction available in the underlying soils controls evaporation, snow melting, and other temperature–related processes (Sellers et al. 1997).

To maintain energy balance under stable climate, the Earth must radiate nearly the same amount of energy (about 235 W m^{-2}) back to space. If the Earth was to emit 235 W m^{-2}, temperature at the Earth surface will be around −19°C (255 K) based on the energy balance requirements, which is the temperature observed at the altitude of 5–6 km above the Earth's surface (Peixoto and Oort 1992). However, the

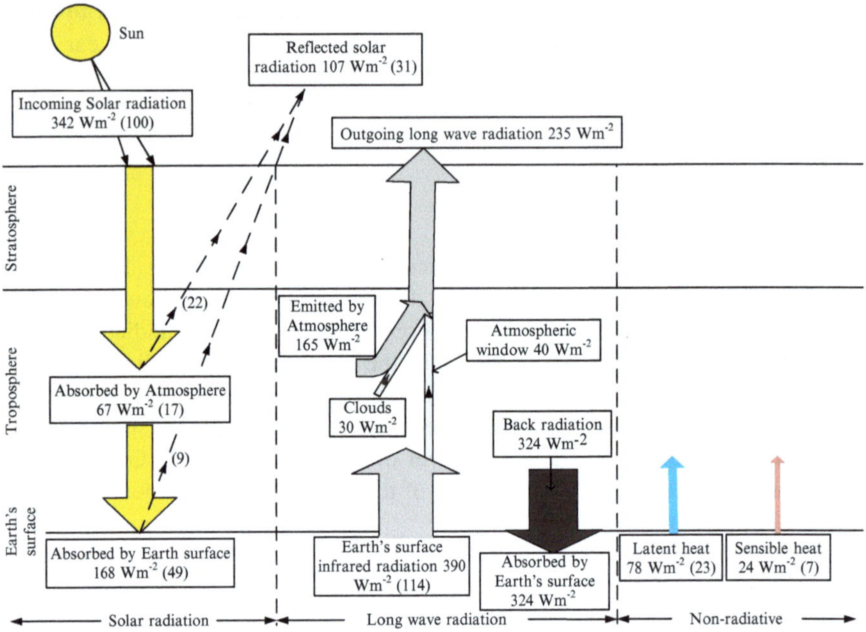

Fig. 1.1 The flow of energy from the Sun to Earth and between the Earth's surface and the atmosphere. *Numbers in brackets* are percent of solar radiation (Redrawn with modification from Kiehl and Trenberth 1997; Baede et al. 2001; Le Treut et al. 2007)

Earth being colder than the Sun, is warmed and it radiates about 390 W m^{-2} (or about 114% of solar radiation from the Sun) as long waves (Fig. 1.1), primarily IR (i.e., radiant heat, Le Treut et al. 2007). The surface IR of 390 W m^{-2} corresponds to blackbody emission at 14°C on Earth's surface, decreasing with altitude, and reaching mean temperature of −58°C at the top of the atmosphere (i.e., troposphere, the layer closest to the Earth) about 15 km above the Earth's surface. Much of this thermal radiation is absorbed in the atmosphere, and radiated back to the Earth (324 W m^{-2}), causing it to be warmer than as if the direct solar radiations were the only source of energy (Le Treut et al. 2007). As a result, the observed global temperature at the Earth surface is about 14°C, 33°C higher than expected, which is attributed to absorption and re-emission of this long wave radiation in the form of IR by the atmosphere, as a result, radiant heat is trapped near the Earth's surface and keeps it warm. This process is referred to as the natural greenhouse effect. Two other mechanisms which transfers heat from Earth's surface to the atmosphere are sensible heat (24 W m^{-2}), and latent heat (i.e., water vapor (78 W m^{-2})) which eventually release its heat through condensation in the atmosphere (Kiehl and Trenberth 1997; Le Treut et al. 2007).

The outgoing long wave radiation leaving the atmosphere is estimated at 235 W m^{-2}, representing 74% of the incoming solar radiation (Fig. 1.1). Some of the radiation leaving the atmosphere originates from the Earth's surface and transmitted relatively unimpeded through the atmosphere (40 W m^{-2}) in the areas where there are no cloud cover, and is present in part of the spectrum known as atmospheric window (8.0–12.0 μm). The atmosphere emits about 165 W m^{-2}, and clouds about 30 W m^{-2}. Thus, very little radiation is transmitted directly to the space as if the atmosphere were transparent.

Presence of greenhouse gases (GHGs), also known as trace gases, those which account for less than 1% of the total volume of dry air in the atmosphere plays important role in Earths' energy budget by absorbing and re-emitting IR radiation emitted by Earth surface, preventing it from escaping to the space. The most important gases causing the greenhouse effect are water vapor (H_2O), carbon dioxide (CO_2), methane (CH_4), nitrous oxide (N_2O) and ozone (O_3). As a result, most of the radiant heat flows back and forth between Earth's surface and atmosphere, and absorbed in the atmosphere to keep the Earth's surface warm (Le Treut et al. 2007). The exchange of energy between earth surface and atmosphere maintains global temperature near the surface at global average of 14°C, and decreases with increase in height above the earth surface. The GHGs occur naturally in the atmosphere and are responsible for natural greenhouse effect, which makes life possible, as we know it. The GHGs that are not naturally occurring include hydrofluorocarbons (HFCs), perfluorocarbons (PFCs; a class of organofluorine compounds that have all hydrogen (H) replaced with fluorine (F)), and sulfur hexafluoride (SF_6) generated from a variety of industrial processes. Without this natural greenhouse effect, the life on Earth would not be possible (Le Treut et al. 2007). The most abundant constituents of the atmosphere- nitrogen (78.1% volume), oxygen (20.9% volume), and argon (0.93% volume) have minimal interaction with the incoming solar radiation and no interaction with IR radiation emitted by the Earth. On the other hand, clouds exert greenhouse effect; however, this effect is offset by their reflectivity such that clouds tend to have a cooling effect on climate (Le Treut et al. 2007).

Atmospheric concentrations of GHGs have changed naturally throughout the Earth's history. However, anthropogenic activities in post-industrial era have resulted in enhanced emission of four major GHGs namely CO_2, CH_4, N_2O, and O_3. In addition, new, manmade GHGs gases generally termed as high-global warming potential gases: HFCs, PFCs, and SF_6 have been added to the atmosphere in the post-industrial era (Baede et al. 2001). These compounds are the most potent GHGs because they have very large heat-trapping capacity and some of them have extremely long atmospheric lifetimes (Table 1.1). Once emitted can remain in the atmosphere for centuries, making their accumulation almost irreversible. The atmospheric concentration of CO_2 has increased due to fossil fuel use in transportation and power generation, cement manufacturing, deforestation and accelerated processes of organic matter decomposition. The CH_4 has increased because of agricultural activities, natural gas distribution, and landfills. Wetlands also release CH_4 naturally. The N_2O has increased as a result of agricultural soil management and N fertilizer use, livestock waste management, mobile and stationary fossil fuel, combustion, and industrial processes. Soils and oceans also emit N_2O naturally.

Table 1.1 Global warming potentials of important anthropogenic greenhouse gases

Greenhouse gas	Lifetime (years)	Radiative efficiency ($W\ m^{-2}\ ppb^{-1}$)	Time horizon (years) 20	100	500
Carbon dioxide (CO_2)	1.2	1.4×10^{-5}	1	1	1
Methane (CH_4)	12	3.7×10^{-4}	62	23	7
Nitrous oxide (N_2O)	114	3.03×10^{-3}	275	298	153
Controlled by the Montreal Protocol					
CFC-11 (CCl_3F)	55	0.25	4,500	3,400	1,400
CFC-12 (CCl_2F_2)	116	0.32	7,100	7,100	4,100
CFC-13 ($CClF_3$)	640	0.25	10,800	14,400	16,400
Hydrofluorocarbons (HFCs)					
HFC-23 (CHF_3)	270	0.19	1,200	14,800	12,200
HFC-134a (CH_2FCF_3)	14	0.16	3,830	1430	435
HFC-152a (CH_3CHF_2)	1.4	0.09	237	124	38
Sulfur hexafluoride (SF_6)	3,200	0.52	16,300	22,800	32,600

(Source: Forster et al. 2007)

The HFCs are man-made chemicals developed as alternatives to ozone-depleting substances for industrial and consumer products, for example, HFC-134a, used in automobile air-conditioning and refrigeration. The PFCs are chemicals primarily produced from aluminum production and semi-conductor manufacture, while SF_6 is gas used for insulation and protection of current interruption in electric power transmission and distribution equipment. Other GHGs include ozone (O_3) continually produced and destroyed in the atmosphere by chemical reaction. Anthropogenic activities have increased ozone in the troposphere (i.e., the atmospheric layer closest to the Earth) through the release of gases such as CO, hydrocarbons, and NO which chemically reacts to produce ozone (Forster et al. 2007). Changes in atmospheric water vapor and O_3 are climate feedbacks due to indirect effect of anthropogenic activities. Because GHGs absorb IR radiation, changes in their atmospheric concentration alters the energy balance of the climate system. Increase in atmospheric GHGs concentrations produces net increase in absorption of energy of the Earth, leading to warming of Earth's surface.

The characteristic absorption of N_2O in the IR radiation range of the atmospheric window of the Earth makes it a potent GHG. Since 1980s, a scientific consensus has emerged that human activities through increasing the concentration of GHGs in the atmosphere have intensified the natural greenhouse effect and set in motion a global warming trend (Fig. 1.2; IPCC 2001, 2007; USEPA 2007). For example, CO_2 in the atmosphere has increased from about 280 ppm in pre-industrial era (1750) to the current 389 ppm (WMO 2011). Similarly, concentrations of CH_4 and N_2O have increased from 700 and 270 ppb in pre-industrial era to current levels of 1,808 and 324 ppbv, respectively (WMO 2011). This change represents an abundance of 139, 258, and 120 for CO_2, CH_4, and N_2O, respectively, relative to year 1750. The mean growth rate estimates of N_2O concentration in the atmosphere is at 0.8 ppb $year^{-1}$ over the past 10 years (WMO 2011). Therefore, anthropogenic activities have

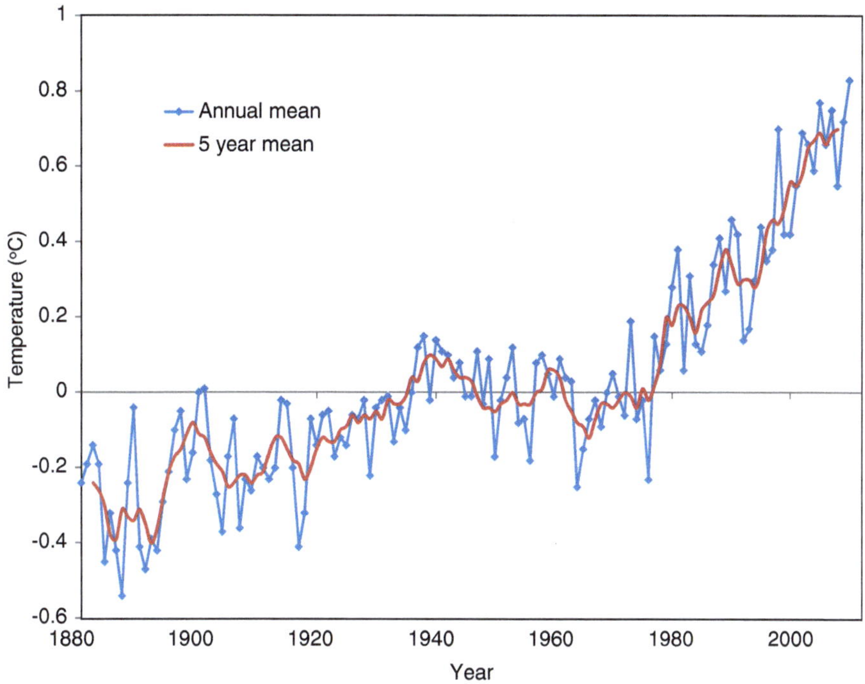

Fig. 1.2 Global surface temperature from 1880 to 2010 based on the National Aeronautics and Space Administration Goddard Institute for Space Sciences (Data source: http//data.giss.nasa.gov/gistemp/graphs)

dramatically altered the chemical composition of the global atmosphere with great implications for current and future climate. Atmospheric theory predicts that changes in the concentration of trace gases will have dramatic consequences for the habitability of the earth, which may include (i) food insecurity, (ii) loss of biodiversity and ecosystems change (iii) destruction of the stratospheric ozone layer due to increase in N_2O and halogenated compounds, (iv) increase in the amount of tropospheric ozone due to increased emissions of NO_x, CO, and hydrocarbons. With the current trends, the earth is likely to warm by 3–5°C for the next century (Le Treut et al. 2007). This is as much as it has warmed since last ice age. Such a warming would have adverse impacts on ecosystems because ecosystems will not be able to adjust to such a rapid temperature changes.

Although the atmospheric concentrations of CH_4 and N_2O are much lower than that of CO_2, they each make a disproportionate contribution to atmospheric anthropogenic greenhouse effect in relation to their concentrations in the atmosphere. Methane contributes some 15%, and N_2O about 6% of the effect, making them the second and third most important GHGs after CO_2. This is because CH_4 has a global warming potential 23 times and N_2O is 298 times that of CO_2 on 100-year timescale (Table 1.1).

1.3 Climate Forcing

Human beings are now the prime drivers of global changes, which include climate change, reduced water quality, loss of diversity, and degraded ecosystem services – i.e., the services provided by nature for free, including fertile soils, clean air, pollination, and water purification (Rockström et al. 2009). The anthropogenic drivers include growing population, and changing per capita consumption patterns that affect food, feed, fiber production, land use change, and energy use. Climate change may be due to natural internal processes or external forcings, or persistent anthropogenic changes in the composition of the atmosphere or in land use (Fig. 1.3). Perturbations to energy balance of the earth system drives the climate change. Factors that affect climate change can be separated into forcings and feedbacks. The concept of climate forcing was introduced in the First Scientific Assessment Report (FAR) of the Intergovernmental Panel on Climate Change (IPCC 1990), and in the original framework and later text climate forcing and radiative forcing have been used synonymously. It is an energy imbalance (or radiative imbalance) imposed on the climate system either externally or by human activities. Examples of forcings include change in volcanic emissions,

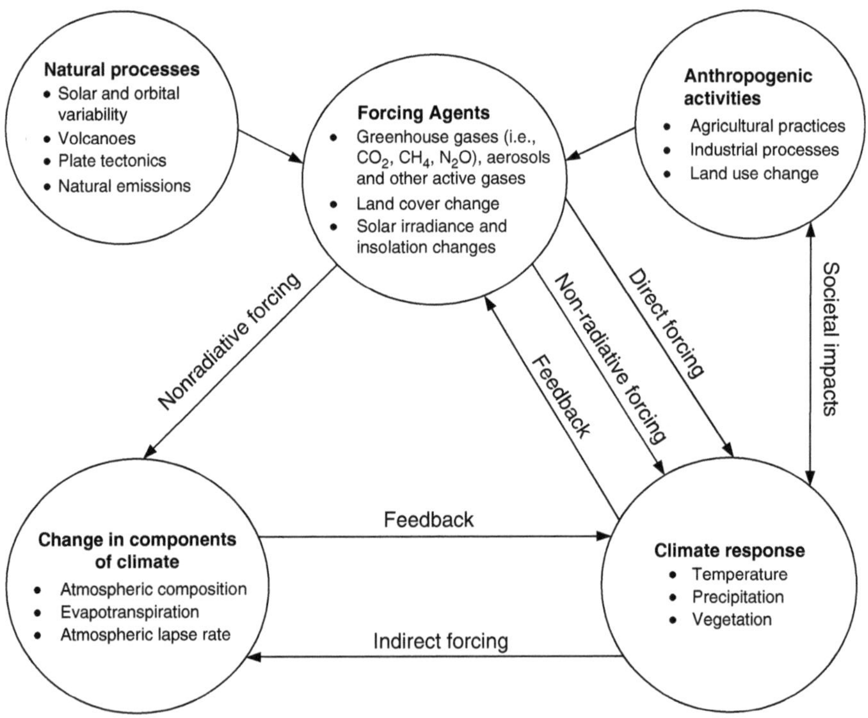

Fig. 1.3 Framework of radiative forcing, climate response and feedbacks of the climate change variables (Modified from NRC 2005; Liepert 2010)

land use modifications, changes in emissions of GHGs, aerosols and solar energy output (Fig. 1.3). The processes that amplify or dampen the initial forcing determine the relationship of radiative imbalance and climate response (Liepert 2010), and they are referred to as climate feedbacks. Climate feedback is the internal climate processes that amplifies or diminish the climate response to the forcing (NRC 2005). Positive feedback processes amplify response of the climate forcing. Examples of feedbacks include changes in atmospheric water vapor caused by increased warming associated with increase in GHGs concentration, which then amplify the warming through the greenhouse properties of water vapor. Warming causes ice to melt, revealing a darker land and water surfaces that absorbs more heat and causing more warming. Climate forcing can be radiative or nonradiative, and radiative forcing can be either direct or indirect. Direct radiative forcings influences the radiative budgets directly (e.g., added GHGs) which absorbs and emit IR radiations. Indirect radiative forcings causes radiative imbalances by first changing climate system components, which then lead to changes in radiative fluxes. One example is increase in evapotranspiration flux. Nonradiative forcings creates energy imbalance that have no direct effects on radiation (NRC 2005).

1.3.1 Radiative Forcing

Radiative forcing (RF) is a concept developed to provide a conceptual framework of understanding and quantifying Earths' energy budget modifications and their potential impacts on surface temperatures response (Hansen et al. 1997; Myhre et al. 2001). The effectiveness of climate forcings has practical importance because of the need to assess and compare the climate impacts of different changing atmospheric constituents. Strategies to reduce global warming can benefit from understanding of potential of different forcings in altering global temperature (Hansen et al. 2005). The RF quantifies the anthropogenic and natural influences on the climate system (Unger et al. 2010). The RF is a measure of how the energy balance of the Earth-Atmosphere system responds when factors that affect climate are altered. Factors affecting climate can be either natural or anthropogenic in origin. Natural factors include solar changes and volcanic eruption. In addition to cyclic changes in solar radiation that follow 10–11 year cycle, solar output has increased gradually in the industrial era, causing a small positive RF, warmer climate, and atmospheric abundance of stratospheric ozone (Forster et al. 2007). Volcanic eruptions can create a short-term (2–3 years) negative forcing through the temporary increase in sulfate aerosol in the stratosphere. Aerosol particles reflect and/or absorb solar and IR radiation in the atmosphere. Aerosols also change the cloud properties. Anthropogenic factors include increase in GHGs, increase in tropospheric ozone, decrease in stratospheric ozone, increase in aerosol particles in atmosphere, change in land cover and atmospheric contrails produced by aircrafts. The positive RF associated with increase in GHGs is the best understood of the anthropogenic RF, and they are positive because of absorption of IR

Table 1.2 Simplified formulas for calculating radiative forcing due to carbon dioxide, methane, nitrous oxide, and halocarbons

Greenhouse gas	Formula for radiative forcing, RF (W m^{-2})	Abundance of GHG in year 1750	Constant
Carbon dioxide (CO_2)	$RF = \alpha In\left(\frac{C}{C_0}\right)$	$C_0 = 278$ ppmv	$\alpha = 5.35$
Methane (CH_4)	$RF = \beta(\sqrt{M} - \sqrt{M_0}) - [f(M,N) - f(M_0N_0)]$	$M_0 = 700$ ppbv	$\beta = 0.036$
Nitrous oxide (N_2O)	$RF = \in (\sqrt{N} - \sqrt{N_0}) - [f(M_0N) - f(M_0N_0)]$	$N_0 = 270$ ppbv	$\epsilon = 0.12$
CFC-11	$RF = \lambda(X - X_0)$	$X_0 = 0$	$\lambda = 0.25$
CFC-12	$RF = \omega(X - X_0)$	$X_0 = 0$	$\omega = 0.32$

Modified from Ramaswamy et al. (2001)

$f(M,N) = 0.47In\left[1 + 2.01 \times 10^{-5}(MN)^{0.75} + 5.31 \times 10^{-15}M(MN)^{1.52}\right]$

C is CO_2 in ppm, M is CH_4 in ppbv, N is N_2O in ppbv and X is CFC in ppbv

radiation. Changes in land cover, principally through forest clearing for cropland has modified the reflective properties of land. Anthropogenic activities have also modified the reflective properties of ice and snow. The overall anthropogenic impact has increased the reflection of solar radiation from the earth. The larger the forcing is, the larger the disruption of steady global surface temperature. It is defined as the change in incoming energy minus outgoing energy at the top of the atmosphere in response to a factor that changes energy balance with reference to unperturbed values (i.e., net energy flux difference at the top of troposphere caused by the imposed change in the pollutant relative to unperturbed initial state) (Ramaswamy et al. 2001; Forster et al. 2007). Mathematically, RF can be expressed as (Ramaswamy et al. 2001):

$$\Delta T_s = \lambda RF \tag{1.1}$$

where λ is the climate sensitivity parameter, ΔT_s is the global mean temperature response, *RF* is radiative forcing. A number of radiative forcings of different origin and sign may exist together and to a first order, they can be added to yield a total forcing. Table 1.2 presents the simplified formulas for calculating RF for most common GHGs.

The year 1750 is a baseline for pre-industrial era benchmark for investigating changes in climate system associated with anthropogenic perturbations. The RF is quantified as a rate of energy change per unit area of a globe measured at the top of the atmosphere (troposphere). It is expressed in Watts per square meter (W m^{-2}). It is the term used to denote externally imposed perturbation in the radiative energy budget of the earths' climate system. In an equilibrium climate state, the average net radiation at the top of the atmosphere is zero. A change in either solar or infrared radiation changes the net radiation. The corresponding imbalance is radiative forcing. When RF from a factor or a group of factors is positive, the energy of the earth-atmosphere system will increase, leading to a warming. On the other hand, negative forcing leads to decrease in energy and cooling of the system (Serreze 2010).

Table 1.3 Concentrations and radiative forcing of the anthropogenic gases since industrialization to 2010

	2005	1750–2005[a]	2010	1750–2010[b]
Component	Concentration	Radiative forcing ($W\ m^{-2}$)	Concentration	Radiative forcing ($W\ m^{-2}$)
Carbon dioxide (CO_2)	377.1 ppmv	1.66 ± 0.17	390 ppmv	1.791
Methane (CH_4)	1,783 ppbv	0.48 ± 0.05	1,808 ppbv	0.504
Nitrous oxide (N_2O)	318.1 ppbv	0.16 ± 0.02	324 ppbv	0.175
CFC_{12} (CCl_2F_2)	541.7 pptv	0.173	530.1 pptv	0.170
CFC_{11} (CCl_3F)	250.9 pptv	0.063	240.2 pptv	0.060
15 minor long-lived halogenated gases[c]	–	0.092	–	0.105
Total	–	2.64	–	2.81

Notes: [a]Data from IPCC 2007

[b]Data extracted from Butler (2011).

[c]15 minor long-lived halogenated gases are: CFC-113 (CCl_3FCClF_2), CCl_4, CH_3CCl_3, HCFC-22 ($CHClF_2$), HCFC-141b (CH_3CCl_2F), HCFC-142b (CH_3CClF_2), HFC-134a (CH_2FCF_3), HFC-152a (CH_3CHF_3), HFC-23 (CHF_3), HFC-143a (CH_3CF_3), HFC-125 (CHF_2CF_3), SF_6, halon-1211 ($CBrClF_2$), halon-1301 ($CBrF_3$) and halon-2402 ($CBrF_2CBrF_2$)

Radiative forcing assesses and compares the anthropogenic and natural drivers of climate change. It is an important concept for prediction of surface temperature change, particularly for comparative studies of different forcings (Hansen et al. 1997). This concept originates from the studies of climatic response to changes in solar insolation and CO_2 using simple radiative convective models. It is applicable for assessment of climatic impact of long-lived GHGs (Ramaswamy et al. 2001). The importance of this concept is that it enables quantification and assessment of various factors that shift the energy balance relative to other factors. Radiative forcing is a simple measure for qualifying and ranking many different influences on climate change since it does not attempt to represent the overall climate response. It has advantage of being easier to calculate and to compare than estimates of the climate response, since climate response and sensitivity to external forcing are not easily quantifiable (Hansen et al. 2005; Colman and McAvaney 2011). The natural RF agents are solar changes and volcanoes. Anthropogenic activities contribute to RF through changing the amount of GHGs, aerosols and cloudiness. However, RF does not indicate how the climate will actually change due to particular forcing factor. Table 1.3 presents the RF of some of the common anthropogenic GHGs. The values reflect the forcing relative to start of industrial era in 1750s. The forcings of the greenhouse gases (e.g., CO_2, N_2O and CH_4) are the best understood of those due to human activities and are all positive because each of these gases absorbs outgoing IR radiation in the atmosphere. Among the anthropogenic long-lived GHGs, increases in CO_2 concentrations have caused the largest forcing during post-industrial era (Table 1.3). Change in the mixing ratio of N_2O from 270 ppbv in 1750 to 319 ppbv in 2005 results in RF of $+0.16 \pm 0.02\ Wm^{-2}$ (IPCC 2007;

Table 1.3). The RF of N_2O has increased by 11% since 1998. As CFC-12 levels slowly decline in response to regulation of ozone depleting gases (IPCC/TEAP 2005), N_2O with its current emission trends has taken over the third place in the ranking of the long-lived GHGs radiative forcings (Table 1.3). Increase in tropospheric ozone concentration also contributes to warming, while decrease in stratospheric ozone concentration contributes to cooling. Aerosol particles influence RF both through reflection and absorption of solar and IR in the atmosphere. Some aerosols can cause negative forcing, resulting into cooling, while others causes positive forcing contributing to warming. The overall effect of aerosols warming is negative (i.e., cooling), which is further intensified by changes they cause in cloud properties.

The RF is a useful tool for estimating the relative global impact of different climate change mechanisms (Ramaswamy et al. 2001); and is particularly useful in estimating the relative equilibrium of average temperature change due to different forcing agents. However, RF alone cannot assess the potential climate change associated with the emissions, since it does not take into account the lifetime of the forcing agents. With growing scientific insight, several attempts have been made since FAR to improve the RF concept by re-evaluating its definition (Forster et al. 2007) by including in it several other feedback processes. In addition, other simple metrics quantify the contribution of individual compounds to climate change. The Global Warming Potential (GWP) is the most established index that has been used in all past climate assessments (IPCC 1990, 2007; Ramaswamy et al. 2001).

1.3.2 Global Warming Potentials

Three characteristics of a gas determine its contribution to greenhouse effect, namely, its: (i) absorptivity for IR radiation, (ii) atmospheric lifetime (Table 1.1), and (iii) concentration in the atmosphere (CAST 2004). Gas absorbs IR radiation in specific energy bands influenced by its chemical properties. In addition, there is an overlap in the radiation-energy absorption bands for different GHGs. Therefore, the net effect of increasing the concentration of any single gas depends on inherent characteristics of the gas and its interaction with the energy absorption bands of other GHGs. The lifetime determines how long the emissions of GHGs into the atmosphere will contribute to global warming. The absorptivity and atmospheric lifetime characteristics of gas(es) is generally combined into a single index called Global Warming Potential (GWP). The concentrations of GHGs in the atmosphere play a significant role in influencing its GWP. If GHG is not present at a sufficiently high concentration, or does not have a potential to reach sufficiently high concentration, then it is not important even if its GWP is high. Only a few trace gases have the combination of the three characteristics to contribute significantly to global warming (Table 1.1; Forster et al. 2007).

When dealing with multiple GHGs, it is necessary to have a simple means of describing the relative abilities of emissions of each GHG to affect RF and therefore, the climate. One of the useful approaches is to express any estimates relative to the trace gas of primary concern, which is CO_2. The lifetime of some GHGs implies that their commitment to possible climate impacts for long periods of time (e.g., decades to centuries) and it necessitates the inclusion of potential in the formulation of the concept (Shine et al. 1990). Thus, GWP is a tool that compares the climate impact of a unit of gas emitted to the atmosphere in a pulse to the impact of a reference gas (i.e., CO_2). Therefore, a unit mass of CO_2 is assigned a GWP of 1. The definition of climate impact depends on the purpose. The impact can be described based on economic (e.g., cost versus benefit) or physically (as in global warming), hence GWP can be used in a variety of functions. The GWP was developed as a metric under Kyoto Protocol with which a multigas emission strategy of limiting anthropogenic climate change can be politically and internationally implemented. Calculation of the GWP of a GHG requires the knowledge of its radiative efficiency and global atmospheric lifetime. Radiative efficiency is defined as the change in net radiation at the tropospause by a given change in GHG concentration or mass. It has units of $\mathrm{W\,m^{-2}\,ppb^{-1}}$ or $\mathrm{W\,m^{-2}\,kg^{-1}}$ (Daniel et al. 2011). The conversion from mass per volume (i.e., $\mathrm{mg\,m^{-3}}$) to ppmv and vice versa is temperature dependent (usually 20–25°C). At an ambient pressure of 1 atm, the conversion can be done according to Eq. 1.2:

$$ppmv = \frac{\rho \times (273.15 + ^\circ \mathrm{C})}{(12.187 \times MW)} \tag{1.2}$$

where *ppmv* is the concentration of the gas expressed in parts per million by volume, ρ is the density of the gas (i.e., mass per unit volume, $\mathrm{mg\,m^{-3}}$), (273.15 + °C) is the temperature expressed in Kelvin, 12.187 is a constant of proportionality, and *MW* is the molar mass of the gas.

The absolute global warming potential (AGWP) for time horizon t' defined as (Daniel et al. 2011):

$$AGWP_x(t') = \int_0^{t'} F_x \cdot x(t)dt \tag{1.3}$$

where *Fx* is the radiative efficiency of species x, $x(t)$ is the decay with time of a unit pulse of compound x, and t' is the time horizon under consideration. The GWP compares the integrated effect of various GHGs on climate and expressed as (Forster et al. 2007; Ramaswamy et al. 2001; Daniel et al. 2011):

$$GWP_x(t) = \frac{AGWP_x(t')}{AGWP_{CO_2}(t')} = \frac{\int_0^{t'} F_x \exp\left(\frac{-t}{\tau_x}\right)dt}{\int_0^{t'} F_{CO_2} R(t)dt} \tag{1.4}$$

where F_{CO2} is the radiative efficiency of CO_2, $R(t)$ is the response function that describes the decay of an instantaneous pulse of CO_2, and the decay of the pulse of compound x has been rewritten assuming it obeys a simple exponential decay curve with a response time of τ_x. The underlying assumption is that the global warming is proportional to climate forcing and that the GHGs are independent of each other.

The pulse response terms lead to a dependence of GWPs on the integration time horizon; GHGs that decay more quickly (slowly) than the reference (CO_2) have GWPs that decrease (increase) with increase in the time horizon. The integration of the forcings over the time horizon of usually 20, 100 or 500 years is important for the political implementation (Ramaswamy et al. 2001; Forster et al. 2007). The choice of time horizon used in calculating GWPs is not determined purely by climate science considerations. Rather, the choice often depends on what information is useful to decision makers, based in part on the time horizon of the impacts and on the values they consider most important. In an effort to account for the impact of the choice of time horizon, typically three time horizons have been considered (20, 100, and 500 years) when reporting GWPs. The GWPs have also been developed to evaluate the effect of sustained emissions (Johnson and Derwent 1996; Berntsen et al. 2005). For short-lived gases, such as ozone, the GWP can be difficult to estimate. In addition, for aerosol forcings, the framework of GWP is not suitable because of (i) the importance of their regional distribution (ii) the short lifetime, and (iii) the impacts of anthropogenic aerosols on clouds and rainfall formation.

The GWP index has three major advantages over most other indices used to measure the contribution of GHGs to global warming: (i) transparency, (ii) simplicity, and (iii) widespread acceptance. However, it also has several disadvantages (IPCC 2009). There is also a growing recognition of the limitations of the GWP metric especially when the impacts of short- and long-lived pollutants need to be considered together (Johansson et al. 2008; Tanaka et al. 2009; Fuglestvedt et al. 2010). Thus, it has been argued whether it is the time to consider other metrics that might be even more useful (Shine 2009). For example, there has been interest in including the economics of emissions mitigation into a climate metric by applying the cost-benefit and cost-effective approaches (e.g., Manne and Richels 2001; Johansson et al. 2006). A range of alternatives has been presented to overcome some of the GWP limitations, but none has yet been widely accepted as a suitable replacement of the GWP.

1.4 Nitrous Oxide Emissions

Nitrous oxide is a highly stable, long-lived gas found in the atmosphere at approximately 1/1,000th the concentration of CO_2, but its contribution to RF is about 6.24% (Table 1.3; Forster et al. 2007). Natural N_2O emissions are estimated at 11.0 Tg N_2O-N year^{-1} (Denman et al. 2007). Soils under natural vegetation, oceans, and atmospheric chemistry are responsible for 6.6, 3.8 and 0.6 Tg N_2O-N year^{-1}, respectively (Ehhalt et al. 2001; Denman et al. 2007). Current anthropogenic emissions are estimated at 6.7 Tg N_2O-N year^{-1} (Denman et al. 2007). Agriculture is responsible for 2.8 Tg N_2O-N year^{-1} and emissions associated with fossil fuel combustion, industrial activities, biomass burning, atmospheric deposition, and human excreta accounts for 2.2 Tg N_2O-N year^{-1}. Emissions from rivers, estuaries, and costal zones attributed to anthropogenic activities contribute about 1.7 Tg N_2O-N year^{-1}. The only known sinks for the atmospheric N_2O is the photolytic reactions in the stratosphere, which is estimated to remove approximately 12.3 ± 2.5 Tg N_2O-N year^{-1} (Kroeze et al. 1999; Ehhalt et al. 2001). The atmospheric concentrations of N_2O are currently increasing at the rate of 0.8 ppbv year^{-1} (WMO 2011), which is equivalent to a net addition of 3.9 Tg N_2O year^{-1}.

1.4.1 Pre-industrial Era Nitrous Oxide Emissions

The atmosphere is considered as reservoir for N_2O, where this gas resides for a specific lifetime. Other reservoirs are soils and oceans. Nitrous oxide can be transferred from one reservoir to another through a process described as flux. Fluxes into atmospheric reservoir are known as sources, while fluxes out of atmospheric reservoir are sinks. Each reservoir has an overall budget, which represents the balance sheet of all sources and sinks. Because sources of N_2O to the atmosphere are essentially the processes that release this gas into the air, it is appropriate to use the term emission to describe the movement of N_2O into the atmosphere.

Changes in atmospheric composition before the contemporary observation began are best estimated by measurements of air occluded in polar ice, either in bubbles embedded in impermeable ice or channels in the overlying firn or the glacial layer (Flückiger et al. 2002; Stauffer et al. 2002; MacFarling Meure et al. 2006). The analysis of polar ice cores to determine the N_2O atmospheric concentrations provides information on its annual emissions and aids in identifying the role of anthropogenic activities on global N cycle and global N_2O distribution. All the known sources of N_2O are at the Earth's surface. The natural source is microbial processes of the terrestrial soils and oceans (Bouwman et al. 1993). In the pre-industrial era the annual N_2O emission and removal were nearly balanced (Forster et al. 2007), hence the atmospheric concentration remained nearly uniform (Fig. 1.4).

The pre-industrial source of N_2O is estimated at 11 (8–13) Tg N_2O-N year^{-1} (Ehhalt et al. 2001; Kroeze et al. 1999; Ruddiman 2003). The pre-industrial emissions are considered mainly natural, and increase in atmospheric N_2O may not have started before 1850s (Khalil and Rasmussen 1988; Kroeze et al. 1999), indicating that human impact on N_2O emissions was relatively small before the onset of industrial era. Pre-agricultural N_2O emission from soil was 6–7 Tg N_2O-N year^{-1} (Bouwman et al. 1993), deep oceans 3–4 Tg N_2O-N year^{-1} (Nevison et al. 1995), and other aquatic and atmospheric deposition sources contributed $<$1.0 Tg N_2O-N year^{-1} (Seitzinger and Kroeze 1998).

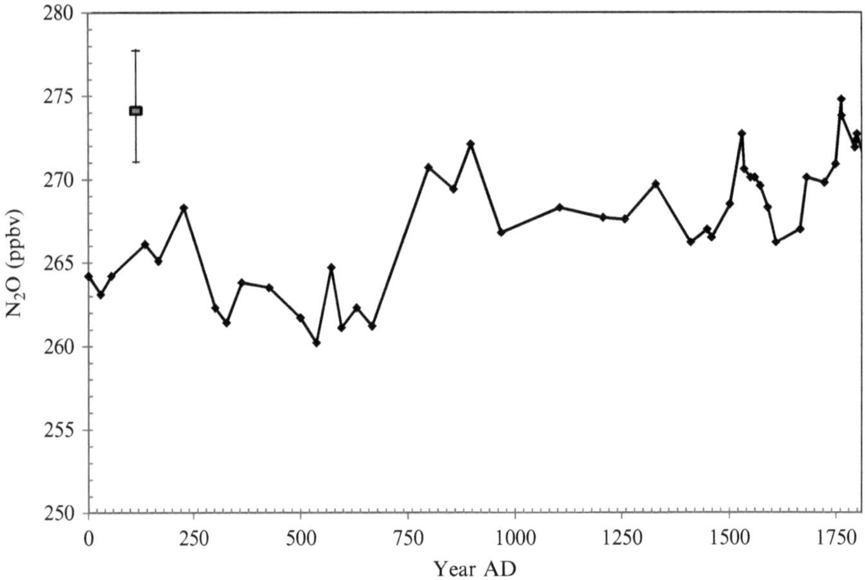

Fig. 1.4 Record of pre-industrial era (AD 0–1800) N_2O emissions as revealed by the Law Dome ice cores records (Adapted from Meure et al. 2006)

Khalil and Rasmussen (1988) analyzed ice cores from northern and southern polar regions to determine the N_2O concentration in pre-industrial and ancient atmospheres ranging from about 150 to 3,000 years B.P. and concluded that the pre-industrial and ancient N_2O concentration in the atmosphere remained at 285 ± 1 ppbv. Their data are consistent with the findings of other researchers, who also deduced similar concentrations from the analysis of air trapped in bubbles of polar ice (e.g., Flückiger et al. 2002; Stauffer et al. 2002; MacFarling Meure et al. 2006). These data show relatively little changes in mixing ratios over 1,800 years of pre-industrial atmospheric N_2O concentration, when human activities had not affected its global N_2O distribution. Data presented by Flückiger et al. (2002) have higher resolution than that of Khalil and Rasmussen (1988) and extends the known ice record to 11,000 years B.P. The data showed that N_2O concentration started to decrease slightly in the early Holocene and reached minimum values of about 260 ppbv at about 8,000 years B.P., before a slow increase to the pre-industrial concentrations of about 265 ppbv (Flückiger et al. 2002). Lower concentrations during Holocene may have been driven mostly by changes in emissions from terrestrial biosphere due to colder temperatures. The Holocene N_2O minimum concentration corresponds to a total decrease of 0.41 Tg N_2O-N year^{-1}, i.e., 3.7% of total pre-industrial global source of ~11.0 Tg N_2O-N year^{-1}, corresponding to concentration of 270 ppbv (IPCC 2001). Measurements of atmospheric N_2O concentrations in the ice from Law Dome, Antarctica showed that variations of up to 10 ppbv (or 14%) occurred between AD 0–1800 (MacFarling Meure et al. 2006; Fig. 1.4). Other researchers have estimated the pre-industrial N_2O concentrations of 270 ppbv

(Machida et al. 1995; Flückiger et al. 1999; IPCC 2001; Holland et al. 2005). About two thirds of the pre-industrial N_2O sources are attributed to soil, and one third to the oceans (IPCC 2001).

During the last glacial/interglacial transition, atmospheric N_2O values increased from 200 ppb during the last glacial period to 265 ppb near the beginning of the Holocene (Flückiger et al. 1999; Leuenberger and Siegenthaler 1992). The cause of this increase is difficult to establish due to complicated spatial and temporal scales involved in the global N budget. It is suggested that the observed increase during the last termination may have been the result of an increase in terrestrial N_2O emissions due to more equitable climate in the early Holocene. During the Younger Dryas event (about 11,000 years before present (BP)), the N_2O oscillation coincided with significant decline in atmospheric CH_4 (Flückiger et al. 1999). The cause of decline in CH_4 during this pre-historic period has been attributed to changes in terrestrial wetlands CH_4 emissions that have been shown to be tightly coupled to the global hydrologic cycle (Brook et al. 1996). Terrestrial N_2O emissions are also strongly influenced by soil type and moisture content (Bouwman et al. 1993). These factors also control CH_4 emissions, such that qualitatively, soil emissions of CH_4 and N_2O should respond to climate change in similar patterns. However, the magnitude of the change may be controlled by a range of other factors. The strong degree of covariance between N_2O and CH_4 records for the last deglacial period suggest that the primary cause of N_2O fluctuations could be the result of terrestrial N_2O emissions, even though changes in stratospheric photolysis or oceanic emissions cannot be ruled out (Sowers 2001).

1.4.2 Post-industrial Nitrous Oxide Emissions

Direct measurements of global atmospheric N_2O concentrations have been monitored since late 1970s (Weiss 1981; Prinn et al. 1990; IPCC 1990; Holland et al. 2005). Earlier records have shown steady rise since the onset of industrial era after a period of relative stability during pre-industrial era (i.e., before 1750s) at about 270 ppbv (Figs. 1.4 and 1.5; MacFarling Meure et al. 2006). The atmospheric N_2O concentrations were relatively constant prior to nineteenth century, and the decrease in N_2O emissions caused by conversion of forests to agricultural land was almost offset by increasing emission intensity from agricultural ecosystems (Kroeze et al. 1999). Industrialization in the later part of nineteenth century caused major changes in agriculture and urbanization. While economic growth induced the demand for food and shift towards meat and milk, and mechanization intensification in agriculture. It is clear that the transformation of agriculture in Europe and North America induced a fast intensification of the use of N from biological fixation and animal manure, which caused increased emissions of N_2O from crop and livestock production systems. The increase in atmospheric N_2O concentration began to occur in the mid-nineteenth century. The atmospheric concentration increased slowly but persistently at a rate of 0.1–0.2 ppbv year^{-1} during the period between 1850 and 1950, and then increased rapidly to ~0.7 ppbv year^{-1} around

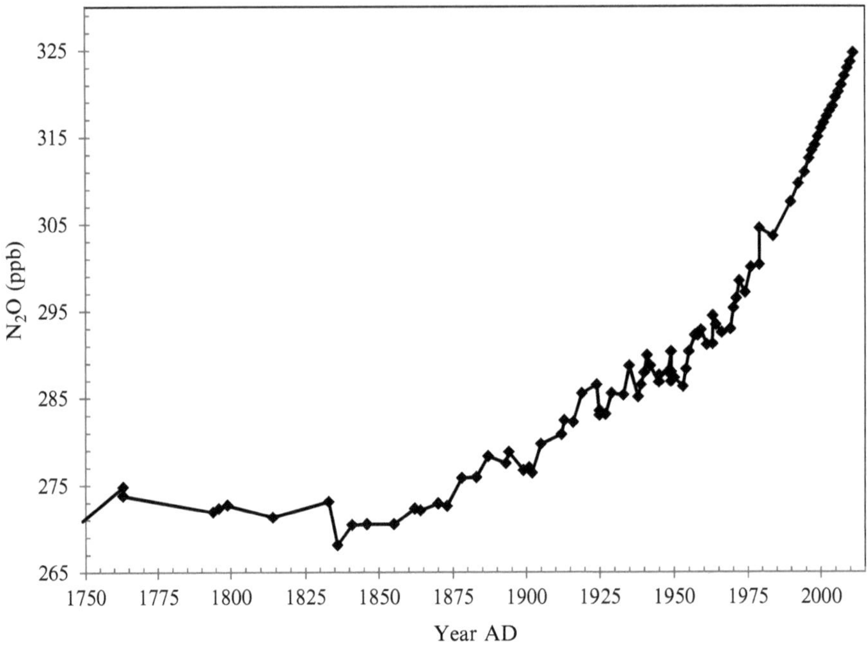

Fig. 1.5 Record of industrial era (AD 1750–1996) N_2O emissions as revealed by the Law Dome ice cores records and continuous monitoring (Adapted from Meure et al. 2006; World Meteorological Organization [http://www.wmo.int])

1990 (Machida et al. 1995; MacFarling Meure et al. 2006; IPCC 2007; Fig. 1.5). It is estimated that the N_2O emission from agricultural ecosystems increased from 0.6 to 1.6 Tg N_2O-N year^{-1} between 1700 and 1900, with most of increases occurring during nineteenth century (Kroeze et al. 1999). This increase is mainly due to expansion of land used for agriculture and intensification of N cycling in agriculture and livestock production. Global input of synthetic fertilizer N increased dramatically starting from 1930s', and this is responsible for dramatic increase in atmospheric N_2O concentration in the twentieth century. Global population also increased from about 1.6 billion in 1930 to about 6 billion in 1999 and 7 billion in 2011.

Anthropogenic activities, mainly intensification of agriculture facilitated by manufacture of synthetic N fertilizers and also increased fossil fuel burning and the land use change, and are the main causes of dramatic increase of N_2O emissions starting from 1950 (Fig. 1.5). The atmospheric abundance of N_2O was 314 ppb in 1998, corresponding to 16.4 Tg N_2O-N year^{-1} (Ehhalt et al. 2001). The global average mole fraction of N_2O in 2005 was 319 ppbv. The growth rate has been approximately constant at ~0.8 ppbv year^{-1} since 1980 with more than one-third of total emissions being anthropogenic, mainly agriculture. There has been a significant multiyear variance in observed growth of N_2O concentrations, but the reasons for these trends are not fully understood yet (IPCC 2001). The World

Meteorological Organization (WMO 2011) estimated the global N_2O abundance in 2010 at 323 ppbv. The mean decadal growth rate has been estimated at 0.8 ppbv. Anthropogenic sources accounts for ~40% of the total N_2O emissions (IPCC 2007; WMO 2011).

The annual source of N_2O from the Earth's surface has increased by almost 50% over the pre-industrial levels, mainly as result of human activity (Hirsch et al. 2006). Global emissions were estimated at 18.0 Tg N $year^{-1}$, of which, 11.0 Tg N $year^{-1}$ was estimated as natural sources in 2005 (Denman et al. 2007). Results from several studies that quantified N_2O emissions from coastal upwelling areas continental shelves, estuaries and rivers suggest that these coastal areas are contributing 0.3–6.6 Tg N_2O-N $year^{-1}$ or about 7–61% of the total oceanic N_2O emissions (Bange et al. 1996; Nevison et al. 2004; Kroeze et al. 2005). Agriculture contributes 1.7–4.8 Tg N_2O-N $year^{-1}$ or about 50% of the anthropogenic N_2O emissions (Denman et al. 2007).

1.4.3 Nitrous Oxide and Climate Change

Climate is the long-term mean seasonal reoccurrence of weather pattern mainly described by temperature, precipitation, and wind over a period ranging from months to thousands or millions of years (Le Treut et al. 2007; Liepert 2010). The classical period, as defined by the WMO is 30 years. The interactions of these variables with atmosphere, hydrosphere, biosphere, cryosphere, and the land surface are integral components of the climate system (Folland et al. 2001; Le Treut et al. 2007). Statistically significant variation either in the mean state of the climate or in its variability, persisting for an extended period (typically decades or longer) is called climate change (Folland et al. 2001; Le Treut et al. 2007). The most fundamental measure of Earth's climate is the global mean annual surface air temperature, which can be assessed with reasonable reliability back to about 1880 (Fig. 1.2). The global surface temperature varies considerably from year to year and exhibits decadal-scale variations, which is a reflection of natural climate variability (Le Treut et al. 2007; Serreze 2010). There is also a general upward trend in global surface temperature. For example, the rise in temperature from 1880 to 2008 amounts to about 0.7°C (Fig. 1.2) with pronounced reductions in mass of many of the world's glaciers, and sea-level rise. This increase is, in part, the manifestation of rising concentrations of atmospheric GHGs, mainly due to fossil fuel burning, increased use reactive N in agriculture and land use change.

The earth's climate has changed continuously over the entire lifetime of the planet due to natural causes. However, human activities in the recent past have produced effects powerful enough to overwhelm the natural mechanisms of climate adjustment and now dominate the changes of climate (IPCC 2007). Increases in human population necessitating, deforestation and intensification of agriculture, fossil-fuel and biomass burning have altered the global N cycle (and those of C and H_2O), leading to an acceleration of emission and deposition of nitrogenous trace gases over the last two centuries (Mayewski et al. 1990; Galloway et al. 1995;

Vitousek et al. 1997; Holland et al. 1999). About 50% of annual global reactive N emissions are generated directly or indirectly by human activities (Vitousek et al. 1997; IPCC 2007). The N_2O is chemically stable nitrogenous trace gas, which persists in the atmosphere for 118 years (Prather and Hsu 2010). Each molecule of N_2O has heat-trapping effects of about 298 times more powerful than that of CO_2 on per-molecule basis (IPCC 2006, 2007). For easy comparison N_2O fluxes are expressed in units of N and in CO_2-carbon equivalents (C eq.), applying relevant GWP of 298 (Eq. 1.5). For this reason, its emission has a long-term influence on climate. It becomes well mixed throughout the atmosphere much faster than it is removed, and its global concentration can be accurately estimated from data at few locations (IPCC 2007).

$$1\text{kg } N_2O - N = 44/28 \times 298 \times 12/44 \text{ C eq.} = 127.71\text{kg C eq.} \tag{1.5}$$

where 44/28 is conversion from N_2O-N to N_2O, GWP of 1 kg N_2O = 298 kg CO_2 equivalents and 12/44 is the conversion from CO_2 equivalents to C equivalents.

1.4.4 *Nitrous Oxide and Atmospheric Ozone Depletion*

In addition to its potential global warming, the interests in atmospheric chemistry of N_2O have also been stimulated by the recognition that this trace gas also plays important roles in the stratosphere chemistry. The photochemical degradation of N_2O in the stratosphere leads to ozone-depleting nitric oxide (NO), nitrogen dioxide (NO_2), and to other important free radical reservoir species (e.g., HNO_3, $ClONO_2$; Crutzen and Schmailzl 1983; Montzka et al. 2011). For the A2 IPCC scenario, the twenty-first century ozone decrease of up to 8% in the 20–40 km altitude region from changes in N_2O alone has been predicted (Portmann and Solomon 2007). Observation that significant amounts of N_2O are added to the atmosphere through N fertilization, fossil fuels, industrial processes and biofuel burning have raised considerable environmental concern and have emphasized the need for better understanding of the geochemistry of tropospheric N_2O.

The Ozone Depletion Potential (ODP) is a metric used for ozone depletion assessment. It has similar qualities as GWP. The integrated impact of a given ozone-depleting substance is estimated relative to CFC-11. The ODPs has widespread use in international agreements such as The Montreal Protocol or The Kyoto Protocol, and in national regulatory discussions. The CFC-11 was widely used industrial compound in the 1970s and 1980s, and has thus been chosen as a convenient reference gas (Fisher et al. 1990).

Comparison of the ODP-weighted anthropogenic emissions of N_2O with those of ozone-depleting substances indicated that N_2O emissions are currently the single most important chemical that depletes ozone (Ravishankara et al. 2009). The 2-D modeling suggested that N_2O is causing ozone depletion in current atmosphere of about 3 Dobson units (DU, one DU is 2.69×10^{16}

O_3 molecules cm^{-2}) compared with about 18 DU caused by other ozone depleting substances (ODS) (Daniel et al. 2010). In the current atmosphere, CFCs lead to much more ozone depletion than does N_2O because of the large historic emissions and long lifetimes of the CFCs, which is expected to decrease in future due to mandatory phase outs. On the contrary, the N_2O emission is not controlled by the Montreal Protocol (MP) and its anthropogenic emissions continues to increase. Based on 2-D model of Solomon et al. (1998), the current estimates of N_2O ODP are at 0.017 (Ravishankara et al. 2009). This value is comparable to the ODP of many hydrochlorofluorocarbons (HCFCs), such as HCFC-123, -124, -225ca, and -225cb (0.02, 0.022, 0.025, and 0.033, respectively), that are currently regulated and being phased out under the MP on substances that deplete the ozone layer (Ajavon et al. 2011). Although N_2O has a relatively small ODP, future changes in emissions and atmospheric concentrations of N_2O could have a significant effect on ozone compared with emissions of other controlled ODSs because of the future larger magnitude of N_2O anthropogenic emissions. The magnitude of past and future N_2O ODP-weighted emissions leads to concerns that include influences on the timing of the recovery of ozone, the "background" ozone level, the distribution of stratospheric ozone depletion, and the possibility of future decreases in ozone due to increasing N_2O (Ravishankara et al. 2009; Wuebbles et al. 2009). When ODS emissions are weighted by their steady state ODPs, N_2O emissions represent the single most important anthropogenic emissions today for ozone depletion (Ravishankara et al. 2009). Unmitigated, these emissions are also expected to remain the most significant throughout the twenty-first century. The doubling the concentration of N_2O in the atmosphere would result in a 10% decrease in the ozone layer, and this would increase the ultraviolet radiation reaching the earth by 20% (Crutzen and Ehhalt 1977). With a relatively long atmospheric lifetime for N_2O, there are justifiable reasons for concern.

The primary source of the stratospheric NO_x is N_2O emission both natural and anthropogenic sources (Wuebbles 2009). The influence of atmospheric N_2O concentration changes on ozone concentrations has been examined (Kinnison et al. 1988; Randeniya et al. 2002; Chipperfield and Feng 2003). The N_2O is very stable in the troposphere where it is emitted, transported to stratosphere, where it releases chemically active compounds that destroy stratospheric ozone through nitrogen oxide catalyzed reaction processes. Approximately 90% of N_2O sinks is the decomposition in the stratosphere by photolysis and photo-oxidation, while the reactions with exited O_2 molecules in the stratosphere consumes the remaining 10% (IPCC 2001; Minschwaner et al. 1993). These reactions are represented by Eqs. 1.6, 1.7, and 1.8 (Machida et al. 1995):

$$N_2O + h\nu \rightarrow N2 + O(^1D) \tag{1.6}$$

$$N_2O + O(^1D) \rightarrow 2NO \tag{1.7}$$

$$N_2O + O(^1D) \rightarrow N_2 + O_2 \quad (1.8)$$

Nitrogen oxides (NO_x i.e. NO + NO_2) are known to catalytically destroy ozone by the following reactions shown in Eqs. 1.8, 1.9, and 1.10 (Crutzen 1970; Johnston 1971):

$$NO + O_3 \rightarrow NO_2 + O_2 \quad (1.9)$$

$$O + NO_2 \rightarrow NO + O_2 \quad (1.10)$$

$$\text{Net reaction}: \; O + O_3 \rightarrow 2O_2 \quad (1.11)$$

Despite the long time recognition of N_2O impact on stratospheric ozone, N_2O is not currently considered as an ozone-depleting substance (ODS) in the same sense as chlorine-, fluorine- and bromide-containing gases, and it is not regulated under the MP (UNEP 2000). Although N_2O ODP is still small, the large anthropogenic emission of N_2O makes up for its lower ODP, making N_2O emissions the single most important of the anthropogenic ODP emission today (Ravishankara et al. 2009). Recent estimates of N_2O emissions under various GHGs mitigation requirements continues to show that N_2O emissions are likely to be higher than they are today, even under most stringent mitigation control requirements (Vuuren et al. 2007).

1.5 Conclusions

Solar radiation powers the global climate. Sun emits energy mainly in the form of short wave radiation. Part of this energy is reflected back into space, while part of it reaches the Earth surface where it is absorbed. The radiation that is absorbed warms the surface the warmer earth surface in turn emits long wave radiation, which is mostly reflected back and forth between the surface and atmosphere. This process dictates the Earth radiation budget. The energy received by Earth must be equal to the energy it emits. In order to achieve the balance, Earth either heats up or cools down when radiation balance is disturbed. The concept of radiative forcing is used to quantify and assess the relative importance of factors that influence radiative balance of the Earth.

The greenhouse gases collectively are the most important forcing factors. They are permeable to short wave radiation from the Sun but impermeable to long wave radiation from the earth. Therefore, increase in their atmospheric concentration is one of the major factors that cause the global warming. Nitrous oxide is a potent greenhouse gas, which also plays significant role in ozone depletion. Its global average mixing ratios has increased from 270 ppmv at the beginning of industrial era in 1750s to 324 ppbv in 2011, and continues to increase at 0.8 ppmv year^{-1} mainly due to enhanced anthropogenic activities. Increase in atmospheric concentration of N_2O and other greenhouse gases have resulted into enhanced greenhouse effect, which is the cause of global warming. The rest of this book discusses the current scientific understanding of anthropogenic perturbations responsible for increase in global average mixing ratio of N_2O with a particular emphasis on agricultural land uses and options for mitigating this increase.

Discussion questions.

1. Describe the natural greenhouse effect. What is the role of N_2O emissions on global warming?
2. What is radiative forcing? How is it related to global warming?
3. What are the merits and limitation of using GWP as an indicator of the radiative forcing?
4. Large volcanic eruption ejects massive sulfur dioxide to the atmosphere causing short term cooling in the climate. Explain.
5. Based on this observation, some scientists suggested that we mimic this effect in order to mitigate global warming. Is it a good suggestion? What are the potential negative and positive of this suggestion?
6. Describe the role of agriculture and world soils in mitigating N_2O emissions.
7. Are there other alternatives to the Haber-Bosch process towards the attempt to feeding the global population projected to be 9.2 billion by 2050? Explain.

References

Adel, A. (1939) Note on the atmospheric oxides of nitrogen. Astro-phys. J. **90**, 627

Ajavon, A.-L.N., Newman, P.A., Pyle, J.A., Ravishankara, A.R. (2011) Scientific assessment of ozone depletion: 2010. Pursuant to Article 6 of the Montreal Protocol on substances that deplete the ozone layer, Global Ozone Research and Monitoring Project. World Meteorological Organization Global Ozone Research and Monitoring Project. Report No. 52. World Meteorological Organization (WMO), Geneva, Switzerland

Baede, A.P.M., Ahlonsou, E., Ding, Y., Schimel, D. (2001) The climate System: an overview. In: Houghton, J.T., Ding, Y., Griggs, D.J., Noguer, M., van der Linden, P.J., Dai, X., Maskell, K., Johnson, C.A. (eds.) Climate Change 2001: The Scientific Basis. Contribution of Working Group I to the Third Assessment Report of the Intergovernmental Panel on Climate Change. pp. 21–84. Cambridge University Press, Cambridge, UK and New York, USA

Bange, H.W., Rapsomanikis, S., Andreae, M.O. (1996) Nitrous oxide in coastal waters. Global Biogeochem. Cycles **10**(1), 197–207

Berntsen, T.K., Fuglestvedt, J.S., Joshi, M.M., Shine, K.P., Stuber, N., Ponater, M., Sausen, R., Hauglustaine, D.A., Li, L. (2005) Response of climate to regional emissions of ozone precursors: sensitivities and warming potentials. Tellus B **57**(4), 283–304. doi:10.1111/j.1600-0889.2005.00152.x

Bouwman, A.F., Fung, I., Matthews, E., John, J. (1993) Gobal analysis of the potential for N_2O production in natural soils. Global Biogeochem. Cycles **7**(3), 557–597

Brook, E.J., Sowers, T., Orchardo, J. (1996) Rapid variations in atmospheric methane concentration during the past 110,000 years. Science **273**(5278), 1087–1091

Butler, H.J. (2011) The National Oceanic and Atmospheric Administration (NOAA) annual greenhouse gas index. NOAA Earth System Research Laboratory Boulder, Colorado. http://www.esrl.noaa.gov/gmd/aggi/ Accessed December, 2011

CAST (2004) Climate change and greenhouse gas mitigation: Challenge and opportunities for agriculture. Council for Agricultural Science and Technology (CAST), Ames, IA.

Chipperfield, M.P., Feng, W. (2003) Comment on: Stratospheric Ozone Depletion at northern mid-latitudes in the 21st century: The importance of future concentrations of greenhouse gases nitrous oxide and methane. Geophys. Res. Lett. **30**(7), 1389–1392. doi:10.1029/2002gl016353

Colman, R.A., McAvaney, B.J. (2011) On tropospheric adjustment to forcing and climate feedbacks. Clim. Dyn. **36**(9–10), 1649–1658. doi:10.1007/s00382-011-1067-4

Crutzen, P.J. (1970) Influence of nitrogen oxides on atmospheric ozone content. Q. J. R. Meteorol. Soc. **96**(408), 320–325

Crutzen, P.J. (1972) SST's - A threat to the Earth's Ozone Shield. Ambio **1**(2), 41–51

Crutzen, P.J. (1974) Estimates of possible variations in total ozone due to natural causes and human activities. Ambio **3**(6), 201–210

Crutzen, P.J., Ehhalt, D.H. (1977) Effects of nitrogen fertilizers and combustion on the stratospheric ozone layer. Ambio **6**(2/3), 112–117

Crutzen, P.J., Schmailzl, U. (1983) Chemical budgets of the stratosphere. Planet. Space Sci. **31**(9), 1009–1032

Daniel, J.S., Fleming, E.L., Portmann, R.W., Velders, G.J.M., Jackman, C.H., Ravishankara, A.R. (2010) Options to accelerate ozone recovery: ozone and climate benefits. Atmos. Chem. Phys. **10**(16), 7697–7707. doi:10.5194/acp-10-7697-2010

Daniel, J.S., Velders, G.J.M., Morgenstern, O., Toohey, D.W., Wallington, T.J., Wuebbles, D.J. (2011) A Focus on information and options for policymakers. In: Ajavon, A.-L.N., Newman, P. A., Pyle, J.A., Ravishankara, A.R. (eds.) Scientific Assessment of Ozone Depletion: 2010. World Meteorological Organization Global Ozone Research and Monitoring Project. Report No. 52. pp. 5.1-5.56. World Meteorological Organization, Geneva, Switzerland

Denman, K.L., Brasseur, G., Chidthaisong, A., Ciais, P., Cox, P.M., Dickinson, R.E., Hauglustaine, D., Heinze, C., Holland, E., D. Jacob, Lohmann, U., Ramachandran, S., Dias, P.L.d.S., Wofsy, S.C., Zhang, X. (2007) Couplings Between Changes in the Climate System and Biogeochemistry. In: Solomon, S., Qin, D., Manning, M., Chen, Z., Marquis, M., Averyt, K.B., M.Tignor, Miller, H.L. (eds.) Climate Change 2007: The Physical Science Basis. Contribution of Working Group I to the Fourth Assessment Report of the Intergovernmental Panel on Climate Change. pp. 501–587. Cambridge University Press, Cambridge, United Kingdom and New York, NY, USA

Ehhalt, D., Prather, M., Dentener, F.J., Derwent, R., Dlugokencky, E.J., Holland, E.A., Isaksen, I., Katima, J., Kirchhoff, V., Matson, P.A. (2001) Atmospheric chemistry and greenhouse gases. In: Houghton, J.T., Ding, Y., Griggs, D.J., Noguer, M., vander Linden, P.J., Dai, X., Maskell, K., Johnson, C.A. (eds.) Climate Change 2001: The Scientific Basis. Contribution of Working Group I to the Third Assessment Report of the Intergovernmental Panel on Climate Change pp. 239–287. Cambridge University Press, Cambridge, United Kingdom and New York, NY, USA

Fisher, D.A., Hales, C.H., Filkin, D.L., Ko, M.K.W., Sze, N.D., Connell, P.S., Wuebbles, D.J., Isaksen, I.S.A., Stordal, F. (1990) Model calculations of the relative effects of CFCs and their replacements on stratospheric ozone. Nature **344**(6266), 508–512. doi:10.1038/344508a0

Flückiger, J., Dallenbach, A., Blunier, T., Stauffer, B., Stocker, T.F., Raynaud, D., Barnola, J.M. (1999) Variations in atmospheric N_2O concentration during abrupt climatic changes. Science **285**(5425), 227–230. doi:10.1126/science.285.5425.227

Flückiger, J., Monnin, E., Stauffer, B., Schwander, J., Stocker, T.F., Chappellaz, J., Raynaud, D., Barnola, J.-M. (2002) High-resolution Holocene N_2O ice core record and its relationship with CH_4 and CO_2. Global Biogeochem. Cycles **16**(1), 1010–1018. doi:10.1029/2001gb001417

Folland, C.K., Karl, T.R., Christy, J.R., Clarke, R.A., Gruza, G.V., Jouzel, J., Mann, M.E., Oerlemans, J., Salinger, M.J., Wang, S.-W. (2001) Observed climate variability and change.

In: Houghton, J.T., Ding, Y., Griggs, D.J., Noguer, M., van der Linden, P.J., Dai, X., Maskell, K., Johnson, C.A. (eds.) Climate Change 2001: The Scientific Basis. Contribution of Working Group I to the Third Assessment Report of the Intergovernmental Panel on Climate Change pp. 99–182. Cambridge University Press, Cambridge, United Kingdom and New York, NY, USA

Forster, P., Ramaswamy, V., Artaxo, P., Berntsen, T., Betts, R., Fahey, D.W., Haywood, J., Lean, J., Lowe, D.C., Myhre, G., Nganga, J., Prinn, R., Raga, G., Schulz, M., Dorland, R.V. (2007) Changes in Atmospheric Constituents and in Radiative Forcing In: Solomon, S., Qin, D., Manning, M., Chen, Z., Marquis, M., Averyt, K.B., M.Tignor, Miller, H.L. (eds.) Climate Change 2007: The Physical Science Basis. Contribution of Working Group I to the Fourth Assessment Report of the Intergovernmental Panel on Climate Change. pp. 129–234. Cambridge University Press, Cambridge, United Kingdom and New York, NY, USA

Fuglestvedt, J.S., Shine, K.P., Berntsen, T., Cook, J., Lee, D.S., Stenke, A., Skeie, R.B., Velders, G.J.M., Waitz, I.A. (2010) Transport impacts on atmosphere and climate: Metrics. Atmos. Environ. **44**(37), 4648–4677. doi:10.1016/j.atmosenv.2009.04.044

Galloway, J.N., Schlesinger, W.H., Levy, H., Michaels, A., Schnoor, J.L. (1995) Nitrogen-fixation: Anthropogenic enhancement-environmental response. Global Biogeochem. Cycles **9**(2), 235–252

Hansen, J., Sato, M., Ruedy, R. (1997) Radiative forcing and climate response. J. Geophys. Res.-Atmos. **102**(D6), 6831–6864

Hansen, J., Sato, M., Ruedy, R., Nazarenko, L., Lacis, A., Schmidt, G.A., Russell, G., Aleinov, I., Bauer, M., Bauer, S., Bell, N., Cairns, B., Canuto, V., Chandler, M., Cheng, Y., Del Genio, A., Faluvegi, G., Fleming, E., Friend, A., Hall, T., Jackman, C., Kelley, M., Kiang, N., Koch, D., Lean, J., Lerner, J., Lo, K., Menon, S., Miller, R., Minnis, P., Novakov, T., Oinas, V., Perlwitz, J., Perlwitz, J., Rind, D., Romanou, A., Shindell, D., Stone, P., Sun, S., Tausnev, N., Thresher, D., Wielicki, B., Wong, T., Yao, M., Zhang, S. (2005) Efficacy of climate forcings. J. Geophys. Res. **110**(D18), D18104. doi:10.1029/2005jd005776

Hirsch, A.I., Michalak, A.M., Bruhwiler, L.M., Peters, W., Dlugokencky, E.J., Tans, P.P. (2006) Inverse modeling estimates of the global nitrous oxide surface flux from 1998–2001. Global Biogeochem. Cycles **20**(1), GB1008. doi:10.1029/2004gb002443

Holland, E.A., Dentener, F.J., Braswell, B.H., Sulzman, J.M. (1999) Contemporary and pre-industrial global reactive nitrogen budgets. Biogeochemistry **46**(1–3), 7–43

Holland, E.A., Lee-Taylor, J., Nevison, C., Sulzman, J. (2005) Global N Cycle: Fluxes and N_2O mixing ratios from human activity. In. Oak Ridge National Laboratory Distributed Active Archive Center, Oak Ridge, Tennessee,

IPCC (1990) Climate Change: The Intergovernmental Panel on Climate Change (IPCC) Scientific Assessment In: Houghton, J.T., Jenkins, G.J., Ephraums, J.J. (eds.) Report prepared for IPCC by Working group I. p. 340 pp. Cambridge University Press, Cambridge, New York, Port Chester, Melbourne, Sydney

IPCC (2001) Climate Change 2001: The Scientific Basis. Contribution of Working Group I to the Third Assessment Report of the Intergovernmental Panel on Climate Change. Houghton, J.T., Ding, Y., Griggs, D.J., Noguer, M., vander Linden, P.J., Dai, X., Maskell, K., Johnson, C.A. (eds.). Cambridge University Press, Cambridge, United Kingdom and New York, NY, USA

IPCC (2006) 2006 IPCC Guidelines for National Greenhouse Gas Inventories, Prepared by the National Greenhouse Gas Inventories Programme. Eggleston, H.S., Buendia, L., Miwa, K., Ngara, T., Tanabe, K. (eds.). Institute for Global Environmental Strategies (IGES), Hayama, Japan

IPCC (2009) Summary Report of the IPCC Expert Meeting on the Science of Alternative Metrics, 18–20 March 2009. available: www.ipcc.ch/meetings/session30/doc13.pdf. In., p. 75. Oslo, Norway

IPCC (2007) Climate Change 2007: The Physical Science Basis. Contribution of Working Group I to the Fourth Assessment Report of the Intergovernmental Panel on Climate Change. In: Solomon, S., Qin, D., Manning, M., Chen, Z., Marquis, M., Averyt, K.B., Tignor, M., Miller, H.L. (eds.). p. 996. Cambridge University Press, Cambridge, United Kingdom and New York, NY, USA

IPCC/TEAP (2005) Safeguarding the Ozone Layer and the Global Climate System: Issues Related to Hydrofluorocarbons and Perfluorocarbons. In: Metz, B., Kuijpers, L., Solomon, S., Andersen, S.O., Davidson, O., Pons, J., de Jager, D., Kestin, T., Manning, M., Meyer, L. (eds.) IPCC/TEAP Special Report. p. 478. Intergovernmental Panel on Climate Change, Cambridge, UK

Johansson, D.J.A., Persson, U.M., Azar, C. (2006) The cost of using global warming potentials: Analysing the trade off between CO_2, CH_4 and N_2O. Clim. Change **77**(3–4), 291–309. doi:10.1007/s10584-006-9054-1

Johansson, D.J.A., Persson, U.M., Azar, C. (2008) Uncertainty and learning: implications for the trade-off between short-lived and long-lived greenhouse gases. Clim. Change **88**(3–4), 293–308. doi:10.1007/s10584-007-9381-x

Johnson, C.E., Derwent, R.G. (1996) Relative radiative forcing consequences of global emissions of hydrocarbons, carbon monoxide and NOx from human activities estimated with a zonally-averaged two-dimensional model. Clim. Change **34**(3–4), 439–462. doi:10.1007/bf00139301

Johnston, H. (1971) Reduction of stratospheric ozone by nitrogen oxide catalysts from supersonic transport exhaust. Science **173**(3996), 517–522

Khalil, M.A.K., Rasmussen, R.A. (1988) Nitrous oxide trends and global mass balance over the last 3000 years. Ann. Glaciol. **10**, 73–79

Kiehl, J.T., Trenberth, K.E. (1997) Earth's annual global mean energy budget. Bull. Amer. Meteorol. Soc. **78**(2), 197–208

Kinnison, D., Johnston, H., Wuebbles, D. (1988) Ozone calculations with large nitrous oxide and chlorine changes. J. Geophys. Res.-Atmos. **93**(D11), 14165–14175

Kroeze, C., Dumont, E., Seitzinger, S.P. (2005) New estimates of global emissions of N_2O from rivers, estuaries and continental shelves. Environm. Sci. **2**(2–3), 159–165

Kroeze, C., Mosier, A., Bouwman, L. (1999) Closing the global N_2O budget: A retrospective analysis 1500–1994. Global Biogeochem. Cycles **13**(1), 1–8. doi:10.1029/1998gb900020

Le Treut, H., Somerville, R., Cubasch, U., Ding, Y., Mauritzen, C., Mokssit, A., Peterson, T., Prather, M. (2007) Historical Overview of Climate Change. In: Solomon, S., Qin, D., Manning, M., Chen, Z., Marquis, M., Averyt, K.B., Tignor, M., Miller, H.L. (eds.) Climate Change 2007: The Physical Science Basis. Contribution of Working Group I to the Fourth Assessment Report of the Intergovernmental Panel on Climate Change. pp. 93–127. Cambridge University Press, Cambridge, United Kingdom and New York, NY, USA

Leuenberger, M., Siegenthaler, U. (1992) Ice age atmospheric concentration of nitrous oxide from an Antarctic ice core. Nature **360**(6403), 449–451

Liepert, B.G. (2010) The physical concept of climate forcing. Wiley Rev: Climate change **1**(6), 786–802. doi:10.1002/wcc.75

MacFarling Meure, C., Etheridge, D., Trudinger, C., Steele, P., Langenfelds, R., van Ommen, T., Smith, A., Elkins, J. (2006) Law Dome CO_2, CH_4 and N_2O ice core records extended to 2000 years BP. Geophys. Res. Lett. **33**(14), L14810. doi:10.1029/2006gl026152

Machida, T., Nakazawa, T., Fujii, Y., Aoki, S., Watanabe, O. (1995) Increase in the atmospheric nitrous oxide concentration during the last 250 years. Geophys. Res. Lett. **22**(21), 2921–2924. doi:10.1029/95gl02822

Manne, A.S., Richels, R.G. (2001) An alternative approach to establishing trade-offs among greenhouse gases. Nature **410**(6829), 675–677. doi:10.1038/35070541

Mayewski, P.A., Lyons, W.B., Spencer, M.J., Twickler, M.S., Buck, C.F., Whitlow, S. (1990) An ice-core record of atmospheric response to anthropogenic sulfate and nitrate. Nature **346**(6284), 554–556

Minschwaner, K., Salawitch, R.J., McElroy, M.B. (1993) Absorption of solar-radiation by O_2 - Implications for O_3 and lifetimes of N_2O, $CFCL_3$, and CF_2CL_2. J. Geophys. Res.-Atmos. **98**(D6), 10543–10561

Montzka, S.A., Reimann, S., Engel, A., Krüger, K., O'Doherty, S., Sturges, W.T. (2011) Ozone-depleting substances (ODSs) and related chemicals In: Ajavon, A.-L.N., Newman, P.A., Pyle, J.A., Ravishankara, A.R. (eds.) Scientific Assessment of Ozone Depletion: 2010. World Meteorological Organization Global Ozone Research and Monitoring Project. Report No. 52. pp. 1.1-1.108. World Meteorological Organization, Geneva, Switzerland

Myhre, G., Myhre, A., Stordal, F. (2001) Historical evolution of radiative forcing of climate. Atmos. Environ. **35**(13), 2361–2373

Nevison, C.D., Lueker, T.J., Weiss, R.F. (2004) Quantifying the nitrous oxide source from coastal upwelling. Global Biogeochem. Cycles **18**(1), GB1018. doi:10.1029/2003gb002110

Nevison, C.D., Weiss, R.F., Erickson, D.J. (1995) Global oceanic emissions of nitrous oxide. J. Geophys. Res.-Oceans **100**(C8), 15809–15820. doi:10.1029/95jc00684

National Research Council (NRC) (2005) Radiative Forcing of Climate Change: Expanding the Concept and Addressing the Uncertainties. Jacob, D.J., Avissan, J., Bond, G.C., Gaffin, S., Kiel, J., Lean, J., Lohmann, U., Mann, M.E., Pieke, R.A., Ramanathan, V., Russell, L.M. (eds.). The National Academies Press, Washington, D.C.

Peixoto, J.P., Oort, A.H. (1992) Physics of climate. American Institute of Physics, New York, USA

Pinty, B., Lavergne, T., Kaminski, T., Aussedat, O., Giering, R., Gobron, N., Taberner, M., Verstraete, M.M., Voßbeck, M., Widlowski, J.L. (2008) Partitioning the solar radiant fluxes in forest canopies in the presence of snow. J. Geophys. Res. **113**(D4), D04104. doi:10.1029/2007jd009096

Portmann, R.W., Solomon, S. (2007) Indirect radiative forcing of the ozone layer during the 21st century. Geophys. Res. Lett. **34**(2), L02813. doi:10.1029/2006gl028252

Prather, M.J., Hsu, J. (2010) Coupling of nitrous oxide and methane by global atmospheric chemistry. Science **330**, 952–954

Prinn, R., Cunnold, D., Rasmussen, R., Simmonds, P., Alyea, F., Crawford, A., Fraser, P., Rosen, R. (1990) Atmospheric emissions and trends of nitrous oxide deduced from 10 years of ALE-GAUGE data. J. Geophys. Res.-Atmos. **95**(D11), 18369–18385

Ramanathan, V., Cicerone, R.J., Singh, H.B., Kiehl, J.T. (1985) Trace gas trends and their porential role in climate change. J. Geophys. Res.-Atmos. **90**(ND3), 5547–5566

Ramaswamy, V., Boucher, O., Haigh, J., Hauglustaine, D., Haiwood, J., Mayhre, G., Nakajima, T., Shi, G.Y., Solomon, S. (2001) Radiative forcing of climate change. In: Houghton, J.T., Ding, Y., Griggs, D.J., Noguer, M., vander Linden, P.J., Dai, X., Maskell, K., Johnson, C.A. (eds.) Climate Change 2001: The Scientific Basis. Contribution of Working Group I to the Third Assessment Report of the Intergovernmental Panel on Climate Change, vol. . pp. 349–416. Cambridge University Press, Cambridge, United Kingdom and New York, NY, USA

Randeniya, L.K., Vohralik, P.F., Plumb, I.C. (2002) Stratospheric ozone depletion at northern mid latitudes in the 21(st) century: The importance of future concentrations of greenhouse gases nitrous oxide and methane. Geophys. Res. Lett. **29**(4), 1051. doi:10.1029/2001gl014295

Ravishankara, A.R., Daniel, J.S., Portmann, R.W. (2009) Nitrous Oxide (N_2O): The Dominant Ozone-Depleting Substance Emitted in the 21st Century. Science **326**(5949), 123–125. doi:10.1126/science.1176985

Rockström, J., Steffen, W., Noone, K., Persson, A., Chapin, F.S., Lambin, E.F., Lenton, T.M., Scheffer, M., Folke, C., Schellnhuber, H.J., Nykvist, B., de Wit, C.A., Hughes, T., van der Leeuw, S., Rodhe, H., Sorlin, S., Snyder, P.K., Costanza, R., Svedin, U., Falkenmark, M., Karlberg, L., Corell, R.W., Fabry, V.J., Hansen, J., Walker, B., Liverman, D., Richardson, K., Crutzen, P., Foley, J.A. (2009) A safe operating space for humanity. Nature **461**(7263), 472–475. doi:10.1038/461472a

Ruddiman, W.F. (2003) Plows, Plagues and Petroleum: How Humans Took Control of the Climate. Princeton University Press, Princeton

Schilt, A., Baumgartner, M., Schwander, J., Buiron, D., Capron, E., Chappellaz, J., Loulergue, L., Schupbach, S., Spahni, R., Fischer, H., Stocker, T.F. (2010) Atmospheric nitrous oxide during the last 140,000 years. Earth Planet. Sci. Lett. **300**(1–2), 33–43. doi:10.1016/j.epsl.2010.09.027

Seitzinger, S.P., Kroeze, C. (1998) Global distribution of nitrous oxide production and N inputs in freshwater and coastal marine ecosystems. Global Biogeochem. Cycles **12**(1), 93–113. doi:10.1029/97gb03657

Sellers, P.J., Dickinson, R.E., Randall, D.A., Betts, A.K., Hall, F.G., Berry, J.A., Collatz, G.J., Denning, A.S., Mooney, H.A., Nobre, C.A., Sato, N., Field, C.B., Henderson-Sellers, A. (1997) Modeling the exchanges of energy, water, and carbon between continents and the atmosphere. Science **275**(5299), 502–509. doi:10.1126/science.275.5299.502

Serreze, M.C. (2010) Understanding Recent Climate Change. Conserv. Biol. **24**(1), 10–17. doi:10.1111/j.1523-1739.2009.01408.x
Shine, K.P. (2009) The global warming potential-the need for an interdisciplinary retrial. Clim. Change **96**(4), 467–472. doi:10.1007/s10584-009-9647-6
Shine, K.P., Derwent, R.G., Wuebbles, D.J., Morcrette, J.-J. (1990) Radiative forcing of climate In: Houghton, J.T., Jenkins, G.J., Ephraums, J.J. (eds.) Climate change: The scientifc assessment report prepared for Intergovernmental Panel on Climate Change (IPCC) Working Group I, vol. report. pp. 41–68. Cambridge University Press, Cambridge, Great Britain, New York, NY, USA and Melbourne, Australia
Solomon, S., Portmann, R.W., Garcia, R.R., Randel, W., Wu, F., Nagatani, R., Gleason, J., Thomason, L., Poole, L.R., McCormick, M.P. (1998) Ozone depletion at mid-latitudes: Coupling of volcanic aerosols and temperature variability to anthropogenic chlorine. Geophys. Res. Lett. **25**(11), 1871–1874
Sowers, T. (2001) N_2O record spanning the penultimate deglaciation from the Vostok ice core. J. Geophys. Res.-Atmos. **106**(D23), 31903–31914
Stauffer, B., Fluckiger, J., Monnin, E., Schwander, M., Barnola, J.M., Chappellaz, J. (2002) Atmospheric CO_2, CH_4 and N_2O records over the past 60,000 years based on the comparison of different polar ice cores. Ann. Glaciol. **35**, 202–208
Tanaka, K., O'Neill, B.C., Rokityanskiy, D., Obersteiner, M., Tol, R.S.J. (2009) Evaluating Global Warming Potentials with historical temperature. Clim. Change **96**(4), 443–466. doi:10.1007/s10584-009-9566-6
United Nations Environment Programme (UNEP) (2000) The Montreal Protocol on Sobstances that Deplete the Ozone Layer as either adjusted and/or amended in London 1990, Copenhagen 1992, Vienna 1995, Montreal 1997 & Beijing 1999. UNEP Ozone Secretariat.
Unger, N., Bond, T.C., Wang, J.S., Koch, D.M., Menon, S., Shindell, D.T., Bauer, S. (2010) Attribution of climate forcing to economic sectors. Proc. Natl. Acad. Sci. U. S. A. **107**(8), 3382–3387. doi:10.1073/pnas.0906548107
United States Environmental Protection Agency (USEPA) (2007) Inventory of U.S. Greenhouse gas emissions and sinks: 1990–2005. US Environmental Protection Agency.
Vitousek, P.M., Aber, J.D., Howarth, R.W., Likens, G.E., Matson, P.A., Schindler, D.W., Schlesinger, W.H., Tilman, G.D. (1997) Human alteration of the global nitrogen cycle: Sources and consequences. Ecol. Appl. **7**(3), 737–750
Vuuren, D.P., Elzen, M.G.J., Lucas, P.L., Eickhout, B., Strengers, B.J., Ruijven, B., Wonink, S., Houdt, R. (2007) Stabilizing greenhouse gas concentrations at low levels: an assessment of reduction strategies and costs. Clim. Change **81**(2), 119–159. doi:10.1007/s10584-006-9172-9
Wang, W.C., Yung, Y.L., Lacis, A.A., Mo, T., Hansen, J.E. (1976) Greenhouse Effects due to Man-Made Perturbations of Trace Gases. Science **194**(4266), 685–690
Weiss, R.F. (1981) The temporal and spatial distribution of tropospheric nitrous oxide. J. Geophys. Res. Oceans **86**(NC8), 7185–7195
World Meteorological Organization (WMO) (2011) The State of Greenhouse Gases in the Atmosphere Based on Global Observations through 2010. World Meteorological Organization Greenhouse Gases Bulletin No. 7. World Meteorological Organization. http://www.wmo.int/gaw
Wuebbles, D.J. (2009) Nitrous Oxide: No Laughing Matter. Science **326**(5949), 56–57. doi:10.1126/science.1179571
Wuebbles, D.J., Youn, D., Patten, K., Wang, D., Martínez-Avilés, M. (2009) Metrics for ozone and climate: Three-dimensional modeling studies of ozone depletion potentials and indirect global warming potentials, in Twenty Years of Ozone Decline. In: Zerefos, C., Contopoulos, G., Skalkeas, G. (eds.) Proceedings of the Symposium for the 20th Anniversary of the Montreal Protocol. pp. 297–326. Springer Publishing, Dordrecht, The Netherlands

Chapter 2
Global Nitrogen Cycle

Contents

Abstract Human activity has significantly altered nitrogen (N) cycle. Prior to the industrialization and agricultural revolution, the vast majority of reactive N (Nr) on the Earth was created by microorganisms through biological N fixation, and this creation was balanced over geological time by denitrification. Over the past century, development of new agricultural practices to satisfy growing global demand for food has drastically disrupted the N cycling. Anthropogenic creation of Nr through Haber-Bosch process now rivals the rate of natural terrestrial N fixation. The unintentional creation of Nr during fossil fuel combustion also adds to the global N cycle. The anthropogenic influences on the N cycling vary across regions of the globe. To date, the largest changes in N cycling have occurred in developed countries in temperate zone. However, this is changing as fertilizer use and fossil fuel combustion increase in South Asia and tropical South America. The significant consequences of anthropogenic acceleration of N cycle is the eutrophication of estuaries and coastal waters, leading to hypoxic zones in many areas, as well as increased global inventories of potent greenhouse gas N_2O.

David Ussiri and R. Lal, *Soil Emission of Nitrous Oxide and its Mitigation*,
DOI 10.1007/978-94-007-5364-8_2, © Springer Science+Business Media Dordrecht 2013

Keywords Reactive nitrogen • Haber-Bosch process • Nitrogen fixation • Fossil fuel combustion • Denitrification

Abbreviations

ATP	Adenosine triphosphate
DNA	Deoxyribonucleic acid
NPP	Net primary productivity
BNF	Biological nitrogen fixation
C-BNF	Cultivation-induced biological nitrogen fixation
Nr	Reactive nitrogen
OM	Organic matter
C:N	Carbon to nitrogen ratio
EDGAR	Emission database for global atmospheric research
UV	Ultraviolet
GHG	Greenhouse gas
UV	Ultraviolet
SOM	Soil organic matter
NO_x	Nitrogen oxides—NO + NO_2
NO_y	Collective term used to describe all oxidized products of N except N_2O

2.1 Introduction

Nitrogen (N), the fifth most abundant element in the earth, is essential to all organisms. Its availability controls ecosystem functions and biogeochemistry. In living tissues, N is essential for the synthesis of nucleic acids and protein, the two most important polymers of life. N is also an integral part of enzymes that drive the metabolic machinery of every living cell. Furthermore, N is part of chlorophyll, the green pigment of plant responsible for photosynthesis. It drives all key metabolic processes involved in growth and energy transfer. Therefore, N is a key element that controls composition, dynamics, diversity, and functioning of terrestrial, freshwater, and marine ecosystems (Vitousek et al. 1997). Changes in the availability of N together with phosphorus (P) an essential component of deoxyribonucleic acid (DNA), adenosine triphosphate (ATP) and phospholipids molecules of cell membranes have controlled the magnitude and activity of the biosphere through geologic times. Nature and its biodiversity can only exist because of the availability of reactive N (Nr) in the system, even if limited. Requirements of N are enormous for life, and it exists in more chemical forms than most other elements (Table 2.1), and undergoes myriads of unique chemical transformations. Nearly all transformations are mediated by microorganisms as part of their metabolism, either to obtain N for synthesis of structural components, to harness energy for their growth or as alternative electron acceptor

Table 2.1 Nitrogenous compounds and their role in biosphere N cycling

Compound	Chemical formula	Natural sources	N cycling processes involved
Nitrogen gas	N_2	Nitrification Denitrification	Biological N fixation Denitrification
Nitrate/Nitrite Nitric acid/Nitrous acid	NO_3^-/NO_2^- HNO_3/HNO_2	Nitrification Denitrification Lightning	Dry deposition Wet deposition Leaching to ground water, rivers, oceans Atmospheric reactions Plant uptake
Nitric oxide	NO	Lightning Nitrification Denitrification	Dry deposition Atmospheric reactions
Nitrogen dioxide	NO_2		Atmospheric reactions Nitrite reduction in acid soils
Nitrous oxide	N_2O	Nitrification Denitrification	Stratospheric reactions Greenhouse effects
Nitrogen pentoxide	N_2O_5		
Organic nitrogen	Variable	Biomass production Organic matter	Dry deposition Volatilization Organic matter decomposition
Ammonia/ammonium	NH_3/NH_4^+	Biological N fixation	Plant uptake Nitrification Fertilizer production Dry deposition Wet deposition Leaching to ground water, rivers and oceans Biomass combustion Volatilization Stratospheric reactions Organic matter decomposition

when growing anaerobically. The large number of N compounds in the biosphere, and their different physical and chemical properties, makes its flows and transformation highly complex and enormously variable.

Nitrogen availability often limits net primary production in agricultural as well as natural ecosystems (De Vries et al. 2006). N cycling and storage in soils varies considerably depending on ecosystem type and land use. In agriculture ecosystems, N cycling is dominated by N fertilization and crop removal, while in natural ecosystems N cycling is affected by climatic, edaphic, and landscape conditions, and sum of N inputs through deposition and biological N fixation (BNF). Temperate forests are naturally N limited, whereas tropical rain forests are often N-rich ecosystems. However, during last few decades the situation has changed markedly due to increased atmospheric N deposition in temperate forests of Europe and North

America (De Vries et al. 2007). Signs of N saturation have been widely reported, which includes accelerated growth and significant reactive N losses via nitrate leaching an N trace gas emissions (Pilegaard et al. 2006; Dise et al. 2009).

While it is highly abundant on earth—(~5 $\times$ 10^9 Tg N contained in the atmosphere, ocean, marine biota, terrestrial ecosystems, soil organic matter and sedimentary rocks), only <2% of it is available to organisms (Mackenzie 2003). The remainder is tied up either in sedimentary rocks, or as triple-bonded dinitrogen (N_2) gas in the atmosphere. The latter is a highly inert gas, with an atmospheric residence time of nearly 1 million years. Over evolutionary history, only few species of Bacteria and Archaea have evolved the ability to convert N_2 to reactive N (Nr), i.e., N bonded to H, C, or O (e.g., NO_y, NH_y, or organic N). Conversion of N_2 to Nr that is available to organisms requires high energy (Galloway 1998). Therefore, even with the adaptations to use Nr efficiently, fixed inorganic N, generally in the form of nitrates (NO_3^-) or ammonium (NH_4^+), often limit the net primary productivity (NPP) of terrestrial and marine ecosystems (Vitousek and Howarth 1991; Falkowski 1997; Vitousek et al. 2002; Falkowski et al. 2008).

Biochemical transformations of N are possible because it can be at valence ranging from reduced, −3 (as in ammonia (NH_3)) to fully oxidized +5 (as in nitrate (NO_3^-)). The conversion between different forms of N is mainly mediated by a range of microorganisms—which capitalize on the potential for transformations of N among different valence states and use the energy released by changes in reduction-oxidation (redox) potential to maintain their life processes (Rosswall 1982; Falkowski 1997). Biogeochemistry of N is also to lesser extent mediated by long-term recycling through the geosphere (Berner 2006). Microorganisms are essential to maintaining the balance between reduced and oxidized forms of N in the ecosystems. Prior to human interference, the biochemistry of N was almost entirely driven by redox reactions primarily mediated by microorganisms (Falkowski 1997; Falkowski et al. 2008). These reactions are: (1) biological N_2 fixation (BNF)—a microbially mediated reductive process that transforms N_2 to NH_4^+, (2) nitrification—a 2–step oxidation process that oxidizes NH_4^+ to NO_3^-, and (3) denitrification—an anaerobic oxidation of organic matter (OM) which uses NO_3^- or nitrite (NO_2^-) as electron acceptors, ultimately releasing N_2. Collectively these microbial reactions drive the cycling of N. The BNF is carried out by symbiotic and free living prokaryotes, which can break the triple bond of N_2. Organisms that fix N_2 symbiotically with legumes are collectively known as rhizobia diazotrophs, whereas those that fix N_2 in the absence of legumes are known as free-living rhizosphere diazotrophs (Russele 2008). Ammonium is oxidized into NO_3^- during nitrification. The NO_3^- produced may be reduced to N_2 via denitrification or NH_3 through dissimilatory NO_3^- reduction (Chap. 3). These processes form the major part of inorganic N cycle in soil. Ammonium and NO_3^- can be used by most living cells to produce organic forms of N. The latter is mineralized and transferred again into NH_3. Therefore, NH_4^+/NH_3 form a link between organic and inorganic N cycle. The N cycle consists of transfer of N between Earth surface reservoirs of the atmosphere, the oceans, the land, and life. It is essential to life and is normally considered as biological cycle, which includes a wide variety of processes involving the uptake and release of various forms of N by terrestrial and marine organisms and transformation of oxidation states by bacteria. Because there is much less N stored in life (i.e., 0.03×10^{18} mol N compared to atmospheric N—142×10^{18} mol N (Berner 2006)), transfer of N between terrestrial or aquatic ecosystems to atmosphere have little effect on the level of atmospheric N_2.

Over geologic history, most Nr has been formed by BNF. However, the discovery of Haber-Bosch process for industrial creation of NH_3 from atmospheric N_2 and H_2 during the twentieth century has replaced BNF as the dominant terrestrial process of creating Nr. The Haber-Bosch process has facilitated the production of agricultural fertilizers and dramatically increased global agricultural productivity through the so-called "Green Revolution." Without the availability of nitrogenous fertilizers produced by the industrial Haber-Bosch process, the large increase in food production over the twentieth century, which has sustained the increase in global population, would not have been possible. Food and energy productions have increased the rate of anthropogenic creation of Nr at a record pace over the past few centuries. This has altered the N cycling by increasing both the availability and mobility of N globally. The outcomes of these alterations are the numerous environmental changes, including the increase in water and air pollution, enrichment of atmospheric concentration of greenhouse gas (GHG), loss of biodiversity due to increase in Nr to the ecosystems, increase in smog, depletion of stratospheric ozone, deposition of acidic substances, and increase in productivity of freshwaters, marine waters, and terrestrial ecosystems. Anthropogenic distortion of the N cycle, combined with changes in phosphorus flows has shifted the state of lake and marine ecosystems from clear to turbid water (Zillen et al. 2008). In addition, hypoxia (defined as O_2 concentrations <2 mg L^{-2} in aquatic environments) is globally a significant problem with over 400 sites suffering from its effects (Zillen et al. 2008; Diaz and Rosenberg 2008). Hypoxia causes severe ecosystem disturbances and alters nutrient biogeochemical cycles in aquatic ecosystems. It also affects food chain dynamics, fish habitats and fisheries (Bonsdorff 2006). The adverse impacts of excessive Nr are associated with its chemical transformations as it moves within the environment. For example, nitric oxide (NO) released from fossil fuel combustion is oxidized in atmosphere to nitric acid (HNO_3), and deposited into an ecosystem where it acidifies soils, leaches into water, and cause eutrophication. It can also be denitrified and releases nitrous oxide (N_2O), a potent GHG, which also destroys the ozone layer.

2.2 Historical Perspective of the Nitrogen Cycle

The N cycle refers to circulation of N compounds through the Earth's atmosphere, hydrosphere, biosphere, and pedosphere. At various points in this cycle, Nr becomes involved in processes that can affect human health and the environment in both positive and negative ways. The human understanding of the N cycle began with its discovery and functions. The element was discovered in the late eighteenth century through the work of several chemists, most notable being Carl Wilhelm Scheele (1742–1786), a Swedish chemist, Laurent Lavoisier (1743–1794), a French chemist, and Daniel Rutherford (1749–1819) a Scottish botanist. Jean Antoine Claude Chaptal (1756–1832) of France formally named the 7th element of the Periodic Table in 1790 as "nitrogene." Subsequent advancements in analysis and experimental agronomy confirmed its presence in the biosphere. Later in 1850s, it was recognized that N was a common element in plant and animal tissues, is essential to plant growth, is constantly cycling between organic compounds, and is an effective fertilizer. Despite its ubiquitous presence in plants and animal tissues, its role in vigorous plant growth and its constant cycling was not recognized until at the beginning of the second half of the nineteenth century. In addition, its source remained uncertain. Lightning and atmospheric deposition were thought to be the most important sources (Smil 2001).

In 1838, Jean-Baptiste Boussingault demonstrated that legumes could restore Nr into the soil and suggested that they must fix atmospheric N directly. It was 50 more years before the puzzle was solved. Marcelin Berthelot (1827–1907) was the first to suggest that something in the soil, rather than something in leguminous plants, was carrying out the process of N fixation. The breakthrough came later when Herman Hellgiel (1831–1895) and Herman Wilfarth (1853–1904) demonstrated that leguminous plants grown in sterile soil with no N addition performed poorly, but those grown in sterile medium inoculated with loamy soil inoculum performed better. In 1888, Herman Hellriegel and Herman Willfrath published the work on microbial communities in which they stated that legumes themselves do not possess the ability to assimilate free N in the air but the active participation of living soil microorganisms is absolutely necessary. They indicated that there are symbiotic relationships between legumes and soil microorganisms (Smil 2001).

The conversion of NH_3 to NO_3^- by microorganisms had been demonstrated earlier, but it was Théofile Schloesing (1824–1919) who conducted an experiment in 1877 to prove the bacterial origin of nitrification. He demonstrated that all of the ammonia in sewage is converted into nitrate within a few days. He further observed that addition of chloroform terminated the process. Russian microbiologist Sergei Nikolaevich (1856–1953) isolated two nitrifiers in 1889—*Nitrosomonas* and *Nitrobacter* responsible for nitrification. Also, Uysse Gayon (1849–1929) and his assistants isolated two bacteria that could reduce NO_3^- to N_2 and named the process denitrification. Unlike nitrification, which is dominated by just a few bacteria genera, denitrification is carried by a variety of microorganisms capable of using oxygen (O_2) in NO_3^- for their respiration as they feed on organic substrates. The most common denitrifying genera include *Pseudomonas*, *Bacillus*, and *Alkaligenes*. Discovery of the nitrification and denitrification towards the end of nineteenth century completed identification of key components of the N cycle.

Over the twentieth century, the knowledge of Nr synthesis, including the Haber-Bosch process, an industrial synthesis of NH_3 from nitrogen and hydrogen transformed the world and cycling of Nr through ecosystems reservoirs and its environmental impact, has increased significantly. Two anthropogenic activities in particular, which have greatly increased Nr availability, are food and energy production. Food production creates additional Nr on purpose to increase agronomic yield. However, energy production creates and mobilizes Nr as an incidental reaction. In addition to industrial fertilizers, cultivation of legumes and paddy rice (*Oriza sativa*) produces anthropogenic Nr, since legumes can self-fertilize via symbiotic N_2 fixing organisms, while rice cultivation creates necessary anaerobic environment to promote the BNF.

2.3 Global Nitrogen Inventory

About 91.3% of N in Earth's system is in its core and the mantle, and is out of circulation (Table 2.2). The release of N from the mantle is negligible and almost equivalent to reincorporation of surface volatiles at subduction zones (Walker 1977). The remaining 8.7% of N is distributed in the biosphere as follows:

(i) Geologic N
About 26.04% is the geologic N, i.e., N incorporated in rocks. It includes N in igneous rocks—where NH_4^+ is frequently present in the lattice of silicate minerals. In some instances, N is introduced into the mineral lattice by ammonium-rich hot spring waters (e.g., buddingtonite). The average for igneous rocks is 25 mg N kg^{-1} (Wedpohl 1972). The N concentration in rocks varies based on rock formation conditions. In general, quartz contains the least NH_4^+ and biotite the most, feldspar falls in between the two. In sedimentary and metamorphic rocks, NH_4^+ is the dominant form of N. In sedimentary rocks, N is derived from organic sediments, which lost the amino groups during conversion of sediments to rock, or retained by adsorption. The average N content of sedimentary rocks is 490 mg kg^{-1}, but the range of concentration is broad—70 mg kg^{-1} limestones, 150 mg kg^{-1} sandstones and greywackes, and 600 mg kg^{-1} in shales (Wedpohl 1972). Some C-rich sedimentary rocks contain large quantities of undestroyed organic N, which can constitute 75–85% of total N (Holloway et al. 2001). Metamorphism generally destroys organic materials and drives off NH_4^+ along with other volatiles. The N content of metamorphic rocks is generally comparable with that of igneous rocks. Geologic N is frequently ignored in N budgets. However, evidence indicates that weathering of certain bedrocks can contribute significant inputs of N to marine and terrestrial ecosystems, and this does affect the long-term accumulation of soil N (Holloway et al. 1998, 2001). This gives rise to less known cycle of N that is much slower than the biological N cycle, generally termed as geological N cycle (Berner 2006), with much longer N turnover times of many millions of years. The discriminating feature of this cycle is that it involves the participation of rocks. The principal processes involved are: (i) burial of organic N and traces of NH_4^+ substituting for potassium (K^+) in clay minerals into sedimentary rocks, and ultimate transformation to silicate minerals (Boyd 2001), (ii) weathering of

Table 2.2 Nitrogen inventory for the earth

Source	Quantity of N (Tg)	Reference	Comments
Lithosphere			
Core	1.3×10^{8}	Stevenson (1982)	
Mantle	5.6×10^{10}	Wedpohl (1972)	
Lithosphere total	5.613×10^{10}	–	
Geologic			
Igneous NH_3	5.8×10^{8}	Wedpohl (1972)	
N_2	7.0×10^{7}	Wedpohl (1972)	
Sedimentary rocks	1.0×10^{9}	Mackenzie (2003)	
Ancient sediments (NH_3 and organic N)	7.5×10^{8}	Wedpohl (1972)	
Coal	1.52×10^{5}	Brown and Skipsey (1986)	Assuming 1.5% N
Geologic total	1.4×10^{9}	–	
Soil			
Soil organic matter	$1.3–1.4 \times 10^{5}$	Batjes (1996)	Top 1 m depth
Soil litter	$1.9–3.3 \times 10^{3}$	Söderlund and Svensson (1976)	
Clay fixed NH_4^+	2.1×10^{4}	Stevenson (1982)	
Soil total	$1.63–1.36 \times 10^{5}$	–	
Terrestrial biota			
Plant biomass	$1.0–1.4 \times 10^{4}$	Mackenzie (2003)	
Animal biomass	2.0×10^{2}	Delwiche (1970)	
Biomass total	$1.1–1.4 \times 10^{4}$	–	
Atmospheric			
Molecular N_2	3.95×10^{9}	Mackenzie (2003)	
Tropospheric N_2O	1.5×10^{3}	Nevison and Holland (1997)	
Tropospheric NH_x	1.74	Rosswall (1982)	
Tropospheric NO_y	0.7	Rosswall (1983)	
Tropospheric organic N	1	Söderlund and Svensson (1976)	
Atmospheric total	4.0×10^{9}	–	
Ocean			
Molecular N_2	2.4×10^{7}	Kennish (1994)	Calculated based on mean water concentrations
DOM	5.3×10^{5}	Söderlund and Svensson (1976)	
Particulate OM	$3–4 \times 10^{3}$	Söderlund and Svensson (1976)	
NO_3^-	5.7×10^{5}	Mackenzie (2003)	
NO_2	5.0×10^{2}	Söderlund and Svensson (1976)	
NH_y	7×10^{3}	Söderlund and Svensson (1976)	
N_2O	2.0×10^{2}	Söderlund and Svensson (1976)	
Living biomass	5.0×10^{2}	Mackenzie (2003)	
Ocean Total	2.4×10^{7}	–	

sedimentary organic matter and liberation of N to surface biological systems (Holloway and Dahlgren 2002), emission of N_2 and other nitrogenous gases from volcanic and metamorphic degassing (Giggenbach and Matsuo 1991). Transfer of N between rocks and the surface reservoirs may involve conversion between several N oxidation states.

(ii) Atmospheric N_2:
Approximately 73.93% is N_2 gas, including 99.99% of the atmospheric N and 95% of marine N. A triple bonded N_2 must be transformed into reactive form to become bioavailable (Galloway 1998). The atmospheric N_2 concentration of well-mixed air is 78.084% with a mass of 3.87×10^9 Tg N (Table 2.2; Trenberth and Guillemot 1994; Schlesinger 1997).

(iii) Reactive N (Nr):
The remaining 0.03% of N is reactive or fixed N (i.e., Nr), in the atmosphere, land, and water. The atmospheric N_2O concentration is 323 ppb, with a mass of 2.42×10^3 Tg N_2O-N (Trenberth and Guillemot 1994; Montzka and Reimann 2010). The terrestrial biomass N is estimated at 3.5×10^3 Tg and soil organic matter (SOM) N ranges from 0.95 to 1.40×10^5 Tg N (Batjes 1996; Schlesinger 1997). By comparison, $\sim 1.0 \times 10^4$ Tg N is held in plant biomass, and $\sim 2.0 \times 10^3$ Tg in the microbial biomass (Davidson 1994). Therefore, on ecosystem scale, soils are the main reservoir for N. The C:N ratio of terrestrial biomass and SOM are ~160 and 15, respectively. At any time, the pool of inorganic N (NH_4^+, NO_3^-, and NO_2^-) is rather small despite the large annual flux through this pool, mainly due to uptake by plants and organisms.

2.4 Nitrogen Fixation

The most abundant form in the surface of the earth is N_2 gas. It is the least reactive species and is not available to most of organisms due to the strength of the triple bond in N_2, which makes this molecule practically inert. Most of N that is available to biota has been derived from this N pool by fixation, either by lightning, free-living, and symbiotic microbes, or by anthropogenic processes. Nitrogen fixation is the transformation of the highly abundant but biologically unavailable atmospheric N_2 to one of the reactive forms—NH_3, NH_4^+, NO_2^- NO_3^-, nitrogen oxides (NO, NO_2), N_2O_5, and N_2O or organic N. The fixation of N_2 to reactive forms of N occurs via four major high-energy routes: (a) lightning fixation (b) biological fixation (c) synthetic N fertilizer production, and (d) high temperature combustion fixation.

2.4.1 Lightning

Lightning strike produces high electrical energy in the atmosphere which overcomes the energy barrier to break the triple bond of molecular N_2 ($N \equiv N$) and double bonds of molecular O_2 ($O{=}O$), whereby the two molecules combines to form NO in the atmosphere. Lightning produces momentary conditions of high pressure and temperature in the lightning channel, which causes thermal dissociation of O_2 and N_2 and allow them to combine and form NO. The NO is photochemically converted to NO_y (Eq. 2.1)

$$N_2 + O_2 + \text{ electrical energy} \rightarrow 2NO \rightarrow 2NO_y \tag{2.1}$$

NO_y is a collective term used to describe all oxidized products of N except N_2O (Eq. 2.2)

$$\begin{aligned} NO_y &= NO_x + NO_2^- + HNO_2 + HNO_3 + HO_2NO_2 + NO_3 + 2N_2O_5 \\ &\quad + \text{ peroxyacetyl nitrate;} \\ NO_x &= NO + NO_2 \end{aligned} \tag{2.2}$$

These compounds have an average atmospheric residence time of <10 days before they are deposited on the Earth's surface either as wet deposition in precipitation or dry deposition in aerosol or gas associated with particulate matter. Lightning, therefore, introduces Nr into ecosystems, especially over the tropical regions.

About 75–90% of the total NO_x formed by lightning is NO depending on the temperature and ozone (O_3) concentration at the time of flash. However, NO reacts quickly with O_3 in the ambient air to produce NO_2 (Shepon et al. 2007) and the equilibrium is quickly achieved between NO and NO_2 (Price et al. 1997). The NO_x produced by lightning are readily carried by convective updrafts to the upper troposphere where its lifetime is considerably longer than in the lower troposphere (Labrador et al. 2005). The link between lightning and nitrogen oxides was recognized in 1820s but its role in controlling O_3 concentration in troposphere was not recognized until 1970s. Lightning NO_x is closely linked with hydroxyl (OH^-) radical production and, therefore, the potential to affect the atmosphere's oxidizing efficiency (Labrador et al. 2004).

The global rate of N fixation by lightning is not well known. Estimates of the global lightning-produced NO_x ranges from 1.0 Tg N year^{-1} (Levine et al. 1981) to 100 Tg N year^{-1} (Franzblau and Popp 1989). The calculated total annual deposition of N (NO_x) from pre-industrial atmosphere estimates lightning fixation at 20 Tg N year^{-1} (Lyons et al. 1990). Recent estimates ranges from 1.0 to 20.0 Tg N year^{-1} (Price et al. 1997; Ridley et al. 2004). The Emission Database for Global Atmospheric Research (EDGAR) estimates of NO_x produced by lightning is at 12.2 Tg N year^{-1} (Olivier et al. 1998). The reasons for the large uncertainty reported in literature are the poorly understood aspect of lightning phenomenon itself, including the charge separation process, the amount of energy deposited per flash, its partitioning among the cloud-to-ground, intra-cloud and inter-cloud flashes and the aspects related to NO_x such as NO molecules produced per flash or per unit energy (Labrador et al. 2005). Estimates of lightning Nr creation are highly uncertain. The IPCC third assessment report used estimates based on

OxComp model, which estimated Nr at 5.0 Tg N year^{-1}, and ranging from 3.0 to 13.0 Tg N year^{-1} (Ehhalt et al. 2001). In this chapter, a global estimate of 5.4 Tg N year^{-1} is used for both pre-industrial and current Nr estimates (Lelieveld and Dentener 2000; Galloway et al. 2004). Although a small quantity, it creates NO_x high in the troposphere where it has longer residence time and more likely to contribute to tropospheric O_3 formation. It may also influence the oxidation capacity of the atmosphere before it is removed via atmospheric deposition. In contrast, the NO_x emitted at the earth's surface mainly contributes to atmospheric Nr deposition.

2.4.2 Biological Nitrogen Fixation

The BNF is a reductive process that transforms N_2 to NH_3. BNF is a microbially mediated process that occurs in several types of bacteria and blue-green algae (cyanobacteria and cyanophyta). BNF occurs when other forms of inorganic N are unavailable in solution. The lack of inorganic N provokes physiological and biochemical changes in organisms which gives them the ability to fix atmospheric N_2. In the case of cyanobacteria, the lack of inorganic N leads to differentiation of specialized cells called 'heterocysts' while in legumes, it provokes formation of root nodules (Boyd 2001). This biologically irreversible reaction is catalyzed by an extremely conserved heterodimeric enzyme complex nitrogenase, which is inhibited by O_2 (Postgate 1998). Nitrogenase enzyme complex consists of two separate enzymes—dinitrogenase reductase and dinitrogenase—which are involved in conversion of N_2 to NH_4^+. Because N fixation is very expensive in terms of energy, this capability is only an advantage when inorganic N is scarce, and the N fixing capability lies dormant during times of plentiful N supply. The microbes can be free-living (non-symbiotic) or in a symbiotic association with roots of higher plants in the terrestrial ecosystems. The majority of symbiotic BNF occurs via a root nodule symbiosis. In this relationship, the plant maintains N fixer by acting as a C source while excreting the fixed N for the benefit of the plant. Legumes and *Rhizobium* spp. are the best-known example of such a symbiotic relationship, and play a large role in BNF in agricultural soils. Non-symbiotic N fixation includes fixation by the true free-living N fixers as well as by autotrophic and heterotrophic organisms not in direct symbioses with vascular plants, including cyanobacteria symbiotic in lichens, bryophytes, or associative N_2 fixing systems. In addition, actinomycetes (*Frankia*) can fix N as free-living microorganisms or in symbiosis with number of non-leguminous vascular species. The free living N fixing microorganisms are also known as rhizosphere diazotrophs—these are most active near plant roots where decomposable C supply is high, but are not found within living tissues. These microorganisms are ubiquitous in soils and are highly diverse. The process uses the enzyme nitrogenase to convert N_2 to NH_3 (Eqs. 2.3 and 2.4).

$$N_2 + 8H^+ + 8e^- + 16ATP \rightarrow 2NH_3 + H_2 + 16ADP + 16P_i \quad (2.3)$$

$$N_2 + 6H^+ + 6e^- \rightarrow 2NH_3 \quad (2.4)$$

The reduction of N_2 is exergonic reaction, but it requires formidable activation energy to break the triple bond and a catalyst to overcome the energy barrier. The heterodimeric enzyme complex, nitrogenase catalyzes the reaction by hydrolyzing 16 ATP molecules per molecule of N_2 fixed (Canfield et al. 2010). The BNF can occur in both managed and unmanaged ecosystems. In managed ecosystems, cultivation of legumes enhances BNF. In the pristine biosphere, N fixation by terrestrial BNF is the dominant source of newly fixed N in the terrestrial ecosystems (Cleveland et al. 1999).

Quantification of the magnitude of natural terrestrial BNF–Nr at a large scale is weak, mostly due to uncertainty in estimates of BNF rates at plot scale, methodological differences, and non-uniform spatial coverage of important N fixing species and locational biases in its study (Cleveland et al. 1999). In the tropical regions (i.e., Africa, Asia and South America), where BNF is likely the important source of Nr there is a little data on terrestrial estimates of BNF rates. Estimates of the global BNF rates are also highly uncertain, mainly because of scant relevant data for natural vegetated ecosystems. In compilation of the rates of natural BNF, symbiotic BNF rates for several biome types are based on few published rates at plot scale within each particular biome (Cleveland et al. 1999). Most studies present BNF estimates as global values based on few but broad components such as forests, grasslands, and other (Paul and Clark 1997). Such coarse divisions average large land areas that contain significant variation in datasets and in biome types, which diminishes the usefulness and credibility of the data.

Cultivation–induced BNF (C-BNF) occurs in several agricultural ecosystems including crops, pasture, and fodder legumes (Smil 1999). Total global C-BNF is estimated at 33 Tg N $year^{-1}$ and ranges from 25 to 41 Tg N $year^{-1}$ (Burns and Hardy 1975; Galloway et al. 1995; Smil 1999). Increase in soybean (*Glycine max*) and meat production since 2000 may have increased global C-BNF. Estimates of about 60–70 Tg N $year^{-1}$ are reported for leguminous crops in agricultural fields (Herridge et al. 2008). A more conservative magnitude of global C-BNF for the year 2005 is estimated at 40 Tg N $year^{-1}$ (Galloway et al. 2008), although there is substantial uncertainty in this value and more precise data are needed.

Synthesis of the available data for native vegetated terrestrial biomes indicate estimates of global natural BNF rate at 168.5 Tg N $year^{-1}$, of which about 60% occurs in tropical rainforests, arid shrub lands, and tropical savannah (Cleveland et al. 1999). Further, BNF on the land is estimated at 140 Tg N $year^{-1}$ or about 10 kg N ha^{-1} $year^{-1}$ on earth's surface (Burns and Hardy 1975). Asymbiotic N fixation ranges from 1.0 to 5.0 kg N ha^{-1} $year^{-1}$ or an average of 44.0 Tg N $year^{-1}$ as a global total. The remainder is assumed to come from symbiotic fixation in higher plants. However, BNF supplies only 12% of the N that is assimilated in natural land plants each year. The remaining 88% is derived from internal recycling and the decomposition of dead materials in soils.

Estimates from the ocean are less reliable due to the importance of *Trichodesmium* (cyanobacteria) and *Richelia intracellularis* (a diazotrophic endosymbiont occurring in large diatoms) as N_2 fixers. The suggested global pelagic BNF rate of 110.0 ± 40.0 Tg N $year^{-1}$ is based on extrapolation of the North Atlantic data (Gruber and Sarmiento 1997). Benthic BNF is estimated at 15.0 ± 10.0 Tg N $year^{-1}$ (Capone 1983).

2.4.3 Industrial Nitrogen Fixation

In 1908, Fritz Haber discovered how NH_3, a chemically reactive, highly usable form of N could be synthesized by reacting atmospheric N_2 with H_2 in the presence of Fe at high pressure and temperature. Carl Bosch subsequently developed this process on an industrial scale, for which he was awarded the Nobel Prize in 1931. This industrial process, known as Haber-Bosch N fixation, uses natural gas methane (CH_4) to produce H_2, which is then combined with N_2 to form gaseous NH_3 (Eqs. 2.5 and 2.6) under conditions of high temperatures (500–600°C) and pressure with metallic Fe catalyst:

$$CH_4 + 6H_2O \rightarrow 3CO_2 + 12H_2 \quad (2.5)$$

$$4N_2 + 12H_2 \rightarrow 8NH_3 \quad (2.6)$$

The resultant NH_3 is refined and compounded to make the nitrogenous fertilizers that modern agriculture depends on. Ammonia synthesis is important because it is the primary ingredient in nitrogenous fertilizers, without which, modern agriculture would be impossible. Virtually this process produces all of N fixed industrially (IFA 2000). The Haber-Bosch process, sometimes called the most important technological advancement of the twentieth century, has significantly lowered the energy requirements and is substantially cheaper, and therefore, forms the basis of an alternative expanding supply of Nr. It has also boosted the production of many expensive or rare compounds such as dyes, artificial fibers, and has its greatest impact on production of explosives and fertilizers (Smil 2001). Chemical fertilizers are one of the major facets of the Green Revolution, which has resulted into many fold increases in crop yields (Mann 1999). Synthetic inorganic fertilizer N production supplies >80 Tg N $year^{-1}$ to agricultural soils (Schlesinger and Hartley 1992; Matthews 1994; Potter et al. 2010). About 40% of the large annual increase in crop production during the post-green revolution period is attributed to the increase in use of synthetic fertilizer N (Brown 1999). It is estimated that in 2008, Haber-Bosch fertilizer N supported about 48% of world population (Erisman et al. 2008). In addition, fertilizer N is required for bioenergy and biofuel production, which is contributing about 10% of global energy requirement, and 1.5% of fuel (Erisman et al. 2008).

About 80% of Nr manufactured by the Haber-Bosch process is used in the agricultural fertilizes (Galloway et al. 2008). However, a large portion of this is lost to the environment, where it cascades through atmospheric, terrestrial, aquatic, and marine pools before eventually being denitrified to N_2 or stored as fossil and Nr. Volatilization of NH_3 or leaching of NO_3^- to adjacent natural ecosystems is environmentally important. Its deposition leads to unintentional fertilization and loss of terrestrial biodiversity (Hesterberg et al. 1996). Transfer of Nr from terrestrial to coastal systems has led to algal blooms and decline in the quality of surface and ground waters. In atmosphere, Nr alters the balance of GHG, enhances tropospheric O_3, decreases stratospheric O_3, increases soil acidification, and stimulates formation of secondary particulate matter in the atmosphere. All these have negative effects on people and the environment (Erisman et al. 2008).

2.4.4 High Temperature Combustion

High temperature pyrolysis (>1,800°C) and pressure formed during fossil fuel combustion, primarily via the internal combustion engines, provides the energy to convert N_2 to NO through reaction with O_2 (Eq. 2.7):

$$N_2 + O_2 + \text{ fossil energy} \rightarrow 2NO \tag{2.7}$$

Some of NO is oxidized into NO_2, and mixture of NO and NO_2 form NO_x. Significant NO_x emissions are also derived from oxidation of fixed N compounds in the fuel: up to 50% of fossil fuel $NO_{x,}$ emissions can be fuel NO_x especially when heavy oil or coal is burned. Fossil fuel combustion releases about 20.0 Tg N year^{-1} in the form of NO_x globally (Muller 1992). Some of this is derived from the organic N contained in the fuels but considered as new sources fixed to the biosphere since this would have remained inaccessible in the earth's crust (Galloway et al. 1995).

Van Aardene et al. (2001) estimated that 0.6 Tg N year^{-1} was created in the form of NO_x during fossil fuel combustion in 1890, primarily from coal combustion. Based on population and other factors such as energy requirement, it is estimated that NO_x emission from fossil fuel at the onset of industrialization in1860 was 0.3 Tg N year^{-1} (Galloway et al. 2004). Most of NO_x emissions occur in the form of NO, which photochemically equilibrates with NO_2 within a few minutes. Estimated anthropogenic emissions of NO_x from fossil fuel for the year 2000 were 33.0 Tg N year^{-1}, of which, 40% originated from transportation sector (Ehhalt et al. 2001). The American and European emissions are relatively stable but emissions from East Asia have been increasing by nearly 4% annually (Kato and Akimoto 1992). Asian emissions from fossil fuel will drive an overall increase in NO_x emissions during the twenty-first century (van Aardenne et al. 2001).

2.5 Nitrogen Cycling

The global N cycle can be divided into three compartments: atmospheric, terrestrial, and marine cycle. Terrestrial and marine ecosystems together forms biosphere cycle (Galloway 2003). The atmospheric N cycling is the simplest because the direct influence of the biota is limited. Thus, chemical and physical processes primarily control the transformations in the atmosphere. In addition, the cycle of oxidized inorganic N (NO_y) is in most cases decoupled from that of the reduced N (NH_x). However, biosphere N cycling (terrestrial and aquatic ecosystems) is more complex due to microbially mediated N transformations such as BNF, nitrification, and denitrification. Despite the large annual flux through the inorganic pools (NH_4^+, NO_3^-), at any time, this pool in the biosphere is very small due to rapid uptake by organisms.

2.5.1 Historical: Pre-industrial Era Nitrogen Cycling

2.5.1.1 Natural Processes

Pre-industrial input of newly fixed N to terrestrial ecosystems were lightning and BNF. High temperatures occurring during lightning strikes produce NO in the atmosphere from molecular O_2 and N_2. This is subsequently oxidized to NO_2 and then to HNO_3, and within few days removed from atmosphere by wet and dry deposition, thus introducing Nr into ecosystems. The creation of Nr by lightning is the highest in tropical terrestrial regions where convective activity is the largest (Galloway et al. 2004). Deposition of lightning Nr occurs primarily in tropical regions where the lightning NO_x formation predominates (Bond et al. 2002). For this budget analysis, a global estimate of 5.4 Tg N year^{-1} is used (Table 2.3; Galloway et al. 2004). Lightning NO_x is important because it occurs high in troposphere where it has longer atmospheric residence time and likely contributes to tropospheric O_3 formation, which contributes to the oxidizing capacity of the atmosphere.

Based on survey of the existing literature from natural terrestrial ecosystems, Galloway et al. (2004) estimated that prior to large anthropogenic alteration of global N cycling during industrial era, natural terrestrial ecosystems BNF (N-BNF) contributed 100–290 Tg N year^{-1}. Of the 11.5 Mha of natural vegetated land (Mackenzie 2003), it is estimated that 0.76 Mha had been altered by anthropogenic action by the beginning of industrial era in 1860—including land clearing for cultivation and conversion of forests to pastures (Houghton 1999). Therefore, N-BNF was estimated at 120.0 Tg N year^{-1} (Fig. 2.1). The BNF also created 40.0–140.0 Tg N year^{-1} in marine ecosystems (Galloway 1998). However, N fixed in marine ecosystems is not transported to terrestrial ecosystems, except in small quantities volatilized as NH_3 and subsequently deposited on the land, which is difficult to quantify.

Table 2.3 Global atmospheric emissions of NO_x and NH_3

	Historical – 1860		Contemporary	
Source[a]	NO_x	NH_3	NO_x	NH_3
Emissions				
Agriculture				
Savannah burning	0.9	0.2	2.9	1.8
Deforestation	0.2	0.2	1.1	1.4
Agricultural soil losses	0.0	0.0	2.6	
Agricultural crops	–	0.2	–	4.1
Animal waste	0.9	5.9	2.4	24.3
Fertilizers	–	0.0	–	9.7
Landfills	–	0.1	–	3.1
Sub total	2.0	6.6	9.0	44.3
Energy				
Fossil fuel combustion	0.3	0.0	20.4	0.1
Biofuel combustion	0.4	0.7	1.3	2.6
Transport	0.0	–	4.1	–
Industrial processes	0.0	0.0	1.5	0.2
Miscellaneous			0.7	–
Sub total	0.6	0.7	27.2	2.9
Natural				
Lightning	5.4	–	5.4	–
Emissions from natural soils	2.9	–	2.9	–
Natural fires	1.6	1.6	0.8	0.8
Stratospheric sources	0.6	–	0.6	–
Vegetation and wild animals	–	6.0	–	4.6
Ocean	–	5.7	–	5.6
Subtotal	10.5	13.3	9.7	11.0
Emission total	13.1	20.6	45.9	58.2
Deposition total	12.8	18.8	45.8	56.7

[a]Emission data from van Aardenne et al 2001; Galloway et al. 2004, deposition data from Lelieveld and Dentener 2000

2.5.1.2 Human Alteration of Nitrogen Cycle in the Pre-industrial Era

Anthropogenic activities influence the N budgets in two ways by (i) increasing the mobilization of existing Nr, and (ii) creating new Nr. As hunter-gatherers, anthropogenic impact on N cycle was limited to mobilization of existing forms of Nr (such as biomass burning), and the extent of the impact was limited relative to the natural processes. Anthropogenic activities that created Nr during the pre-industrial era were mainly cultivation of legumes and fossil fuel combustion for energy, especially coal. It is estimated that fossil fuel combustion of coal generated 0.6 Tg N $year^{-1}$ in the form of NO_x in 1890 (van Aardenne et al. 2001). Scaling back these estimates by population and other factors such as energy demand, Galloway et al. (2004) estimated that fossil fuel combustion in 1860 created 0.3 Tg N $year^{-1}$ in the

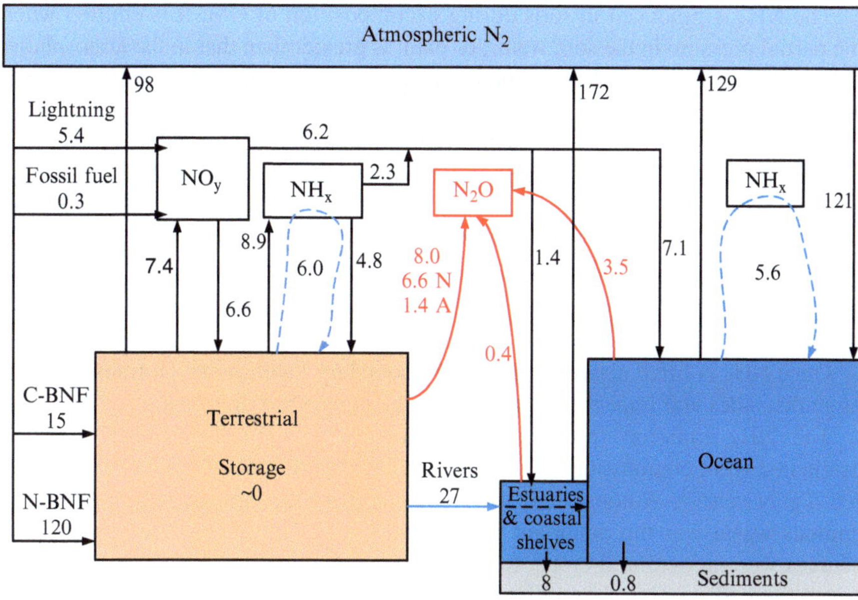

Fig. 2.1 Global N cycle for the pre-industrial era showing of N in Tg year^{-1}. C-BNF, biological N_2 fixation within agroecosystem (agricultural crops), N-BNF, biological N_2 fixation within natural ecosystems (Source: van Aardenne et al 2001; Galloway et al. 2004)

form of NO_x, (mainly through coal combustion at the beginning of industrial era). Advent of legume cultivation ~5,000 years. ago initiated C-BNF on earth. At the beginning of industrial era, C-BNF was ~15.0 Tg N year^{-1} (Galloway et al. 2004; Fig. 2.1).

2.5.1.3 Pre-industrial Era Nitrogen Budget

The NO_x and NH_3 emissions can result from natural processes, agronomic practices, and energy production. For NO_x, natural emissions occur from soil processes, lightning, wild fires, and stratospheric injection. NO is the most commonly emitted species. Once in the atmosphere, it is quickly oxidized to NO_2, which is then oxidized to HNO_3, potentially reacting with NH_3 to form aerosol. Natural NO_x emission totals in 1860 were 10.5 Tg N year^{-1} (Table 2.3). Energy production contributed NO_x emissions of 0.7 Tg year^{-1}, of which, fossil fuel combustion of coal and biofuel combustion were 0.3 and 0.4 Tg N year^{-1}, respectively (Table 2.3). Food production related NO_x were 2.0 Tg year^{-1}, of which, agricultural waste contributed 0.9 Tg N year^{-1}, slash-and-burn of forests accounted for 0.2 Tg N year^{-1} and savanna grass and shrubs produced 0.9 Tg N year^{-1} (Table 2.3). The total NO_x emissions were 13.1 Tg N year^{-1} (Table 2.3; Fig. 2.1). Total NO_x deposition is estimated at 12.8 Tg year^{-1} (Table 2.3), of which 6.6 Tg N year^{-1} was deposited on terrestrial ecosystem and 6.2 Tg N year^{-1} was exported to coastal regions and oceans (Fig. 2.1).

The NH_3 is produced in soils during decomposition of OM. It is emitted when the partial pressure in the soil, water, or plant is greater than that in the atmosphere. Other NH_3 sources include savanna burning, deforestation, animal waste, landfills and sewage disposal, wild fires and oceans. It is the common atmospheric gaseous base. Once in the atmosphere it can be converted into aerosol in acid–base reactions with gases such as HNO_3 (Eq. 2.8) or aerosol such as H_2SO_4 (Eq. 2.9):

$$NH_3 + HNO_3 \rightarrow NH_4NO_{3(s)} \quad (2.8)$$

$$NH_{3(g)} + H_2SO_{4(s)} \rightarrow (NH_4)_2SO_{4(s)} \quad (2.9)$$

Once NH_3 is lifted above the planetary boundary layer, it can be transported to large distances and impact receptors far in the downwind direction.

The NH_3 emission from energy production was 0.7 Tg N $year^{-1}$, which originated from combustion of biofuels (Table 2.3). Agronomic practices released 6.6 Tg N $year^{-1}$. Among this, low temperature NH_3 emission from domestic animals waste was the dominant fraction at 5.3 Tg N $year^{-1}$. Other agricultural sources were agricultural crops, forests, and savanna, which contributed 0.2 Tg N $year^{-1}$ each (Table 2.3). Landfills and sewage disposal contributed 0.1 Tg N $year^{-1}$. Natural NH_3 emissions included natural vegetation and wild animals, natural fires and oceans, which emitted 6.0, 1.6 and 5.7 Tg N $year^{-1}$, respectively (Table 2.3). Total NH_3 emissions were 20.6 Tg N $year^{-1}$, of which, 14.9 Tg N $year^{-1}$ were emitted from the land and 5.6 Tg N $year^{-1}$ was emitted from the ocean (Table 2.3; Fig. 2.3). Among the emissions, the rapid local recycling of NH_3 that were deposited within the same region for the terrestrial and ocean ecosystems were 6.0 and 5.6 Tg N $year^{-1}$, respectively (Fig. 2.1). Total NH_3 deposition was 18.8 Tg N $year^{-1}$ (Table 2.3). Deposition of NH_3 on the continents for non-local sources was 4.8 Tg N $year^{-1}$, and transfer to the oceans was 2.3 Tg N $year^{-1}$ (Fig. 2.1).

The oxidized inorganic N species not included in NO_y is N_2O, which is produced during nitrification and denitrification (Chap. 3). Because of its stability, no significant chemical reaction takes place in the troposphere. However, once in the stratosphere it is converted to NO by ultraviolet (UV) radiation (Eq. 2.10) viz:

$$N_2O + O(^1D) \rightarrow 2NO \quad (2.10)$$

The NO thus produced then destroys stratospheric ozone in a reaction that regenerates it (Eqs. 2.11, 2.12, and 2.13), viz:

$$NO + O_3 \rightarrow NO_2 + O_2 \quad (2.11)$$

$$O_3 \rightarrow O + O_2 \quad (2.12)$$

$$NO_2 + O \rightarrow NO + O_2 \quad (2.13)$$

Net reaction:

$$2O_3 \rightarrow 3O_2 \qquad (2.14)$$

The N_2O emissions in 1860 from terrestrial natural ecosystems were 6.6 Tg N_2O-N year^{-1} (Bouwman et al. 1995). Anthropogenic emissions from animal waste, C-BNF, crop residues and biomass burning were 1.4 Tg N_2O-N year^{-1} (Kroeze et al. 1999), and rivers contributed 0.05 Tg N_2O-N year^{-1} (Seitzinger et al. 2000). Estuaries and coastal shelves contributed 0.02 and 0.4 Tg N_2O-N year^{-1} (Seitzinger et al. 2000) and open ocean emitted 3.5 Tg N_2O-N year^{-1} (Nevison et al. 1995) (Fig. 2.1). Thus, total emissions in 1860 were 12.0 Tg N_2O-N year^{-1}.

In 1860, the natural Nr generation in pre-industrial era was 125.4 Tg N year^{-1}, of which, 120.0 Tg N year^{-1} by terrestrial BNF and 5.4 Tg N year^{-1} by lightning (Fig. 2.1). Anthropogenic Nr generation rates were 15.3 Tg N year^{-1}. The C-BNF accounted for most of anthropogenic Nr (15 Tg N year^{-1}), while coal combustion contributed 0.3 Tg N year^{-1}. Total terrestrial Nr creation was 141 Tg N year^{-1} (Fig. 2.1). The riverine export to estuaries, coastal waters was 27 Tg N year^{-1}; atmospheric deposition to the coastal waters was 1.4 Tg N. In addition, 8.0 Tg N was emitted into the atmosphere as N_2O where it was stored or transported to stratosphere. The remaining 98.0 Tg N is either denitrified back to N_2 or part of it is stored in the terrestrial ecosystems. The atmospheric N_2O data recorded in ice bubbles from the pre-industrial era dating back to 0 AD indicated that N_2O emissions remained constant at approximately 270–280 ppb, which implies that Nr creation and denitrification was approximately equal and Nr was not accumulating in pristine ecosystems prior to anthropogenic perturbations (MacFarling Meure et al. 2006).

2.5.2 Contemporary: Post-industrial Era Nitrogen Cycling

The fundamental anthropogenic change in the contemporary global N cycle is an increase in the transfer from the vast and unavailable forms of N to the terrestrial ecosystem. Human activities have significantly altered the natural N cycle in the post-industrial era through food production and energy use. The most fundamental change in N cycling is the dramatic increase in Nr. Anthropogenic activities have more than doubled the rate of transfer of N from highly abundant but unavailable form—N_2 in the atmosphere to the available forms such as NH_4^+, NO_3^- and NO_2^- (Smil 1999). The major sources of changes are: (1) mobilization and fixation during fuel combustion, (2) industrial fixation for use as fertilizers for crop production, and (3) BNF increase associated with increase in agriculture production of crops that fix N symbiotically.

From 1860 to the present, food production and energy consumption increased steadily on both absolute and per capita basis. The anthropogenic Nr generation also increased from ~15.3 Tg N year^{-1} to 156 Tg N in 1995 (Galloway et al. 2004).

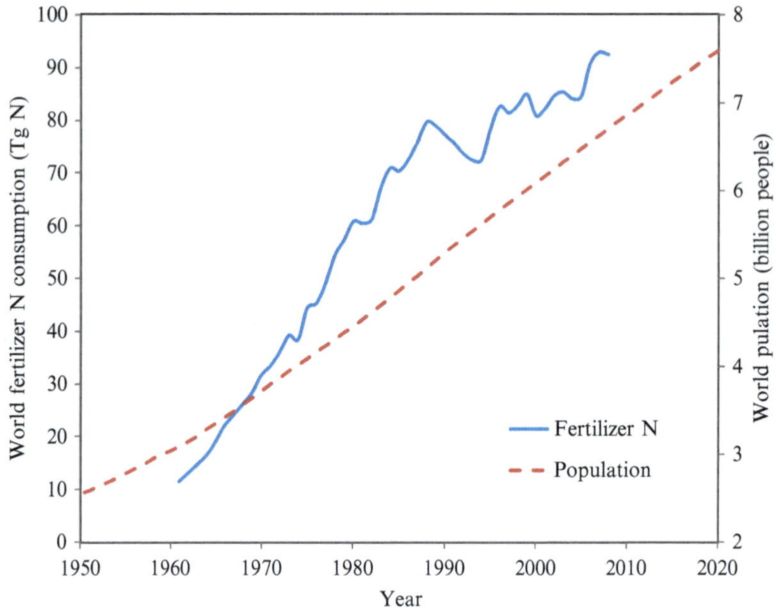

Fig. 2.2 Global population trends from 1950 to 2020 and actual global fertilizer N consumption between 1961 and 2008. (Source: Population data-US Census Bureau, fertilizer consumption data: FAOSTAT database)

In large part, the increase was due to rising agricultural demands for Nr, which was sustained by the Haber-Bosch process. From 1995 to 2005, the Nr increased further to 187.0 Tg N $year^{-1}$, in large part because of 20% and 26% increase in cereal and meat production, respectively (Galloway et al. 2008). Between 1961 and 2008, fertilizer N application to agricultural lands increased at about 1.72 Tg N $year^{-1}$ to match the food demand for the world population which grew by about 72 million $year^{-1}$ (Fig. 2.2). Industrial fixation of N has increased exponentially from near zero in 1940s. Until 1970s, most of industrial N fertilizers were applied in developed countries. At present, however, the fertilizer N use in developed countries has stabilized, but application is increasing dramatically in developing countries (Zhu et al. 2005). The momentum of growth in human population and increase in urbanization ensure that industrial N fixation may continue to grow even at higher rates for decades to come because of the increase in demand for food.

The C-BNF occurred in several agricultural systems, including leguminous crops, pasture, and leguminous fodder (Smil 1999). The C-BNF increased significantly due to expansion of leguminous crops production and improved pastures to meet the demand for fodder crop. The C-BNF estimate for 1995 was 31.5 Tg N $year^{-1}$ (Galloway et al. 2004). The current C-BNF is estimated at 40 Tg N $year^{-1}$ (Galloway et al. 2008). The increase over the past 10 years is because of the increase in area under soybean (*Glycine max*) and fodder crops.

In addition, with the industrial revolution, fossil fuel combustion became an increasingly important energy source for both industrial and domestic uses (Mackenzie 2003). Primary commercial energy production by coal, natural gas, and petroleum consumption resulted in significant increase in NO_x emissions as a waste product of energy generation and high temperature N_2 fixation. In 1995, NO_x emissions from fossil fuel combustion were estimated at 27.2 Tg N year^{-1} (Galloway et al. 2004). However, decrease in NO_x emissions in the developed world recently as a result of control in emissions has led to stabilization of global Nr-NO_x generation at about 25 Tg N year^{-1} from 1995 (Cofala et al. 2007). Other sources of NO_x include agriculture, natural soils, wild fires, and stratospheric deposition.

The NH_3 emission from anthropogenic activities increased from 7.3 Tg N in 1860 to current levels 47.2 Tg N year^{-1}. Finally, NH_3 released from industrial use of Nr is poorly understood. NH_3 from the Haber-Bosch process is used to create a range of industrial products including nylon, plastics, resins, glues, melamine, animal/fish/shrimp feeds supplements and explosives. About 20% of the Haber-Bosch created Nr is utilized in industrial processes and its fate in the environment is not well known.

The data on the anthropogenic generation of Nr have been relatively less uncertain. However, the fate of Nr in the ecosystems is less certain. About 18% of Nr is exported to coastal ecosystems, and its fate is not clearly understood. Some of it is denitrified, releasing N_2O and N_2. About 13% deposits into the ocean from marine atmosphere, and 4% emitted as N_2O. As much as 65% either accumulates in soils, vegetation, and ground water or denitrified into N_2. However, the uncertainty of this component is large.

2.5.2.1 Current Reactive Nitrogen Distribution and Its Fate

Once Nr molecule has been created it remains in the environment for considerable time. The Nr can be highly mobile by virtue of its direct injection into the atmosphere (NO_x, NH_y and N_2O), or conversion to reduced N (NH_4^+) to mobile species (NO_2^- and NO_3^-). Over time, it cascades through the environment and drives different environmental impacts such as eutrophication, acidification in space and time, and affects radiative forcing and therefore climate. Both natural and anthropogenic processes emit NO_x into the atmosphere. Agricultural sources predominantly emit the alkaline NH_x (i.e. the sum of NH_3 and NH_4^+). In the analysis of the fate of Nr introduced in the global ecosystems, Smil (1999) estimated that 50% is removed by harvested crops, 23% is leached N, 6% is volatilized as NH_3, and 6% is released into the atmosphere as NO_x and N_2O. Only 10% of Nr is denitrified and converted into inert N_2 within the ecosystem. The remaining, ~90% is recycled within the ecosystems. The endpoint of cascading is ultimately the emission of N_2 or N_2O to the atmosphere. During the cascade, Nr can influence GHGs exchange with the atmosphere, aerosol production, tropospheric O_3 or increase biological productivity, which require C. All these processes have impact on global climate.

The ultimate fate of NO_x and NH_x is their removal by the wet and dry deposition on terrestrial and aquatic ecosystems. The deposition of Nr induces cascade of effects (Galloway et al. 2004). The vast majority of the land area of northern hemisphere is receiving increased N deposition due to anthropogenic activity. For example, deposition to forests ranges from 5 kg N ha^{-1} $year^{-1}$ in Northern Europe to >60 kg N ha^{-1} $year^{-1}$ in central Europe, and the range for short vegetation is about half (Dise et al. 2009). Ammonium is the dominant form of atmospheric N input in Europe. Ammonium originates from NH_3 emitted primarily from animal husbandry, while NO_3^- originates from N oxides emitted by fossil fuel combustion and automotive exhaust. At remote sites in the southern hemisphere, wet deposition is <1 kg N ha^{-1} $year^{-1}$ (Vitousek et al. 1997). Total pre-industrial inputs to forests were therefore <2 kg N ha^{-1} $year^{-1}$. The NPP of most terrestrial ecosystems are limited by N availability and Nr deposition may enhance ecosystem productivity (Vitousek et al. 1997) with possible consequences to the global C cycle (Prentice et al. 2001). It also leads to imbalance in nutrients cycling leading to change of ecosystem diversity (Bobbink et al. 1998; Phoenix et al. 2006). In fresh water and coastal regions, Nr inputs lead to noxious algal blooms, increase in water turbidity, shifts in food webs and loss of fish stocks (Rabalais 2002). In addition, Nr deposition in both terrestrial and marine ecosystems leads to denitrification and enhanced emissions of N_2O.

Owing to increase in Nr generation due to anthropogenic activities such as fossil fuel combustion and agricultural practices, the global emissions of NO and NH_3 have increased by a factor of 3 since the pre-industrial era. Current global NO_x and NH_3 emissions from anthropogenic sources are estimated at 40 and 53.4 Tg N $year^{-1}$, respectively (Dentener et al. 2006; Lamarque et al 2010). The major part of NO_x emissions (30 Tg N $year^{-1}$) originate from fossil fuel combustion, including road transport (10 Tg $year^{-1}$), shipping, and aviation (6 Tg $year^{-1}$) and energy sector (7.4 Tg $year^{-1}$). These emissions rapidly increased during twentieth century. In addition, about 6.0 Tg N $year^{-1}$ originates from biomass burning. Soil emissions of NO_x are also affected by use of fertilizers, and therefore, difficult to separate between natural and anthropogenically-perturbed component from soils. Regionally, the emission increases have been more substantial especially in North America, Europe, and Asia where increases by up to a factor of 10 have been reported during the last century (van Aardenne et al. 2001). Further increases of emissions and deposition are predicted by 2050 and 2100 (Galloway et al. 2004; Lamarque et al. 2005; Dentener et al. 2006), in particular in developing countries, especially the most important sources of NOx, that is transport and power production. However, landmasses and ocean in the southern hemisphere are less impacted by human activity. However, this will likely change as population grows along with the increase in demand for food. Much of the knowledge on N dynamics is from the temperate regions. Yet, the tropical regions are predicted to experience the most dramatic increase in Nr during the first half of the twenty-first century (Zhu et al. 2005) because of rapid urbanization and intensification of agriculture. Tropical ecosystems generally P and cation deficient, but relatively N-rich compared to temperate ecosystems (Matson et al. 1999). Thus, responses in these ecosystems may be different compared to those in the temperate regions, and result in rapid N losses into water and air, soil cation depletion, and reduced C uptake.

Most of the Nr produced by human activities in NH_3 form is less mobile; it could be absorbed by the soil surface (e.g., NH_4^+). However, once nitrified, it becomes highly mobile (i.e., NO_3^-) and is easily transported into surface waters. It is estimated that ~20 Tg of dissolved N and 20 Tg of particulate N from anthropogenic sources are transported by rivers into coastal ecosystems (Galloway et al. 1995). Modeling estimates suggest that riverine dissolved inorganic N discharge to coastal ecosystems is 16 Tg $year^{-1}$, or nearly 11% of the anthropogenic N produced (Seitzinger and Kroeze 1998). Most of the transport occurs in rivers draining East and South Asia. There are three possible pathways of N discharged into coastal waters: (1) storage in the sediments, (2) transport into the open ocean, and (3) denitrification and conversion into N_2 and N_2O. For the North Atlantic Ocean, a region from which most of the research data are generated, the riverine N is almost entirely denitrified in the coastal and shelves regions (Nixon et al. 1996; Seitzinger and Giblin 1996). On global basis, however, data are too sparse to discern the fate of N transported into coastal waters.

A large proportion of anthropogenic Nr remains within the terrestrial ecosystems. The possible fate of this component is: (1) storage in long-term reservoirs, including soil OM, ground water, and biomass, and (2) denitrification into N_2 and N_2O. Forested ecosystems are accumulating N, as is evidenced by the observations of N saturation syndrome (Stoddard 1994; Aber et al. 1998; Lovett et al. 2000; Corre and Lamersdorf 2004), defined as a condition of plants and soil biotic and abiotic processes to accumulate N that exceeds the biotic demand (Aber et al. 1998). While there is a substantial potential for N storage in the groundwater, biomass and soil OM, there is also a large potential for significant increase in NO_3^- concentration of streams and other surface waters, increased denitrification and increased N_2O emissions (Liu and Greaver 2009).

2.5.2.2 Contemporary Nitrogen Budget

The contemporary NO_x emission estimate is at 45.9 Tg N $year^{-1}$, of which, 9.7 Tg N $year^{-1}$ is from natural sources, while 36.2 Tg N $year^{-1}$ (79%) is anthropogenic. Natural sources include lightning, emissions from natural soils, natural wild fires, and stratospheric sources at 5.4, 2.9, 0.8, and 0.6 Tg N $year^{-1}$, respectively (Table 2.3). Anthropogenic source of NO_x include biofuel combustion at 1.3 Tg N $year^{-1}$, fossil fuel combustion, at 20.4 Tg N $year^{-1}$, transportation at 4.1 Tg N $year^{-1}$, of which ships and aircraft contribute 3.6 and 0.5 Tg N $year^{-1}$, respectively (Table 2.3). Industrial NO_x emission estimate is 1.5 Tg N $year^{-1}$. Agriculture Nr-NO_x is 9.0 Tg N $year^{-1}$, which consists of agricultural waste, forest clearing and burning, savanna grass and shrub clearing and burning and emissions from fertilized agricultural soils at 2.4, 1.1, 2.9, and 2.6 Tg N $year^{-1}$ (Table 2.3).

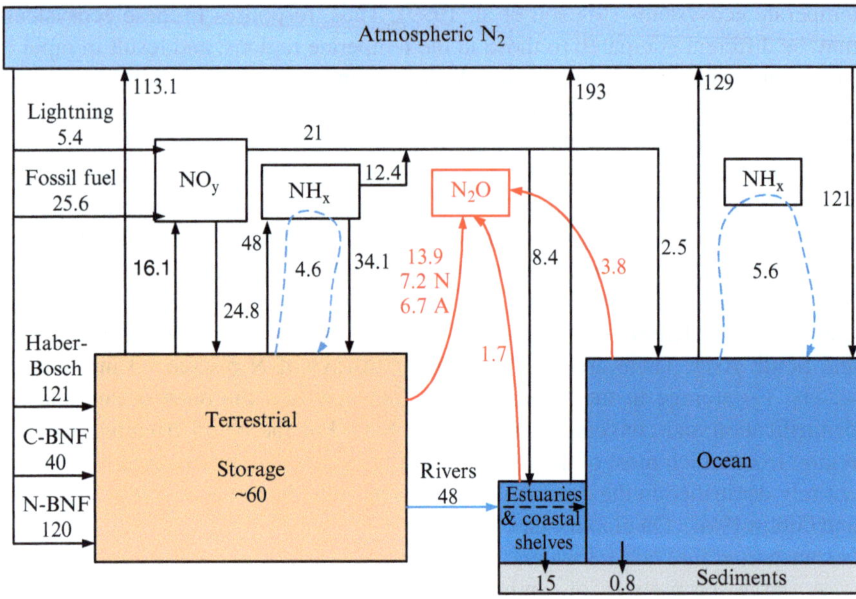

Fig. 2.3 Contemporary global N cycle showing of N in Tg year^{-1}. C-BNF, biological N_2 fixation within agroecosystem (agricultural crops), N-BNF, biological N_2 fixation within natural ecosystems (Source: Galloway et al. 2004, 2008)

All NO_x emissions are of the terrestrial sources. Total NO_y deposition is 45.8 Tg N year^{-1}, 24.8 Tg N year^{-1} to the continents and 21.0 Tg N year^{-1} to oceans (Table 2.3; Fig. 2.3).

The contemporary NH_3 emissions are 58.2 Tg N year^{-1}, of which, anthropogenic emissions are 47.2 Tg N year^{-1} or 81%. Natural NH_3 sources include vegetation and wild animals at 4.6 Tg N year^{-1} and ocean emissions at 5.6 Tg N year^{-1}. Rapid re-deposition within the same regions effectively removes natural NH_3 emissions (green dotted lines in Fig. 2.3) from the ocean. Agriculture is the dominant anthropogenic source, contributing 44.3 Tg N year^{-1} or 94% of the anthropogenic sources. The components of agricultural emissions are low temperature NH_3 emission from domestic animal waste at 22.9 Tg N year^{-1} and fertilizers at 9.7 Tg N year^{-1}. Other agricultural sources include combustion of agriculture waste, forests, savannas, and crops emission at 1.4, 1.4, 1.8 and 4.0 Tg N year^{-1}, respectively. Sewage and landfill emission estimate is 3.1Tg N year^{-1} (Table 2.3). Total NH_3 deposition is 56.7 Tg N year^{-1}, of which, 38.7 Tg N year^{-1} is deposited into the terrestrial ecosystems and 18.0 Tg N year^{-1} into the oceans. The NH_3 which is rapidly re-deposited within the same region of the terrestrial ecosystems (mostly forest) is estimated at 4.6 Tg N year^{-1} (Fig. 2.3).

Global N_2O emissions increased from 12 Tg N_2O-N year^{-1} in 1860 to about 19.4 Tg N_2O-N year^{-1} in 2010 (Fig. 2.3). Emissions from terrestrial soils were estimated at 8.0 Tg N_2O-N year^{-1} in 1860 and increased to 13.9 Tg N_2O-N year^{-1} in 2010, primarily due to increase in N use in agricultural production systems (Denman et al. 2007). The N_2O emissions from estuaries and coastal shelves also increased from 0.4 to 1.7 Tg N_2O-N year^{-1} due to indirect agricultural effects (Seitzinger et al. 2000; Fig. 2.3). Davidson (2009) updated the global N_2O emission to 19.8 Tg, N_2O-N year^{-1} that is higher than the IPCC fourth assessment report based on estimate that 2.0% of manure N is converted to N_2O. The findings have important implication that projected increase in N fertilizer use to support the demand for human food production and higher per capita meat consumption is likely to increase N_2O emissions.

Current BNF by natural terrestrial ecosystems is estimated at 107 Tg year^{-1}, lightning is generating an additional 5.4 Tg N year^{-1}. Anthropogenic activities-mainly agricultural and energy production are generating 156 Tg N year^{-1}. Riverine export of Nr to coastal areas increased from 27 Tg N year^{-1} in 1860 to 48 Tg N year^{-1}. Currently, 268 Tg N year^{-1} of new Nr is introduced into the terrestrial ecosystems, 81 Tg N year^{-1} is transferred to marine environment via atmospheric deposition and riverine export, and 13.9 Tg N year^{-1} is emitted into the atmosphere as N_2O. Among the remaining 175 Tg N year^{-1}, about 113.0 Tg N year^{-1} is denitrified into N_2. About 40% of European Nr input is denitrified (van Egmond et al. 2002). Country estimates include about 40% for the Netherlands (Kroeze et al. 2003) and 32% for USA (Howarth et al. 2002). However, the quantity of Nr that is denitrified into N_2 remains the largest uncertainty in the N cycling due to scarcity of reliable field measurements (Galloway et al. 2004; Boyer et al. 2006). About 60.0 Tg N year^{-1} of Nr is accumulating in the terrestrial ecosystems.

Research efforts on marine BNF has focused on planktonic cyanobacterium *Trichodesmium* spp., (Capone et al. 1997; Karl et al. 2002), although now a variety of marine cyanobacteria and other bacteria are known to fix N_2 biologically in marine environment (Mulholland 2007; Carpenter and Capone 2008). Based on direct rate measurements, *Trichodesmium* accounts for a quarter to half of geochemically derived estimates of marine N_2 fixation (Mahaffey et al. 2005). Estimated pelagic marine N_2 fixation ranging from 87.0 to 200.0 Tg N year^{-1} have been reported (Gruber and Sarmiento 1997; Deutsch et al. 2001; Brandes and Devol 2002; Karl et al. 2002; Lee et al. 2002; Galloway et al. 2004). Because many direct estimates of global pelagic BNF are based on highly spatially, temporally and physiologically limited and variable data and since most of existing geochemical estimates rely on stoichiometric relationships of nutrient standing stocks without consideration of the imbalances between rate estimates of C and N_2 fixation (Mulholland 2007), the uncertainty associated with pelagic BNF is high. For this budget, a conservative value of 121.5 Tg N year^{-1} is used, with an assumption that this value does not vary over time.

In addition, marine ecosystem is currently receiving riverine input of 48 Tg N year^{-1}, atmospheric deposition of NO_y and NH_3 of 33 Tg N year^{-1}. The Nr losses are primarily from denitrification to N_2 at 150–450 Tg N year^{-1} (Codispoti et al. 2001). Estimated particulate OM storage into coastal shelves and ocean is 15 and 0.8 Tg N year^{-1}, respectively (Fig. 2.3). The estimated residence time of Nr in the ocean is 1,500–5,000 years (Codispoti et al. 2001). The atmospheric deposition and riverine inputs of Nr are the two major connection points between terrestrial and oceans. Their importance as sources of Nr to the oceans is increasing with time, as illustrated by 1,860 N budget compared to current N budget.

2.5.2.3 Questions Pertaining to Increased Reactive Nitrogen in Terrestrial and Aquatic Ecosystems

The fate of anthropogenic Nr input into the terrestrial ecosystems is relatively well understood and has low uncertainty. However, the fate of Nr into other ecosystems, including aquatic and atmospheric systems is not well known. It is estimated that 18% of Nr added into the terrestrial ecosystems is exported to rivers, estuaries and coastal waters and is denitrified; 13% is deposited in oceans from marine atmosphere, and 4% is emitted as N_2O (Galloway et al. 2004). Therefore, 65% of Nr added into the terrestrial ecosystem either accumulates in soils, vegetation, and groundwater or denitrified into N_2. The denitrification rates in soils are not well understood, as a result, the quantity of Nr stored in the terrestrial ecosystems is uncertain. The Nr input into terrestrial ecosystems continues to rise, and the impacts of increase in Nr input to terrestrial ecosystems are not fully understood. There is a possibility that as the terrestrial ecosystems becomes saturated with Nr, more of it will be exported into coastal waters, or more may be denitrified, increasing risks of global warming. The most direct effect of increased Nr to the environment is through the formation of N_2O, which is responsible for anthropogenic radiative forcing. Pathways and impact of Nr are also influenced by its chemical form, a factor which highlights the need for a better understanding of the fate of oxidized versus reduced forms of N.

N is affected by and affects the climate, and the net contributions of anthropogenic Nr to climate change are vigorously debated (Sutton et al. 2007). The Nr increases radiative forcing in the troposphere, principally by the production of N_2O and tropospheric O_3. In contrast, it can also have a cooling effect, largely through tropospheric aerosols and stratospheric O_3 decline (Denman et al. 2007). In addition, Nr can interact with the C cycle with the global effects on atmospheric CO_2 and CH_4 (Hungate et al. 2003; Gruber and Galloway 2008). Elevated levels of Nr deposition may stimulate plant growth in N limited ecosystems, and increase CO_2 uptake. The terrestrial sink of CO_2 is influenced by increased Nr availability. The C and N cycles are linked through the ecosystem C:N ratio. Uptake of atmospheric CO_2 is therefore, limited by N availability. Effects of increase in N on C storage need to be assessed, however. The acceleration of N cycling because of global warming is one of the mechanisms to further increase N availability to the ecosystem. This mechanism works through faster decomposition of organic matter and associated N mineralization rates.

An increase in Nr in the atmosphere influences oxidation capacity, radiation balance, and acidity. Many of these effects are specific to individual N compounds, (e.g., oxidative effects of NO_x, radiative balance of N_2O, acidification of some NO_x, and acid neutralizing capacity of NH_x). However, the biogeochemistry of Nr is intriguing because under appropriate conditions, almost any species of Nr can be converted into any other form of Nr. Therefore, one atom of N can have cascading impacts. For example, NO emitted from fossil fuel combustion can change the oxidative capacity of the atmosphere, increase the acidity of aerosols, change the radiative properties of the atmosphere, increase the acidity of precipitation and ecosystems, and enhance the NPP of the ecosystems etc. These cascading effects can continue as the same atoms are converted from one species into another until they are either stored in a long-term reservoir or are denitrified into N_2. Such cascading effects of Nr make its anthropogenic increase of a great interest to biogeochemists, environmentalists, agriculturalists, and atmospheric scientists as well as policy makers.

Effects of Nr input into tropical regions in terrestrial and aquatic ecosystems must be quantified. Due to BNF, P, and cations deficiency, many tropical ecosystems tend to be relatively N-rich, suggesting that increase in Nr may lead to much different response compared to that in the temperate regions, probably leading to more N_2O emissions (Matson et al. 1999).

Rapidly growing biofuel development may create yet new and rapidly changing dimensions of the anthropogenic influence on the global N cycle. In addition to increased emissions of NO_x and NH_x, the increase in use of biofuel to reduce the dependence of fossil fuel as energy source may increase the use of fertilizer N to meet the new demand for fuel crops for use in energy feedstock. Production of biofuel may increase release of Nr into an ecosystem, accentuating contribution to change in the N cycle. It may also lead to increase in N_2O emissions from fertilizers and combusted biomass, which may cancel out the savings in CO_2 emissions due to change from fossil fuel use into biofuel (Crutzen et al. 2008). It can also lead to tropospheric O_3 production. Therefore, before recommending for changes in the energy policy in favor of agro-biofuel, there is a need for better analysis to quantify the true emissions saving based on global warming potential of CO_2 and N_2O.

The increasing population and increase in per capita food and energy consumption will lead to a continued increase in Nr generation and N mobilization in the future because of use of Nr for agricultural production (Fig. 2.2), and increase in energy demand to match the population growth. It is possible to slow down the increase in fertilizer N use by increase in N fertilizer use efficiency, since crops do not utilize about 50% of fertilizer N applied (Smil 1999). There are several interventions to improve N use efficiency and minimize the cascading effects of Nr (Smil 1999). These include: (i) choice of appropriate fertilizing compounds based on soil testing, (ii) maintenance of proper nutrient ratios, and (iii) attention to timing and placement of N fertilizer. In addition, other indirect approaches which can reduce the need for synthetic fertilizers or increase the efficiency of N use are: (i) frequent planting of leguminous crops and optimization of conditions to favor BNF organisms, (ii) maximizing the recycling of organic wastes, (iii) integrated use of organic and synthetic fertilizers, (iv) improving soil management by reducing soil erosion, (v) maintaining adequate soil moisture, and (vi) controlling crop pests (Smil 1999).

2.6 Conclusions

Nitrogen is the important element that controls the productivity and functioning of terrestrial and marine ecosystems. Human activities continue to transform global N cycle at a record pace, reflecting growing demand for reactive N in agriculture and industry and increased combustion of fossil fuels. From the 1860 to the current, the anthropogenic reactive N generation increased dramatically from 15 Tg N $year^{-1}$ to current levels at about 187 Tg N $year^{-1}$. The major driving factor for this growth is the increased agricultural demand for food to match with the population growth. The Nr generation trend is still accelerating, and is unlikely to change in the foreseeable future, because of growing demand of food for growing global population. N is influenced by and influences climate. Reactive N can directly increase radiative forcing in the troposphere through the production of N_2O and tropospheric O_3, but Nr can also have cooling effect through tropospheric aerosol formation and decreasing of stratospheric O_3. In addition, N has strong interaction with C, and elevated Nr deposition may stimulate plant growth in N-limited ecosystems, thereby increasing CO_2 sequestration. Further research is needed to evaluate the overall consequences of increased Nr to the ecosystem. Other aspects of N cycling alteration, which need further research, include: (i) the ultimate fate of Nr in both terrestrial and marine ecosystem need to be refined, (ii) effects increasing Nr on human health.

Suggested Study Questions

1. What is the fate and consequences of the Nr introduced into the global environments?
2. Explain why global cycles of N and C are coupled.
3. The global N cycle can be compartmentalized into atmosphere, terrestrial and marine cycling. Describe each cycle, emphasizing the interconnections between the compartments.
4. Describe the mechanisms responsible for transferring Nr from biosphere to the atmosphere.
5. The global N cycle plays significant role in understanding global N_2O emissions. Explain.
6. Is the Haber-Bosch process a mixed blessing? Explain.

References

Aber, J., McDowell, W., Nadelhoffer, K., Alison, M., Berntson, G., Kamakea, M., McNulty, S., Currie, W., Rustad, L., Fernandez, I. (1998) Nitrogen saturation in temperate forest ecosystems. BioScience **48**(11), 921–934

Batjes, N.H. (1996) Total carbon and nitrogen in the soils of the world. Eur. J Soil Sci. **47**(2), 151–163

Berner, R.A. (2006) Geological nitrogen cycle and atmospheric N_2 over Phanerozoic time. Geology **34**(5), 413–415. doi:10.1130/g22470.1

Bobbink, R., Hornung, M., Roelofs, J.G.M. (1998) The effects of air-borne nitrogen pollutants on species diversity in natural and semi-natural European vegetation. J. Ecol. **86**(5), 717–738

Bond, D.W., Steiger, S., Zhang, R., Tie, X., Orville, R.E. (2002) The importance of NO_x production by lightning in the tropics. Atmos. Environ. **36**(9), 1509–1519

Bonsdorff, E. (2006) Zoobenthic diversity-gradients in the Baltic Sea: Continuous post-glacial succession in a stressed ecosystem. Journal of Experimental Marine Biology and Ecology **330**(1), 383–391. doi:10.1016/j.jembe.2005.12.041

Bouwman, A.F., Vanderhoek, K.W., Olivier, J.G.J. (1995) Uncertainties in the global source distribution of nitrous oxide. J. Geophys. Res.-Atmos. **100**(D2), 2785–2800

Boyd, S.R. (2001) Nitrogen in future biosphere studies. Chem. Geol. **176**(1–4), 1–30. doi:10.1016/s0009-2541(00)00405-8

Boyer, E.W., Alexander, R.B., Parton, W.J., Li, C.S., Butterbach-Bahl, K., Donner, S.D., Skaggs, R.W., Del Gross, S.J. (2006) Modeling denitrification in terrestrial and aquatic ecosystems at regional scales. Ecol. Appl. **16**(6), 2123–2142

Brandes, J.A., Devol, A.H. (2002) A global marine-fixed nitrogen isotopic budget: Implications for Holocene nitrogen cycling. Global Biogeochem. Cycles **16**(4), 1120. doi:10.1029/2001gb001856

Brown, G.C., Skipsey, E. (1986) Energy resources: Geology, supply and demand. Open Univ. Press, Milton Keynes, UK

Brown, L.R. (1999) Feeding nine billion. In: Brown, L.R., Flavin, H. (eds.) State of the World 1999. A world report on progress towards a sustainable society. pp. 115–132. W. Norman & Company, New York

Burns, R.C., Hardy, R.W.F. (1975) Nitrogen Fixation in Bacteria and Higher Plants. Springer-Verlag, New York

Canfield, D.E., Glazer, A.N., Falkowski, P.G. (2010) The Evolution and Future of Earth's Nitrogen Cycle. Science **330**(6001), 192–196. doi:10.1126/science.1186120

Capone, D.G. (1983) Benthic nitrogen fixation. In: Carpenter, E.J., Capone, D.G. (eds.) Nitrogen in the marini environments. pp. 105–137. Acad. Press, New York

Capone, D.G., Zehr, J.P., Paerl, H.W., Bergman, B., Carpenter, E.J. (1997) *Trichodesmium*, a globally significant marine cyanobacterium. Science **276**(5316), 1221–1229. doi:10.1126/science.276.5316.1221

Carpenter, E.J., Capone, D.G. (2008) Nitrogen Fixation in the Marine Environment In: Capone, D.G., Bronk, D.A., Mulholland, M.R., Carpenter, E.J. (eds.) Nitrogen in the Marine Environment. pp. 141–197. Elsevier, San Diego, CA

Cleveland, C.C., Townsend, A.R., Schimel, D.S., Fisher, H., Howarth, R.W., Hedin, L.O., Perakis, S.S., Latty, E.F., Von Fischer, J.C., Elseroad, A., Wasson, M.F. (1999) Global patterns of terrestrial biological nitrogen (N_2) fixation in natural ecosystems. Global Biogeochem. Cycles **13**(2), 623–645

Codispoti, L.A., Brandes, J.A., Christensen, J.P., Devol, A.H., Naqvi, S.W.A., Paerl, H.W., Yoshinari, T. (2001) The oceanic fixed nitrogen and nitrous oxide budgets: Moving targets as we enter the anthropocene? Sci. Mar. **65**, 85–105

Cofala, J., Amann, M., Klimont, Z., Kupiainen, K., Hoglund-Isaksson, L. (2007) Scenarios of global anthropogenic emissions of air pollutants and methane until 2030. Atmos. Environ. **41**(38), 8486–8499. doi:10.1016/j.atmosenv.2007.07.010

Corre, M.D., Lamersdorf, N.P. (2004) Reversal of nitrogen saturation after long-term deposition reduction: Impact on soil nitrogen cycling. Ecology **85**(11), 3090–3104

Crutzen, P.J., Mosier, A.R., Smith, K.A., Winiwarter, W. (2008) N_2O release from agro-biofuel production negates global warming reduction by replacing fossil fuels. Atmos. Chem. Phys. **8**(2), 389–395

Davidson, E.A. (1994) Climate change and soil microbial processes: Secondary effects are hypothesized from better known interacting primary effects. In: Rounsevell, D.A., Loveland, P.J. (eds.) Soil response to climate change. pp. 156–168. Springer, Berlin

Davidson, E.A. (2009) The contribution of manure and fertilizer nitrogen to atmospheric nitrous oxide since 1860. Nature Geosci. **2**(9), 659–662. doi:10.1038/ngeo608

De Vries, W., Reinds, G.J., Gundersen, P., Sterba, H. (2006) The impact of nitrogen deposition on carbon sequestration in European forests and forest soils. Global Change Biol. **12**(7), 1151–1173. doi:10.1111/j.1365-2486.2006.01151.x

de Vries, W., van der Salm, C., Reinds, G.J., Erisman, J.W. (2007) Element fluxes through European forest ecosystems and their relationships with stand and site characteristics. Environ. Pollut. **148**(2), 501–513. doi:10.1016/j.envpol.2006.12.001

Delwiche, C.C. (1970) Nitrogen cycle. Scientific American **223**(3), 136–146

Denman, K.L., Brasseur, G., Chidthaisong, A., Ciais, P., Cox, P.M., Dickinson, R.E., Hauglustaine, D., Heinze, C., Holland, E., Jacob, D., Lohmann, U., Ramachandran, S., Dias, P.L.d.S., Wofsy, S.C., Zhang, X. (2007) Couplings Between Changes in the Climate System and Biogeochemistry. In: Solomon, S., Qin, D., Manning, M., Chen, Z., Marquis, M., Averyt, K.B., M.Tignor, Miller, H.L. (eds.) Climate Change 2007: The Physical Science Basis Contribution of Working Group I to the Fourth Assessment Report of the Intergovernmental Panel on Climate Change. pp. 501–587. Cambridge University Press, Cambridge, United Kingdom and New York, NY, USA

Dentener, F., Drevet, J., Lamarque, J.F., Bey, I., Eickhout, B., Fiore, A.M., Hauglustaine, D., Horowitz, L.W., Krol, M., Kulshrestha, U.C., Lawrence, M., Galy-Lacaux, C., Rast, S., Shindell, D., Stevenson, D., Van Noije, T., Atherton, C., Bell, N., Bergman, D., Butler, T., Cofala, J., Collins, B., Doherty, R., Ellingsen, K., Galloway, J., Gauss, M., Montanaro, V., Muller, J.F., Pitari, G., Rodriguez, J., Sanderson, M., Solmon, F., Strahan, S., Schultz, M., Sudo, K., Szopa, S., Wild, O. (2006) Nitrogen and sulfur deposition on regional and global scales: A multimodel evaluation. Global Biogeochem. Cycles **20**(4), GB4003, doi:4010.1029/2005 GB002672. doi:10.1029/2005gb002672

Deutsch, C., Gruber, N., Key, R.M., Sarmiento, J.L., Ganachaud, A. (2001) Denitrification and N_2 fixation in the Pacific Ocean. Global Biogeochem. Cycles **15**(2), 483–506

Diaz, R.J., Rosenberg, R. (2008) Spreading dead zones and consequences for marine ecosystems. Science **321**(5891), 926–929. doi:10.1126/science.1156401

Dise, N.B., Rothwell, J.J., Gauci, V., van der Salm, C., de Vries, W. (2009) Predicting dissolved inorganic nitrogen leaching in European forests using two independent databases. Sci. Total Environ. **407**(5), 1798–1808. doi:10.1016/j.scitotenv.2008.11.003

Ehhalt, D., Prather, M., Dentener, F., Derwent, R., Dlugokencky, E., Holland, E., Isaksen, I., Katima, J., Kirchhoff, V., Matson, P. (2001) Atmospheric chemistry and greenhouse gases. In: Houghton, J.T., Ding, Y., Griggs, D.J., Noguer, M., van der Linden, P.J., Dai, X., Maskell, K., Johnson, C.A. (eds.) Climate Change 2001: The Scientific Basis. Contribution of Working Group I to the Third Assessment Report of the Intergovernmental Panel on Climate Change. pp. 239–287. Cambridge University Press, Cambridge, UK and New York, USA

Erisman, J.W., Sutton, M.A., Galloway, J., Klimont, Z., Winiwarter, W. (2008) How a century of ammonia synthesis changed the world. Nature Geosci. **1**(10), 636–639. doi:10.1038/ngeo325

Falkowski, P.G. (1997) Evolution of the nitrogen cycle and its influence on the biological sequestration of CO_2 in the ocean. Nature **387**(6630), 272–275

Falkowski, P.G., Fenchel, T., Delong, E.F. (2008) The microbial engines that drive Earth's biogeochemical cycles. Science **320**(5879), 1034–1039. doi:10.1126/science.1153213

Franzblau, E., Popp, C.J. (1989) Nitrogen oxides produced from lightning. J. Geophys. Res.-Atmos. **94**(D8), 11089–11104

Galloway, J.N. (1998) The global nitrogen cycle: changes and consequences. Environ. Pollut. **102**, 15–24

Galloway, J.N. (2003) The global nitrogen cycle. Treatise on geochemistry **8**, 557–583

Galloway, J.N., Dentener, F.J., Capone, D.G., Boyer, E.W., Howarth, R.W., Seitzinger, S.P., Asner, G.P., Cleveland, C.C., Green, P.A., Holland, E.A., Karl, D.M., Michaels, A.F., Porter, J.H., Townsend, A.R., Vorosmarty, C.J. (2004) Nitrogen cycles: past, present, and future. Biogeochemistry **70**(2), 153–226

Galloway, J.N., Schlesinger, W.H., Levy, H., Michaels, A., Schnoor, J.L. (1995) Nitrogen fixation - anthropogenic enhancement environmental response. Global Biogeochem. Cycles **9**(2), 235–252

Galloway, J.N., Townsend, A.R., Erisman, J.W., Bekunda, M., Cai, Z.C., Freney, J.R., Martinelli, L.A., Seitzinger, S.P., Sutton, M.A. (2008) Transformation of the nitrogen cycle: Recent trends, questions, and potential solutions. Science **320**(5878), 889–892. doi:10.1126/science.1136674

Giggenbach, W.F., Matsuo, S. (1991) Evaluation of results from 2nd and 3rd IAVCEI Field Workshops on Volcanic Gases, Mt Usu, Japan, and White island, New-Zealand. Applied Geochemistry **6**(2), 125–141. doi:10.1016/0883-2927(91)90024-j

Gruber, N., Galloway, J.N. (2008) An Earth-system perspective of the global nitrogen cycle. Nature **451**(7176), 293–296. doi:10.1038/nature06592

Gruber, N., Sarmiento, J.L. (1997) Global patterns of marine nitrogen fixation and denitrification. Global Biogeochem. Cycles **11**(2), 235–266

Herridge, D.F., Peoples, M.B., Boddey, R.M. (2008) Global inputs of biological nitrogen fixation in agricultural systems. Plant Soil **311**(1–2), 1–18. doi:10.1007/s11104-008-9668-3

Hesterberg, R., Blatter, A., Fahrni, M., Rosset, M., Neftel, A., Eugster, W., Wanner, H. (1996) Deposition of nitrogen-containing compounds to an extensively managed grassland in central Switzerland. Environ. Pollut. **91**(1), 21–34

Holloway, J.M., Dahlgren, R.A. (2002) Nitrogen in rock: Occurrences and biogeochemical implications. Global Biogeochem. Cycles **16**(4). doi:10.1029/2002gb001862

Holloway, J.M., Dahlgren, R.A., Casey, W.H. (2001) Nitrogen release from rock and soil under simulated field conditions. Chem. Geol. **174**(4), 403–414

Holloway, J.M., Dahlgren, R.A., Hansen, B., Casey, W.H. (1998) Contribution of bedrock nitrogen to high nitrate concentrations in stream water. Nature **395**(6704), 785–788

Houghton, R.A. (1999) The annual net flux of carbon to the atmosphere from changes in land use 1850–1990. Tellus B **51**(2), 298–313

Howarth, R.W., Boyer, E.W., Pabich, W.J., Galloway, J.N. (2002) Nitrogen use in the United States from 1961–2000 and potential future trends. Ambio **31**(2), 88–96

Hungate, B.A., Dukes, J.S., Shaw, M.R., Luo, Y.Q., Field, C.B. (2003) Nitrogen and climate change. Science **302**(5650), 1512–1513

International Fertilizer industry Association (IFA) (2000) Mineral Fertilizer Production and the Environment. Part 1. The Fertilizer Industry's Manufacturing Processes and Environmental Issues. Technical Report No.26. pp. International Fertilizer Industry Association (IFA), Paris, France

Karl, D., Michaels, A., Bergman, B., Capone, D., Carpenter, E., Letelier, R., Lipschultz, F., Paerl, H., Sigman, D., Stal, L. (2002) Dinitrogen fixation in the world's oceans. Biogeochemistry **57**(1), 47–98

Kato, N., Akimoto, H. (1992) Anthropogenic emissions of SO_2 and NO_x in Asia: emission inventories. Atmos. Environ. **26**(16), 2997–3017. doi:10.1016/0960-1686(92)90291-r

Kennish, M.J. (1994) Practical handbook of marine science. CRC Press, Boca Raton, LA

Kroeze, C., Aerts, R., van Breemen, N., van Dam, D., van der Hoek, K., Hofschreuder, P., Hoosbeek, M., de Klein, J., Kros, H., van Oene, H., Oenema, O., Tietema, A., van der Veeren, R., de Vries, W. (2003) Uncertainties in the fate of nitrogen I: An overview of sources of uncertainty illustrated with a Dutch case study. Nutr. Cycl. Agroecosys. **66**(1), 43–69

Kroeze, C., Mosier, A., Bouwman, L. (1999) Closing the global N_2O budget: A retrospective analysis 1500–1994. Global Biogeochem. Cycles **13**(1), 1–8

Labrador, L.J., von Kuhlmann, R., Lawrence, M.G. (2004) Strong sensitivity of the global mean OH concentration and the tropospheric oxidizing efficiency to the source of NO_x from lightning. Geophys. Res. Lett. **31**(6), L06102. doi:10.1029/2003gl019229

Labrador, L.J., von Kuhlmann, R., Lawrence, M.G. (2005) The effects of lightning-produced NO_x and its vertical distribution on atmospheric chemistry: sensitivity simulations with MATCH-MPIC. Atmos. Chem. Phys. **5**, 1815–1834

Lamarque, J.F., Bond, T.C., Eyring, V., Granier, C., Heil, A., Klimont, Z., Lee, D., Liousse, C., Mieville, A., Owen, B., Schultz, M.G., Shindell, D., Smith, S.J., Stehfest, E., Van Aardenne, J., Cooper, O.R., Kainuma, M., Mahowald, N., McConnell, J.R., Naik, V., Riahi, K., van Vuuren, D.P. (2010) Historical (1850–2000) gridded anthropogenic and biomass burning emissions of reactive gases and aerosols: methodology and application. Atmos. Chem. Phys. **10**(15), 7017–7039. doi:10.5194/acp-10-7017-2010

Lamarque, J.F., Kiehl, J.T., Brasseur, G.P., Butler, T., Cameron-Smith, P., Collins, W.D., Collins, W.J., Granier, C., Hauglustaine, D., Hess, P.G., Holland, E.A., Horowitz, L., Lawrence, M.G., McKenna, D., Merilees, P., Prather, M.J., Rasch, P.J., Rotman, D., Shindell, D., Thornton, P. (2005) Assessing future nitrogen deposition and carbon cycle feedback using a multimodel approach: Analysis of nitrogen deposition. J. Geophys. Res.-Atmos. **110**(D19), D19303. doi:10.1029/2005jd005825

Lee, K., Karl, D.M., Wanninkhof, R., Zhang, J.Z. (2002) Global estimates of net carbon production in the nitrate-depleted tropical and subtropical oceans. Geophys. Res. Lett. **29**(19), 1907. doi:10.1029/2001gl014198

Lelieveld, J., Dentener, F.J. (2000) What controls tropospheric ozone? J. Geophys. Res.-Atmos. **105**(D3), 3531–3551

Levine, J.S., Rogowski, R.S., Gregory, G.L., Howell, W.E., Fishman, J. (1981) Simultaneous measurements of NO_x NO and O_3 production in a laboratory discharge: Atmospheric implications. Geophys. Res. Lett. **8**(4), 357–360

Liu, L.L., Greaver, T.L. (2009) A review of nitrogen enrichment effects on three biogenic GHGs: the CO_2 sink may be largely offset by stimulated N_2O and CH_4 emission. Ecol. Lett. **12**(10), 1103–1117. doi:10.1111/j.1461-0248.2009.01351.x

Lovett, G.M., Weathers, K.C., Sobczak, W.V. (2000) Nitrogen Saturation and Retention in Forested Watersheds of the Catskill Mountains, New York. Ecol. Appl. **10**(1), 73–84

Lyons, W.B., Mayewski, P.A., Spencer, M.J., Twickler, M.S. (1990) Nitrate concentrations in snow from remote areas: Implication for the global NO_x flux. Biogeochemistry **9**(3), 211–222

MacFarling Meure, C., Etheridge, D., Trudinger, C., Steele, P., Langenfelds, R., van Ommen, T., Smith, A., Elkins, J. (2006) Law Dome CO_2, CH_4 and N_2O ice core records extended to 2000 years BP. Geophys. Res. Lett. **33**(14), L14810. doi:10.1029/2006gl026152

Mackenzie, F.T. (2003) Our Changing Planet, 3rd ed. pp. Prentice Hall, Upper Saddle River

Mahaffey, C., Michaels, A.F., Capone, D.G. (2005) The conundrum of marine N_2 fixation. Am. J. Sci. **305**(6–8), 546–595

Mann, C.C. (1999) Crop scientists seek a new revolution. Science **283**(5400), 310–314

Matson, P.A., McDowell, W.H., Townsend, A.R., Vitousek, P.M. (1999) The globalization of N deposition: ecosystem consequences in tropical environments. Biogeochemistry **46**(1–3), 67–83

Matthews, E. (1994) Nitrogen fertilizers: Global distribution of consumption and associated emissions of nitrous oxide and ammonia. Global Biogeochem. Cycles **8**(4), 411–439

Montzka, S.A., Reimann, S. (2010) Ozone-depliting substances (ODSs) and related chemicals. In: Ajavon, A.-L.N., Newman, P.A., Pyle, J.A., Ravishankara, A.R. (eds.) Scientific assessment of Ozone depletion: 2010. World meteorological Organization, Global Ozone Research Monitoring, Geneva, Switzerland

Mulholland, M.R. (2007) The fate of nitrogen fixed by diazotrophs in the ocean. Biogeosciences **4**(1), 37–51

Muller, J.F. (1992) Geographical distribution and seasonal variation of surface emissions and deposition velosities of atmospheric trace gases. J. Geophys. Res.-Atmos. **97**(D4), 3787–3804

Nevison, C., Holland, E. (1997) A reexamination of the impact of anthropogenically fixed nitrogen on atmospheric N2O and the stratospheric O-3 layer. J. Geophys. Res.-Atmos. **102**(D21), 25519–25536

Nevison, C.D., Weiss, R.F., Erickson, D.J. (1995) Global oceanic emissions of nitrous oxide. J. Geophys. Res.-Oceans **100**(C8), 15809–15820

Nixon, S.W., Ammerman, J.W., Atkinson, P., Berounsky, V.M., Billen, G., Boicourt, W.R., Church, T.M., Ditoro, D.M., Elmgren, R., Garber, J.H., Giblin, A.E., Jahnke, R.E., Owens, N.J.P., Pilson, M.E.Q., Seitzinger, S. (1996) The fate of nitrogen and phosphorus at the land-sea margin of the North Atlantic Ocean. In: Howarth, R.W. (ed.) Nitrogen cycling in the North Atlantic Ocean and its Watersheds. pp. 141–180. Kluwer, Dordrecht, The Netherlands

Olivier, J.G.J., Bouwman, A.F., Van der Hoek, K.W., Berdowski, J.J.M. (1998) Global air emission inventories for anthropogenic sources of NO_x, NH_3 and N_2O in 1990. Environ. Pollut. **102**, 135–148

Paul, E.A., Clark, F.E. (1997) Soil Microbiology and Boiochemistry. Academic Press, New York

Phoenix, G.K., Hicks, W.K., Cinderby, S., Kuylenstierna, J.C.I., Stock, W.D., Dentener, F.J., Giller, K.E., Austin, A.T., Lefroy, R.D.B., Gimeno, B.S., Ashmore, M.R., Ineson, P. (2006) Atmospheric nitrogen deposition in world biodiversity hotspots: the need for a greater global perspective in assessing N deposition impacts. Global Change Biol. **12**(3), 470–476. doi:10.1111/j.1365-2486.2006.01104.x

Pilegaard, K., Skiba, U., Ambus, P., Beier, C., Bruggemann, N., Butterbach-Bahl, K., Dick, J., Dorsey, J., Duyzer, J., Gallagher, M., Gasche, R., Horvath, L., Kitzler, B., Leip, A., Pihlatie, M.K., Rosenkranz, P., Seufert, G., Vesala, T., Westrate, H., Zechmeister-Boltenstern, S. (2006) Factors controlling regional differences in forest soil emission of nitrogen oxides (NO and N_2O). Biogeosciences **3**(4), 651–661

Postgate, R. (1998) Nitogen fixation. pp. Cambridge University Press, Cambridge, UK

Potter, P., Ramankutty, N., Bennett, E.M., Donner, S.D. (2010) Characterizing the Spatial Patterns of Global Fertilizer Application and Manure Production. Earth Interact. **14**, 1–23. doi:10.1175/2009ei288.1

Prentice, I.C., Farquhar, G.D., Fasham, M.J.R., Goulden, M.L., Wallace, D.W.R. (2001) The carbon cycle and the atmospheric carbon dioxide. In: Houghton, J.T., Y. Ding, D.J. Griggs, M. Noguer,, P.J. van der Linden, X.D., K. Maskell, and C.A. Johnson (eds.) Climate Change 2001: The Scientific Basis. Contribution of Working Group I to the Third Assessment Report of the Intergovernmental Panel on Climate Change. pp. 183–237. Cambridge University Press, Cambridge, United Kingdom and New York, NY, USA

Price, C., Penner, J., Prather, M. (1997) NO_x from lightning .1. Global distribution based on lightning physics. J. Geophys. Res.-Atmos. **102**(D5), 5929–5941

Rabalais, N.N. (2002) Nitrogen in aquatic ecosystems. Ambio **31**(2), 102–112

Ridley, B., Ott, L., Pickering, K., Emmons, L., Montzka, D., Weinheimer, A., Knapp, D., Grahek, F., Li, L., Heymsfield, G., McGill, M., Kucera, P., Mahoney, M.J., Baumgardner, D., Schultz, M., Brasseur, G. (2004) Florida thunderstorms: A faucet of reactive nitrogen to the upper troposphere. J. Geophys. Res.-Atmos. **109**(D17), D17305. doi:10.1029/2004jd004769

Rosswall, T. (1982) Microbial regulation of the biogeochemical nitrogen cycle. Plant Soil **67**(1–3), 15–34

Rosswall, T. (1983) The nitrogen cycle. In: Bolin, J., Cook, R.B. (eds.) The major biogeochemical cycles and their interactions. SCOPE. pp. 46–50. Wiley & Sons, Chichester, UK

Russele, M.P. (2008) Biological dinitrogen fixation in agriculture. In: Schepers, J.S., Raun, W.R. (eds.) Nitrogen in agricultural systems. pp. 281–358. ASA, CSA, SSSA, Madison, WI

Schlesinger, W.H. (1997) Biogeochemistry: An analysis of global change, 2nd ed. 587 pp. Academic Press, New York

Schlesinger, W.H., Hartley, A.E. (1992) A global budget for atmospheric NH_3. Biogeochemistry **15**(3), 191–211

Seitzinger, S.P., Giblin, A.E. (1996) Estimating denitrification in North Atlantic continental shelf sediments. Biogeochemistry **35**(1), 235–260

Seitzinger, S.P., Kroeze, C. (1998) Global distribution of nitrous oxide production and N inputs in freshwater and coastal marine ecosystems. Global Biogeochem. Cycles **12**(1), 93–113

Seitzinger, S.P., Kroeze, C., Styles, R., eacute, e, V. (2000) Global distribution of N_2O emissions from aquatic systems: natural emissions and anthropogenic effects. Chemosphere: Global Change Sci. **2**(3–4), 267–279

Shepon, A., Gildor, H., Labrador, L.J., Butler, T., Ganzeveld, L.N., Lawrence, M.G. (2007) Global reactive nitrogen deposition from lightning NO_x. J. Geophys. Res.-Atmos. **112**(D6), D06304. doi:10.1029/2006jd007458

Smil, V. (1999) Nitrogen in crop production: An account of global flows. Global Biogeochem. Cycles **13**(2), 647–662. doi:10.1029/1999gb900015

Smil, V. (2001) Enriching the Earth: Fritz Harber, Carl Bosch and the Transformation of World Food Production. The MIT Press, Cambridge, Massachusetts

Söderlund, R., Svensson, B.H. (1976) The Global Nitrogen Cycle. Ecol. Bull. (22), 23–73

Stevenson, F.J. (1982) Origin and distribution of nitrogen in soil. In: Stevenson, F.J. (ed.) Nitrogen in agricultural soils. Agronomy Monograph (22), 1–42. American Society of Agronomy, Madison, Wisconsin

Stoddard, J.L. (1994) Long-term changes in watershed retention of nitrogen. In: Baker, L.A. (ed.) Environmental chemistry of lakes and reservoirs. Advances in Chemistry Series. **237**, 223–284. American Chemical Society, Washington, D.C., USA

Sutton, M.A., Nemitz, E., Erisman, J.W., Beier, C., Bahl, K.B., Cellier, P., de Vries, W., Cotrufo, F., Skiba, U., Di Marco, C., Jones, S., Laville, P., Soussana, J.F., Loubet, B., Twigg, M., Famulari, D., Whitehead, J., Gallagher, M.W., Neftel, A., Flechard, C.R., Herrmann, B., Calanca, P.L., Schjoerring, J.K., Daemmgen, U., Horvath, L., Tang, Y.S., Emmett, B.A., Tietema, A., Penuelas, J., Kesik, M., Brueggemann, N., Pilegaard, K., Vesala, T., Campbell, C. L., Olesen, J.E., Dragosits, U., Theobald, M.R., Levy, P., Mobbs, D.C., Milne, R., Viovy, N., Vuichard, N., Smith, J.U., Smith, P., Bergamaschi, P., Fowler, D., Reis, S. (2007) Challenges in quantifying biosphere-atmosphere exchange of nitrogen species. Environ. Pollut. **150**, 125–139. doi:10.1016/j.envpol.2007.04.014

Trenberth, K.E., Guillemot, C.J. (1994) The total mass of the atmosphere. J. Geophys. Res.-Atmos. **99**(D11), 23079–23088

van Aardenne, J.A., Dentener, F.J., Olivier, J.G.J., Goldewijk, C.G.M.K., Lelieveld, J. (2001) A 1° × 1° resolution data set of historical anthropogenic trace gas emissions for the period 1890–1990. Global Biogeochem. Cycles **15**(4), 909–928. doi:10.1029/2000gb001265

van Egmond, K., Bresser, T., Bouwman, L. (2002) The European Nitrogen Case. Ambio **31**(2), 72–78. doi:10.1043/0044-7447(2002)031(0072:tenc)2.0.co;2

Vitousek, P.M., Aber, J.D., Howarth, R.W., Likens, G.E., Matson, P.A., Schindler, D.W., Schlesinger, W.H., Tilman, G.D. (1997) Human alteration of the global nitrogen cycle: Sources and consequences. Ecol. Appl. **7**(3), 737–750

Vitousek, P.M., Hattenschwiler, S., Olander, L., Allison, S. (2002) Nitrogen and nature. Ambio **31**, 97–101

Vitousek, P.M., Howarth, R.W. (1991) Nitrgen limitation on land and in the sea: How it can occur. Biogeochemistry **13**(2), 87–115

Walker, J.C.G. (1977) Evolution of the atmosphere. pp. Macmillan Publishing Co., Inc., New York

Wedpohl, K.H. (1972) Handbook of geochemistry. pp. Springer-Verlag, Berlin, Germany

Zhu, Z.L., Xiong, Z.Q., Xing, G.X. (2005) Impacts of population growth and economic development on the nitrogen cycle in Asia. Sci. China C **48**, 729–737. doi:10.1360/062005-266

Zillen, L., Conley, D.J., Andren, T., Andren, E., Bjorck, S. (2008) Past occurrences of hypoxia in the Baltic Sea and the role of climate variability, environmental change and human impact. Earth-Science Reviews **91**(1–4), 77–92. doi:10.1016/j.earscirev.2008.10.001

Chapter 3
Formation and Release of Nitrous Oxide from Terrestrial and Aquatic Ecosystems

Contents

Abstract Soil nitrogen (N) is in a constant state of flux, moving and changing chemical forms. Nitrification and denitrification are the main processes to remove reactive N (Nr) from the environment. Both are predominantly microbial processes that provide energy to specialized groups of microorganisms. Nitrification oxidizes reduced N, generally NH_3 or NH_4^+ to NO_3^- via nitrite under aerobic conditions. Denitrification is the process under which oxidized N is reduced back into N_2 under anaerobic conditions. Autotrophic nitrification and heterotrophic denitrification are the major N_2O forming processes in terrestrial and aquatic forming in terrestrial and aquatic ecosystems. Major factors regulating nitrification and denitrification are the availability of reactive N, the availability of reductant (mostly labile organic carbon compounds), and oxygen concentration. These three factors are in turn governed by many other factors such as water content, pH, porosity, and the presence of inhibitory compounds, which may act to cause accumulation of ionic (nitrite) or

DOI 10.1007/978-94-007-5364-8_3,

gaseous (nitric oxide, nitrous oxide) intermediates. It has been estimated that soils under terrestrial ecosystems denitrify ~124 Tg N year^{-1} or about 35–40% of total land based reactive N. Arable soils receiving high inputs of N are hot spots for denitrification and dominant sources of anthropogenic N_2O emissions.

Keywords Nitrification • Denitrification • Dissimilatory nitrate reduction to ammonium • Nitrate reductase • Nitrite reductase, nitrous oxide reductase

Abbreviations

ATP	adenosine triphosphate
AMO	ammonia monooxygenase
HAO	hydroxylamine oxidoreductase
NADPH	nicotinamide adenine dinucleotide phosphate oxidase
OM	organic matter
SOM	soil organic matter
DNRA	dissimilatory nitrate reduction to ammonium
CEC	cation exchange capacity
SOC	soil organic carbon
AOB	ammonia oxidizing bacteria
WHO	World Health Organization of the United Nations
US EPA	U.S. Environmental Protection Agency
NT	no tillage

3.1 Introduction

Global N_2O production is largely due to microbial processes. Microbial transformations of nitrification and denitrification contribute about 70% of the annual N_2O budgets worldwide (Mosier 1998; Godde and Conrad 2000). The transformation processes may occur simultaneously within different microsites of the same soil. Knowledge of the underlying processes and microorganisms community structure is important to better understand the regulation processes, improve the global estimates of N_2O with a view of developing more targeted management practices for mitigation of N_2O. Traditionally, autotrophic nitrification and heterotrophic denitrification are major N_2O forming processes in both terrestrial and marine ecosystems. However, these are not the sole production pathways of N_2O. Other pathways includes heterotrophic nitrification, commonly carried out by fungi (Odu and Adeoye 1970), aerobic denitrification by heterotrophic nitrifiers (Robertson and Kuenen 1990) and nitrifier denitrification, i.e. denitrification by autotrophic NH_3 oxidizing bacteria (Ritchie and Nicholas 1972; Wrage et al. 2001). Other processes which may produce N_2O are heterotrophic nitrification (Papen et al. 1989; Laughlin et al. 2008), fungal denitrification,

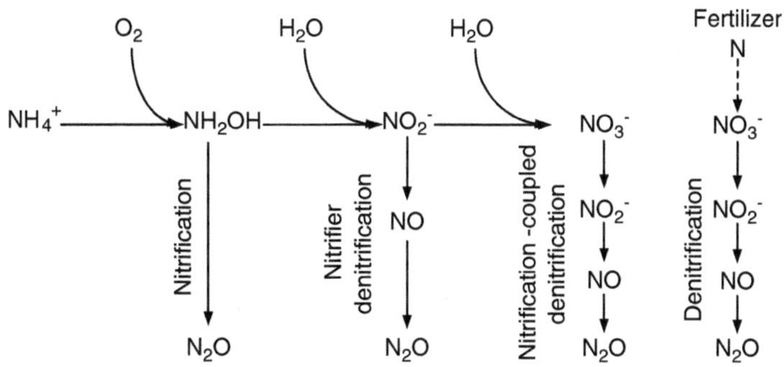

Fig. 3.1 Major pathways for N_2O formation as outlined in this book (Modified from Kool et al. 2009)

where N_2O is the final product, since many fungi lack the N_2O-reductase enzyme to further reduce N_2O to N_2 (Laughlin and Stevens 2002). Co-denitrification (Laughlin and Stevens 2002) by both bacteria and fungi and dissimilatory nitrate reduction to ammonia (DNRA) (Bleakley and Tiedje 1982) may also produce N_2O as one of their byproducts. Major microbial pathways for N_2O production are presented in Fig. 3.1. In addition, chemodenitrification, which is a non-biological reaction of the intermediate products of nitrification with organic or inorganic compounds in soils (Chalk and Smith 1983; Tiedje 1988), may also produce small quantities of N_2O.

Bacteria produce N_2O through nitrification and denitrification processes – which are key transformation processes within the natural N cycle (Chap. 2). These processes are generally accepted as principal sources of N_2O in soils, but almost all microbial processes that involve oxidation or reduction of N through +1 or +2 oxidation state may at least yield trace amounts of N_2O (Conrad 1990; Smith et al. 2001; Freney 1997). Nitrification is the main source of N_2O under aerobic conditions, while denitrification dominates under anoxic conditions. In nitrification, bacteria oxidize N through a 2-step oxidation process. Two groups of nitrifying bacteria are responsible: those that oxidize ammonium (NH_4^+) to nitrite (NO_2^-) and those that oxidize NO_2^- to nitrate (NO_3^-). In this process, N_2O is produced as a byproduct or as an alternate product of ammonium oxidation. Nitrous oxide and NO are minor byproducts of the transformation under O_2 limited conditions when nitrifiers use NO_2^- as terminal electron acceptors. In denitrification bacteria reduce oxidized inorganic forms of N under anoxic/suboxic conditions. This process may form N_2O as intermediate byproduct, or it may consume N_2O. Therefore, the process of denitrification can be either a source or a sink of N_2O, depending on environmental conditions such as oxygen availability, reactive N availability, pH, and temperature (Sorai et al. 2007). Production of N_2O has also been demonstrated among bacteria that respires nitrate to nitrite and those that dissimilate nitrite to ammonium (DNRA) (Stevens et al. 1998; Xiong et al. 2007). A smaller quantity is also formed by chemodenitrification (van Cleemput 1998) and fungal transformation (Laughlin and Stevens 2002).

Nitrous oxide is a by-product of aerobic nitrification and an obligate intermediate in the denitrification pathway, and is emitted by both nitrifiers and denitrifiers. Production and consumption of N_2O is moderated by oxygen partial pressure; nitrification is additionally controlled by the concentration of NH_4^+, while denitrification is also controlled by availability of carbon (C) and NO_3^- (Conrad 1995). Denitrification is the main biological process responsible for returning fixed N to the atmosphere as N_2, thus closing the N cycle (Philippot et al. 2009). This reduction of soluble N to gaseous N represents a loss for agriculture, since it can deplete NO_3^- in the soil, an essential plant nutrient. The denitrification N_2O/N_2 product ratio is variable, and N_2O may even be the dominant end product. However, denitrification also provides a valuable ecosystem service by mediating N removal from NO_3^- polluted waters in sediments and other water-saturated soils (Mosier 1998). Denitrifiers can be sinks for N_2O, and the sink activity appears to be stimulated by low availability of mineral N (Chapuis-Lardy et al. 2007; Conen and Neftel 2007). Whether soils are a net source or sinks for the atmospheric N_2O depends on the environmental factors such as O_2 partial pressure, which regulates consumption and production (Philippot et al. 2009). But most soils are a net N_2O source (Conrad 1996). Flushes of N_2O can occur when previously well- aerated soils becomes moistened or saturated from precipitation or irrigation, or when frozen soils thaw during snowmelt.

3.2 Nitrification

Nitrification is the biological oxidation of reduced N, generally in the form of ammonia (NH_3) or ammonium (NH_4^+) to nitrite (NO_2^-) and ultimately to nitrate (NO_3^-). It is mediated by chemoautotrophic bacteria belonging to family Nitrobacteriaceae, which derive their energy from oxidation of either NH_4^+ or NO_2^-. The NH_4^+ ions are released from ammonification reactions of soil organic matter (SOM), organic forms of N fertilizers or added as NH_3/NH_4^+ inorganic fertilizers. Since the rate of conversion of NO_2^- to NO_3^- is faster than the conversion of NH_4^+ to NO_2^-, it is unlikely that NO_2^-, which is toxic to plants, accumulates under most soil and climatic conditions. Nitrification links the most oxidized (NH_3) and most reduced (NO_3^-) forms of N redox cycle and helps to determine the overall distribution of these important nutrients in the ecosystems (Ward 2008). The nitrification process produces hydrogen ions (H^+) thereby reducing the pH (Eq. 3.4). Application of stable isotope approaches to distinguish between nitrification and denitrification has provide the greater understanding of the potential of nitrification as a source of N_2O, with the evidence that under certain conditions ammonia oxidizing bacteria can significantly contribute to N_2O emissions from soils (Kool et al. 2009; Shaw et al. 2006; Bateman and Baggs 2005).

In soils, nitrification determines the form of N present and therefore how N is absorbed, utilized or dispersed into the environment, and has large implications for plant productivity and environmental quality. Nitrate is also susceptible to denitrification, but it is highly available to plants and is often the major uptake form of N.

The NH_4^+ and NO_2^- rarely occur at significant concentrations in the oxygenated habitats, including the majority of agricultural soils, since it is rapidly converted into NO_3^- which may accumulate in the soil solution and water column to a high concentration. The rapid conversion of NH_4^+ to NO_3^- in the soil limits the effectiveness of much of the applied N fertilizers. Nearly 90% of all added N fertilizers are applied in the NH_4^+ form, which is mostly nitrified within 4 weeks after application (Sahrawat 1982).

In freshwater and marine ecosystems, NH_4^+ is also rapidly recycled between heterotrophic and N_2 fixing organisms and photosynthetic community which utilizes it as a source of N. The NH_4^+ cations have a tendency to bind electrostatically with negatively charged surfaces of soil and functional groups of soil SOM. This binding is sufficient to limit N losses by leaching. However, its conversion to anions (NO_2^- or NO_3^-) makes N highly mobile, providing a higher potential for N to be leached beyond the root zone. Therefore, nitrification influences the movement of N through generally negatively charged soil matrix and determine the fate of N in soil. The NO_3^- is more likely to move rapidly by mass flow to plant roots, leach out of root zone or denitrified (Norton 2008). However, NO_3^- accumulates in deep ocean and seasonally in deep waters of lakes where there is no demand for inorganic N by phytoplankton (Ward 2008). Therefore, nitrification strongly influences the fate of N in the soil and aquatic environments. Nitrification also produces nitrogenous gases (NO and N_2O) important in greenhouse effect and ozone atmospheric chemistry (Godde and Conrad 2000). Nitrification therefore, lowers the effectiveness of N fertilization and can have serious environmental implications when excess N enters the natural environments (Jarvis 1996).

The nitrification process is carried out by nitrifying autotrophic and heterotrophic bacteria and NH_3 oxidizing archaea, which are ubiquitous components of all soil microorganism populations. There are two functionally distinct groups of nitrifiers: (1) NH_3 oxidizing bacteria and archaea, and (2) NO_2^- oxidizing bacteria. A new group of NH_3 oxidizing nitrifiers of Archaeal origin utilizing similar pathway to that known in NH_3 oxidizing bacteria was recently discovered and verified by cultivation (Konneke et al. 2005; Schleper et al. 2005). It is now estimated that ammonia oxidizing archaea are abundant in marine (Wuchter et al. 2006) and also prevalent in soils (Leininger et al. 2006). The major sources of reduced forms of soil N (NH_3/NH_4^+) are excretion and mineralization of organic matter (OM) derived from animal and plants, and application of ammonia-based fertilizers. In aquatic ecosystems, NH_4^+ generally originates from internal cycling and atmospheric deposition. Nitrifiers have slow turnover rates and possess the ability to survive long periods in a state of dormancy (Belser 1979). Other bacterial species such as *Nitrocystus* and *Nitrospira*, and also some heterotrophic fungi groups, such as *Aspergillus flavus* can also carry out nitrification in some ecosystems (Subbarao et al. 2006).

Nitrification is controlled by the O_2 partial pressure, NH_3/NH_4^+ concentration and pH (Tiedje 1988). In soils, these factors are affected by presence of plant roots. The O_2 partial pressure can be altered by respiration of roots and root-associated microorganisms. Roots consumption of water and penetration creates channels

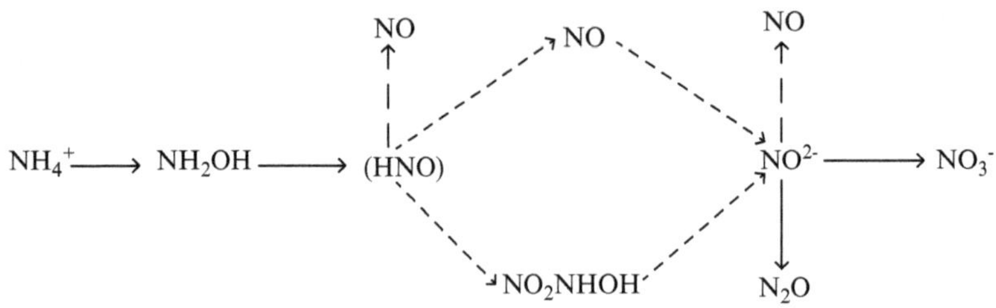

Fig. 3.2 Pathways for chemoautotrophic nitrification. *Dashed lines* represent the unconfirmed pathways

(macro/biopores) for gaseous transfer. Plants also release readily available organic compounds in the soil solution through rhizodeposition, which become major source of microbial nutrients in the rhizosphere.

3.2.1 Autotrophic Nitrification

Chemoautotrophic nitrifiers are aerobes that synthesize their cell constituents from CO_2. The driving force for CO_2 reduction is the production of adenosine triphosphate (ATP) during oxidation of NH_4^+ to NO_2^- or NO_2^- to NO_3^-. Pathways for chemoautotrophic nitrification, N_2O and NO production are presented in Fig. 3.2.

3.2.1.1 Ammonia Oxidation

Most of the aerobic nitrification that occurs in natural habitats is performed by obligate autotrophic bacteria and archaea and in some cases mixotrophic (i.e., microorganisms that can use combination of energy – autotrophic and heterotrophic) (Norton 2008; Ward 2008). Most of the existing knowledge of autotrophic NH_3 oxidation was derived from studies on cultivated strains of genus *Nitrosomonas* which oxidizes NH_3 to NO_2^- (Chain et al. 2003) and *Nitrobacter* which oxidizes NO_2^-. Chemoautotrophic nitrifiers are aerobes that synthesize their cell constituents from CO_2. The driving force for CO_2 reduction is the production of adenosine triphosphate (ATP) during oxidation of NH_4^+ to NO_2^- or NO_2^- to NO_3^-. Chemoautotrophic bacteria or archaea responsible for oxidizing NH_3 to NO_2^- gains metabolic energy from this process. Ammonia oxidizing bacteria are highly specialized for their chemolithotrophic metabolism and lacks gene encoding catabolic pathway for the use of organic substrate (Chain et al. 2003) although small quantities of pyruvate and amino acids may be incorporated into their biomass (Prosser 1989). The oxidation of NH_3 to NO_2^- requires two unique enzymes – ammonia monooxygenase (AMO) and hydroxylamine oxidoreductase (HAO) (Arp et al. 2002). AMO is the major protein found in the intracytoplasmic membranes of all autotrophic ammonia oxidizing bacteria. It is required for all NH_3 oxidizing organisms (Rotthauwe et al. 1997; Norton et al. 2002). This molecule and the gene that encode it, is sometimes the markers for this physiological group of organisms (Norton 2008). Discovery of its gene encoding in archaea was the basis for the discovery and subsequently isolation of NH_3-oxidizing archaea (Konneke et al. 2005). HAO is a periplasmic enzyme anchored in the membrane (Arp et al. 2002).

The bacteria responsible for NO_2^- oxidation gains energy from the process and may assimilate C either autotrophically or heterotrophically. The nitrite oxidizing bacteria identified in soils or waste water system belong to genera *Nitrobacter* or *Nitrospira* (Bartosch et al. 2002). *Nitrobacter winogradsky* and *N. hamburgesis* have been the model organisms for the biochemistry in the nitrite oxidizing bacteria. Nitrite is oxidized by the membrane bound nitrite oxidoreductase enzyme, with O_2 coming from H_2O (Spieck et al. 1996). The majority of energy produced through NH_3 and NO_2^- oxidation is used for the fixation of CO_2 via the Calvin cycle involving the ribose biphosphatase carboxylase/oxygenase enzyme (Prosser 1989).

The cultured autotrophic nitrifiers depend on CO_2 as their major C source and C fixation is carried out under the Calvin Cycle. It is estimated that CO_2 fixation accounts for nearly 80% of the energy budget of an autotroph (Forrest and Walker 1971; Kelly 1978). The use of NH_4^+ or NO_2^- as a sole source of reducing power for autotrophic growth is less efficient. About 35 mol. of NH_4^+ or 100 mol. of NO_2^- must be oxidized to support fixation of single mole of CO_2. Therefore, nitrification is a low-yield energy source, which accounts for slow growth of nitrifiers even under optimum conditions (Wood 1986).

Oxidation of NH_3 to NO_2^- occurs in two steps. In the first step, NH_3 is oxidized to hydroxylamine (Eq. 3.1), and in the second step hydroxylamine is oxidized to NO_2^-. Oxidation of NH_3 to hydroxylamine is facilitated by AMO, a copper-containing enzyme that is a membrane-bound protein (Basu et al. 2003). This reaction is endergonic and requires O_2 and a source of reducing equivalents. Gaseous NH_3 is required as its substrate as demonstrated by pH dependence of the reaction rate (Suzuki et al. 1974; Ward 1986). In AMO catalyzed reaction, one atom of O from O_2 is reduced with two electrons from the substrate (NH_3) with the insertion of O atom into substrate (i.e., formation of NH_2OH). Synthesis and activity of AMO enzyme in the *Nitrosomonas europea* respond directly to NH_3 concentration (Stein et al. 1997).

The transformation of NH_3 to NO_2^- occurs according to Eq. 3.1, 3.2, 3.3 and 3.4,

$$NH_3 + O_2 + 2H^+ + 2e^- \rightarrow NH_2OH + H_2O \quad (3.1)$$

$$NH_2OH + H_2O \rightarrow NO_2^- + 5H^+ + 4e^- \quad (3.2)$$

$$4H^+ + O_2 \rightarrow H_2O \quad (3.3)$$

$$2NH_3 + O_2 \rightarrow NO_2^- + H_2O + 2H^+ \quad (3.4)$$

Intermediate product hydroxylamine is further oxidized by HAO to NO_2^- (Eq. 3.2). Molecular O_2 from water is consumed by terminal cytochrome oxidase as a result of electron transport, and generates ATP for cellular metabolism (Eq. 3.3). The overall reaction of NH_3 oxidation process consumes molecular O_2 and produces H^+ and NO_2^- (Eq. 3.4). It is an energy yielding reaction, which produces sufficient ATP to support reverse electron transport for CO_2 fixation. Hydroxylamine oxidation yields 4 electrons, two of them are shunted back to AMO for activation of NH_3 (Ward 2008). The remaining two electrons are available for cells reducing reaction needs (i.e., generation of NADPH). Depending on conditions and O_2 concentrations, nitric oxide (NO) nitrous oxide (N_2O) or molecular nitrogen (N_2) can be produced as secondary products in the autotrophic NH_3 oxidation by both terrestrial and marine ecosystems (Zart and Bock 1998; Schmidt et al. 2004). Ammonia oxidizing archaea do not possess the hydroxylamine reductase gene, so the pathway for ammonia oxidation is different from that outlined for bacteria. The only cultured ammonia oxidizing archaea—*Nitrostipumilus maritimus* depends on CO_2 as its only C source, and presence of low levels of organic C has an inhibitory effect. The pathway for CO_2 fixation in archaea is currently unknown (Ward 2008).

3.2.1.2 Nitrite Oxidation

The biochemistry of NO_2^- oxidation involves two electrons transfer with no intermediates. Nitrite oxidoreductase enzyme facilitates the transfer of O_2. The additional O_2 atom in NO_3^- is derived from water (Eq. 3.5) and molecular O_2 that is involved in net reaction (Eq. 3.7) results from electron transport involving cytochrome oxidase (Eq. 3.6).

$$NO_2^- + H_2O \rightarrow NO_3^- + 2H^+ \quad (3.5)$$

$$4H^+ + O_2 \rightarrow H_2O \quad (3.6)$$

$$NO_2^- + O_2 \rightarrow NO_3^- \quad (3.7)$$

The energy yield of NO_2^- oxidation is lower than that of NH_3 oxidation necessitating the oxidation of large quantities of NO_2^- in order to fix small quantity of CO_2. Overall, therefore, the biochemical impact of nitrification is much greater on N cycle than on the C cycle. No bacteria have been found that can convert NH_3 to NO_3^- directly (Hooper et al. 1997). Recent evidence suggests that archaea of the phylum Crenarchaeota are also capable of NH_3 oxidation, which contain genes encoding key enzymes of this process and are widespread in soil environments (Treusch et al. 2005; Leininger et al. 2006).

The production of N_2O by autotrophic NH_4^+ oxidizers results from reductive process in which the organisms use NO_2^- as an electron acceptor, especially when O_2 is limiting (Poth and Focht 1985). This process allows organisms to conserve O_2

for the oxidation of NH_4^+, from which they gain energy for growth and regeneration, and avoiding the potential accumulation of toxic levels of NO_2^-. For the autotrophic nitrifying bacteria such as *Nitrosomonas* spp., N_2O can be either the product of ammonia and hydroxylamine oxidation or it may result from nitrite reduction under anaerobic conditions, which is referred to as nitrifiers denitrification (Poth and Focht 1985). *Thiosphaera pantothrapha* and *Alcaligenes fecalis* are special organisms that can accomplish both nitrification and denitrification and generate N_2O in aerobic conditions (Robertson and Tiedje 1987). *Nitrosomonas europea* can also produce N_2O during the oxidation of NH_4^+ or NH_2OH to NO_2^- (Yoshida and Alexander 1971).

Both the organisms NH_4^+ and NO_2^- oxidizers are found together in the soil. As a result, NO_2^- never accumulates in the soil except under the conditions of low pH which inhibits NO_2^- oxidizers. Although decreased O_2 and increase in acidity enhance the production of N_2O but these factors also retard the oxidation of NH_4^+. Therefore, it is difficult to predict the net effect of decreased O_2 and increased acidity on N_2O production. The optimum temperature for nitrifiers activity ranges from 2 to 40°C. The overall nitrification process is controlled primarily by NH_4^+ and O_2 concentrations Since O_2 supply is moderated by soil moisture, the effect of soil water content on N transformation probably reflects its impact on O_2 diffusion in the normal soil moisture ranges.

3.2.1.3 Taxonomy of Bacteria and Archaea Involved in Autotrophic Nitrification

The classic nitrifying species distinctions are based on cell shape and arrangement of internal membranes (Watson et al. 1986; Koops and Moller 1992). However, these have been superseded by the evolutionary relationships deduced from ribosomal sequences. Based on molecular sequencing data, NH_3 oxidizing bacteria fall into two principal sub divisions of major phylum in the Proteobacteria – Gammaproteobacteria and Betaproteobacteria which occurs in a wide range of ecosystems (Table 3.1). The Betaproteobacteria subdivision containing genus *Nitrospira* and *Nitrosomonas* is the most extensively studied (Purkhold et al. 2003).

Current taxonomy is based on ribosomal sequences, and phylogeny of nitrifying bacteria and archaea fall under phylum Crenarchaeota (Table 3.1). Although low numbers of a few other NH_4^+ oxidizing chemoautotrophs are also present in many soils, *Nitrobacter* is the only genus known to be involved in the oxidation of NO_2^- to NO_3^-. Thus, it is surprising that this ion is promptly oxidized and rarely accumulates in soil. The notable exceptions include tropical savannah soils (Johansson and Sanhueza 1988) and seasonally dry tropical forest soils (Davidson et al. 1991) that accumulate NO_2^- during dry season. The O_2 is obligatory for the chemoautotrophic oxidation of either NH_4^+ or NO_2^-, and both reactions are coupled to electron transport phosphorylation, thereby providing the energy required for growth and regeneration of the responsible organisms. All members of family Nitrobacteriaceae are aerobes that synthesize their cell constituents from

Table 3.1 Selected strains of chemolithotrophic nitrifiers and their habitats

Compound	Genus	Species	Habitat
Ammonium to nitrite	*Nitrosomonas*	*europea*	Soil, water, sewage
		communis	Soil
		cryotolerans	Marine
		eutropha	Sewage
		halophila	Salt or soda lakes
		marina	Marine
		nitrosa	Eutrophic
		oligotrapha	Freshwater
		ureae	Soils, freshwater
	Nitrosospira	*multiformis*	Soil
	Nitrosospira	*briensis*	Soil
	Nitrosospira	*tenuis*	Soil
	Nitrosococcus	*mobilis*	Soil
	Nitrosococcus	*oceani*	Marne
	Nitrosocpumilus	*maritimus*	Marine
Nitrite to nitrate	*Nitrobacter*	*winogradskyi*	Soil, water
		hambugensis	soil
		vulgaris	soil
		alkalicus	Alkaline soil
	Nitrospina	*gracillis*	Marine
	Nitrococus	*mobilis*	Marine
	Nitrospira	*marina*	Marine
	Nitrospira	*moscoviensis*	Heating pipes

CO_2 by way of the Calvin reductive pentose phosphate cycle. The relatively narrow species diversity of the chemoautotrophs responsible for nitrification in soil is thought to render the process unusually susceptible to external influences (Haynes 1995).

3.2.1.4 Controls of Nitrification Rates in Environment

Soil texture and structure influence N mineralization by affecting the aeration status, distribution of organic matter, physical distribution of organic material, physical chemical and biological properties (Strong et al. 1999). The availability of NH_4^+ to nitrifiers in soil depends on NH_4^+ fixation capacity of clay minerals and the presence of other competing cations on the exchange sites (Wang et al. 2003). In addition, O_2 levels, temperature, and moisture content are critical factors that affect nitrifier activity, making nitrification a complex process in soils.

(a) Substrate availability
Ammonia oxidizing bacteria are widely distributed in soils and require only CO_2, O_2, and NH_4^+ to grow and multiply. In soils, CO_2 is generally available, and O_2 supply is usually in adequate supply in well-drained conditions, except

for brief intermittent periods, and few anaerobic environments such as sediments, bogs and sludge. Therefore, NH_4^+ availability is the factor that most frequently limits the overall rate of nitrification. Ammonia inhibition has been observed at 2.0–5.0 mM NH_4^+ concentrations in agricultural soils (Shi and Norton 2000); and at approximately 5 mM NH_4^+ for ammonia-sensitive ammonia oxidizing bacteria isolated from sewage sludge (Suwa et al. 1994; Norton 2008). Because N_2O is the result of the reductive process, its importance as a product of nitrification increases with decrease in O_2 availability. However, whether the increased importance on N_2O as a product translates to higher yields of N_2O production depends on how much the overall process rate is reduced by the limited availability of O_2.

(b) Oxygen and water potential

O_2 affects nitrification through its roles both as substrate for AMO and as the terminal electron acceptor from cytochrome C oxidase. Oxygen availability is controlled by the interaction of O_2 consumption and diffusion from the soil surface through the air filled pores spaces. Sufficient O_2 diffuses into most of soils that are at field capacity or drier to maintain nitrification, although microsites lacking O_2 may frequently occur inside soil aggregates. Nitrification rates generally decline in soils that remain wetter than field capacity for several days. In addition, higher clay contents may also inhibit nitrification, mainly due to higher water potential (Schjonning et al. 2003). Some ammonia oxidizing bacteria have the ability to use NO_2^- as an alternative electron acceptor when O_2 is limiting. Low O_2 availability may repress nitrite oxidizer activity before ammonia oxidation and result in the accumulation of NO_2^- (Laanbroek and Gerards 1993).

In soils at field capacity and drier moisture regimes, water availability impacts nitrification rates through direct effect on cell physiology and metabolic activity and through indirect effects on substrate availability (Stark and Firestone 1995).

(c) Temperature

The optimum temperature for nitrification is environmental-specific. The optimum temperatures for nitrification in pure cultures range from 25 to 35°C (Focht and Verstraete 1977). Nitrification in soils shows similar temperature optimum at 25–35°C. Soils from temperate climates and nitrifiers isolated from temperate regions have optimum temperatures close to 25°C, while those from tropical climates have optimum temperatures for nitrification around 35°C (Myers 1975). Although nitrifier activity in soils is low at temperatures <15°C, soil nitrifier activity continues even under winter season soil temperatures (2–10°C; Cookson et al. 2002) and as low as 0°C (Focht and Verstraete 1977; Koops et al. 1991).

(d) Acidity and alkalinity

Both ammonia and NO_2^- oxidation are considered optimal at neutral to slightly alkaline soil pH values. Growth rates and activity are significantly reduced outside the relatively narrow pH range around the optimum for the organism (Prosser 1989). However, autotrophic nitrification have been confirmed in soils with pH values from 3.0 (De Boer and Kowalchuk 2001) to 10.0 (Sorokin et al. 2001). Generally, nitrification is rapid in soils with pH $\geq$ 6.0, but slower in soils with pH $\leq$ 5.0. The effect of acidity on NH_3 oxidation is related to the exponential decrease in NH_3 availability with decreasing pH (pKa of NH_3/NH_4^+ is 9.25). Under acidic conditions ammonia occurs as NH_4^+, thus it is not suitable as substrate for AMO (Suzuki et al. 1974). Nevertheless nitrification proceeds in soils with a pH $\leq$ 4.0 (Vitousek et al. 1982). Several theories have been proposed to explain the nitrification in acidic soils including existence of acid-tolerant strains, microsites of relatively higher pH, syntrophic associations with mineralizing organisms, and protection via aggregate formation (Kowalchuk and Stephen 2001).

(e) Carbon to nitrogen ratio and soil organic matter
Nitrification is influenced by the availability of NH_4^+ to the nitrifiers in soils. The C:N ratio (i.e., quality of SOM) determines the heterotrophic bacterial populations and their ability to compete with the nitrifiers for NH_4^+. Nitrifiers are poor competitors for C and N compared with heterotrophic bacteria; thus a high C:N ratio in soils usually leads to microbial immobilization of the NH_4^+-N, effectively suppressing nitrification (Focht and Verstraete 1977; Sahrawat 1996). Most arable soils have a C:N ratios of about 10 where nitrification proceeds at a normal rate and immobilization of NH_4^+-N formed from mineralization is limited (Vitousek et al. 2002). Under forest and grassland soils, however, where SOM often have C:N ratios >100, microbial immobilization of NH_4^+-N is common (Vitousek et al. 1982; Vitousek et al. 2002).

3.2.1.5 Soil Factors Influencing Nitrification

Nitrification is a biological process which can be influenced by soil physical factors including texture and structure, soil chemical factors such as pH, cation exchange capacity (CEC) and SOM concentration. Soil texture and structure influence nitrification by affecting the aeration status, physical distribution of SOM, and accessibility of NH_4^+ to nitrifiers (physical protection) (Strong et al. 1999). Plants modify soil physical and chemical environment as they grow. Environmental factors such as temperature, moisture content, aeration (O_2 and CO_2 concentrations) also play a significant role in nitrification. In soil system, moisture and aeration are inversely related. Therefore, it is difficult to differentiate the extent of their relative roles in nitrification. O_2 concentration is reduced at higher soil moisture, since most of pore spaces are filled with water. Also, high soil moisture restricts diffusion of air into the soil. In addition, plants can modify both soil physical, biological and chemical environment as they grow. Plants can change soil organic C (SOC) content, nutrients availability, aeration and moisture content from time to time. Nitrification releases protons which acidifies the soil. Therefore, natural buffering capacity of soil could influence nitrification rate. Low soil pH reduces the rate of nitrification (De Boer and Kowalchuk 2001).

3.2.2 Heterotrophic Nitrification

Heterotrophic nitrification is defined as the oxidation of reduced N compounds (including organic N) producing NO_2^- or NO_3^-. It is catalyzed by a variety of microorganisms including fungi, actinomycetes, and heterotrophic bacteria, using a wide variety of metabolic pathways (Prosser 1989), and can be significant in some forest soils (Killham 1990). In some organisms, the mechanism is similar to that in autotrophic ammonia oxidizers and is also linked in some strains to aerobic denitrification. The second mechanism, termed as fungal nitrification, is linked to lignin degradation and involves reaction of reduced organic compounds with hydroxyl radicals produced in the presence of hydrogen peroxide and superoxide. There is little evidence that heterotrophs gain energy or other benefit such as growth of organisms from heterotrophic nitrification (Prosser 1989; Jetten et al. 1997), and cellular rates of heterotrophic nitrification activity are significantly lower than that of autotrophs. However, the process is important in acid soils or where C:N ratios and heterotrophy biomass are high (Killham 1990; Prosser 2005).

3.2.3 Chemolithotrophic Ammonia Oxidation

The chemolithoautotrophic bacteria or archaea responsible for the oxidation of ammonia or ammonium to nitrite gain metabolic energy from this oxidation process. The ammonia oxidizing bacteria (AOB) then use this energy for the biosynthesis of all their cellular requirements from CO_2 and inorganic nutrients. The AOB have been identified from genera *Nitrosomonas*, *Nitrospira* and *Nitrosococcus*. The AOB are characteristically autotrophs although small amounts of organic C (pyruvate and amino acids) may be incorporated into their biomass (Prosser 1989). These bacteria are highly specialized for their chemolithotrophic metabolism, which is consistent with their lack of genes encoding catabolic pathways for the use of organic substrates (Chain et al. 2003). These organisms are characterized by intracytoplasmic membrane containing the system responsible for oxidation and energy capture.

3.3 Denitrification

Denitrification is the reduction of nitrate (NO_3^-) or nitrite (NO_2^-) to the gases nitric oxide (NO) N_2O, and di-nitrogen (N_2), which is considered to occur under anaerobic condition (Morley et al. 2008). It is the last step in the N cycle, where fixed N is returned to the atmospheric pool of N_2. Denitrification is a microbial respiratory process in which NO_3^- or NO_2^- substitute for O_2 as an alternative terminal electron acceptor when O_2 supply is limited for aerobic respiration. It is a heterotrophic

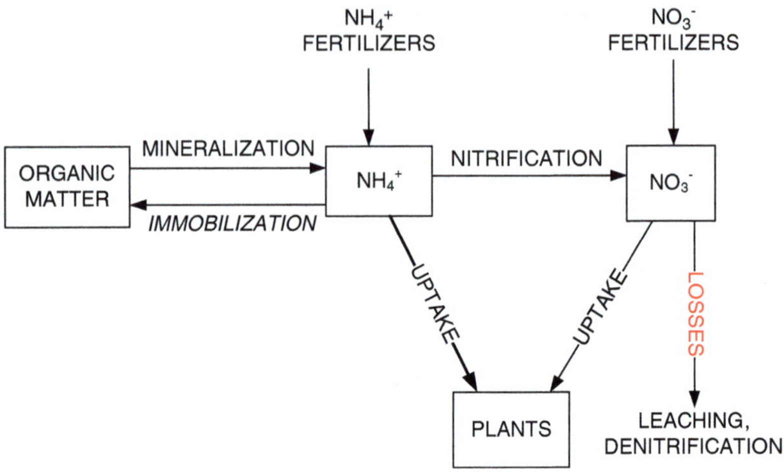

Fig. 3.3 An overview of the key processes and transformations influencing N in soils (Modified from Norton 2008)

process in which N oxides serve as the terminal electron acceptor for organic C metabolism (Codispoti et al. 2001). It consists of sequential reduction of NO_3^-, and the ultimate end product of denitrification is N_2, with various N-oxides as intermediate (Fig. 3.3). In much broader sense, however, from biogeochemical perspective, any process that reduces nitrogenous compounds to gaseous nitrogen (N_2) as the end-products can be considered as denitrification (Devol 2008). Thus, metabolism processes such as autotrophic, anammox, and inorganic reactions between metal oxides and ammonium that may produce N_2 would be denitrifying process. Denitrification was discovered in the second half of the nineteenth century by several researchers: (1) observation of systematic disappearance of NO_3^- from water by R.A. Smith in 1867, and (2) bacteria responsible for NO_3^- loss from soil were identified by E. Meusel in 1875. In 1882, Gayon and Dupetit introduced the term denitrification, and isolated pure cultures of denitrifying bacteria in 1886 (Coyne 2008). Substantial progress has been made since 1990s to understand biochemistry and genetic of denitrification. Some of it has been summarized in number of comprehensive reviews (Zumft 1997; Philippot 2002; Philippot et al. 2002, 2007).

It is estimated that soils of terrestrial ecosystems denitrify approximately 124 Tg N $year^{-1}$, which is about 35–40% of total N from land-based reactive N (Nr) sources (Seitzinger et al. 2006). Agricultural fields, which receive high N loading and poor soil drainage tends to be hot spots for denitrification (Hofstra and Bouwman 2005). Other major sinks for land-based Nr are rivers, estuaries and coastal waters.

In soils and aquatic ecosystems, denitrification occurs when O_2 concentration is reduced to near zero. The suboxic conditions initiate a series of respiratory processes through which alternate electron acceptors are utilized during the oxidation of OM. In normal soils and aquatic ecosystems, the dominant electron acceptor, after O_2 is NO_3^-. Chemical distribution of electron acceptors in natural ecosystems generally follows the thermodynamic energy yields, i.e. $O_2 > NO_3^- > Mn(IV) > Fe(III) > SO_4^-$ (Luther et al. 1998; Devol 2008). As a rule, denitrification requires anaerobic or suboxic conditions (Coyne 2008). Almost all heterotrophic denitrifiers are facultative anaerobes (Tiedje 1988). The expression of denitrification in the organism cell is triggered by the environmental parameters of low O_2 tension and availability of N-oxides. Denitrification is part of the bioenergetic apparatus of a bacterial cell where the N oxides —NO_3^-, NO_2^-, NO and N_2O serve in place of O_2 as terminal acceptors for electron transport phosphorylation (Zumft 1997).

For the biological denitrification to occur, microorganisms possessing the metabolic capacity must be present. In addition, three essential conditions must exist: (1) suitable electron donors such as organic compounds, (2) anaerobic conditions or restricted O_2 availability (3) presence of N-oxides (NO_3^-, NO_2^- NO or N_2O) as electron acceptors. The main limitation of co-occurrence of these factors is that the production of nitrate requires oxic while denitrification requires suboxic conditions. The denitrification process is promoted under anaerobic conditions, high levels of NO_3^- and readily available source of C. Denitrification can occur at oxic/suboxic interfaces, with the interface being a separation in either space or time or both. The wide range of environments in which the denitrification occurs reflect the variety of physical conditions that bring aerobically produced NO_3 in contact with denitrifiers in suboxic environments. Denitrification occurs in microsites within well-drained soils in forests, grasslands, and agricultural lands. It also occurs in partially to fully water-saturated soils, ground water aquifers, surface, hyporheic and riparian sediments. In addition, denitrification can occur in continental shelf sediments, permanent column in suboxic bottom waters of river reaches and seasonally varying suboxic bottom waters of lakes, estuaries, continental shelves and enclosed seas throughout the water column in suboxic river riches, and in oxygen minimum zones at intermediate water depths in the oceans.

The microorganisms capable of denitrifying, including bacteria, archaea and some fungi, are ubiquitous, and thus denitrification occurs widely throughout the terrestrial, marine, and freshwater ecosystems where combined conditions of NO_3 or NO_2 availability, low O_2 concentrations, and sufficient OM occur. In agricultural soils, denitrification occurs in waterlogged soils in which the redox potential falls below 400 mV. Molecular O_2 is a potent denitrification inhibitor. Strict anaerobic condition is not required, however. The upper limit of O_2 concentration for the occurrence of denitrification is ~5 μM (0.5 mg O_2 L^{-1}), and generally more denitrification occurs at O_2 concentrations <0.2 mg O_2 L^{-1} (Seitzinger et al. 2006; Devol 2008).

Denitrification can be viewed either as beneficial or detrimental process. In agricultural, forest and grassland soils, denitrification can lead to losses of valuable N from soil system, decrease the efficiency of fertilizer use, and reduce productivity of the land. Nitrates play significant role as essential plant nutrient, and denitrification leads to considerable N losses in agricultural soils. Losses increase with increase in rate of fertilization. Between 0 and 25% of fertilizer N can be lost as N_2 or N_2O (Mogge et al. 1999). Denitrification causes fertilizer N losses from field

(Ryden and Lund 1980). In contrast, the removal of excess NO_3^- from the soil by denitrifiers in land based, wastewater treatment systems is considered beneficial process that can aid in protection of ground and surface waters (Kim and Burger 1997; Barton et al 1999). The NO_3^- is susceptible to leaching and contaminates surface and groundwater sources. Denitrification is a means to remove N compounds from wastewaters and minimizing groundwater and surface waters contamination by wastewater treatment systems (Knowles 1982; Schipper and Vojvodić-Vuković 2000). It also plays significant role in removing N oxides from the environments that may be adversely affected by available N (Smith et al. 1976). It can block other environmentally harmful respiratory processes such as sulfate reduction and methanogenesis by competing for available electron donors (Coyne 2008). Nitrates can also become a pollutant of groundwater and surface water, causing major problems for the supply of drinking water. High NO_3^-concentrations in water bodies is potentially harmful to human and animals health. The Word Health Organization (WHO) and US Environmental Protection Agency (US EPA) has stipulated a safe upper limit for drinking water for human consumption at 45 mg NO_3^- L^{-1} and is 10 mg L^{-1} NO_3^--N (WHO 2007; US EPA 2009). Denitrification is fundamentally important in the biogeochemical cycle because it is a major route by which inorganic oxidized N compounds in the ecosystem return to the atmospheric N_2 pool (Bowden 1986). In both cases, denitrification produces NO and N_2O trace gases that contribute to ozone depletion and global warming (Wuebbles 2009; Ravishankara et al. 2009). N_2O emission by denitrification is the net result of the balance between production and reduction by denitrifying bacteria.

3.3.1 Respiratory Denitrification

Respiratory denitrifying bacteria are generally an aerobic bacteria which have the alternative capacity to reduce nitrogen oxides when O_2 becomes limiting (Tiedje 1988). In the laboratory incubations it has been demonstrated that the respiratory denitrifying bacteria do not require strict anaerobic media, rather, the growth is more reliable and rapid when initially aerobic media is inoculated with the denitrifying culture. The culture will consume O_2 and then gradually allow culture to shift to denitrifying metabolism. If an aerobic inoculum is transferred to a strictly anaerobic medium, growth will be delayed or prevented due to inability of the inoculum to generate energy to synthesize the required denitrifying enzymes (Tiedje 1988). Respiratory denitrification is carried out by heterotrophic bacteria that oxidize OM for energy and biosynthesis. The process requires NO_3^- or NO_2^- which serves as the terminal electron acceptor. NO_3^- is usually the starting point in agricultural soils. However, any other N oxides (NO_2^-, NO, or N_2O) can also serve as substrate (Coyne 2008). Reduction of N-oxides is typically coupled to electron transport phosphorylation (Knowles 1982; 1996). The capacity for respiratory denitrification is widespread among bacteria and is distributed across various taxonomic subclasses within Proteobacteria (Zumft 1997). The main criterion for a respiratory denitrifier is that the growth yield should be enhanced proportional to the amount of N-oxide present, and that the increase should be greater than that provided if the N-oxide simply served as an electron sink (Tiedje 1988). For the complete pathway, NO_3^- with an oxidation state for N at +5 is sequentially reduced to molecular N_2 with the oxidation state of 0, and can be represented by Eq. 3.9 (Coyne 2008):

$$NO_3^- \rightarrow NO_2^- \rightarrow NO \rightarrow N_2O \rightarrow N_2 \tag{3.9}$$

The following stoichiometric equation for denitrification is often used with glucose as a substrate (Eq. 3.10):

$$5CH_2O + 4NO_3^- + 4H^+ \rightarrow 2N_2 + 5CO_2 + 7H_2O \tag{3.10}$$

The conversion of NO_3- to N_2 can be complete, but a small and variable portion of N is often emitted as N_2O gas. In agriculture soils, emissions are sporadic, occurring before, during and after growing seasons. Flushes of N_2O can also occur when well aerated soils becomes moistened or saturated from precipitation or irrigation, or when frozen soils thaw during snowmelt. There has been a serious debate over the status of NO as true intermediate of denitrification (Averill and Tiedje 1982). However, occurrence of NO as an intermediate product of denitrification has been proven (Zumft 1993; Ye et al. 1994). The sequential reduction of NO_3^- to N_2 is facilitated by four enzymes systems, namely nitrate reductase (NAR), nitrite reductase (NIR), nitric oxide reductase (NOR), and nitrous oxide reductase (NOS) (Zumft and Korner 1997). A denitrifier does not need to have the capacity to reduce all possible intermediates. In most cases denitrification is often truncated (Tiedje 1988). In the marine ecosystems, the intermediates NO_2^- and N_2O escape the cells and are frequently found in the denitrifying environments, and can accumulate externally (Codispoti et al. 2005; Tiedje 1988). The NO is a free radical and very reactive, and toxic to most bacteria (Zumft 1997). Therefore, nitrite reductase and nitric oxide reductase are controlled interdependently at both transcriptional and enzyme activity levels in order to minimize the accumulation of NO (Zumft 1997), and their levels are generally undetectable under low O_2 (<0.5 nM) marine environments (Ward and Zafiriou 1988).

3.3.1.1 Denitrifying Organisms

Denitrification is a metabolic capacity distributed among microbial groups in Archaea, Proteobacteria and eukaryotic fungi (Zumft 1997). Many soil prokaryotes can denitrify and exhibit a variety of reduction pathways for N-oxides. The proportion of denitrifiers represent nearly 5% of the total soil microbial community, thereby out-ranking other microbial groups involved in the N cycle such as nitrifiers (Henry et al. 2004, 2006). Not all microorganisms can reduce all possible N-oxides, however. Some microorganisms produce N_2, while others produce only N_2O.

Table 3.2 The diversity of archaea, bacteria and fungi genera harboring denitrifying species

Microorganism	Genus	Species
Archaea	*Haloarcula*	*marismortui*
	Halobacterium	*denitrificans*
	Pyrobaculum	*aerophilum*
Bacteria		
Actinomycetes	*Corynebacterium*	*nephridii*
	Streptomyces	*thioluteus*
Firmicutes	*Bacillus*	*azotoformans*
	Paenibacillus	*terrae*
Bacteroides	*Flavobacterium*	*denitrificans*
	Flexibacter	*canadiensis*
Aquifaceae	*Hydrogenobacter*	*thermophilus*
Proteobacteria-Alphaproteobacteria	*Agrobacterium*	spp.
	Azosprillum	*lipoferum*
	Bradyrhizobium	*japonicum*
	Rhizobium	sp.
	Brucella	*melitensis*
	Pseudovibrio	*denitrificans*
	Rhodobacter	*sphaeroides*
Betaproteobacteria	*Alcaligenes*	*fecalis*
	Aquasprillum	*magnetotcticum*
	Azonexus	*caeni*
	Nitrosomonas	*europea*
		eutropha
	Thibacillus	*denitrificans*
Gammaproteobacteria	*Halomonas*	*desierata*
		campisalis
	Pseudomonas	*fluorescens*
	Thioalkalivibrio	*denitrificans*
Epsilonproteobacteria	*Nitratifractor*	*salsuginis*
		tegarcus
	Thiomcrospira	*denitrificans*
Fungi	*Fusarium*	*oxysporum*

Furthermore, some microorganisms cannot reduce NO_3^-. Therefore, they only use NO_2^-. In addition, some NO_3^- reducing bacteria reduce the produced NO_2^- into ammonium and organic N, and not into nitric oxide (NO) (i.e., dissimilatory NO_3^- reduction). Thus, a true denitrifying microorganisms: (1) produce N_2O and/or N_2 as the major end product of NO_3^- or NO_2^- reduction, and (2) the reduction must result into increased growth rather than using NO_3^- or NO_2^- as an electron sink (Mahne and Tiedje 1995).

The distribution of the denitrifying trait among microorganisms cannot be predicted by the taxonomic affiliation, since distantly related microorganisms can denitrify, while closely related strains may exhibit different respiratory pathways. Over 60 genera of denitrifying microorganisms have been identified within bacteria, archaea, fungi, and other eukaryotes (Bremner 1997; Shoun et al. 1992; Pina-Ochoa et al. 2010) (Table 3.2). Some denitrifiers can undertake nitrate or nitrite reduction under conditions not typically associated with conventional denitrification. For example, several bacteria isolated from soils and sediments are capable of denitrifying in the presence of O_2 (Patureau et al. 2000). Among the phylogenetically diverse groups of denitrifiers, several bacteria are also involved in other steps in N cycle, e.g., *Nitrospira* and *Nitrosomonas*, which are capable of nitrification (Shaw et al. 2006). Similarly, many N fixing rhizobia can denitrify (O'hara and Daniel 1985).

Most studies aimed at quantifying N_2O production are focused on bacterial denitrification despite the fact that fungi can denitrify in woodlands and also grasslands (Laughlin and Stevens 2002; Blagodatskiy et al. 2008). In most cases the main product of fungal denitrification is N_2O because many fungi lack the N_2O reductase enzyme (Takaya 2009). It has been shown that Ectomycorrhizal fungal species possess the ability to produce N_2O, suggesting that this group of fungi may have significant but yet unexplored role in N_2O production in forest ecosystems (Prendergast-Miller et al. 2011).

3.3.1.2 Factors Affecting Denitrification

Composition of microbial population exerts a dominant control on denitrification and also N_2O emissions, and these remains relatively constant over time (Bedard-Haughn et al. 2006). In general, denitrification proceeds in soils and broad range of conditions than would be predicted on the basis of the biochemistry of the process and the physiology of denitrifiers. Denitrification is promoted by high soil moisture, neutral soil pH, high soil temperature, low rate of O_2 diffusion and presence of soluble OM and NO_3^-.

(a) Soil water and aeration

Soil water content together with the rate of O_2 consumption determines the O_2 availability (Tiedje 1988). Oxygen is the dominant environmental moderator of denitrification. It is generally assumed that nitrogen oxide reductase enzymes are repressed by O_2 and when O_2 is removed, the reductase enzymes are activated within a period of 40 min to 3 h (Payne 1973). In the presence of O_2 denitrification and synthesis of denitrifying enzymes is inhibited (Smith and Tiedje 1979). The potential denitrification rate under aerobic conditions is typically 0.3–3% of the anaerobic rate. Increase denitrification rate is associated with restricted soil aeration at high soil water content in soils. Studies have indicated the soil water threshold value of 60% water-filled pore space (WFPS) above which the denitrification rates increases sharply with increase in soil water content to about 90% WFPS. Below 60% WFPS, the denitrification is not related to soil water content However, O_2 sensitivity of denitrification varies among bacteria from extreme sensitive in *Paracoccus denitrificans* to less sensitive in *Thiosphaera patotropha* which maintains >25% of their denitrifying activity in aerobic conditions (Robertson et al. 1988). O_2 exerts its moderation effects through synthesis and activity of the denitrifying enzymes. Typical threshold O_2 concentration for synthesizing denitrifying enzymes is <10 μmol O_2 L^{-1} (Tiedje 1988), the sensitivity enzymes synthesis follows the sequence N_2O reductase > NO reductase > NO_2^- reductase > NO_3^- reductase, but it also varies among denitrifier organisms. In some organisms e.g. *Achromobacter cyclocastes*, both NO_3^- and N_2O reductase are synthesized at all O_2 concentrations but only fully induced when O_2 concentration is <2 μM (Coyne and Tiedje 1990). The threshold O_2 content at which denitrification occurs in soils and water is a complex interaction of diffusion, respiration, and global regulation of enzyme transcription and expression by O_2 within the denitrifiers present in the environment (Coyne 2008; Devol 2008). In soils, flooding has immediate effect of reducing aeration and directing electron flow to denitrifying enzymes.

The O_2 status of a habitat microsite is controlled by the rate of O_2 supply to the site and the rate of O_2 consumption by respiration. In soils, water restricts diffusion by increasing the distance to air channels and forcing O_2 to flow through water where diffusion coefficient is much smaller, rather than through gas phase, where diffusion coefficient is much greater (Tiedje 1988). Therefore, moisture status has significant impact on denitrification, largely through its influence on O_2 diffusion. Main factors which control moisture status are rainfall, plants-through evapotranspiration, and soil texture-which influence water holding capacity.

(b) Temperature

Denitrification occurs in temperature ranging from 4 to 60°C. At temperature of 1°C little NO_3^- is lost even from saturated soil. Temperature must typically rise to 5–7°C before significant NO_3^- reduction occurs. One of the effects of declining temperature is increased solubility of O_2 which may play a role in decreasing denitrification activity (Coyne 2008).

(c) Carbon

The role of carbon is to provide the electron donor for nitrate reduction. It also indirectly affects the O_2 status of aerobic soils. Available C increases heterotrophic respiration in aerobic soils decreasing O_2 concentrations and creating anaerobic microsites for denitrifiers. Under denitrifying conditions, C may not be limiting factor, since O_2-stressed cells excrete more available C. Therefore, C limitation to denitrification is very rare (Tiedje 1988). Available C influences respiration. It has been demonstrated that denitrification occurs in "hot spots" which are created by decaying SOM that generates anaerobic microsites within aerobic soils (Barton et al. 1999). This hypothesis has been used to explain large spatial variability in denitrification and N_2O emission typically observed in soils.

(d) Soil nitrates

Nitrate is required as an electron acceptor for denitrification. The availability of NO_3^- to denitrifiers depends on the rate of NO_3^- transport from bulk soil solution to the denitrifying microsite and the rate of consumption in the soil. NO_3^- availability often limits denitrification in soils not receiving additional N inputs via fertilizers or biological N fixation. NO_3^- has been shown to limit denitrification in unfertilized grasslands (Groffman et al. 1993; Tenuta and Beauchamp 1996) and undisturbed forest soils (Henrich and Haselwandter 1997). Diffusion of NO_3^- from the aerobic nitrifying microsite to anaerobic denitrifying microsite can potentially limit denitrification under low N loading (Myrold and Tiedje 1985).

3.3.1.3 Effects of Soil Management

In agricultural soils, land management practices such as tillage, compaction, fertilization, liming, irrigation, herbicide use and pesticide use have significant effects on microbial activity in soil. Studies have shown that denitrification losses are greater under no till (NT) than in plowed soils. The differences may be due to higher water contents under NT soils. Liming acid soils may have beneficial effects on overall denitrification rates.

Nitrogen fertilizers promote denitrification activity in soils, and substantial amount of fertilizer added N is lost through denitrification. Fertilization also affects the N_2O to N_2 ratio from denitrification, and N_2O emission increases due to increased input of fertilization (Skiba and Smith 2000). The rate of denitrification in fertilized soils is influenced by fertilizer type, application rate, timing and application method. The combination of high N application rates and poor drainage leads to high denitrification activity. Fertilization can also cause secondary effect that affect the denitrification, such as soil pH change. Denitrification is higher at neutral pH than acidic conditions. Mulvaney et al. (1997) reported higher emission of N_2O and N_2 after application of alkaline hydrolyzing fertilizers than acidic fertilizers with the following order NH_3 > urea > $(NH_4)_2HPO_4$ > $(NH_4)_2SO_4 \approx NH_4H_2PO_4$. Organic fertilizers such as manures and sewage sludge often promote denitrification more than mineral fertilizers (Enwall et al. 2005; Dambreville et al. 2006). This is probably due to additional supply of readily available organic C.

3.3.2 Dissimilatory Nitrate Reduction to Ammonium (DNRA)

Dissimilatory NO_3^- reduction to ammonia (DNRA), sometimes referred to as nitrate ammonification, is an anaerobic microbial pathway of N cycle that transform NO_3^- first to NO_2^-, and then to NH_4^+. There is a growing body of evidence showing that this process can occur in range of ecosystems including upland tropical forests (Silver et al. 2001; Rutting et al. 2008), paddy rice soils (Yin et al. 2002), and grasslands (Rutting et al. 2010). The evidence also suggests that this process may not be strictly confined to highly reducing and high C:N conditions as traditionally understood (Wan et al. 2009; Schmidt et al. 2011).

DNRA is catalyzed by facultative and obligate fermentative bacteria. The major criterion used to identify DNRA is the production of ammonium from NO_3^- in excess of the reduced N needed for growth (Tiedje 1988). The organisms involved in DNRA are fermentative, and uses NO_3^- mainly as electron sink. The process occurs in highly reducing environments capable of maintaining sustained anaerobic metabolism (Tiedje 1988). These conditions are thought to occur only in flooded environments. Upland humid tropical forests are good candidates for DNRA activity, due to high N availability and rapid N cycling (Silver et al. 2001). The conditions for DNRA are similar to those for denitrification (e.g. low redox potential, available NO_3^- and labile C). However, DNRA is favored by high ratio of available C to electron acceptors. The direct and rapid reduction of NO_3^- to NH_4^+ via DNRA has important implications for ecosystem N retention and loss.

3.3.3 *Chemodenitrification*

This occurs under acidic conditions (pH < 5), where inorganic and organic compounds reacts with N oxides and yield free energy. Some of these reactions are catalyzed by microbes, while others are known to occur spontaneously and catalyzed by abiotic agents (Tiedje 1988). The commonly known chemodenitrification is the acid-catalyzed destruction of NO_2^- which can become significant at pH < 5.0 (Eq. 3.10 and 3.11).

$$3HNO_3 \rightarrow 2NO + HNO_2 + H_2O \tag{3.10}$$

$$2HNO_2 \rightarrow NO + NO_2 + H_2O \tag{3.11}$$

Although N_2, N_2O and NO have all been reported as products of chemodenitrification, the predominant product is NO. Since NO is minor product of biological denitrification, its occurrence can be used as preliminary indication of chemodenitrification. Another chemodenitrification mechanism involves the oxidation of organic N by NO_2^- forming N_2 gas (Eq. 3.12), generally known as van Slyke reaction, which generally occurs in frozen soils, where NO_2^- is concentrated by the salting-out effect (Tiedje 1988).

$$R_2C \rightarrow NOH + HNO_2 \rightarrow R_2C \rightarrow O + N_2O + H_2O \tag{3.12}$$

Chemodenitrification is considered minor process on a global N cycle, occurring only under specific conditions.

3.3.4 *Nitrifier Denitrification*

Nitrifier denitrification by ammonia oxidizing bacteria has been acknowledged in pure cultures (Hooper 1968; Ritchie and Nicholas 1972) and it has been suggested that nitrifier denitrification could be a universal trait in betaproteobacterial ammonium oxidizing bacteria in soil (Shaw et al. 2006). Studies suggest that nitrifier denitrification may contribute significantly to N_2O production from soils but definite proof remains elusive due to methodological constraints.

3.3.5 Measurements of Total Denitrification Rates

Quantification of the rates of nitrate removal is of interest in a variety of ecosystems. Denitrification to N_2 is the only natural process of permanent removal of excess N from ecosystems. Direct quantification of total denitrification is hampered by the very high natural background of atmospheric N_2. Quantification of denitrification is also hindered by high spatial and temporal variation in the process, especially in terrestrial ecosystems. Denitrification has become a key regulator of water and air quality at regional and global scales (Galloway et al. 2003). A wealth of methods has been developed in the past decades for quantification of total denitrification (Groffman et al. 2006). Unfortunately, none is without drawbacks and even today, there is no method that can be used at the field scale or at high temporal resolution.

The most common method is the acetylene (C_2H_2) inhibition method (Balderston et al. 1976), by which the terminal step of denitrification, i.e. the reduction of N_2O to N_2 is inhibited by C_2H_2. Major drawbacks are that (i) it is not easy to achieve 100% diffusion of C_2H_2 to the active denitrification sites, (ii) nitrification is also inhibited by C_2H_2, and (iii) C_2H_2 interacts with NO in oxic environments (Bollmann and Conrad 1997). To overcome these problems a completely new concept of replacing the background N_2 during soil core incubations with a noble gas (e.g. with a He:O_2 mixture) has been developed and facilitates direct measurements of N_2O and N_2 (Scholefield et al. 1997; Butterbach-Bahl et al. 2002). However, the high capital investment in equipment and the time-consuming flushing procedure to remove N_2 limit wide adoption of the method. Despite the drawbacks, acetylene inhibition method is considered to be the most practical method for measuring in situ denitrification rates in soils (Davidson et al. 1990).

The use of stable isotope analysis either in tracer studies with isotopically enriched tracer compounds or at the natural abundance level offer promising alternatives. Application of $^{15}NO_3^-$ and $^{15}NH_4^+$ containing fertilizer and monitoring ^{15}N-labelled N_2O and N_2 provides a suitable tracer or denitrification to N_2 for agricultural N fertilized soils. The method can be used to generate reliable estimates of denitrification rates in soils and sediments (Groffman et al. 2006). For N poor environments, the ^{15}N tracers can artificially stimulate N turnover, microbial immobilization or dissimilatory reduction of NO_3^- to NH_4^+. In these environments, natural abundance of N and O isotopes may offer an alternative, as due to kinetic isotope fractionation the intermediates and the end product of denitrification become increasingly depleted in ^{15}N, whereas the remaining soil NO_3^- becomes increasingly enriched in ^{15}N and ^{18}O. If substrate is not limiting, large kinetic N isotope fractionation factors of up to −40‰ can be observed during denitrification

(Groffman et al. 2006). However, if denitrification is limiting or rates are small, as in the case for N-poor ecosystems, the apparent N isotope fractionation is too small to provide unambiguous interpretation of the data.

Dual-isotope labeling with ^{15}N and ^{18}O-enriched NO_3^- can identify nitrification or denitrification as source of N_2O; this information is desirable for the models (Wrage et al. 2005). However, at low pH, NO_2^-, intermediate of nitrification and denitrification, rapidly undergoes O-isotope exchange with water (Casciotti et al. 2007), which may lead to misinterpretation of the results when stoichiometric relationships in the different N_2O formation pathways are assumed, and is very likely the cause of O-isotope exchange between N_2O and water, (Kool et al. 2009). The most recent approach of quantifying denitrification rates and differentiating between nitrification and denitrification as sources of N_2O is the analysis of N_2O isotopic isomers. Intramolecular physicochemical site differences between terminal and central N atom lead to differences in N isotope ratios between the two positions during N_2O formation and consumption. Differences in this so-called ^{15}N site preference have been attributed to N_2O production during nitrification and denitrification, respectively (Perez et al. 2001). However, microbial populations have a larger impact on the isotopic and isotopic isomer signatures of N_2O than the production pathway itself (Sutka et al. 2003). The ^{15}N isotopic abundance of soil-emitted NO was determined for the first time (Li and Wang 2008). The authors found very ^{15}N-depleted NO with $\delta^{15}N$ values down to −50‰ and could identify both nitrification and denitrification as sources of soil-emitted NO. Despite great research efforts, many challenges of quantifying total denitrification and differentiating between N_2O produced by nitrification from that produced by denitrification remain.

3.3.5.1 Annual Denitrification Rates from Agricultural and Forest Soils

Annual denitrification rates for agricultural soils ranging from 0 to 239 kg N ha^{-1} yr^{-1} and mean 13.0 kg N ha^{-1} yr^{-1} have been reported in literature (Barton et al. 1999 and references therein). Highest rates were observed for the N fertilized and irrigated fields. Application of inorganic N fertilizers increases rates of denitrification more than organic N fertilizers, probably due to greater NO_3^- availability. Annual denitrification rates for forest soils ranges from 0 to 40 kg N ha^{-1} yr^{-1}, mean 2.2 kg N ha^{-1} yr^{-1} (Davidson et al. 1990; Barton et al. 1999). Denitrification rates are generally higher under deciduous than coniferous forest soils, which is attributed to greater N and C availability through more frequent litter inputs and higher soil pH (Davidson et al. 1990). Gaseous N losses are lower during the winter and highest in spring or early summer (Henrich and Haselwandter 1997). Disturbances such as forest clearing or forest clear-felling tends to increase denitrification rates for both coniferous and deciduous forests which may last for relatively short period – up to 4 years (Dutch and Ineson 1990).

3.4 Nitrous Oxide Production by Nitrification and Denitrification

N_2O is produced and emitted from soils during nitrification under aerobic conditions (i.e. up to 50% of soil pore spaces filled with water) and denitrification of NO_3^- under anaerobic conditions (e.g. >75% of the soil pore spaces filled with water) (Mosier et al. 1996). These two microbial processes contribute about 90% of the N_2O emissions from agricultural ecosystems (Smith et al. 1997). In agriculture, N lost as N_2O during nitrification is relatively low, between 2 and 4% (Duxbury and McConnaughey 1986), making this pathway of limited significance in terms of N_2O emissions. However, the N losses as N_2O from denitrification are high and occurs very rapidly in soils that are wet but not waterlogged (Linn and Doran 1984), sufficiently warm to support microbial activity. About 60–80% of soil NO_3-N can be converted into gaseous form (N_2O, NO and N_2) under anaerobic conditions (Mosier et al. 1996). The rate of N_2O production by denitrification also depends on nitrate supply and availability SOM, which determine the degree of heterotrophic bacteria activity involved in denitrification.

Production of N_2O in the soil is affected by numerous factors ranging from physical, chemical, biological to environmental. N_2O production in the soil is influenced by soil moisture content, mineral N availability, O_2 supply, SOC supply, soil pH, and redox potential. Like all biogenic processes, the microbial processes of nitrification and denitrification vary highly in magnitude with variations of environmental conditions. Soil temperature and soil moisture regimes are major drivers on hourly to inter-annual timescales for observed temporal changes of N_2O emissions (Gasche and Papen 1999; Butterbach-Bahl et al. 2004). The effect of soil temperature for N_2O emission is generally direct. Temperature influence enzyme kinetics and metabolic turnover rates of nitrifiers and denitrifiers with the optimum temperatures at approximately 30–35ºC. However, soil moisture effect is mostly indirect, since soil moisture affects the rate of O_2 diffusion into the soil profile and thus also determines if soil is predominantly aerobic or anaerobic (Smith 1980). The oxidation status of soil determines whether an oxic process of nitrification or anoxic process of denitrification predominates. Both processes can produce N_2O, but the production of N_2O by denitrification process is higher than that produced from the nitrification process. The optimum soil water filled pore space (WFPS) for N_2O production is 60–65%. At WFPS values >70–80%, emissions are strongly reduced, since N_2O in the soil profile will be further reduced to N_2, while under high soil nitrate concentrations N_2O can be the end product of denitrification (Conrad 1996, 2002; Firestone and Davidson 1989). High soil moisture content stimulates microbial respiration, restricts O_2 diffusion in soil, and increases NO_3^- diffusion to microsites. Therefore denitrification can be high after rainfall events, spring snowmelt, or after irrigation.

3.5 Conclusions

Nitrification is the biological oxidation of reduced forms of N to NO_3^- or NO_2^-. The conversion of cations (NH_4^+) to anions influence the mobility of N through generally negatively charged soil matrix. Therefore, NO_3^- moves rapidly by mass flow, and leach out of root zone. In waste treatment, however, nitrification is a desirable process, since NO_3^- is less toxic than NH_4^+, and is easily removed from the waste stream through denitrification. Denitrification is a microbial process by which N-oxides are alternative electron acceptor when O_2 is limiting. Under agricultural soils, the process results into loss of crop available N. Therefore, denitrification is of a major concern in agriculture. In aquatic ecosystem, however, denitrification is beneficial because it improves water quality, since it reduces the N loading which is responsible for eutrophication. Both nitrification and denitrification processes are of environmental concern, since they are the biological processes responsible for emissions of N_2O and NO, which contributes to greenhouse effect and ozone depletion. Nitrification and denitrification are microbial processes, which depend on environmental factors to drive its magnitude, temporal dynamics and end products.

3.6 Future Research Needs

Measurements of nitrification and denitrification and the associated N_2O and N_2 production are still scarce. Therefore, there is a need for long-term data on nitrification and denitrification to provide thorough understanding of the variability of nitrification and denitrification processes as affected by environmental constraints. Single site observations also need to be supplemented by more holistic view on denitrification and N_2O emissions at landscape scales. It is also apparent that the existing knowledge on how nitrification, denitrification, and the associated N_2O emission from soils may change in changing climate is limited. Traditional N_2O mitigation strategies have focused on increasing N use efficiency. However, there is a great potential for lowering N_2O emissions by developing management options to lower net N_2O emissions by enhancing its reduction to N_2 rather than eliminating denitrification altogether. Such options include application of copper-based fertilizers to increase Cu-based enzyme. Therefore, future research needs to focus on the following:

- Increase in field data on denitrification from various ecosystems,
- Understanding the mechanisms controlling nitrification and denitrification is needed to improve modeling and prediction of N_2O emission at landscape and regional level and also for understanding of plant-soil link and its importance as controller of soil N and C substrate availability,
- Need for improved numerical description of denitrification and nitrification in process models.
- Increasing the understanding of regulation of nitrification and denitrification and development of specific management strategies that will enhance conversion of N_2O to N_2 as a mitigation option for N_2O emission from fertilized fields.

Study Questions

1. Describe nitrification-coupled denitrification giving specific ecosystems and conditions which favors this biological process.
2. How does dissimilatory nitrate reduction differ from assimilatory nitrate reduction?
3. Excessive use of N fertilizers in agriculture may cause algal bloom in estuaries draining agricultural watersheds. Describe.
4. How can the loss of N_2O be minimized in agro-ecosystems?
5. Describe the importance of nitrification and denitrification to the environment.

References

Arp, D.J., Sayavedra-Soto, L.A., Hommes, N.G. (2002) Molecular biology and biochemistry of ammonia oxidation by *Nitrosomonas europaea*. Arch. Microbiol. **178**(4), 250–255. doi:10.1007/s00203-002-0452-0

Averill, B.A., Tiedje, J.M. (1982) The chemical mechanism of microbial denitrification. Febs Lett. **138**(1), 8–12

Balderston, W.L., Sherr, B., Payne, W.J. (1976) Blockage by acetylene of nitrous oxide reduction in *Pseudomonas perfectomarinus*. Appl. Environ. Microbiol. **31**(4), 504–508

Barton, L., McLay, C.D.A., Schipper, L.A., Smith, C.T. (1999) Annual denitrification rates in agricultural and forest soils: a review. Aus. J. Soil Res. **37**(6), 1073–1093. doi:10.1071/sr99009

Bartosch, S., Hartwig, C., Spieck, E., Bock, E. (2002) Immunological detection of nitrospira-like bacteria in various soils. Microbial Ecol. **43**(1), 26–33. doi:10.1007/s00248-001-0037-5

Basu, P., Katterle, B., Andersson, K.K., Dalton, H. (2003) The membrane-associated form of methane mono-oxygenase from *Methylococcus capsulatus* (Bath) is a copper/iron protein. Biochem. J. **369**, 417–427. doi:10.1042/bj20020823

Bateman, E.J., Baggs, E.M. (2005) Contributions of nitrification and denitrification to N_2O emissions from soils at different water-filled pore space. Biol. Fert. Soils **41**(6), 379–388. doi:10.1007/s00374-005-0858-3

Bedard-Haughn, A., Matson, A.L., Pennock, D.J. (2006) Land use effects on gross nitrogen mineralization, nitrification, and N_2O emissions in ephemeral wetlands. Soil Biol. Biochem. **38**(12), 3398–3406. doi:10.1016/j.soilbio.2006.05.010

Belser, L.W. (1979) Population ecology of nitrifying bacteria. Annu. Rev. Microbiol. **33**, 309–333

Blagodatskiy, S.A., Avksent'ev, A.A., Davydova, M.A., Blagodatskaya, E.V., Kurakov, A.V. (2008) Nitrous oxide production in soils and the ratio of the fungal to bacterial biomass. Eurasian Soil Sci. **41**(13), 1448–1455. doi:10.1134/s1064229308130140

Bleakley, B.H., Tiedje, J.M. (1982) NIitrous oxide production by organisms other than nitrifiers or denitrifiers. Appl. Environ. Microbiol. **44**(6), 1342–1348

Bollmann, A., Conrad, R. (1997) Acetylene blockage technique leads to underestimation of denitrification rates in oxic soils due to scavenging of intermediate nitric oxide. Soil Biol. Biochem. **29**(7), 1067–1077

Bowden, W.B. (1986) Gaseous nitrogen emissions from undisturbed terrestrial ecosystems: An assessment of their impacts on local and global nitrogen budgets. Biogeochemistry **2**(3), 249–279

Bremner, J.M. (1997) Sources of nitrous oxide in soils. Nutr. Cycl. Agroecosys. **49**(1–3), 7–16. doi:10.1023/a:1009798022569

Butterbach-Bahl, K., Kock, M., Willibald, G., Hewett, B., Buhagiar, S., Papen, H., Kiese, R. (2004) Temporal variations of fluxes of NO, NO_2, N_2O, CO_2, and CH_4 in a tropical rain forest ecosystem. Glob. Biogeochem. Cycles **18**(3), GB3012. doi:10.1029/2004gb002243

Butterbach-Bahl, K., Willibald, G., Papen, H. (2002) Soil core method for direct simultaneous determination of N_2 and N_2O emissions from forest soils. Plant Soil **240**(1), 105–116

Casciotti, K.L., Bohlke, J.K., McIlvin, M.R., Mroczkowski, S.J., Hannon, J.E. (2007) Oxygen isotopes in nitrite: Analysis, calibration, and equilibration. Anal. Chem. **79**(6), 2427–2436. doi:10.1021/ac061598h

Chain, P., Lamerdin, J., Larimer, F., Regala, W., Lao, V., Land, M., Hauser, L., Hooper, A., Klotz, M., Norton, J., Sayavedra-Soto, L., Arciero, D., Hommes, N., Whittaker, M., Arp, D. (2003) Complete genome sequence of the ammonia-oxidizing bacterium and obligate chemolithoautotroph Nitrosomonas europaea (vol 185, pg 2759, 2003). J. Bacteriol. **185**(21), 6496–6496. doi:10.1128/jb.185.21.2003.6496

Chalk, P.M., Smith, C.J. (1983) Chemodenitrification. In: Freney, J.R., Simpson, J.R. (eds.) Gaseous loss of nitrogen from plant-soil systems. Development in plant soil sciences, **9**. 65–89.

Chapuis-Lardy, L., Wrage, N., Metay, A., Chotte, J.L., Bernoux, M. (2007) Soils, a sink for N_2O? A review. Global Change Biol. **13**, 1–17

Codispoti, L.A., Brandes, J.A., Christensen, J.P., Devol, A.H., Naqvi, S.W.A., Paerl, H.W., Yoshinari, T. (2001) The oceanic fixed nitrogen and nitrous oxide budgets: Moving targets as we enter the anthropocene? Sci. Mar. **65**, 85–105

Codispoti, L.A., Yoshinari, T., Devol, A.H. (2005) Suboxic respiration in the oceanic water-column. In: del Giorgio, P.A., Williams, P.J.l.B. (eds.) Respiration in Aquatic Ecosystems. pp. 225–247. Oxford University Press, Oxford, U.K.

Conen, F., Neftel, A. (2007) Do increasingly depleted delta N-15 values of atmospheric N_2O indicate a decline in soil N_2O reduction? Biogeochemistry **82**(3), 321–326. doi:10.1007/s10533-006-9066-y

Conrad, R. (1990) Flux of NO_x between soil and atmosphere: Importance and soil microbial metabolism. In: Revsbech, N.P., Sorensen, J. (eds.) Denitrification in Soil and Sediments. pp. 105–128. Plenum, New York

Conrad, R. (1995) Soil microbial processes involved in production and consumption of atmospheric trace gases. Adv. Microb. Ecol. **14**, 207–250

Conrad, R. (1996) Soil microorganisms as controllers of atmospheric trace gases (H_2, CO, CH_4, OCS, N_2O, and NO). Microbiol. Rev. **60**(4), 609–640

Conrad, R. (2002) Microbial and biogeochemical background of production and consumption of NO and N_2O in soil. In: Gasche, R., Papen, H., Rennenberg, H. (eds.) Trace Gas Exchange in Forest Ecosystem. pp. 3–33. Springer, New York, USA

Cookson, W.R., Cornforth, I.S., Rowarth, J.S. (2002) Winter soil temperature (2-4 °C) effects on nitrogen transformations in clover green manure amended or unamended soils; a laboratory and field study. Soil Biol. Biochem. **34**(10), 1401–1415

Coyne, M.S., Tiedje, J.M. (1990) Induction of denitrifying enzymes in oxygen-limited *Achromobacter cycloclastes* continuous culture. Fems Microbiol. Ecol. **73**(3), 263–270. doi:10.1111/j.1574-6968.1990.tb03949.x

Coyne, M.S. (2008) Biological denitrification. In: Schepers, J.S., Raun, W.R. (eds.) Nitrogen in agricultural soils. Agronomy Monograph No. 49, pp. 201–253. ASA-CSSA-SSSA, Madison, WI

Dambreville, C., Hallet, S., Nguyen, C., Morvan, T., Germon, J.C., Philippot, L. (2006) Structure and activity of the denitrifying community in a maize-cropped field fertilized with composted pig manure or ammonium nitrate. Fems Microbiol. Ecol. **56**(1), 119–131. doi:10.1111/j.1574-6941.2006.00064.x

Davidson, E.A., Myrold, D.D., Groffman, P.M. (1990) Denitrification in temperate forest ecosystems. In: Gessel, S.P., Lacate, D.S., Weetam, G.F., Powers, R.F. (eds.) Sustained productivity of forest soils. Proceedings of 7th North America Forest Soils Conference. pp. 196–220. University of British Columbia, Vancouver, Britis Columbia

Davidson, E.A., Vitousek, P.M., Matson, P.A., Riley, R., Garciamendez, G., Maass, J.M. (1991) Soil emissions of nitric-oxide in a seasonally dry tropical forest of Mexico. J. Geophys. Res.-Atmos. **96**(D8), 15439–15445

De Boer, W., Kowalchuk, G.A. (2001) Nitrification in acid soils: microorganisms and mechanisms. Soil Biol. Biochem. **33**(7–8), 853–866

Devol, A.H. (2008) Denitrification inluding anammox. In: Capone, D.G., Bronk, D.A., Muholland, M.R., Carpenter, E.J. (eds.) Nitrogen in the Marine Environment. pp. 263–301. Elsevier Amsterdam, The Netherland

Dutch, J., Ineson, P. (1990) Denitrification of an upland forest site. Forestry **63**(4), 363–377. doi:10.1093/forestry/63.4.363

Duxbury, J.M., McConnaughey, P.K. (1986) Effect of fertilizer source on denitrification and nitrous oxide emissions in a maize field. Soil Sci. Soc. Am. J. **50**(3), 644–648

Enwall, K., Philippot, L., Hallin, S. (2005) Activity and composition of the denitrifying bacterial community respond differently to long-term fertilization. Appl. Environ. Microbiol. **71**(12), 8335–8343. doi:10.1128/aem.71.12.8335-8343.2005

Firestone, M.K., Davidson, E.A. (1989) Microrbiolgical basis of NO and N_2O production and consumption in soil. In: Andreae, M.O., Schimel, D.S. (eds.) Exchange of Trace Gases between Terrestrial Ecosystems and the Atmosphere, Life Sciences Research Report **47**, 7–21.

Focht, D.D., Verstraete, W. (1977) Biochemical ecology of nitrification and denitrification. In: Alexander, M. (ed.) Advances in Microbial Ecology **1**, 135–214. Plenum Press, New York

Forrest, W.W., Walker, D.J. (1971) The generation and utilization of energy during growth. Adv. Microb. Physiol. **5**, 213–274

Freney, J.R. (1997) Emission of nitrous oxide from soils used for agriculture. Nutr. Cycl. Agroecosys. **49**(1–3), 1–6

Galloway, J.N., Aber, J.D., Erisman, J.W., Seitzinger, S.P., Howarth, R.W., Cowling, E.B., Cosby, B.J. (2003) The nitrogen cascade. Bioscience **53**(4), 341–356

Gasche, R., Papen, H. (1999) A 3-year continuous record of nitrogen trace gas fluxes from untreated and limed soil of a N-saturated spruce and beech forest ecosystem in Germany. 2. NO and NO_2 fluxes. J. Geophys. Res.-Atmos. **104**(D15), 18505–18520

Godde, M., Conrad, R. (2000) Influence of soil properties on the turnover of nitric oxide and nitrous oxide by nitrification and denitrification at constant temperature and moisture. Biol. Fert. Soils **32**(2), 120–128

Groffman, P.M., Altabet, M.A., Böhlke, J.K., Butterbach-Bahl, K., David, M.B., Firestone, M.K., Giblin, A.E., Kana, T.M., Nielsen, L.P., Voytek, M.A. (2006) Methods for measuring denitrification: Diverse approaches to a difficult problem. Ecol. Appl. **16**(6), 2091–2122

Groffman, P.M., Rice, C.W., Tiedje, J.M. (1993) Denitrification in a tallgrass Prairie landscape. Ecology **74**(3), 855–862. doi:10.2307/1940811

Haynes, R.J. (1995) Nitrous oxide budget Enc. Environ. Biol. **2**, 579–585

Henrich, M., Haselwandter, K. (1997) Denitrification and gaseous nitrogen losses from an acid spruce forest soil. Soil Biol. Biochem. **29**(9–10), 1529–1537. doi:10.1016/s0038-0717(97) 00010-2

Henry, S., Baudoin, E., Lopez-Gutierrez, J.C., Martin-Laurent, F., Baumann, A., Philippot, L. (2004) Quantification of denitrifying bacteria in soils by nirK gene targeted real-time PCR. J. Microbiol. Meth. **59**(3), 327–335. doi:10.1016/j.mimet.2004.07.002

Henry, S., Bru, D., Stres, B., Hallet, S., Philippot, L. (2006) Quantitative detection of the nosZ gene, encoding nitrous oxide reductase, and comparison of the abundances of 16S rRNA, narG, nirK, and nosZ genes in soils. Appl. Environ. Microbiol. **72**(8), 5181–5189. doi:10.1128/ aem.00231-06

Hofstra, N., Bouwman, A.F. (2005) Denitrification in agricultural soils: Summarizing published data and estimating global annual rates. Nutr. Cycl. Agroecosys. **72**(3), 267–278. doi:10.1007/ s10705-005-3109-y

Hooper, A.B.. (1968) A nitrite-reducing enzyme from *Nitrosomonas europaea*: preliminary characterization with hydroxylamine as electron donor. Biochim. Biophys. Acta **162**(1), 49–65

Hooper, A.B.., Vannelli, T., Bergmann, D.J., Arciero, D.M. (1997) Enzymology of the oxidation of ammonia to nitrite by bacteria. Ant. Leew. Int. J. G. Mol. Microb. **71**(1–2), 59–67

Jarvis, S.C. (1996) Future trends in nitrogen research. Plant Soil **181**(1), 47–56

Jetten, M.S.M., Logemann, S., Muyzer, G., Robertson, L.A., deVries, S., vanLoosdrecht, M.C.M., Kuenen, J.G. (1997) Novel principles in the microbial conversion of nitrogen compounds. Ant. Leew. Int. J. G. Mol. Microb. **71**(1–2), 75–93

Johansson, C., Sanhueza, E. (1988) Emission of NO from savanna soils during rainy season. J. Geophys. Res.-Atmos. **93**(D11), 14193–14198

Kelly, D.P. (1978) Bioenergetics of chemolithotrophic bacteria. In: Bull, Meadow (eds.) Companion to microbiology. pp. 363–386. Longman, London

Killham, K. (1990) Nitrification in coniferous forest soils. Plant Soil **128**(1), 31–44

Kim, D.Y., Burger, J.A. (1997) Nitrogen transformations and soil processes in a wastewater-irrigated, mature Appalachian hardwood forest. Forest Ecol. Manag. **90**(1), 1–11. doi:10.1016/s0378-1127(96)03889-3

Knowles, R. (1982) Denitrification. Microbiol. Rev. **46**(1), 43–70

Knowles, R. (1996) Denitrification: microbiology and ecology. Life Support Bioph. Sci. **3**(1–2), 31–34

Konneke, M., Bernhard, A.E., de la Torre, J.R., Walker, C.B., Waterbury, J.B., Stahl, D.A. (2005) Isolation of an autotrophic ammonia-oxidizing marine archaeon. Nature **437**(7058), 543–546. doi:10.1038/nature03911

Kool, D.M., Wrage, N., Oenema, O., Harris, D., Van Groenigen, J.W. (2009) The O-18 signature of biogenic nitrous oxide is determined by O exchange with water. Rapid Comm. Mass Sp. **23**(1), 104–108. doi:10.1002/rcm.3859

Koops, H.P., Bottcher, B., Moller, U.C., Pommereningroser, A., Stehr, G. (1991) Classification of 8 new species of ammonia-oxidizing bacteria: *Nitrosomonas communis* sp-nov, *Nitrosomonas ureae* sp-nov, *Nitrosomonas aestuarii* sp-nov, *Nitrosomonas marina* sp-nov, *Nitrosomonas nitrosa* sp-nov, *Nitrosomonas eutropha* sp-nov, *Nitrosomonas oligotropha* sp-nov and *Nitrosomonas halophila* sp-nov. J. Gen. Microbiol. **137**, 1689–1699

Koops, H.P., Moller, U.C. (1992) The lithotropic ammonia-oxidizing bacteria. In: Ballows (ed.) Prokaryotes **111**,2625–2637. Springer, New York

Kowalchuk, G.A., Stephen, J.R. (2001) Ammonia oxidizing bacteria: A model for molecular microbial ecology. Annu. Rev. Microbiol. **55**, 485–529. doi:10.1146/annurev.micro.55.1.485

Laanbroek, H.J., Gerards, S. (1993) Competition for limiting amounts of oxygen between *Nitrosomonas europaea* and *Nitrobacter winogradskyi* grown in mixed continuous cultures. Arch. Microbiol. **159**(5), 453–459

Laughlin, R.J., Stevens, R.J. (2002) Evidence for fungal dominance of denitrification and codenitrification in a grassland soil. Soil Sci. Soc. Am. J. **66**(5), 1540–1548

Laughlin, R.J., Stevens, R.J., Muller, C., Watson, C.J. (2008) Evidence that fungi can oxidize NH_4^+ to NO_3^- in a grassland soil. Eur. J. Soil Sci. **59**(2), 285–291. doi:10.1111/j.1365-2389.2007.00995.x

Leininger, S., Urich, T., Schloter, M., Schwark, L., Qi, J., Nicol, G.W., Prosser, J.I., Schuster, S.C., Schleper, C. (2006) Archaea predominate among ammonia-oxidizing prokaryotes in soils. Nature **442**, 806–809. doi:10.1038/nature04983

Li, D., Wang, X. (2008) Nitrogen isotopic signature of soil-released nitric oxide (NO) after fertilizer application. Atmos. Environ. **42**(19), 4747–4754. doi:10.1016/j.atmosenv.2008.01.042

Linn, D.M., Doran, J.W. (1984) Effect of water-filled pore space on carbon dioxide and nitrous oxide production in tilled and nontilled soils. Soil Sci. Soc. Am. J. **48**(6), 1267–1272

Luther, G.W., Brendel, P.J., Lewis, B.L., Sundby, B., Lefrancois, L., Silverberg, N., Nuzzio, D.B. (1998) Simultaneous measurement of O_2, Mn, Fe, I^-, and S(-II) in marine pore waters with a solid state voltammetric microelectrode. Limnol. Oceanogr. **43**(2), 325–333

Mahne, I., Tiedje, J.M. (1995) Criteria and methodology for identifying respiratory denitrifiers. Appl. Environ. Microbiol. **61**(3), 1110–1115

Mogge, B., Kaiser, E.A., Munch, J.C. (1999) Nitrous oxide emissions and denitrification N-losses from agricultural soils in the Bornhöved Lake region: influence of organic fertilizers and land-use. Soil Biol. Biochem. **31**(9), 1245–1252

Morley, N., Baggs, E.M., Dorsch, P., Bakken, L. (2008) Production of NO, N_2O and N_2 by extracted soil bacteria, regulation by NO_2^- and O_2 concentrations. Fems Microbiol. Ecol. **65**(1), 102–112. doi:10.1111/j.1574-6941.2008.00495.x

Mosier, A.R. (1998) Soil processes and global change. Biol. Fert. Soils **27**(3), 221–229

Mosier, A.R., Duxbury, J.M., Freney, J.R., Heinemeyer, O., Minami, K. (1996) Nitrous oxide emissions from agricultural fields: Assessment, measurement and mitigation. Plant Soil **181**(1), 95–108. doi:10.1007/bf00011296

Mulvaney, R.L., Khan, S.A., Mulvaney, C.S. (1997) Nitrogen fertilizers promote denitrification. Biol. Fert. Soils **24**(2), 211–220

Myers, R.J.K. (1975) Temperature effects on ammonification and nitrification in a tropical soil. Soil Biol. Biochem. **7**(2), 83–86. doi:10.1016/0038-0717(75)90003-6

Myrold, D.D., Tiedje, J.M. (1985) Diffusional constraints on denitrification in soil. Soil Sci. Soc. Am. J. **49**(3), 651–657

Norton, J.M. (2008) Nitrification in agricultural soils. In: Schepers, J.S., Raun, W.R. (eds.) Nitrogen in agricultural soils. Agronomy Monograph No. 49, pp. 173–199. ASA-CSSA-SSSA, Madison, WI

Norton, J.M., Alzerreca, J.J., Suwa, Y., Klotz, M.G. (2002) Diversity of ammonia monooxygenase operon in autotrophic ammonia-oxidizing bacteria. Arch. Microbiol. **177**(2), 139–149. doi:10.1007/s00203-001-0369-z

O'hara, G.W., Daniel, R.M. (1985) Rhizobial denitrification: A review. Soil Biol. Biochem. **17**(1), 1–9

Odu, C.T.I., Adeoye, K.B. (1970) Heterotrophic nitrification in soils - a preliminary investigation. Soil Biol. Biochem. **2**, 41–45

Papen, H., Vonberg, R., Hinkel, I., Thoene, B., Rennenberg, H. (1989) Heterotrophic nitrification by *Alcaligenes faecalis*: NO_2^-, NO_3^-, N_2O, and NO production in exponentially growing cultures. Appl. Environ. Microbiol. **55**(8), 2068–2072

Patureau, D., Zumstein, E., Delgenes, J.P., Moletta, R. (2000) Aerobic denitrifiers isolated from diverse natural and managed ecosystems. Microbial Ecol. **39**(2), 145–152. doi:10.1007/s002480000009

Payne, W.J. (1973) Reduction of nitrogenous oxides by microorganisms. Becterial. Reviews **37**(4), 409–452

Perez, T., Trumbore, S.E., Tyler, S.C., Matson, P.A., Ortiz-Monasterio, I., Rahn, T., Griffith, D.W. T. (2001) Identifying the agricultural imprint on the global N_2O budget using stable isotopes. J. Geophys. Res.-Atmos. **106**(D9), 9869–9878

Philippot, L. (2002) Denitrifying genes in bacterial and Archaeal genomes. Bioch. Biophys. Acta **1577**(3), 355–376

Philippot, L., Hallin, S., Borjesson, G., Baggs, E.M. (2009) Biochemical cycling in the rhizosphere having an impact on global change. Plant Soil **321**(1–2), 61–81. doi:10.1007/s11104-008-9796-9

Philippot, L., Hallin, S., Schloter, M. (2007) Ecology of denitrifying prokaryotes in agricultural soil. Adv. Agron. **96**, 249–305. doi:10.1016/s0065-2113(07)96003-4

Philippot, L., Piutti, S., Martin-Laurent, F., Hallet, S., Germon, J.C. (2002) Molecular analysis of the nitrate-reducing community from unplanted and maize-planted soils. Appl. Environ. Microbiol. **68**(12), 6121–6128. doi:10.1128/aem.68.12.6121-6128.2002

Pina-Ochoa, E., Hogslund, S., Geslin, E., Cedhagen, T., Revsbech, N.P., Nielsen, L.P., Schweizer, M., Jorissen, F., Rysgaard, S., Risgaard-Petersen, N. (2010) Widespread occurrence of nitrate storage and denitrification among Foraminifera and Gromiida. Proc. Natl. Acad. Sci. USA **107**(3), 1148–1153. doi:10.1073/pnas.0908440107

Poth, M., Focht, D.D. (1985) N-15 Kinetic analysis of N_2O production by Nitrosomonas europaea: an examination of nitrifier denitrification. Appl. Environ. Microbiol. **49**(5), 1134–1141

Prendergast-Miller, M.T., Baggs, E.M., Johnson, D. (2011) Nitrous oxide production by the ectomycorrhizal fungi *Paxillus involutus* and *Tylospora fibrillosa*. Fems Microbiol. Lett. **316**(1), 31–35. doi:10.1111/j.1574-6968.2010.02187.x

Prosser, J.I. (1989) Autotrophic nitrification in bacteria. Adv. Microb. Physiol. **30**, 125–181

Prosser, J.I. (2005) Nitrification. In: Hillel, D. (ed.) Encyclopedia of soils in the environment, vol. 3. pp. 31–39. Elsevier Academic Press, Oxford,

Purkhold, U., Wagner, M., Timmermann, G., Pommerening-Roser, A., Koops, H.P. (2003) 16S rRNA and amoA-based phylogeny of 12 novel betaproteobacterial ammonia-oxidizing isolates: extension of the dataset and proposal of a new lineage within the nitrosomonads. Int. J. Syst. Evol. Micr. **53**, 1485–1494. doi:10.1099/ijs.0.02638-0

Ravishankara, A.R., Daniel, J.S., Portmann, R.W. (2009) Nitrous oxide (N_2O): The ominant ozone-depleting substance emitted in the 21st Century. Science **326**(5949), 123–125. doi:10.1126/science.1176985

Ritchie, G.A.F., Nicholas, D.J. (1972) Identification of sources of nitrous oxide produced by oxidative and reductive processes in *Nitrosomonas europaea*. Biochem. J. **126**(5), 1181–1191
Robertson, G.P., Tiedje, J.M. (1987) Nitrous oxide sources in aerobic soils: Nitrification, denitrification and other biological processes. Soil Biol. Biochem. **19**(2), 187–193
Robertson, L.A., Kuenen, J.G. (1990) Combined heterotrophic nitrification and aerobic denitrification in *Thiosphaera pantotropha* and other bacteria. Ant. Leew. Int. J. G. Mol. Microb. **57**(3), 139–152
Robertson, L.A., Vanniel, E.W.J., Torremans, R.A.M., Kuenen, J.G. (1988) Simultaneous nitrification and denitrification in aerobic chemostat cultures of *Thiosphaera pantotropha*. Appl. Environ. Microbiol. **54**(11), 2812–2818
Rotthauwe, J.H., Witzel, K.P., Liesack, W. (1997) The ammonia monooxygenase structural gene amoA as a functional marker: Molecular fine-scale analysis of natural ammonia-oxidizing populations. Appl. Environ. Microbiol. **63**(12), 4704–4712
Rutting, T., Huygens, D., Muller, C., Cleemput, O., Godoy, R., Boeckx, P. (2008) Functional role of DNRA and nitrite reduction in a pristine south Chilean *Nothofagus* forest. Biogeochemistry **90**(3), 243–258. doi:10.1007/s10533-008-9250-3
Rutting, T., Clough, T.J., Muller, C., Lieffering, M., Newton, P.C.D. (2010) Ten years of elevated atmospheric carbon dioxide alters soil nitrogen transformations in a sheep-grazed pasture. Global Change Biol. **16**(9), 2530–2542. doi:10.1111/j.1365-2486.2009.02089.x
Ryden, J.C., Lund, L.J. (1980) Nature and extent of directly measured denitrification losses from some irrigated vegetable crop production units. Soil Sci. Soc. Am. J. **44**(3), 505–511
Sahrawat, K.L. (1982) Nitrification in some tropical soils. Plant Soil **65**(2), 281–286
Sahrawat, K.L. (1996) Nitrification inhibitors, with emphasis on natural products, and the persistence of nitrogen in the soil. In: Ahmad, N. (ed.) Nitrogen economy in tropical soils. pp. 379–388. Kluwer Academic Publishers, Dordrecht
Schipper, L.A., Vojvodić-Vuković, M. (2000) Nitrate removal from groundwater and denitrification rates in a porous treatment wall amended with sawdust. Ecol. Engn. **14**(3), 269–278
Schjonning, P., Thomsen, I.K., Moldrup, P., Christensen, B.T. (2003) Linking soil microbial activity to water- and air-phase contents and diffusivities. Soil Sci. Soc. Am. J. **67**(1), 156–165
Schleper, C., Jurgens, G., Jonuscheit, M. (2005) Genomic studies of uncultivated archaea. Nat. Rev. Microbiol. **3**(6), 479–488. doi:10.1038/nrmicro1159
Schmidt, I., van Spanning, R.J.M., Jetten, M.S.M. (2004) Denitrification and ammonia oxidation by Nitrosomonas europaea wild-type, and NirK- and NorB-deficient mutants. Microbiol.-SGM **150**, 4107–4114. doi:10.1099/mic.0.27382-0
Schmidt, C.S., Richardson, D.J., Baggs, E.M. (2011) Constraining the conditions conducive to dissimilatory nitrate reduction to ammonium in temperate arable soils. Soil Biol. Biochem. **43**(7), 1607–1611. doi:10.1016/j.soilbio.2011.02.015
Scholefield, D., Hawkins, J.M.B., Jackson, S.M. (1997) Use of a flowing helium atmosphere incubation technique to measure the effects of denitrification controls applied to intact cores of a clay soil. Soil Biol. Biochem. **29**(9–10), 1337–1344
Seitzinger, S., Harrison, J.A., Bohlke, J.K., Bouwman, A.F., Lowrance, R., Peterson, B., Tobias, C., Van Drecht, G. (2006) Denitrification across landscapes and waterscapes: A synthesis. Ecol. Appl. **16**(6), 2064–2090
Shaw, L.J., Nicol, G.W., Smith, Z., Fear, J., Prosser, J.I., Baggs, E.M. (2006) Nitrosospira spp. can produce nitrous oxide via a nitrifier denitrification pathway. Environ. Microbiol. **8**(2), 214–222. doi:10.1111/j.1462-2920.2005.00882.x
Shi, W., Norton, J.M. (2000) Effect of long-term, biennial, fall-applied anhydrous ammonia and nitrapyrin on soil nitrification. Soil Sci. Soc. Am. J. **64**(1), 228–234
Shoun, H., Kim, D.H., Uchiyama, H., Sugiyama, J. (1992) Dinitrification by fungi. Fems Microbiol. Lett. **94**(3), 277–281. doi:10.1111/j.1574-6968.1992.tb05331.x
Silver, W.L., Herman, D.J., Firestone, M.K. (2001) Dissimilatory Nitrate Reduction to Ammonium in Upland Tropical Forest Soils. Ecology **82**(9), 2410–2416
Skiba, U., Smith, K.A. (2000) The control of nitrous oxide emissions from agricultural and natural soils. Chemosphere **2**(3–4), 379–386

Smith, J.H., Gilbert, R.G., Miller, J.B. (1976) Redox potentials and denitrification in a cropped potato processing waste-water disposal field. J. Environ. Qual. **5**(4), 397–399
Smith, K.A. (1980) A model of the extent of anaerobic zones in aggregated soils and its potential application to estimates of denitrification. J. Soil Sci. **31**(2), 263–277
Smith, K.A., McTaggart, I.P., Tsuruta, H. (1997) Emissions of N_2O and NO associated with nitrogen fertilization in intensive agriculture, and the potential for mitigation. Soil Use Manage. **13**(4), 296–304
Smith, M.S., Tiedje, J.M. (1979) Phases of denitrification following oxygen depletion in soil. Soil Biol. Biochem. **11**(3), 261–267. doi:10.1016/0038-0717(79)90071-3
Smith, P., Goulding, K.W., Smith, K.A., Powlson, D.S., Smith, J.U., Falloon, P., Coleman, K. (2001) Enhancing the carbon sink in European agricultural soils: including trace gas fluxes in estimates of carbon mitigation potential. Nutr. Cycl. Agroecosys. **60**(1–3), 237–252. doi:10.1023/a:1012617517839
Sorai, M., Yoshida, N., Ishikawa., M. (2007) Biogeochemical simulation of nitrous oxide cycle based on the major nitrogen processes. J. Geophys. Res.-Biogeo. **112**, G01006 doi:10.1029/2005JG000109
Sorokin, D., Tourova, T., Schmid, M.C., Wagner, M., Koops, H.P., Kuenen, J.G., Jetten, M. (2001) Isolation and properties of obligately chemolithoautotrophic and extremely alkali-tolerant ammonia-oxidizing bacteria from Mongolian soda lakes. Arch. Microbiol. **176**(3), 170–177
Spieck, E., Muller, S., Engel, A., Mandelkow, E., Patel, H., Bock, E. (1996) Two-dimensional structure of membrane-bound nitrite oxidoreductase from Nitrobacter hamburgensis. J. Struct. Biol. **117**(2), 117–123
Stark, J.M., Firestone, M.K. (1995) Mechanisms for soil moisture effects on activity of nitrifying bacteria. Appl. Environ. Microbiol. **61**(1), 218–221
Stein, L.Y., Arp, D.J., Hyman, M.R. (1997) Regulation of the synthesis and activity of ammonia monooxygenase in Nitrosomonas europaea by altering pH to affect NH_3 availability. Appl. Environ. Microbiol. **63**(11), 4588–4592
Stevens, R.J., Laughlin, R.J., Malone, J.P. (1998) Soil pH affects the processes reducing nitrate to nitrous oxide and di-nitrogen. Soil Biol. Biochem. **30**(8–9), 1119–1126
Strong, D.T., Sale, P.W.G., Helyar, K.R. (1999) The influence of the soil matrix on nitrogen mineralisation and nitrification. IV. Texture. Aus. J. Soil Res. **37**(2), 329–344
Subbarao, G.V., Ito, O., Sahrawat, K.L., Berry, W.L., Nakahara, K., Ishikawa, T., Watanabe, T., Suenaga, K., Rondon, M., Rao, I.M. (2006) Scope and strategies for regulation of nitrification in agricultural systems-challenges and opportunities. Crit. Rev. Plant Sci. **25**(4), 303–335. doi:10.1080/07352680600794232
Sutka, R.L., Ostrom, N.E., Ostrom, P.H., Gandhi, H., Breznak, J.A. (2003) Nitrogen isotopomer site preference of N2O produced by Nitrosomonas europaea and Methylococcus capsulatus Bath. Rapid Comm. Mass Sp. **17**(7), 738–745. doi:10.1002/rcm.968
Suwa, Y., Imamura, Y., Suzuki, T., Tashiro, T., Urushigawa, Y. (1994) Ammonia-oxidizing bacteria with different sensitivities to $(NH_4)_2SO_4$ in activated sludges. Water Res. **28**(7), 1523–1532
Suzuki, I., Dular, U., Kwok, S.C. (1974) Ammonia or ammonium ion as substrate for oxidation by *Nitrosomonas europaea* cells and extracts. J. Bacteriol. **120**(1), 556–558
Takaya, N. (2009) Response to hypoxia, reduction of electron acceptors, and subsequent survival by Filamentous fungi. Biosci. Biotechnol. Biochem. **73**(1), 1–8. doi:10.1271/bbb.80487
Tenuta, M., Beauchamp, E.G. (1996) Denitrification following herbicide application to a grass sward. Can.J. Soil Sci. **76**(1), 15–22
Tiedje, J.M. (1988) Ecology of denitrification and the dissimilatory nitrate reduction to ammonium. In: Zehnder, J.B. (ed.) Biology of anaerobic microorganisms. pp. 179–244. Wiley, New York
Treusch, A.H., Leininger, S., Kletzin, A., Schuster, S.C., Klenk, H.P., Schleper, C. (2005) Novel genes for nitrite reductase and Amo-related proteins indicate a role of uncultivated mesophilic crenarchaeota in nitrogen cycling. Environ. Microbiol. **7**(12), 1985–1995. doi:10.1111/j.1462-2920.2005.00906.x
U.S. Environmental Protection Agency (US EPA) (2009) National Primary Drinking Water Regulations. Report No. EPA 816-F-09-0004. United States Enveronmental Protection Agency.

van Cleemput, O. (1998) Subsoils: chemo- and biological denitrification, N_2O and N_2 emissions. Nutr. Cycl. Agroecosys. **52**(2–3), 187–194

Vitousek, P.M., Gosz, J.R., Grier, C.C., Melillo, J.M., Reiners, W.A. (1982) A comparative analysis of potential nitrification and nitrate mobility in forest ecosystems. Ecol. Monogr. **52**(2), 155–177. doi:10.2307/1942609

Vitousek, P.M., Hattenschwiler, S., Olander, L., Allison, S. (2002) Nitrogen and nature. Ambio **31**(2), 97–101. doi:10.1639/0044-7447(2002)031[0097:nan]2.0.co;2

Wan, Y.J., Ju, X.T., Ingwersen, J., Schwarz, U., Stange, C.F., Zhang, F.S., Streck, T. (2009) Gross nitrogen transformations and related nitrous oxide emissions in an intensively used calcareous soil. Soil Sci. Soc. Am. J. **73**(1), 102–112. doi:10.2136/sssaj2007.0419

Wang, W.J., Dalal, R.C., Moody, P.W., Smith, C.J. (2003) Relationships of soil respiration to microbial biomass, substrate availability and clay content. Soil Biol. Biochem. **35**(2), 273–284

Ward, B.B. (1986) Nitrification in marine environments. In: Prosser, J.I. (ed.) Nitrification. pp. 157–184. IRL Press, Oxford

Ward, B.B. (2008) Nitrification in marine systems. In: Capone, D.G., Bronk, D.A., Muholland, M. R., Carpenter, E.J. (eds.) Nitrogen in the Marine Environment. vol. 5, pp. 199–261. Elsevier, Amsterdam, The Netherland

Ward, B.B., Zafiriou, O.C. (1988) Nitrification and nitric-oxide in the oxygen minimum of the eastern tropical north pacific. Deep Seas Res. Pt A **35**(7), 1127–1142

Watson, S.W., Bock, E., Valois, F.W., Waterbury, J.B., Schlosser, U. (1986) Nitrospira marina gen-nov sp-nov - a chemolithotrophic nitrite-oxidizing bacterium. Arch. Microbiol. **144**(1), 1–7

World Health Organization (WHO) (2007) Nitrate and nitrite in drinking-water. World Health Organization of the United Nations.

Wood, P.M. (1986) Nitrification as an energy source In: Prosser, J.I. (ed.) Nitrification. pp. 39–62. IRL Press, Oxford

Wrage, N., van Groenigen, J.W., Oenema, O., Baggs, E.M. (2005) A novel dual-isotope labelling method for distinguishing between soil sources of N_2O. Rapid Comm. Mass Sp. **19**(22), 3298–3306. doi:10.1002/rcm.2191

Wrage, N., Velthof, G.L., van Beusichem, M.L., Oenema, O. (2001) Role of nitrifier denitrification in the production of nitrous oxide. Soil Biol. Biochem. **33**(12–13), 1723–1732

Wuchter, C., Abbas, B., Coolen, M.J.L., Herfort, L., van Bleijswijk, J., Timmers, P., Strous, M., Teira, E., Herndl, G.J., Middelburg, J.J., Schouten, S., Damste, J.S.S. (2006) Archaeal nitrification in the ocean. Proc. Natl. Acad. Sci. USA **103**, 12317–12322. doi:10.1073/pnas.0600756103

Wuebbles, D.J. (2009) Nitrous Oxide: No Laughing Matter. Science **326**(5949), 56–57. doi:10.1126/science.1179571

Xiong, Z.Q., Xing, G.X., Zhu, Z.L. (2007) Nitrous oxide and methane emissions as affected by water, soil and nitrogen. Pedosphere **17**(2), 146–155

Ye, R.W., Averill, B.A., Tiedje, J.M. (1994) Denitrification: Production and consumption of nitric oxide. Appl. Environ. Microbiol. **60**(4), 1053–1058

Yin, S.X., Chen, D., Chen, L.M., Edis, R. (2002) Dissimilatory nitrate reduction to ammonium and responsible microorganisms in two Chinese and Australian paddy soils. Soil Biol. Biochem. **34**(8), 1131–1137. doi:10.1016/s0038-0717(02)00049-4

Yoshida, T., Alexander, M. (1971) Hydroxylamine oxidation by *Nitrosomonas europea*. Soil Sci. **111**(5), 307–312

Zart, D., Bock, E. (1998) High rate of aerobic nitrification and denitrification by Nitrosomonas eutropha grown in a fermentor with complete biomass retention in the presence of gaseous NO_2 or NO. Arch. Microbiol. **169**(4), 282–286

Zumft, W.G. (1993) The biological role of nitric oxide in bacteria. Arch. Microbiol. **160**(4), 253–264

Zumft, W.G. (1997) Cell biology and molecular basis of denitrification. Microbiol. Mol. Biol. R. **61**(4), 533–616

Zumft, W.G., Korner, H. (1997) Enzyme diversity and mosaic gene organization in denitrification. Ant. Leew. Int. J. G. Mol. Microb. **71**(1–2), 43–58

Chapter 4
Nitrous Oxide Fluxes Measurement

Contents

Abstract Accurate field measurements of the fluxes of N_2O are important for reliable up-scaling and model prediction of regional and global N_2O emissions. Gaseous flux chambers are the most common field measurement technique used and represent the smallest scale, sampling an area $\geq 1\ m^2$. Their operating principle is simple, they can be highly sensitive, the cost can be low and field requirements small. Chamber techniques can be applied to field trials with multiple small plots. Chamber measurements are usually made weekly and rarely more frequently than once daily. Other techniques include mass balance, and micrometeorological methods. Mass balance technique is suitable for small, defined source areas, while micrometeorological methods are suitable for measuring gas emissions on large landscape scales. Eddy covariance is the preferred technique for this scale. Relaxed eddy accumulation retains the attraction of eddy covariance by providing a direct point measurement but removes the need for a fast response gas sensor by substituting it with a fast solenoid valve sampling system. Assumptions of these methods and precautions in their application are discussed in this chapter. For N_2O

David Ussiri and R. Lal, *Soil Emission of Nitrous Oxide and its Mitigation*,
DOI 10.1007/978-94-007-5364-8_4,

flux, spatially extensive and continuous measurements from agroecosystems are required, but number and frequency of measurements needed makes this difficult to achieve. Use of robust simulation models that incorporate all major regulatory processes to estimate the emissions can be the alternative of field measurements.

Keywords Gas flux chambers • Mass balance • Eddy covariance • Eddy accumulation • Flux gradient • Empirical models • Process-based models

Abbreviations

BAE	Biosphere atmosphere change GHG greenhouse gas
EF	Emission factor
EC	Eddy covariance
IPCC	Intergovernmental Panel on Climate Change
OC	Organic carbon
SOM	Soil organic matter
PVC	Polyvinyl chloride
GC	Gas chromatography
ALE	Atmospheric Lifetime Experiment
GAGE	Global Atmospheric Gases Experiment
AGAGE	Advanced Global Atmospheric Gases Experiment
NOAA	National Oceanic and Atmospheric Administration
ESRL	Earth System Research Laboratory

4.1 Introduction

Greenhouse gases (GHG) fluxes are measured at scales ranging from a few grams of soil to several square km of land area. Each scale of measurement and method has contributed to the current understanding of biosphere atmosphere exchange (BAE) of GHG (Denmead 2008). Quantification of soil-atmosphere exchange rates of GHG fluxes is necessary when evaluating environmental impact of land and land uses. It is also a prerequisite for improving the accuracy of the GHG inventory and assessing the viability of mitigation options. Emission inventory can be used directly to establish source categories, identify trends, and examine the impact of different policies approach. Quantitative information on gaseous fluxes for a range of ecosystems is also required in order to evaluate their contribution and assess potential mitigation options. Monitoring fluxes is also important in developing and verifying the empirical as well as process-based modeling approaches that provide emission estimates at plot scale and beyond and assist in advice on policy issues.

Microbial processes of nitrification and denitrification play an important role in soil N_2O production (Firestone and Davidson 1989; Chap. 3). N_2O poses much greater challenges to adequately design monitoring and observation protocols at plot, farm and ecosystem level. In addition, quantitative modeling and up-scaling techniques are needed. N_2O emissions, typically exhibit a high degree of spatial and temporal variability (Freibauer and Kaltschmitt 2003) due to the dependence on soil, land management, and environmental factors. For example, N_2O emissions pulses over a few hours to days as triggered by freezing thawing, soil rewetting or fertilization can dominate annual fluxes at a given site. Extreme events lasting a few hours to several days can contribute up to 80% of the annual N_2O emissions from an ecosystem. Managed soils, particularly fertilized, have wider site-to-site variation in magnitude of N_2O emissions than unmanaged soils. Soil type, land management including nitrogen additions, concentration of organic matter, temperature, and precipitation influence the N_2O fluxes that are measured in the field (Skiba and Smith 2000). Furthermore, these factors interact with each other. It is obvious that the relationships between these factors and N_2O fluxes are highly non-linear and difficult to predict. High degree of variability present a major challenge for accurate assessment of gaseous fluxes and extrapolating N_2O emissions from point measurements to an area basis. N_2O fluxes are difficult to predict at a plot scale, and extrapolation of fluxes to annual emissions requires large number of measurement to obtain representative annual flux estimate (Skiba et al. 2009).

The variability is due, in part, to the underlying processes responsible for N_2O production. Proximal soil factors drive the microbial processes of N_2O production and consumption at micro scale. Substrate availability, O_2 availability, soil moisture, and soil temperature exert direct control on hourly and diurnal variation of N_2O fluxes (Freibauer and Kaltschmitt 2003). The proximal physical and chemical factors are themselves controlled by biological drivers supplying substrate, such as readily degradable organic carbon (OC), and O_2 demand of decomposing organisms and roots (Freibauer and Kaltschmitt 2003). The N_2O emissions are also influenced by environmental and climatic factors such as precipitation and air temperature (Smith et al. 2002). Farm management contributes to the emissions but the driving forces of the microbial activity intimately link also climate weather, site properties (Smith et al. 1998; Skiba and Smith 2000). Field-level estimates of N_2O are required for evaluating the environmental impact of farming practices and for developing techniques for N_2O mitigation (Hellebrand and Kalk 2001; Smith et al. 2002). Such estimates are also useful to developing inventories with a known level of uncertainty (Saggar et al. 2004). Field measurements of N_2O fluxes are also challenging because of the low ambient concentrations and high spatial and temporal variability.

A number of methods and approaches can be used to determine the rate of soil surface-atmosphere exchange of N_2O. These include simple and widely used enclosure methods (static chambers), sub-surface methods, mass balance, backward Lagrarian dispersion, micrometeorological methods with various degrees of complexity (eddy covariance, flux gradient, eddy accumulation, and relaxed eddy accumulation) and the emission factor (EF) method developed by Intergovernmental Panel on Climate Change (IPCC). Most of global understanding of GHG fluxes and their control by physical, chemical, and microbial processes has largely arisen from flux chamber measurements. The micrometeorological methods have the advantage of being non-intrusive and achieving better spatial integration of fluxes but they are generally expensive to use to integrate temporal variations in emission over the year. The IPCC–EF also provides a simple method for calculating N_2O emissions from agriculture where N input in each component is multiplied by emission factors. In addition, there are a number of possible approaches to produce improved estimates by using models. Advancements in development of high frequency instruments, that detect extremely small concentration changes has improved the knowledge of BAE at the field/landscape scale and at a high temporal resolution. This chapter describes some commonly used techniques for monitoring BAE of N_2O, emphasizing the principles behind, strengths, and weaknesses associated with each technique. Although the emphasis is placed on N_2O emissions, most of the techniques are also capable of monitoring other GHGs such as CO_2 and CH_4 also.

4.2 Measurement Methods

4.2.1 Flux Chambers

Static chambers are most commonly used tools for measuring GHG fluxes from soil. Chambers are cylinders or boxes randomly inserted into the soil to form airtight enclosure. They are cheap, straightforward to use and can be employed under a wide range of conditions (Conen and Smith 1998). Commonly, static chambers seal a certain volume of atmosphere above the soil surface area for a predetermined time to allow GHG to accumulate to a concentration above the ambient, that can be determined by gas chromatography (GC) or directly by infrared analysis. The surface area of soil monitored ranges from ~0.1 m^2 (Ambus and Christensen 1994) to ~0.5 m^2 (Clayton et al. 1994), depending on the dimensions of the gas chambers. Alternatively, for the experiments where there is a need to control environmental variables, intact soil cores of 0.1 m diameter and up to 0.1 m length are collected in metal liners and placed in robust PVC chambers glued to PVC base (Saggar et al. 2004; Tate et al. 2007). Gas fluxes are monitored at regular intervals after temporarily sealing the chambers with a gas-tight seal using self-sealing lids and greased O-ring.

The chamber technique is based on the increase (or decrease in case of sink) in gas concentration within the enclosed headspace to a concentration that is determined by GC (Hedley et al. 2006). Gas fluxes (*F*) are then calculated using linear regression and the ideal gas laws (Ussiri et al. 2009, Eq. 4.1) viz:

$$F = \left(\frac{\Delta g}{\Delta t}\right)\left(\rho\frac{V}{A}\right)\left(\frac{273}{T+273}\right)k \tag{4.1}$$

where $\Delta g/\Delta t$ is the average rate of change in gas concentration inside the chamber (mg m^{-2} min^{-1}), ρ is the gas density, V is the volume of the chamber (m^3), A is the surface area circumscribed by the chamber (m^2), T is the temperature in the chamber (°C) and k is the time conversion factor.

Gas chambers operate on simple principle and they offer flexibility, portability, and are of low cost to fabricate and maintain. The general principle is to restrict the volume of air with which the gas exchange occurs to magnify changes in concentration of gas in the headspace. The gas chamber method is a bottom up approach most commonly used worldwide for monitoring gas fluxes under different land management systems, including farmland (Jacinthe and Dick 1997; Ussiri et al. 2009), pastures (Carran et al. 1995; Luo et al. 2008) and forests (von Arnold et al. 2005; Pihlatie et al. 2010). This approach is essential for capturing emissions that are highly variable in space and time, resulting from episodic nature of emissions from land. The approach is also suitable for understanding the processes that regulate N_2O fluxes from soil.

Gas chambers are portable, compact, easy to install and maintain and can be readily adapted for taking measurements. The technique has contributed most to the current understanding of the magnitude and spatiotemporal variability of N_2O fluxes and soil and environmental variables regulating N_2O fluxes from soils. Most researchers have applied manual static chambers combined with gas chromatography (GC) and electron capture detector (ECD) or infrared gas analysis and monitor in time steps of days to weeks. Despite their low cost and ease of use under a wide range of conditions, the chambers can cover only small area of soil surface. Therefore, a large number of chambers are required to provide a representative estimate of the GHG fluxes. Also, the necessary temporal interpolation between measurements introduces an additional uncertainty to annual N_2O flux estimates. In addition, chambers alter the physical status of the observed surface. The accuracy and representativeness of chamber-based measurements are also limited by the high spatial variability of N_2O emission (Ambus and Christensen 1994). A combination of weekly and event-driven sampling can produce a robust annual N_2O flux estimates and improve the accuracy of estimated annual fluxes.

An optimal size of chamber needed to provide robust flux estimates is not known. The results from 2-week intensive sampling using 20 small chambers (25 cm diameter by 30 cm height, i.e. ~0.015 m^3) and 8 large chambers (1 m × 0.5 m × 0.3 m high, i.e. ~0.15 m^3) installed in dairy grazed pastures indicated no significant differences in N_2O fluxes among two chamber sizes. However, spatial variability was higher from small than large chambers (Saggar et al. 2008). The similarity in N_2O fluxes from large and small chambers supports the conclusion that the small chambers method provides an accurate and representative soil-atmosphere-exchange of GHG but higher variability from smaller chambers emphasizes the spatial variability of factors controlling N_2O fluxes. However, comparison of CH_4 uptake by the two types of chambers described above using three small chambers and three large chambers under forest clear-cut site indicated a 0.4-fold lower CH_4 uptake by small than large chambers (Tate et al. 2006). Variation in moisture content and limited number of replications, rather than chamber sizes are among as the main cause of CH_4 uptake variations, since the two types of chambers were located 0.5 km apart (Tate et al 2006). Chambers are classified as flow through or closed chambers based on whether or not they are open to the atmosphere, respectively.

4.2.1.1 Flow-Through Chambers

Flow-through chambers are designed to maintain a constant flow of outside air through the headspace and the difference in concentration between the air entering and leaving the headspace is measured. The flux density of gas at the surface (F; kg m^{-2} s^{-1}) is calculated from the relationship (Eq. 4.2):

$$F = v\frac{\left(\rho_{g0} - \rho_{gi}\right)}{A} \tag{4.2}$$

where v (m^3 s^{-1}) is the volumetric flow rate of the gas being monitored, ρ_{g0} is the gas concentration in the air leaving the chamber, ρ_{gi} is the gas concentration in the air entering the chamber, and A is the surface area covered by the chamber (m^2). Flow-through chambers have the advantage of controlled flow rate, which eliminates the risk of inhibiting the surface flux through too high gas concentration in the head-space and changing head-space pressure. The major disadvantage of flow through chamber is the small magnitude of concentration increase when gas fluxes are low.

4.2.1.2 Closed Chambers

Closed chambers are designed to be airtight so that there is no replacement of air in the chamber headspace, and gas concentration increases continuously. These types of chambers are most commonly used for flux monitoring because the greater concentration changes are easier to detect. The rate of increase is monitored and the flux density is calculated from the function described in Eq. 4.1. These chambers can be static or dynamic. In static chambers, there is no air circulation inside the chambers. Such chambers can be operated manually or automated to incorporate gas sampling and analysis equipment. For manually operated static chambers, the usual practice is to take periodic air samples from the headspace with the gas syringe and measure the gaseous concentration later in the laboratory using a GC.

Dynamic chambers are more complex and are generally automated and equipped with a gas analyzer. Air is circulated in a closed loop between the headspace and a gas analyzer. The circulation permits the rate of concentration increase in the headspace to be monitored to detect any inhibition of flux through a buildup in the headspace concentration. The aim is to have a linear increase indicating a constant flux from the surface. These chambers are designed such that the lids can be opened and closed automatically at a controlled time intervals through pneumatic control systems.

4.2.1.3 Advantages of Gas Flux Chambers Method

The method is simple in concept and operation, and, in the case of static closed chamber, economic and does not require power in the field and nor on-line gas analyzers. Samples can be collected with a syringe, stored in vials, and transported to be analyzed remotely in the laboratory using a high sensitivity GC or other gaseous analyzers off-line. In addition, chambers do not require large experimental areas. Therefore, they permit process studies and experiments with many treatments. The automated chambers can also be designed to provide continuous records of gas fluxes from the same location.

4.2.1.4 Disadvantages of Gas Flux Chambers Method

Sealed static chambers exclude fluctuations of ambient pressure caused by wind turbulence, and which promotes gaseous mixing of soil air with the atmosphere (Hutchinson and Mosier 1981). This effect tends to enhance the transfer of GHG such as N_2O, which is otherwise driven by diffusion only inside the chamber (Kimball and Lemon 1971; Conen and Smith 1998). Low frequency field measurements may not be able to capture the intensive emission pulses and thereby significantly underestimate seasonal annual N_2O emissions (Scott et al. 1999). Other drawbacks associated with gas chambers technique includes: (i) increase in gas concentration in the chamber headspace, which may affect the gas fluxes. This is particularly a problem for the closed chambers. In this type of chambers, one gas sample is taken at certain time interval. As a result, the flux changes from linear to nonlinear. Livingston and Huchinson (1995) described a protocol for inferring fluxes from both linear and non-linear concentration changes, (ii) pressure changes between inside and outside of the chamber can create an artifact gas fluxes by inducing viscous flows from soil to headspace and vice-versa. This can occur in closed chambers where vigorous mixing is employed to ensure that a representative sample is collected from the chamber. This is a problem also in closed dynamic chambers, where air is pumped between chamber and gas analyzer. To ensure a pressure equilibration, Hutchinson and Mosier (1981) proposed an incision of a small orifice or tube to the chamber, small enough such that the air exchange between the chamber headspace and atmosphere is negligible, (iii) small surface area covered by the chamber and large variability in soil gaseous fluxes causes large coefficients of variation for chamber measurements, necessitating large number of replications, (iv) chambers must be well established before measurements commence to avoid effects from initial soil disturbances, and (v) gas fluxes have marked diurnal cycles, and there is always a lag between time of production in different regions of soil profile and emissions at the soil surface. Therefore, measurements at one time of the day may be misleading.

4.2.2 Sub-Surface Method

Net N_2O fluxes from soil surface determined by headspace sampling using chamber techniques are not necessarily the representative of the extent of gas production/ or consumption beneath the surface. Entrapment of gas bubbles within the soil structure and denitrification may reduce the fluxes of N_2O measured by headspace sampling. Measuring the concentration of gases in the soil sub-surface helps determine the location of sources and sinks, and establish the variation with time and causes of variation (Clark et al. 2001). This information helps to elucidate the underlying soil processes needed for developing credible mechanistic models (Li and Kelliher 2005). Transport of gases within the soil and between the soil and atmosphere is driven by vertical concentration gradients. Sub-surface measurements technologies for determining N_2O emissions rely on equilibration of gas composition between the soil pore space and that in small headspace probe, by diffusion through a perforated silicon membrane (Gut et al. 1998; Clark et al. 2001; Li and Kelliher 2005). The perforated stainless steel tubes are sealed at the ends with rubber septa and are installed in the trenches. The method causes only limited soil disturbance and is considered unaffected by soil moisture content. The method appears promising for pasture soils as measurements can be made as the animals are grazing, and allowing the inclusion of grazing animals as one of the controlling variables. Further testing and comparisons are needed among measurements from commonly used chambers and micrometeorological approaches.

4.2.3 Mass Balance Approaches

Mass balance methods generally equate the rate of gaseous emission in a control volume with the difference between the rate at which the gas is carried out and in the control volume by the wind (Denmead 2008). These techniques are suitable for small-defined source areas, typically tens to few thousands of square meters in extent. Emissions are calculated from the difference in the rates at which the gas is carried into control volume above the source area and out by the wind. The mean horizontal flux of emitted gas F_h (kg m^{-2} s^{-1}) at any level z on a downwind boundary can be expressed as (Eq. 4.3), viz:

$$F_{h,z} = \overline{u_{z(\rho_{g,z}-\rho_{g,u})}} \tag{4.3}$$

where u_z is the horizontal wind speed at distance z, $\rho_{g,z}$ and $\rho_{g,u}$ are the concentrations of the gas at z on the downwind and upwind boundaries. The overbar in Eq. 4.3 indicates time average. The net flux Q across a plane of unit width on the downwind can be obtained by integrating Eq. 4.4:

$$Q = \int_0^Z \overline{u_z\left(\rho_{g,z} - \rho_{g,u}\right)}\, dz \tag{4.4}$$

where Z is the top of the plume of emitted gas. Q can be evaluated by numerical integration (i.e., trapezoidal rule). The surface flux density in the emitting region F_0 (kg m^{-2} s^{-1}) can be expressed as (Eq. 4.5):

$$F_0 = Q/X \tag{4.5}$$

The required primary data are the profiles of gas concentration on the downwind boundaries and wind speed profile, the wind direction and upwind background gas concentration. The mass balance approach can be applied on a closed or an open system. The closed system techniques have been used to determine gaseous emissions from landfill, treated fields, and small confined animal herds (Denmead et al. 1998, 2000).

The major strength of mass balance methods is that it fills the gap between gas chamber method and micrometeorological approaches that require large uniform source areas. Mass balance methods are appropriate for measuring gas fluxes from small well defined source areas. Closed systems are suitable for both homogenous and heterogeneous source distributions. They are appropriate for determining NH_3 and N_2O from fertilizer applications, or CH_4 and NH_4 from small-scale grazing operations (Prasertsak et al. 2001).

4.2.4 Micrometeorological Approach

Commonly known as top-down approach, the micrometeorological methods uses flux gradient technique. The approach is conceptually ideal for measuring GHG emission over large ecologically uniform areas (10^2–10^4 m^2) without changing the physical condition of observed surface, i.e. non-intrusive (Li et al. 2008). The techniques can reduce the spatial variability problems inherent in static chamber techniques (Lapitan et al. 1999). The typical experimental area for micrometeorological flux measurements is several hundreds of meters in lateral extent. The approach is particularly useful for heterogeneous ecosystems such as grazed lands and soft surfaces where compaction by walking to the flux chambers may release gases into the chamber and affect the measured fluxes (e.g. peat, wetlands or dung heaps). The approach requires sampling of the turbulent air above the surface of interest and determination of concentration gradients of gas of interest with height (flux-gradient, FG) or concentration fluctuation at one height (eddy covariance, EC). Due to the nature of turbulent flow and mixing in the near-surface air layer, trace gas concentrations vary rapidly in time, and concentration gradients are usually small at the measurement height (Wagner-Riddle et al. 2005).

The conservation of equation principles provides the framework for measuring and interpreting gaseous flux data under the micrometeorological techniques (Baldocchi et al. 1988). The conservation equation states that the time rate of

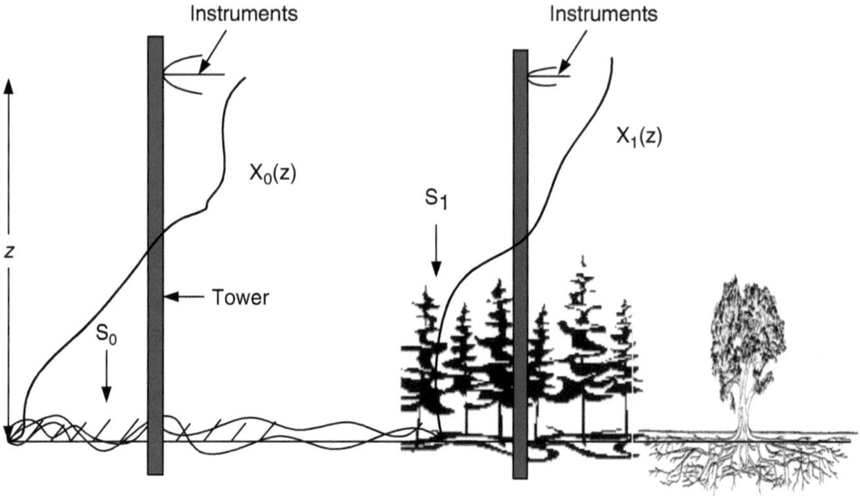

Fig. 4.1 Schematic representation of the continuity equation

change of the mean mixing ratio or concentration of a gas constituent at a fixed point in space is: (i) balanced by the mean horizontal and vertical advection, (ii) mean horizontal and vertical divergence or convergence of turbulent flux, and (iii) molecular diffusion (D) and by any source or sink (S), viz (Eq. 4.6):

$$\frac{\partial \overline{x}}{\partial t} = -\frac{\overline{u}\partial \overline{x}}{\partial x} - \frac{\overline{v}\partial \overline{x}}{\partial y} - \frac{\overline{w}\partial \overline{x}}{\partial z} - \frac{\partial \overline{u'\chi'}}{\partial x} - \frac{\partial \overline{v'\chi'}}{\partial y} - \frac{\partial \overline{w'\chi'}}{\partial z} + D + S \tag{4.6}$$

where χ is the mixing ratio of gas constituent (i.e. mass of gas per mass of dry air), u, v, and w are the stream wise, lateral, and vertical wind velocity components, respectively, that operate in respective longitudinal (x), lateral (y), and vertical (z) directions. The mean covariances between wind velocity components and χ represent turbulent fluxes, D is molecular diffusion, and S is a source/sink term. Over-bars indicates time averaging and prime (′) indicates fluctuations from the mean. Schematic descriptions of various aspects of the equation are presented in Fig. 4.1.

The assumption underlying micrometeorological techniques is that the fluxes are nearly constant with height and the concentrations change vertically but not horizontally (Denmead 2008). In addition, no sources or sinks exist in the atmosphere above the surface and concentration of gas does not vary significantly with time (Baldocchi et al. 1988). With these assumptions, then Eq. 4.6 is reduced to:

$$\frac{\overline{\partial w'x'(z)}}{\partial z} = D = -v\frac{\partial^2 \overline{\chi}(z)}{\partial z^2} \tag{4.7}$$

where ν is molecular diffusivity. Since molecular diffusion in the surface boundary layer is negligible in comparison to the turbulent flux, integrating Eq. 4.6 with respect to z produces Eq. 4.7:

$$\overline{wz}(z) = F = -\frac{\nu \partial \overline{\chi}(0)}{\partial z} \tag{4.8}$$

Therefore, in well-adjusted internal boundary layer, mean vertical turbulent covariance (F) is constant and does not vary with height, and is equal to gradient-diffusion flux at the surface. Vertical turbulent flux over horizontally homogenous surface under steady-state condition can be expressed as:

$$F = -\overline{\rho_\alpha}\overline{w'\chi'} \tag{4.9}$$

where ρ_α is the density of dry air. The flux is directed downward when $F < 0$ and upwards when F is >0 (positive).

The flux at a particular height (z) is, therefore, the result of many upwind ground level sources. The contribution of sources at different distances from the sensor is predicted by using footprint analyses, which use theories of atmospheric dispersion to predict trajectories of parcels of air transported by wind. Surface roughness and thermal stability are important influences on the footprint.

Difficulties in measuring GHG concentration for micrometeorological techniques flux quantification generally arises from: (i) slow time response instruments, and (ii) need for detection of small concentration differences or fluctuations against a large background concentration (Wagner-Riddle et al. 2005). Three common methods that fall under this category include: (1) eddy covariance (EC), (2) eddy accumulation (EA), and (3) flux gradient methods.

4.2.4.1 Eddy Covariance Technique

The EC technique is a direct measure of the turbulent flux of a scalar across horizontal wind streamlines (Paw et al. 2000). Ideally, the meteorological conditions controlling the state of the turbulence should not vary over the course of measurements. The technique requires simultaneous fast and accurate measurement of both the vertical velocity and trace gas concentration (Stull 1988). Sonic anemometers are used to produce air motion data with the high resolution. Diode laser absorption spectroscopy is one of the commonly used sensors with the time response and sensitivity required for the direct EC flux measurements (Werle 1998). The EC technique is becoming popular for ecosystem assessment of gaseous fluxes because: (1) It is a scale-appropriate method and allows the assessment of whole ecosystem gaseous exchange, (2) It produces a direct measurement of net gaseous exchange across canopy-atmosphere interface. This is achieved by using micrometeorological theory to interpret measurements of covariance between vertical wind velocity and scalar concentration fluctuations (Baldocchi et al. 1988),

(3) The area sampled, called the flux footprint, possesses longitudinal dimensions ranging between 100 m to several km (Schmid 1994), (4) The technique is capable of measuring ecosystem gaseous exchange across a spectrum of timescales ranging from hours to years (Baldocchi et al. 2001). The EC technique is best applied when three conditions are met: (i) flat terrain, (ii) steady environmental conditions, and (iii) extending upwind for an extended distance of the underlying vegetation (Baldocchi 2003). Violation of these conditions can cause systematic errors in interpretation of EC measurements, which accumulate when integrated over time to produce daily and annual estimates (Baldocchi 2003; Pattey et al. 2006).

The intention of EC measurement is to make a direct estimate of the rate of vertical flux transport of the gaseous component in question. The instantaneous flux density (Fg) is estimated based on vertical wind speed (w) and gas density (ρ_g) as in Eq. 4.10:

$$F_g = \rho_g w \tag{4.10}$$

The mean flux density is calculated by Eq. 4.11.

$$\overline{F_g} = \overline{w\rho_g} \tag{4.11}$$

The technique averages the instantaneous fluxes over a period long enough to encompass all the effective transporting eddy sizes. Usually w and c values are measured with fast response instruments at frequencies of 10 Hz or higher and the average is formed over sampling periods of 15 min to 1 h. The averaging sampling times approximate to particular applications. The vertical velocity is usually measured with a 3-D sonic anemometer, which provides additional information on turbulence levels, the momentum flux, horizontal wind speeds in the x and y directions, the wind direction and the flux of sensible heat between the surface and the atmosphere. Fast response gas analyzers for measuring ρ_g are of two types: (a) open-path and (b) closed-path.

Open path instruments are currently used for measuring water vapor and CO_2 concentrations. Open path analyzers require correction of gaseous fluxes for sensible heat and water vapor. Therefore, both sensible heat and water vapor need to be monitored simultaneously with gas flux.

Closed path analyzers use tunable diode lasers, which can measure CO_2, CH_4, and N_2O. The air stream is piped through heat exchanger to bring it to a uniform temperature before analysis. This removes the need for correction for sensible heat. Water vapor is removed by the use of counter-flow drier, which apparently does not attenuate the gas fluctuations (Laville et al. 1999). If water is not removed, corrections for its flux must be applied.

Piping the air stream to the analyzer introduces a lag between the measurement of w and ρ_g, which must be accounted for. Piping the air stream may dampen the gas fluctuations. Gases can also be absorbed or desorbed on the walls of the tubing (Ibrom et al. 2007), which can dampen the fluctuations in concentration of transported gases, leading to an underestimation of their true flux. Ibrom et al. (2007) proposed a method of calculating correction factors for handling this situation.

Advantages and Disadvantages of Eddy Covariance

The EC is a direct measurement, which gives the vertical flux at the point of measurement. It is independent of atmospheric stability, and does not require some of the simplifying assumptions, which are inbuilt into other micrometeorological approaches. The practical problems in EC measurement for N_2O fluxes include: (a) dealing with the effects of simultaneous fluxes of heat and water vapor either by measuring them and apply the corrections to the apparent values of gaseous flux or by preconditioning air samples before measuring their concentrations, (b) accounting for lags between measuring vertical wind speeds and gas concentrations, (c) accounting for possible damping of gas fluctuating by sampling down tubes, (d) mounting at a correct height in relation to sensor path lengths, (e) correctly alignment of the sonic anemometer, vertically and with reference to wind direction, (f) minimizing distortion from neighboring instruments and support structures, and (g) using appropriate sampling frequencies and the averaging times.

4.2.4.2 Eddy Accumulation

The EA method does not require fast response gas analyzer, which makes it particularly appropriate for trace gases. However, rapid measurement of w is still required. The principle behind EA is sampling of "updraft" and "douwndraft" air into two separate bins at the rate proportional to w. Gas concentrations in the up and down bins are then measured with high resolution gas analyzers. The EA requires fast response solenoid valves rather than fast response gas analyzer required for EC approach. The gas flux is given by Eq. 4.12:

$$F_g = \overline{w_u \rho_{g,u}} - \overline{w_d \rho_{g,d}} \tag{4.12}$$

the subscript u and d denotes the up and down sampling periods.

4.2.4.3 Relaxed Eddy Accumulation

In addition to a fast solenoid valve, in relaxed eddy accumulation (REA), a fast-response anemometer measures w and controls a simple valve system. Air is sampled at a constant rate and diverted into the up and down bins depending on the direction of w. The gas flux is estimated by Eq. 4.13:

$$\overline{F_g} = b\sigma_w \left(\overline{\rho_{g,u}} - \overline{\rho_{g,d}} \right) \tag{4.13}$$

where b is a coefficient and σ_w is the standard deviation of w.

4.2.4.4 Advantages and Disadvantages of Eddy Accumulation

The EA method is a direct measurement of gas flux at a point. It allows both on-line gas sensor in the field or a high precision mass spectrometer or GC in a laboratory. Off-line gas analysis is advantageous because samples can be pre-conditioned before analysis to remove the effect of heat and water vapor on the flux measurement. The precision needed for measuring ρ_g and w is the same as that for EC applications. However, measurement of w must be of high accuracy. Data correction cannot be applied with later processing.

4.2.4.5 Flux Gradient Methods

In this technique, gaseous fluxes are calculated as the product of eddy diffusivity and the vertical concentration gradient of the gas or as a product of a transfer coefficient and the difference in gas concentration between two heights. The vertical transport of gases in the surface layer is given by Eq. 4.14:

$$F_g = \frac{K_g \partial \overline{\rho_g}}{\partial z} \tag{4.14}$$

where Kg is turbulent diffusivity or eddy turbulent for gas.

Eddy diffusion arises from random movement of parcels of air. K_g is determined by wind speed, height, aerodynamic roughness, and stability of the atmosphere.

Three approaches commonly used for determining gas fluxes are aerodynamic method (Prueger and Kustas 2005), tracer technique (Denmead 2008) and energy balance (Bowen ratio) method (Denmead 2008). Detailed description for these approaches are beyond the scope of this chapter, the reader is recommended to refer to the cited references for more detailed description, equations employed to calculate the gaseous fluxes, strengths and limitations.

4.2.5 Comparison of Chamber and Micrometeorological Approaches

Scaling up to the field and regional scale is usually based on data from small flux chambers. Several studies have been conducted to establish the validity of this approach. The N_2O fluxes from grasslands and arable lands, chambers strategically placed within the footprint of the micrometeorological tower are generally in reasonable agreement with eddy covariance (Christensen et al. 1996). However, for CH_4 fluxes from rice paddies discrepancies of a factor of 2–3 between chamber and micrometeorological method, are reported, the chambers giving lower emissions (Kanemasu et al. 1995). These differences suggest that more comparisons need to be carried out and that chambers may well be suited for relatively firm surfaces, but not those of low bulk density or under completely waterlogged conditions.

For N_2O flux measurements, tunable diode lasers and quantum cascade lasers offers a methodology for flux measurements by eddy covariance technique (Kroon et al. 2007, 2010), but sometimes inadequate sensitivity of these instruments to detect small fluxes limit their use for long-term monitoring purposes.

4.3 Modeling Approaches

Because of large variability of N_2O fluxes, models are important tools for extrapolation over large regions even continents. Models estimate emissions based on spatially explicit input data such as soil information and meteorological records enclosing the spatial heterogeneity of the landscape. A model is a simplified representation of a more complex process or a system. The main aim of designing a model is to produce the simplest representation that exhibit all the relevant behavior and dependencies of the investigated system. Specific purposes of models are to: (i) increase the knowledge and reduce the uncertainty associated with gaseous emissions estimates (ii) provide a tool to estimate gaseous emissions at times and locations where the emission data are not available, and (iii) provide a framework for answering a "what if" questions about a possible changes of current processes, i.e. to perform simulations. Environmental models generally represent climate system, hydrological processes, ecosystem, and biogeochemical cycles. Once designed, the model is usually validated against the experimental data.

Models fall into three broad categories: (1) empirical (2) process-based, and (3) meta models. Empirical models use relationships that fit the observed behavior of the system, but do not describe any of the underlying processes. They are developed based on identified main driving variables for particular systems, which can be directly measured, such as soil and/or climatic parameters (Conen et al. 2000). Empirical models differ in complexity (Mosier et al. 1998; Bouwman et al. 2002; Lesschen et al. 2011). Process-based models in contrast, simulate the underlying physical, chemical, or biological processes. Process-based models can, therefore, predict results outside the measured data range providing that the underlying assumptions made in the model are still valid. In addition to predicting system behavior, process-based models can provide insight into why the system behaves in a particular way. Process-oriented models for predicting N_2O fluxes, therefore, must consider the ecosystem N cycle in some detail to include N aspects such as N mineralization, N assimilation by plants and microbes, N leaching, microbially-driven N transformations and interaction of N cycle with C cycle, as well as ecosystem biophysical drivers (Frolking et al. 1998; Del Grosso et al. 2000; Li 2000). Process-based models are complex, they frequently require large amounts of input data, they can be conceptually intensive, and independent verification of the emissions can be challenging. Meta models are integrated assessment tools developed based on application of detailed process-based models for assessing the impact of agricultural activities on the environment (de Vries et al. 2007; Leip 2011). The advantage of using models is that they can be applied to any scale depending on the availability of input data matching the requirements of the model. In addition, the more elaborate empirical and process-based models can be used to test a wide range of scenarios including the assessment of possible mitigation measures (Leip et al. 2011).

4.3.1 Empirical Models

Empirical models are developed based on the functional relationships observed from the experimental measurements. The relationship is then defined by a mathematical function. Their advantage is that they can be designed to require only simple and easily obtained input data. The disadvantage of empirical models is that it is often difficult to generalize beyond the range of database used to develop the model. Some empirical models include:

4.3.1.1 IPCC Emission Factor Method

Under the United Nations Framework Convention for Climate Change (UNFCCC), member countries are required to submit national GHG inventories. Estimating the sources and sinks of GHG emissions is needed for calculating the national inventories. Methodologies for agricultural soils fall under three Tiers: (i) Tier 1 consists of simple equations and emission factors based on IPCC guidelines, (ii) Tier 2 uses IPCC guidelines default equations modified for country-specific local climate, soil, management and other conditions, and (iii) Tier 3 methods are based on more complex models and inventory systems, with more disaggregated data to capture variability in local conditions (IPCC 1997, 2006; Lokupitiya and Paustian 2006). Most countries use the default methodology of IPCC to calculate anthropogenic emissions from agricultural soils, including those from fertilizers animal waste, N fixed and crop residues (Lokupitiya and Paustian 2006). The direct N_2O emissions from agricultural soils are calculated by multiplying total soil N input from various sources such as synthetic fertilizer N, and excretal N from grazing animals by an appropriate emission factors (EFs). These guidelines are based on an evaluation of hundreds of field measurements that show fertilization causes 1% of N released as N_2O from agricultural soils. A similar method is used to account for indirect N_2O emissions from volatilization, runoff, and leaching. Countries may use default EFs, but are encouraged to develop their own country-specific EFs.

No account is taken for the variations in N_2O emissions from different climate, seasonal, soil and management factors, and site specific controls are ignored. The IPCC methodology can be refined by disaggregating the EFs by various factors including animal species, soil types, and climate regimes (Clough et al. 2007)

Table 4.1 Regression analyses of N_2O emission from various land uses

Response variable	Land use	Model variables	R^2	References
Daily N_2O fluxes	Grazed pasture, New Zealand	Soil water content, inorganic N, soil temperature	0.44, 0.57	Carran et al. (1995)
N_2O production rate	Incubated soils at constant moisture and temperature	Soil mineral N, sand, substrate induced respiration rate	0.63–0.97	Godde and Conrad (2000)
Daily N_2O fluxes	Fertilized corn, no till, reduced till and plow till, Ohio, USA	Precipitation, air temperature, gravimetric water content, soil temperature	−0.18–0.65	Ussiri et al. (2009)
Log N_2O fluxes	Managed grassland across Europe	Soil temperature, WFPS, monthly rainfall	0.50	Flechard et al. (2007)

proposed a modification of IPCC methodology to account for the effects of applying nitrification inhibitors with autumn and late winter early spring fertilizer N applications. Crutzen et al. (2008) argued that the IPCC methodology seriously underestimates N_2O emissions from agriculture. Their estimates using known global removal rates and concentration growth of N_2O show an overall EF of 3–5% whereas the EF estimated for direct and indirect concentration growth of N_2O using IPCC methodology cover only part of these emissions. The IPCC approach is also limited by: (i) uncertainty in emission factors, (ii) uncertainty in indirect emissions, (iii) limited data on the type and amount of N excreted by grazing animals and (iv) spatial and temporal variability of N_2O emissions. The IPCC default methodology is, therefore, only a first approximation of actual emissions, being too simplistic and generalized, and ignoring all specific controls (Brown et al. 2001; Saggar et al. 2004). It does not allow mitigation options to be assessed because it does not adequately account for regulatory controls.

4.3.1.2 Regression Analysis

Large number of analyses have been performed relating N_2O emissions in temperate soils, climate and management properties (Table 4.1). The regression models are developed by determining relationships between numbers of input variables that produce the closest estimate of N_2O emissions, using the method of least squares. Regression models provide statistical relationships between variables and do not necessarily imply any underlying causal or process-based relationship. This means regression models should not be extrapolated beyond the range of the datasets for which these have been developed.

The strength of a regression model is often assessed by using the coefficient of determination (R^2) which indicates the fraction of the total variability in the responsive variable explained by the model. Given the R^2 values, most regression models can explain half of the observed variability. However, R^2 always increases with addition of extra explanatory variables even if of no relationship. Therefore, a different statistical parameter such as adjusted R^2 (R_{adj}^2) should be used to compare models with differing numbers of explanatory variables.

The regression analyses from various land uses are listed in Table 4.1. They differ by: (i) the time of N_2O emissions considered (daily to annual), (ii) the variables included in the model, and (iii) the data used to derive the model (field measurements from single site versus field measurements from multiple sites). The regression models consider only small subset of factors controlling N_2O emissions and cannot be used to examine the effects of any other factors on emissions. For example, some of the models presented in Table 4.1 can account for the effect of total N inputs but none of the models consider the effects of the timing of N inputs.

4.3.1.3 Boundary Line Approach

Boundary line analysis (BLA) is a technique used for defining bivariate relationships for processes that are limited by multiple factors (Farquharson and Baldock 2008). The BLA considers the outer envelope of a data scatter rather than fitting a regression line through the scatter using a least squares estimate. This is applicable when there is lack of relationship, and the regression line through the data scatter is of little use. The dependence of N_2O emissions on a specified variable, if no other factors are limiting, can be established using the boundary line approach. This approach considers how the response variable is limited by the different input variables. Graphically, the outer boundary envelope in a scatter plot of response variable (N_2O emission) against each explanatory variable (e.g. water filled pore space (WFPS) or mineral N) is assumed to define the boundary where no other variable limit the response. Points from a sufficiently large dataset, which lie on the outer limit of each scattergram represent the effect of that variable on N_2O. The boundary line is fitted to these points and describes the fractional reduction (F_v) in the dependent variable (v) below the maximum caused by the specified variable is limiting the response.

Once the boundary line has been established, the effect of any value of the specified variable on N_2O emission can be obtained. This method requires a large collection of experimental observations to ensure that sufficient instances of the response being limited by only a single variable (for each explanatory variable and over the range of values) are included. This method was devised to explain plant growth response to nutrient limitations (Webb 1972). (Conen et al. 2000) devised a boundary line model for N_2O emissions in grasslands and cropped sites based on soil mineral N (NH_4^+ and NO_3^-) concentrations in topsoil, soil WFPS and temperature. The model predicted the range of the N_2O fluxes within broad categories (1–10, 10–100, 100–1,000 g N_2O-N ha^{-1} day^{-1}) reasonably well for grasslands but underestimated the emissions from vegetable crops.

4.3.2 Process-Based Modeling

To account for the large spatial and temporal variability of GHG fluxes on site and regional and national scales, the development and use of process-oriented models is the most promising tool (Butterbach-Bahl et al. 2004). They can also reduce the uncertainty associated with national and global GHG estimations while enhancing understanding of the complex interactions of biogeochemical processes involved in trace gases production (Barnsley 2007). Models generally simulate the GHG exchange at a given site based on the underlying processes, i.e. the dominant physico-chemical, plant and microbial processes involved in ecosystem C and N cycling and associated GHG exchange (Li et al. 2000). The underlying assumption in process-oriented modeling is that the N_2O emission is controlled by comparable factors across the climatic zones and land uses e.g. microbial C and N turnover, temperature, moisture and substrate responses, and that by capturing the major biogeochemical processes within an ecosystem it is possible to predict the temporal variability of N_2O fluxes. Therefore, such models require a thorough process understanding of the coupled C and N cycles, even though the level of process description may vary between the models currently in use (e.g., Li et al. 2000). Process-based models not only allow the simulation of GHG emissions at a range of scales from farm to national and global levels, but also the exploration of potential mitigation strategies (Giltrap et al. 2010). Generally, two approaches namely bottom-up and top-down, are employed in modeling for estimating the emissions. In bottom-up approach, data from small scale are used to up-scale to larger regions. In bottom-up, emissions are estimated from measured mixing ratios at atmospheric monitoring stations, and inverse modeling techniques, generally based on three-dimensional atmospheric transport and chemistry models. The atmospheric inversions integrate over large regions and generally provide emissions estimates mainly on continental to global scale (e.g., Hirsch et al. 2006; Huang et al. 2008; Corazza et al. 2011).

A number of process-oriented simulation models, which attempt to describe the complex interaction of water, N and C cycles of the terrestrial ecosystems have been developed. The common features within the process-oriented N_2O emission simulation models include soil-air atmosphere and climate interactions, plant growth, C and N cycling, and land use and management. In the N cycling component, the contributions of N_2O from both nitrification and denitrification are estimated. The differences in structure and functionality of the commonly used models are summarized in Table 4.2. Among these models, the DeNitrification DeComposition (DNDC, Li et al. 1992a, b) and Daily time-step version of CENTURY ecosystem model (DAYCENT; Parton et al. 1996, 2001; Del Grosso et al. 2000) have been used to provide site specific and regional scale estimates of N_2O emissions.

There are only few studies that have compared field-scale models for estimating N_2O fluxes from agricultural soils using field observed data (Frolking et al. 1998; Li et al. 2005). Frolking et al. (1998) compared the predictions of four process based

Table 4.2 Structure and functionality of major field scale N_2O simulation models

Modeled process	DNDC	DAYCENT	Ecosys	NLOSS	ExpertN	WNMN	NASA CASA
Mineralization	Function of C:N of 7 SOM pools, T, WFPS	Function of T, WFPS, soil texture, SOM C and N	Function of C:N	Function of C, C:N	Function of C:N of 3 SOM pools, T, water content	Function of C:N, pH, T, water content, clay content	Function of T, WFPS, SOM C:N, soil texture
Crop N uptake	Optimal growth curve	Plant growth controlled by climate and soil N	Plant growth	Optimal growth	Optimum N uptake curve, reduced by suboptimal N availability	Mass flow of N to the roots	Maintain target C:N ratio
Nitrification	Function of NH_4^+, T, WFPS, pH	Function of NH_4^+, T, WFPS, pH; 2% lost as N2O	Nitrifier dynamics, N_2O: dynamic	Nitrifier dynamics, N_2O = fixed proportion (0.25%)	Function of dissolved NH_4^+, T, water content, pH; maximum rate =10% of NH_4^+ pool per day, 0.5% lost as N_2O	First order kinetics, N_2O fixed proportion (0.1–0.5%)	Not explicitly modeled; 2% of N mineralized
Denitrification	Microbial growth kinetics controls N reduction	Function of WFPS, NO_3^-, soil respiration	Denitrifier dynamics	Denitrifier dynamics and anaerobic bacteria driven	Function of dissolved NH_4^+, T, WFPS	Function of WFPS, threshold driven and first-order kinetics	Not explicitly modeled; $NO:N_2O:N_2$ ratio a function of WFPS
Nitrate leaching	Function of water flow	Function of water flow, NO_3^- concentration	Function of water flow, NO_3^- concentration	Function of water flow, NO_3^- concentration	Function of water flow, nitrate concentration	Function of water flow, NO_3^- concentration	Not modeled

Ammonia volatilization	Function of WFPS, T, NH_3: NH_4^+ at pH and T controlled equilibrium in soil solution	Not modeled	Not modeled	Function of WFPS, T, NH_3: NH_4^+ at pH and T, controlled equilibrium in soil solution	Not modeled	Function of WFPS, T, NH_3: NH_4^+ at pH	Not explicitly modeled
Gases produced	NH_3, N_2O, N_2	NH_3, NO, N_2O, N_2	NH_3, N_2O, N_2	NH_3, N_2O, N_2	N_2O, N_2	NH_3, N_2O, N_2	NH_3, NO, N_2O, N_2
C cycling	8 C pools	8 C pools	6 C pools	8 C pools	3–7 C pools	5 C pools	4 C pools (Active-(microbial biomass, labile), slow, passive)
Gases simulated	CO_2, CH_4	CO_2, CH_4	CO_2, CH_4	CO_2	CO_2, CH_4	CO_2	CO_2, CH_4
References	Li et al. (1992a, b, 2000)	Parton et al. (1996, 1998, 2001), Del Grosso et al. (2000)	Grant (2001)	Riley and Matson (2000)	Engel and Priesack (1993), Priesack et al. (2001), Kaharabata et al. (2003)	Li et al. (2005, 2007)	Potter et al. (1997), Frolking et al. (1998)

T = temperature

models – DAYCENT, DNDC, ExpertN and the daily version of NASA CASA with the field measurement of N_2O fluxes from five sites – unfertilized semi-arid rangelands in Colorado, fertilized grasslands in Scotland, and two fertilized cultivated fields in Germany. Simulations were run for two consecutive years and compared with the field observations. All models simulated daily and annual N_2O emission fluxes. The models produced similar results for the general cycling of N through the ecosystems but varied in the partitioning of gaseous N losses between N_2O, NO, NH_3 and N_2. Both DAYCENT and ExpertN predicted N_2O emissions close to observed emissions for all sites, and particularly DAYCENT tended to perform better than ExpertN. However, DNDC and the CASA models are less consistent in their predictions with overestimates in the Colorado site, underestimates of emissions in the Scottish site and comparable to measured emissions in the German sites. The study revealed that accurate partitioning of gaseous N loss into NO, N_2O and N_2 is a challenge for all models, and accurate simulation of soil water content dynamics with denitrification and N_2O flux is a key component of each model (Frolking et al. 1998). Since the model comparison by Frolking et al. (1998), considerable improvements of the performances of DNDC and DAYCENT have been made, and updated comparisons are needed.

Li et al. (2005) used WNMM with three different N gas modules to simulate soil denitrification fluxes and N_2O emissions from two irrigated sites in North China Plains. The three gas modules were MNMM, DAYCENT (or NGAS) and DNDC. Gas modules are the actual model components, which simulate N_2O emissions from the processes of soil nitrification and denitrification. Soil water, temperature, organic matter (OM) decomposition and other N transformations and crop growth are simulated by WNMM platform. All three-gas modules generate generally similar soil mineral N dynamics in the 0–20 cm top soil, but the relatively simpler daily time-step WNMM gas module outperforms both DAYCENT and DNDC approaches for predicting soil denitrification fluxes and N_2O emissions (Li et al. 2005). Abdalla et al. (2010) used DayCent and DNDC models to estimate N_2O emissions from fertilized humid pasture, and investigate the impacts of future climate change on N_2O fluxes and biomass production. Comparison between simulated and measured flux indicated that DNDC overestimated the measured flux with relative deviations of +132 and +258% due to overestimation of the effects of SOC. After calibration for Irish conditions, DayCent simulated N_2O fluxes more consistently than did DNDC (Abdalla et al. 2010)

Process-based models can be linked to large-scale economic models to capture the complex interplay between market, environmental policies, and the economic behavior of different agents such as farmers, consumers, and processors at country, regional or global scale. This kind of analysis use robust data to perform integrated multi-sector modeling system. These kinds of simulation tools enables assessment of environmental policy impacts on wide range of environmental problems such as climate change, air pollution, groundwater pollution or GHG emissions (Leip et al. 2008; Oenema et al. 2009; Velthof et al 2009; Leip 2011).

4.3.3 Modeling Issues for Further Consideration

4.3.3.1 Uncertainty

Since model is a simplification of reality, the model prediction value and the measured value may differ. The difference between measured value and predicted value is known as "error term." There are two major sources of error: (a) error introduced by model simplification, and (b) error introduced by uncertainties of the input data. In addition, experimental data used to validate the model also contain measurement errors. Understanding which sources of error predominate is important in improving the model. Sensitivity analysis is useful for indicating the inputs having the greatest effect on the model results so that the main effort can be directed into obtaining reliable values for those inputs. Monte Carlo simulation, a probability distribution for each input data, is used to randomly generate a large number of input datasets. The model outputs from these randomly generated datasets are then analyzed statistically. The error introduced by model simplifications is estimated by comparing the model predictions with the experimental results. Subtracting the uncertainties associated with input data from model error produces the uncertainty arising from the model.

Uncertainty in such emission inventories and mitigation/feedback studies is associated with the uncertainties in input parameters as well as with those in model parameters. However, uncertainties in input parameters are mainly addressed using Monte Carlo techniques (Kesik et al. 2006; Werner et al. 2007), and model parametric uncertainty is often neglected (Van Oijen et al. 2005). This is a shortcoming, which needs to be properly addressed in future work. Nevertheless, biogeochemical models offer a great opportunity to improve understanding of ecosystem processes and GHG exchange and they can play an important role in identifying and predicting consequences and feedbacks of global changes (climate and land use change) for ecosystem functioning and BAE of GHG.

4.3.3.2 Scaling Up

Simulation of N_2O emissions over spatial and temporal scales provides regional or national estimates. However, three potential problems with up-scaling are: (a) different processes may be important at different scales, (b) input data availability is reduced at larger scales, and (c) model input and model output undergo change of support which affects the relationship between them (Heuvelink 1998).

Regionalization of N_2O emissions is achieved by coupling of the process-based models to geographical information system (GIS) databases which provides the relevant information needed for initializing (soil and vegetation properties, management) and driving (meteorological conditions) the models (Kesik et al. 2006). Such approaches partly neglect landscape processes, such as lateral flow and transport of nutrients and sediments via leaching or erosion. It needs to be addressed for more reliable regional and national emission estimations.

National inventories for N_2O emissions from soils using DNDC or DAYCENT models have been calculated for US, UK, China, Germany, India, or Europe. On a global scale, the GIS coupled Forest-DNDC model has been used to estimate N_2O emissions from soils under a tropical rain forest (Werner et al. 2007). Increasingly biogeochemical models have been used to study potential strategies for mitigating GHG emissions from soils on site as well as on regional scales (e.g., Li et al. 2006) or to improve the understanding of how future changes in climate or land use may feedback on BAE of GHG (Parton et al. 2007). Uncertainty in such emission inventories and mitigation/feedback studies is associated with those in input parameters as well as with those in model parameters.

At the large scale (national to continental), very simple models such as IPCC emission factors, which use different source of N input as a sole explanatory variable, have predicted N_2O emissions that are comparable to other approaches within large uncertainties (Leip 2010; Li et al. 2001). However, they fail to depict the variation of N_2O emissions with inter-annual variability of climate and regional patterns. Nevertheless, the explanatory power of models for site-level N_2O fluxes is often low. Several Europe-wide models have been used to estimate GHG fluxes from agriculture soils (De Vries et al 2011a, b; Corazza et al 2011). Generally, estimation of N_2O fluxes at regional scale remains a challenge due to lack of robust data needed for good validation of the model estimates.

4.4 Global and Regional Atmospheric Nitrous Oxide Monitoring

The Atmospheric Lifetime Experiment (ALE) started monitoring of global greenhouse gases including N_2O in 1978 at four globally distributed sites namely Agrigole (Ireland), Ragged Point (Barbados), Cape Matula (American Samoa), and Cape Grim (Tasmania). In 1980, Cape Meares (Oregon, USA) came on line, and later replaced by Trinidad Head (California, USA) in 1995, while Mace Heard replaced Agrigole in 1987. Based on the changes in instrumentation, the program also evolved into Global Atmospheric Gases Experiment (GAGE) and then into Advanced Global Atmospheric Experiment (AGAGE). The current monitoring station locations for ALE/GAGE/AGAGE global network program are Cape Grim, Tasmania, Cape Mata Tula, American Samoa, Ragged Point, Barbados, Mace Head, Ireland, Trinidad Head, California, Jungfraujoch, Switzerland and Zappelin Mount, Norway. The ALE and GAGE data have all been recalculated according to AGAGE calibration standards to create a unified ALE/GAGE/AGAGE dataset (Prinn et al. 2000). National Oceanic and Atmospheric Administration (NOAA) Earth System Research Laboratory (ESRL) operates global atmospheric baseline observatories with stations at Barrow (Alaska), Trinidad Head, American Samoa, South Pole (Antarctica) and Summit (Greenland). A major limitation in the global dataset was the lack of an international primary scale standard for N_2O that would allow inter-calibrating the various existing N_2O observational sites and networks. A standardized scale for atmospheric observation for calibrating N_2O measurements internationally were recently established (Hall et al. 2007). In 1990s, more national stations came to operation under the frame of CarboEurope, and N_2O was included in the multispecies observation network in Europe. However, N_2O measurements are less standardized than those of CO_2 are (Corazza et al. 2011).

4.4.1 Top-Down Modeling of Nitrous Oxide Emissions

Top-down or inverse modeling uses measurements from atmospheric GHG network monitoring to estimate GHG emissions at regional and/or global scale. Almost the entire earth surface acts as a source of N_2O and these sources are very homogenously spread in space. However, extreme N_2O emission events such as during freeze-thaw periods, after fertilization or rewetting dry soils make them very variable at synoptic timescales in a presently unpredictable manner. Together with the long atmospheric residence time of 122 ± 24 years as derived using atmospheric measurements (Volk et al. 1997), these features produce complex atmospheric N_2O patterns and very small atmospheric N_2O concentration gradients signals. In addition, there are N_2O sources in the atmosphere, which makes simulation of stratospheric sink and stratosphere-troposphere N_2O exchange difficult. All these factors complicate the top-down modeling. Based on the atmospheric lifetime of 122 ± 24 years, the global total sinks of N_2O are estimated at 12.5 ± 2.5 Tg N_2O-N year^{-1}. The current increase of atmospheric mixing ratios of 0.73 ± 0.06 ppbv year^{-1} is equivalent to imbalance between sources and sinks (Hirsch et al. 2006). Hirsch et al. (2006) estimated the pre-industrial N_2O sources at about 11.0 Tg N_2O-N year^{-1}.

Atmospheric measurements combined with inverse atmospheric models provide independent "top-down" emission estimates of N_2O emissions. Inverse modeling also can be important tool for regional emission estimates, independent verification of international agreements on emission reduction, and improving our knowledge about GHG cycles (Corazza et al. 2011). Hirsch et al. (2006) were the first to use NOAA-CMDL flask data to inversely model N_2O source/sink patterns for four to six regions of the globe with atmospheric transport model. Their work demonstrated the feasibility to calculate robust long-term average N_2O fluxes for large global regions. The major constraint of this approach is the unknown temporal and spatial variation in stratosphere-troposphere exchange of N_2O. Furthermore, top-down approaches generally estimates total emissions, hence, good bottom-up estimates of natural sources are needed for better estimates of anthropogenic sources. Using new comprehensive dataset of continuous measurement from various European monitoring stations and NOAA data for Europe observations Corazza et al. (2011) demonstrated that atmospheric observation combined with inverse modeling can significantly decrease the uncertainties of N_2O emissions at country and regional level.

4.5 Conclusions and Research Priorities

The N_2O emissions from soils are spatially and temporally variable. This variability can be attributed in large part to the underlying microbial processes responsible for N_2O production in soil, although other soil factors may also play a significant role. Number of methods and approaches has been used to determine N_2O fluxes from soils. Static chambers are the most commonly used tool worldwide, and most of the current understanding of N_2O emissions, especially under agroecosystems was derived from studies involving static chamber technique. Micrometeorological measurements may provide spatially integrated N_2O emissions at the field scale, watershed, and even larger scales in the near future. Most of the existing regional, national, and global estimates of N_2O emissions are obtained based on some form of IPCC-EF estimates, which have limited application because of: (i) uncertainty in emission EF, (ii) uncertainty in indirect emissions estimates based on this approach, and (iii) spatial and temporal variability of N_2O emissions. Although attempts have been made to improve the IPCC EF method, overall, the approach seems unsatisfactory for producing reliable global estimates for N_2O emission. Other approaches, such as regression models, are also constrained by similar limitations. Since 1990s, process-oriented ecosystem modeling has emerged as a promising approach to deal with some of the limitations and improve the understanding of GHG production and emissions at regional, national and global scales. However, more research needs to be undertaken to develop reliable databases for model parameterization, simulations, and validations. Overall, there are no agreements between complex empirical model estimates, process-based models, and inverse model estimates of N_2O emissions from agricultural soils. The reason may be combination of insufficient or non-representative experimental observations for model calibrations, the overestimation of sensitivity to important model parameters and poor spatial input data. More progress is needed to improve the performance of the models. The key to improving N_2O emission estimates lies in soil borne fluxes. This requires scientific development of cheap, reliable, and operational flux measurement techniques to improve the density of observations in critical ecosystems and seasons prone to high N_2O emissions, as well as improved process models. In addition, there is a strong knowledge gap, which needs to be filled. The key developments and gaps in knowledge include the following:

1. New molecular tools are now available to link soil microbial biodiversity with soil function and can provide an overview of the distribution of functional microbial groups in soils of different land uses, and assign trace gas emissions to the active microbial population
2. Need for instruments development to facilitate N_2O flux measurements at the field and landscape scale and provide long-term measurements at large spatial scale and high temporal resolution at key sites.
3. New methods to study denitrification rates to N_2 and isotope studies to elucidate the microbial pathway responsible for N_2O production and removal need to be developed, deployed, and used at the field and regional scale.

4. Insufficient data to up-scale N_2O emissions to the national and global scale, also inclusion of 'new' crops data (i.e., bioenergy crops) should be addressed.
5. Gap in knowledge of the contribution and quantification of plants, especially trees, in producing and transporting N_2O soil to atmosphere (Chaps. 5 and 7).
6. Improvement in modeling to incorporate prediction of consequences and feedbacks of global changes (climate and land use change) for ecosystem functioning and biosphere–atmosphere trace gas exchange.
7. Ecosystem level N_2O observation network such as that which has emerged in Europe, provides a way around for the implementing robust monitoring and source of reliable data for modeling. Such efforts should be replicated in other regions especially Africa and Asia, where the reliable field data are scarce.

Suggested Questions

1. Describe the approach for monitoring N_2O fluxes from a diary animal farm. What factors will need considerations in selecting the emission monitoring in such a land use?
2. What will be a method of choice for long-term monitoring N_2O fluxes from pastures used for grazing?
3. Describe the necessary steps for obtaining estimated annual N_2O emissions from 250 ha farm by process-oriented modeling technique.
4. Describe the main obstacles for application of modeling for estimating N_2O emissions.

References

Abdalla, M., Jones, M., Yeluripati, J., Smith, P., Burke, J., Williams, M. (2010) Testing DayCent and DNDC model simulations of N_2O fluxes and assessing the impacts of climate change on the gas flux and biomass production from a humid pasture. Atmos. Environ. **44**(25), 2961–2970. doi:10.1016/j.atmosenv.2010.05.018

Ambus, P., Christensen, S. (1994) Measurement of N_2O emission from a fertilized grassland: an analysis of spatial variability. J. Geophys. Res.-Atmos. **99**(D8), 16549–16555

Baldocchi, D., Falge, E., Gu, L.H., Olson, R., Hollinger, D., Running, S., Anthoni, P., Bernhofer, C., Davis, K., Evans, R., Fuentes, J., Goldstein, A., Katul, G., Law, B., Lee, X.H., Malhi, Y., Meyers, T., Munger, W., Oechel, W., U, K.T.P., Pilegaard, K., Schmid, H.P., Valentini, R., Verma, S., Vesala, T., Wilson, K., Wofsy, S. (2001) FLUXNET: A new tool to study the temporal and spatial variability of ecosystem-scale carbon dioxide, water vapor, and energy flux densities. Bull. Am. Meteorol. Soc. **82**(11), 2415–2434

Baldocchi, D.D. (2003) Assessing the eddy covariance technique for evaluating carbon dioxide exchange rates of ecosystems: past, present and future. Global Change Biol. **9**(4), 479–492

Baldocchi, D.D., Hicks, B.B., Meyers, T.P. (1988) Measuring biosphere-atmosphere exchanges of biologically related gases with micrometeorological methods. Ecology **69**(5), 1331–1340

Barnsley, M.J. (2007) Environmental modeling: A practical introduction. pp. Taylor & Francis, Boca Raton, Fl

Bouwman, A.F., Boumans, L.J.M., Batjes, N.H. (2002) Modeling global annual N_2O and NO emissions from fertilized fields. Global Biogeochem. Cycles **16**(4). doi:10.1029/2001gb001812

Brown, L., Brown, S.A., Jarvis, S.C., Syed, B., Goulding, K.W.T., Phillips, V.R., Sneath, R.W., Pain, B.F. (2001) An inventory of nitrous oxide emissions from agriculture in the UK using the IPCC methodology: emission estimate, uncertainty and sensitivity analysis. Atmos. Environ. **35**(8), 1439–1449. doi:10.1016/s1352-2310(00)00361-7

Butterbach-Bahl, K., Kesik, M., Miehle, P., Papen, H., Li, C. (2004) Quantifying the regional source strength of N-trace gases across agricultural and forest ecosystems with process based models. Plant Soil **260**(1–2), 311–329

Carran, R.A., Theobald, P.W., Evans, J.P. (1995) Emission of nitrous oxide from some grazed pasture soils in New Zealand. Aus. J. Soil Res. **33**(2), 341–352

Christensen, S., Ambus, P., Arah, J.R.M., Clayton, H., Galle, B., Griffith, D.W.T., Hargreaves, K.J., Klemedtsson, L., Lind, A.M., Maag, M., Scott, A., Skiba, U., Smith, K.A., Welling, M., Wienhold, F.G. (1996) Nitrous oxide emission from an agricultural field: Comparison between measurements by flux chamber and micrometerological techniques. Atmos. Environ. **30**(24), 4183–4190

Clark, M., Jarvis, S., Maltby, E. (2001) An improved technique for measuring concentration of soil gases at depth in situ. Comm. Soil Sci. Plan. Anal. **32**(3–4), 369–377

Clayton, H., Arah, J.R.M., Smith, K.A. (1994) Measurement of nitrous oxide emissions from fertilized grassland using closed chambers. J. Geophys. Res.-Atmos. **99**(D8), 16599–16607

Clough, T.J., Di, H.J., Cameron, K.C., Sherlock, R.R., Metherell, A.K., Clark, H., Rys, G. (2007) Accounting for the utilization of a N_2O mitigation tool in the IPCC inventory methodology for agricultural soils. Nutr. Clycl. Agroecosys. **78**(1), 1–14. doi:10.1007/s10705-006-9069-z

Conen, F., Dobbie, K.E., Smith, K.A. (2000) Predicting N_2O emissions from agricultural land through related soil parameters. Global Change Biol. **6**(4), 417–426

Conen, F., Smith, K.A. (1998) A re-examination of closed flux chamber methods for the measurement of trace gas emissions from soils to the atmosphere. Eur. J. Soil Sci. **49**(4), 701–707

Corazza, M., Bergamaschi, P., Vermeulen, A.T., Aalto, T., Haszpra, L., Meinhardt, F., O'Doherty, S., Thompson, R., Moncrieff, J., Popa, E., Steinbacher, M., Jordan, A., Dlugokencky, E., Bruhl, C., Krol, M., Dentener, F. (2011) Inverse modelling of European N_2O emissions: assimilating observations from different networks. Atmos. Chem. Phys. **11**(5), 2381–2398. doi:10.5194/acp-11-2381-2011

Crutzen, P.J., Mosier, A.R., Smith, K.A., Winiwarter, W. (2008) N_2O release from agro-biofuel production negates global warming reduction by replacing fossil fuels. Atmos. Chem. Phys. **8**(2), 389–395

de Vries, W., Butterbach-Bahl, K., van der Gon, H.D., Oenema, O. (2007) Impact of atmospheric nitrogen deposition on the exchange of carbon dioxide, nitrous oxide and methane from European forests. In: Reay, D.S., Hewitt, C.N., Smith, K.A., Grace, J. (eds.) Greenhouse Gas Sinks. pp. 249–283

de Vries, W., Leip, A., Reinds, G.J., Kros, J., Lesschen, J.P., Bouwman, A.F. (2011a) Comparison of land nitrogen budgets for European agriculture by various modeling approaches. Environ. Pollut. **159**(11), 3254–3268. doi:10.1016/j.envpol.2011.03.038

de Vries, W., Leip, A., Reinds, G.J., Kros, J., Lesschen, J.P., Bouwman, A.F., Grizzetti, B., Bouraoui, F., Butterbach-Bahl, K., Bergamaschi, P., Winiwarter, W. (2011b) Geographical variation in terrestrial nitrogen budgets across Europe. In: Sutton, M.A., Howard, C.M., Erisman, J.W., Billen, G., Bleeker, A., Grennfelt, P., Grinsven, H.v., Grizzetti, B. (eds.) The European Nitrogen Assessment: Sources, Effects and Policy Perspectives. pp. 317–344. Cambridge University Press, Cambridge, UK

Del Grosso, S.J., Parton, W.J., Mosier, A.R., Ojima, D.S., Kulmala, A.E., Phongpan, S. (2000) General model for N_2O and N_2 gas emissions from soils due to denitrification. Global Biogeochem. Cycles **14**(4), 1045–1060

Denmead, O.T. (2008) Approaches to measuring fluxes of methane and nitrous oxide between landscapes and the atmosphere. Plant Soil **309**(1–2), 5–24. doi:10.1007/s11104-008-9599-z

Denmead, O.T., Harper, L.A., Freney, J.R., Griffith, D.W.T., Leuning, R., Sharpe, R.R. (1998) A mass balance method for non-intrusive measurements of surface-air trace gas exchange. Atmos. Environ. **32**(21), 3679–3688

Denmead, O.T., Leuning, R., Jamie, I., Griffith, D.W.T. (2000) Nitrous oxide emissions from grazed pastures: measurements at different scales. Chemosphere: Global Change Sci. **2**(3–4), 301–312

Engel, T., Priesack, E. (1993) Expert-N, a building block system of nitrogen models as a resource for advice, research, water management and policy In: Eijsackers, H.J.P., Hamers, T. (eds.) Integrated Soil and Sediment Research: A Basis for Proper Protection. pp. 503–507. Kluwer Academic Publishers, Dordrecht, Netherlands

Farquharson, R., Baldock, J. (2008) Concepts in modelling N_2O emissions from land use. Plant Soil **309**(1–2), 147–167. doi:10.1007/s11104-007-9485-0

Firestone, M.K., Davidson, E.A. (1989) Microbiological basis of NO and N_2O production and consumption in the soil. In: Andreae, M.O., Schimel, D.S. (eds.) Exchange of trace gases between terrestrial ecosystems and the atmosphere. pp. 7–21. John Wiley & Sons, Chichester, UK

Flechard, C.R., Ambus, P., Skiba, U., Rees, R.M., Hensen, A., van Amstel, A., Pol-van Dasselaar, A.V., Soussana, J.F., Jones, M., Clifton-Brown, J., Raschi, A., Horvath, L., Neftel, A., Jocher, M., Ammann, C., Leifeld, J., Fuhrer, J., Calanca, P., Thalman, E., Pilegaard, K., Di Marco, C., Campbell, C., Nemitz, E., Hargreaves, K.J., Levy, P.E., Ball, B.C., Jones, S.K., van de Bulk, W.C.M., Groot, T., Blom, M., Domingues, R., Kasper, G., Allard, V., Ceschia, E., Cellier, P., Laville, P., Henault, C., Bizouard, F., Abdalla, M., Williams, M., Baronti, S., Berretti, F., Grosz, B. (2007) Effects of climate and management intensity on nitrous oxide emissions in grassland systems across Europe. Agric. Ecosys. Environ. **121**(1–2), 135–152. doi:10.1016/j.agee.2006.12.024

Freibauer, A., Kaltschmitt, M. (2003) Controls and models for estimating direct nitrous oxide emissions from temperate and sub-boreal agricultural mineral soils in Europe. Biogeochemistry **63**(1), 93–115. doi:10.1023/a:1023398108860

Frolking, S.E., Mosier, A.R., Ojima, D.S., Li, C., Parton, W.J., Potter, C.S., Priesack, E., Stenger, R., Haberbosch, C., Dorsch, P., Flessa, H., Smith, K.A. (1998) Comparison of N_2O emissions from soils at three temperate agricultural sites: simulations of year-round measurements by four models. Nutr. Clycl. Agroecosys. **52**(2–3), 77–105

Giltrap, D.L., Li, C.S., Saggar, S. (2010) DNDC: A process-based model of greenhouse gas fluxes from agricultural soils. Agric. Ecosys. Environ. **136**(3–4), 292–300. doi:10.1016/j.agee.2009.06.014

Godde, M., Conrad, R. (2000) Influence of soil properties on the turnover of nitric oxide and nitrous oxide by nitrification and denitrification at constant temperature and moisture. Biol. Fert. Soils **32**(2), 120–128. doi:10.1007/s003740000247

Grant, R.F. (2001) A review of the Canadian ecosystem model ecosys. In: Shaffer, M. (ed.) Modeling carbon and nitrogen dynamics for soil management. CRC Press, Boca Raton, FL

Gut, A., Blatter, A., Fahrni, M., Lehmann, B.E., Neftel, A., Staffelbach, T. (1998) A new membrane tube technique (METT) for continuous gas measurements in soils. Plant Soil **198**(1), 79–88

Hall, B.D., Dutton, G.S., Elkins, J.W. (2007) The NOAA nitrous oxide standard scale for atmospheric observations. J. Geophys. Res.-Atmos. **112**(D9). doi:10.1029/2006jd007954

Hedley, C.B., Saggar, S., Tate, K.R. (2006) Procedure for fast simultaneous analysis of the greenhouse gases: Methane, carbon dioxide, and nitrous oxide in air samples. Comm. Soil Sci. Plan. Anal. **37**(11–12), 1501–1510. doi:10.1080/00103620600709928

Hellebrand, H.J., Kalk, W.D. (2001) Emission of methane, nitrous oxide, and ammonia from dung windrows. Nutr. Clycl. Agroecosys. **60**(1–3), 83–87

Heuvelink, G.B.M. (1998) Uncertainty analysis in environmental modelling under a change of spatial scale. Nutr. Clycl. Agroecosys. **50**(1–3), 255–264

Hirsch, A.I., Michalak, A.M., Bruhwiler, L.M., Peters, W., Dlugokencky, E.J., Tans, P.P. (2006) Inverse modeling estimates of the global nitrous oxide surface flux from 1998–2001. Global Biogeochem. Cycles **20**(1). doi:10.1029/2004gb002443

Huang, J., Golombek, A., Prinn, R., Weiss, R., Fraser, P., Simmonds, P., Dlugokencky, E.J., Hall, B., Elkins, J., Steele, P., Langenfelds, R., Krummel, P., Dutton, G., Porter, L. (2008) Estimation of regional emissions of nitrous oxide from 1997 to 2005 using multinetwork measurements, a chemical transport model, and an inverse method. J. Geophys. Res.-Atmos. **113**(D17). doi:10.1029/2007jd009381

Hutchinson, G.L., Mosier, A.R. (1981) Improved soil cover method for field measurement of nitrous oxide fluxes. Soil Sci. Soc. Am. J. **45**(2), 311–316

Ibrom, A., Dellwik, E., Flyvbjerg, H., Jensen, N.O., Pilegaard, K. (2007) Strong low-pass filtering effects on water vapour flux measurements with closed-path eddy correlation systems. Agric. Forest Meteorol. **147**(3–4), 140–156. doi:10.1016/j.agrformet.2007.07.007

Intergovernmental Panel on Climate Change (IPCC) (1997) Revised 1996 IPCC guidelines for national greenhouse gas inventories. Houghton, J.T., Filho, L.G.M., Lim, B., Treanton, K., Mamaty, I., Bonduki, Y., Griggs, D.J., Callender, B.A. (eds.). pp. IPCC/OECD/EA, Bracknell, UK

Intergovernmental Panel on Climate Change (IPCC) (2006) 2006 IPCC guidelines for national greenhouse gas inventories: Prepared by the National Greenhouse Gas Inventories Programme. Eggleston H.S., B.L., Miwa K., Ngara T. and Tanabe K. (ed.). pp. Institute for Global Environmental Strategies (IGES) Kanagawa, Japan

Jacinthe, P.A., Dick, W.A. (1997) Soil management and nitrous oxide emissions from cultivated fields in southern Ohio. Soil Till. Res. **41**(3–4), 221–235

Kaharabata, S.K., Drury, C.F., Priesack, E., Desjardins, R.L., McKenney, D.J., Tan, C.S., Reynolds, D. (2003) Comparing measured and Expert-N predicted N_2O emissions from conventional till and no till corn treatments. Nutr. Clycl. Agroecosys. **66**(2), 107–118. doi:10.1023/a:1023978830307

Kanemasu, E.T., Flitcroft, I.D., Shah, T.D.H., Nie, D., Thurtell, G.W., Kidd, G., Simpson, I., Lin, M., Neue, H.U., Bronson, K. (1995) Comparison of micrometeorological and chamber measurements of methane emissions from rice fields. Peng, S., Ingram, K.T., Neue, H.U., Ziska, L.H. (eds.). Climate Change and Rice. 91–101

Kesik, M., Bruggemann, N., Forkel, R., Kiese, R., Knoche, R., Li, C.S., Seufert, G., Simpson, D., Butterbach-Bahl, K. (2006) Future scenarios of N_2O and NO emissions from European forest soils. J. Geophys. Res.- Biogeo. **111**(G2), G02018. doi:10.1029/2005jg000115

Kimball, B.A., Lemon, E.R. (1971) Air turbulence effects upon soil gas exchange. Soil Sci. Soc. Am. Proc. **35**(1), 16–21

Kroon, P.S., Hensen, A., Jonker, H.J.J., Zahniser, M.S., van 't Veen, W.H., Vermeulen, A.T. (2007) Suitability of quantum cascade laser spectroscopy for CH4 and N2O eddy covariance flux measurements. Biogeosciences **4**(5), 715–728

Kroon, P.S., Vesala, T., Grace, J. (2010) Flux measurements of CH_4 and N_2O exchanges. Agric. Forest Meteorol. **150**(6), 745–747. doi:10.1016/j.agrformet.2009.11.017

Lapitan, R.L., Wanninkhof, R., Mosier, A.R. (1999) Methods for stable gas flux determination in aquatic and terrestrial ecosystems. In: Bouwman, A.F. (ed.) Approaches to scaling trace gas fluxes in ecosystems. pp. 27–66. Elsevier Science, Amsterdam, The Netherlands

Laville, P., Jambert, C., Cellier, P., Delmas, R. (1999) Nitrous oxide fluxes from a fertilised maize crop using micrometeorological and chamber methods. Agric. Forest Meteorol. **96**(1–3), 19–38

Leip, A. (2010) Quantitative quality assessment of the greenhouse gas inventory for agriculture in Europe. Climatic Change **103**(1–2), 245–261. doi:10.1007/s10584-010-9915-5

Leip, A. (2011) Assessing the environmental impact of agriculture in Europe: The indicator database for European agriculture. In: Guo, L., Gunasekara, A., McConnell, L. (eds.) Understanding Greenhouse Gas Emissions from Agricultural Management. pp. 371–375. American Chemical Society (ACS), Washington, DC

Leip, A., Busto, M., Corazza, M., Bergamaschi, P., Koeble, R., Dechow, R., Monni, S., de Vries, W. (2011) Estimation of N_2O fluxes at the regional scale: data, models, challenges. Curr. Opin. Environ. Sustainability **3**(5), 328–338. doi:10.1016/j.cosust.2011.07.002

Leip, A., Marchi, G., Koeble, R., Kempen, M., Britz, W., Li, C. (2008) Linking an economic model for European agriculture with a mechanistic model to estimate nitrogen and carbon losses from arable soils in Europe. Biogeosciences **5**(1), 73–94

Lesschen, J.P., Velthof, G.L., de Vries, W., Kros, J. (2011) Differentiation of nitrous oxide emission factors for agricultural soils. Environ. Pollut. **159**(11), 3215–3222. doi:10.1016/j.envpol.2011.04.001

Li, C.S. (2000) Modeling Trace Gas Emissions from Agricultural Ecosystems. Nutr. Clycl. Agroecosys. **58**(1), 259–276. doi:10.1023/a:1009859006242

Li, C.S., Aber, J., Stange, F., Butterbach-Bahl, K., Papen, H. (2000) A process-oriented model of N_2O and NO emissions from forest soils: 1. Model development. J. Geophys. Res.-Atmos. **105**(D4), 4369–4384

Li, C.S., Frolking, S., Frolking, T.A. (1992a) A model of nitrous-oxide evolution from soil driven by rainfall events .1. Model structure and sensitivity. J. Geophys. Res.-Atmos. **97**(D9), 9759–9776

Li, C.S., Frolking, S., Frolking, T.A. (1992b) A model of nitrous-oxide evolution from soil driven by rainfall events .2. Model applications. J. Geophys. Res.-Atmos. **97**(D9), 9777–9783

Li, C.S., Salas, W., DeAngelo, B., Rose, S. (2006) Assessing alternatives for mitigating net greenhouse gas emissions and increasing yields from rice production in China over the next twenty years. J. Environ. Qual. **35**(4), 1554–1565. doi:10.2134/jeq2005.0208

Li, C.S., Zhuang, Y.H., Cao, M.Q., Crill, P., Dai, Z.H., Frolking, S., Moore B, III, Salas, W., Song, W.Z., Wang, X.K. (2001) Comparing a process-based agro-ecosystem model to the IPCC methodology for developing a national inventory of N_2O emissions from arable lands in China. Nutr. Clycl. Agroecosys. **60**(1–3), 159–175. doi:10.1023/a:1012642201910

Li, J., Tong, X., Yu, Q., Dong, Y., Peng, C. (2008) Micrometeorological measurements of nitrous oxide exchange above a cropland. Atmos. Environ. **42**(29), 6992–7001. doi:10.1016/j.atmosenv.2008.04.040

Li, Y., Chen, D.L., Zhang, Y.M., Edis, R., Ding, H. (2005) Comparison of three modeling approaches for simulating denitrification and nitrous oxide emissions from loam-textured arable soils. Global Biogeochem. Cycles **19**(3), GB3002. doi:10.1029/2004gb002392

Li, Y., White, R., Chen, D.L., Zhang, J.B., Li, B.G., Zhang, Y.M., Huang, Y.F., Edis, R. (2007) A spatially referenced water and nitrogen management model (WNMM) for (irrigated) intensive cropping systems in the North China Plain. Ecol. Model. **203**(3–4), 395–423. doi:10.1016/j.ecolmodel.2006.12.011

Li, Z., Kelliher, F.M. (2005) Determining nitrous oxide emissions from subsurface measurements in grazed pasture: A field trial of alternative technology. Aus. J. Soil Res. **43**(6), 677–687. doi:10.1071/sr04106

Livingston, G.P., Huchinson, G.L. (1995) Enclosure-based measurement of trace gases: application and sources of error. In: Matson, P.A., Harriss, R.C. (eds.) Biogenic trace gases: measuring emissions from soil and water. pp. 14–51. Blackwell, Oxford, UK

Lokupitiya, E., Paustian, K. (2006) Agricultural soil greenhouse gas emissions: A review of National Inventory Methods. J. Environ. Qual. **35**(4), 1413–1427. doi:10.2134/jeq2005.0157

Luo, J., Ledgard, S.F., de Klein, C.A.M., Lindsey, S.B., Kear, M. (2008) Effects of dairy farming intensification on nitrous oxide emissions. Plant Soil **309**(1–2), 227–237. doi:10.1007/s11104-007-9444-9

Mosier, A., Kroeze, C., Nevison, C., Oenema, O., Seitzinger, S., van Cleemput, O. (1998) Closing the global N_2O budget: nitrous oxide emissions through the agricultural nitrogen cycle. Nutr. Clycl. Agroecosys. **52**(2), 225–248. doi:10.1023/a:1009740530221

Oenema, O., Witzke, H.P., Klimont, Z., Lesschen, J.P., Velthof, G.L. (2009) Integrated assessment of promising measures to decrease nitrogen losses from agriculture in EU-27. Agric. Ecosys. Environ. **133**(3–4), 280–288. doi:10.1016/j.agee.2009.04.025

Parton, W.J., Hartman, M., Ojima, D., Schimel, D. (1998) DAYCENT and its land surface submodel: description and testing. Global Planet. Change **19**(1–4), 35–48. doi:10.1016/s0921-8181(98)00040-x

Parton, W.J., Holland, E.A., Grosso, S.J.D., Hartman, M.D., Martin, R.E., Mosier, A.R., Ojima, D.S., Schimel, D.S. (2001) Generalized model for NO_x and N_2O emissions from soils. J. Geophys. Res.-Atmos. **106**(D15), 17403–17419. doi:10.1029/2001jd900101

Parton, W.J., Morgan, J.A., Wang, G.M., Del Grosso, S. (2007) Projected ecosystem impact of the Prairie Heating and CO_2 Enrichment experiment. New Phytol. **174**(4), 823–834. doi:10.1111/j.1469-8137.2007.02052.x

Parton, W.J., Mosier, A.R., Ojima, D.S., Valentine, D.W., Schimel, D.S., Weier, K., Kulmala, A.E. (1996) Generalized model for N_2 and N_2O production from nitrification and denitrification. Global Biogeochem. Cycles **10**(3), 401–412. doi:10.1029/96gb01455

Pattey, E., Edwards, G., Strachan, I.B., Desjardins, R.L., Kaharabata, S., Riddle, C.W. (2006) Towards standards for measuring greenhouse gas fluxes from agricultural fields using instrumented towers. Can. J. Soil Sci. **86**(3), 373–400

Paw, U.K.T., Baldocchi, D.D., Meyers, T.P., Wilson, K.B. (2000) Correction of eddy-covariance measurements incorporating both advective effects and density fluxes. Bound-Lay. Meteorol. **97**(3), 487–511

Pihlatie, M.K., Kiese, R., Bruggemann, N., Butterbach-Bahl, K., Kieloaho, A.J., Laurila, T., Lohila, A., Mammarella, I., Minkkinen, K., Penttila, T., Schonborn, J., Vesala, T. (2010) Greenhouse gas fluxes in a drained peatland forest during spring frost-thaw event. Biogeosciences **7**(5), 1715–1727. doi:10.5194/bg-7-1715-2010

Potter, C.S., Riley, R.H., Klooster, S.A. (1997) Simulation modeling of nitrogen trace gas emissions along an age gradient of tropical forest soils. Ecol. Model. **97**(3), 179–196. doi:10.1016/s0304-3800(96)01903-5

Prasertsak, P., Freney, J.R., Saffigna, P.G., Denmead, O.T., Prove, B.G. (2001) Fate of urea nitrogen applied to a banana crop in the wet tropics of Queensland. Nutr. Clycl. Agroecosys. **59**(1), 65–73. doi:10.1023/a:1009806826141

Priesack, E., Achatz, S., Stenger, R. (2001) Parametrisation of soil nitrogen transport models by use of laboratory and field data In: M.J., S., Liwang, M., S, H. (eds.) Modeling carbon and nitrogen dynamics for soil management. pp. 461–484. CRC Press, Boca Raton, USA

Prinn, R.G., Weiss, R.F., Fraser, P.J., Simmonds, P.G., Cunnold, D.M., Alyea, F.N., O'Doherty, S., Salameh, P., Miller, B.R., Huang, J., Wang, R.H.J., Hartley, D.E., Harth, C., Steele, L.P., Sturrock, G., Midgley, P.M., McCulloch, A. (2000) A history of chemically and radiatively important gases in air deduced from ALE/GAGE/AGAGE. J. Geophys. Res.-Atmos. **105**(D14), 17751–17792. doi:10.1029/2000jd900141

Prueger, J.H., Kustas, W.P. (2005) Aerodynamic methods for estimating turbulent fluxes. In: Hatfield, J.L., Baker, J.M. (eds.) Micrometeorology in agricultural systems. pp. 407–436. American Society of Agronomy, Madison, WI, USA

Riley, W.J., Matson, P.A. (2000) NLOSS: A mechanistic model of denitrified N_2O and N_2 evolution from soil. Soil Sci. **165**(3), 237–249. doi:10.1097/00010694-200003000-00006

Saggar, S., Andrew, R.M., Tate, K.R., Hedley, C.B., Rodda, N.J., Townsend, J.A. (2004) Modelling nitrous oxide emissions from dairy-grazed pastures. Nutr. Clycl. Agroecosys. **68**(3), 243–255

Saggar, S., Tate, K.R., Giltrap, D.L., Singh, J. (2008) Soil-atmosphere exchange of nitrous oxide and methane in New Zealand terrestrial ecosystems and their mitigation options: a review. Plant Soil **309**(1–2), 25–42. doi:10.1007/s11104-007-9421-3

Schmid, H.P. (1994) Source areas for scalars and scalar fluxes. Bound-Lay. Meteorol. **67**(3), 293–318

Scott, A., Crichton, I., Ball, B.C. (1999) Long-term monitoring of soil gas fluxes with closed chambers using automated and manual systems. J. Environ. Qual. **28**(5), 1637–1643

Skiba, U., Drewer, J., Tang, Y.S., van Dijk, N., Helfter, C., Nemitz, E., Famulari, D., Cape, J.N., Jones, S.K., Twigg, M., Pihlatie, M., Vesala, T., Larsen, K.S., Carter, M.S., Ambus, P., Ibrom, A., Beier, C., Hensen, A., Frumau, A., Erisman, J.W., Bruggemann, N., Gasche, R., Butterbach-Bahl, K., Neftel, A., Spirig, C., Horvath, L., Freibauer, A., Cellier, P., Laville, P., Loubet, B., Magliulo, E., Bertolini, T., Seufert, G., Andersson, M., Manca, G., Laurila, T.,

Aurela, M., Lohila, A., Zechmeister-Boltenstern, S., Kitzler, B., Schaufler, G., Siemens, J., Kindler, R., Flechard, C., Sutton, M.A. (2009) Biosphere-atmosphere exchange of reactive nitrogen and greenhouse gases at the NitroEurope core flux measurement sites: Measurement strategy and first data sets. Agric. Ecosys. Environ. **133**(3–4), 139–149. doi:10.1016/j.agee.2009.05.018

Skiba, U., Smith, K.A. (2000) The control of nitrous oxide emissions from agricultural and natural soils. Chemosphere: Global Change Sci. **2**(3–4), 379–386. doi:10.1016/s1465-9972(00)00016-7

Smith, K.A., Thomson, P.E., Clayton, H., McTaggart, I.P., Conen, F. (1998) Effects of temperature, water content and nitrogen fertilisation on emissions of nitrous oxide by soils. Atmos. Environ. **32**(19), 3301–3309. doi:10.1016/s1352-2310(97)00492-5

Smith, W.N., Desjardins, R.L., Grant, B., Li, C., Lemke, R., Rochette, P., Corre, M.D., Pennock, D. (2002) Testing the DNDC model using N_2O emissions at two experimental sites in Canada. Can. J. Soil Sci. **82**(3), 365–374

Stull, R.B. (1988) An introduction to boundary layer meteorology. Kluwer Academic Publishers, Dordrecht

Tate, K.R., Ross, D.J., Saggar, S., Hedley, C.B., Dando, J., Singh, B.K., Lambie, S.M. (2007) Methane uptake in soils from Pinus radiata plantations, a reverting shrubland and adjacent pastures: Effects of land-use change, and soil texture, water and mineral nitrogen. Soil Biol. Biochem. **39**(7), 1437–1449. doi:10.1016/j.soilbio.2007.01.005

Tate, K.R., Ross, D.J., Scott, N.A., Rodda, N.J., Townsend, J.A., Arnold, G.C. (2006) Post-harvest patterns of carbon dioxide production, methane uptake and nitrous oxide production in a Pinus radiata D. Don plantation. For. Ecol. Manage. **228**(1–3), 40–50. doi:10.1016/j.foreco.2006.02.023

Ussiri, D.A.N., Lal, R., Jarecki, M.K. (2009) Nitrous oxide and methane emissions from long-term tillage under a continuous corn cropping system in Ohio. Soil Till. Res. **104**(2), 247–255. doi:10.1016/j.still.2009.03.001

Van Oijen, M., Rougier, J., Smith, R. (2005) Bayesian calibration of process-based forest models: bridging the gap between models and data. Tree Physiol. **25**(7), 915–927

Velthof, G.L., Oudendag, D., Witzke, H.R., Asman, W.A.H., Klimont, Z., Oenema, O. (2009) Integrated Assessment of Nitrogen Losses from Agriculture in EU-27 using MITERRA-EUROPE. J. Environ. Qual. **38**(2), 402–417. doi:10.2134/jeq2008.0108

Volk, C.M., Elkins, J.W., Fahey, D.W., Dutton, G.S., Gilligan, J.M., Loewenstein, M., Podolske, J.R., Chan, K.R., Gunson, M.R. (1997) Evaluation of source gas lifetimes from stratospheric observations. J. Geophys. Res.-Atmos. **102**(D21), 25543–25564. doi:10.1029/97jd02215

von Arnold, K., Nilsson, M., Hanell, B., Weslien, P., Klemedtsson, L. (2005) Fluxes of CO_2, CH_4 and N_2O from drained organic soils in deciduous forests. Soil Biol. Biochem. **37**(6), 1059–1071. doi:10.1016/j.soilbio.2004.11.004

Wagner-Riddle, C., Thurtell, G.W., Edwards, G.C. (2005) Trace gas concentration measurements for micrometeorological flux quantification. In: Hatfield, J.L., Baker, J.M. (eds.) Micrometeorology in agricultural systems. pp. 321–343. American Society of Agronomy, Madison, WI, USA

Webb, R.A. (1972) Use of boundary line in the analysis of biological data. J. Hortic. Sci. **47**, 309–319

Werle, P. (1998) A review of recent advances in semiconductor laser based gas monitors. Spectrachim. Acta A **54**(2), 197–236

Werner, C., Butterbach-Bahl, K., Haas, E., Hickler, T., Kiese, R. (2007) A global inventory of N_2O emissions from tropical rainforest soils using a detailed biogeochemical model. Global Biogeochem. Cycles **21**(3), GB3010. doi:10.1029/2006gb002909

Chapter 5
Global Sources of Nitrous Oxide

Contents

Abstract Both natural and anthropogenic sources produce N_2O. The primary natural sources of N_2O are upland soils under natural vegetation, oceans, coastal waters, riparian zones, estuaries, and rivers. Riparian zones, rivers, estuaries, and coastal waters are also impacted by anthropogenic activities from agriculture. The anthropogenic source is associated with leaching and export from agricultural soils. Natural sources accounts for 64% of global N_2O emissions. Riparian zones have saturated conditions and microbially available C, which contribute to higher rates of production of N_2O than dry land soils. The oceans are another major source of natural of N_2O emissions to the atmosphere with production of N_2O primarily occurring in the water column. Agriculture accounts for 67–80% of anthropogenic N_2O emissions. The major anthropogenic sources from agriculture include organic and inorganic N added to as fertilizers, cultivation of legumes that fixes atmospheric N_2 biologically. Other anthropogenic sources include industrial processes, biomass burning, and fossil fuel combustion. The current estimated global N_2O emissions are estimated at 19.8 Tg N_2O-N year^{-1}, of which the anthropogenic

David Ussiri and R. Lal, *Soil Emission of Nitrous Oxide and its Mitigation*,
DOI 10.1007/978-94-007-5364-8_5,

emissions are 6.7 Tg N_2O-N year^{-1}. The prediction of future emissions depends on the changing human activities as well as on climate patterns that are shifting because of global climate change.

Keywords Indirect N_2O emissions • Top-down modeling • Bottom-up modeling • N_2O:N_2 ratios • Anthropogenic sources

Abbreviations

SOC	soil organic carbon WFPS water-filled pore spaces
NOAA-GMD	National Oceanic and Atmospheric Administration Global Monitoring Division
ESRL	Earth System Research Laboratory
GEIA	global Emissions Inventory Activity
AOU	apparent oxygen utilization
ALE	Atmospheric Lifetime Experiment
GAGE	Global Atmospheric Gases Experiment
EDGAR	Emissions Database Global Atmospheric Research

5.1 Introduction

Nitrous oxide (N_2O) can be produced from a wide variety of biological sources in soils and water by both natural and anthropogenic processes, some of which cannot be easily quantified. Natural sources of N_2O are mainly microbial processes of nitrification and denitrification in uncultivated soils, oceans, and other aquatic systems and wetlands. Primary anthropogenic sources of N_2O are agricultural soil management, especially the use of manure and synthetic N fertilizers, cultivation of N-fixing crops and forages, manure deposition by livestock in grasslands and/or grazed pastures, and land use change. Other anthropogenic sources include sewage treatment, combustion of fossil fuel, adipic acid and nitric acid production, and biomass burning (US EPA 2010). Human activities can drastically change N_2O emissions directly through activities listed above, and indirectly through atmospheric reactive N deposition, and changes in rainfall and temperature patterns because of global warming.

Some sources can be related to both natural and anthropogenic processes, such as riparian zones, rivers, estuaries, and continental shelves, which may be polluted by agricultural runoff and drainage, and forest and grassland fires which can be human initiated (e.g. land clearing) or by lightning ignition. To develop effective mitigation strategies, it is important to quantify both natural and anthropogenic emissions and to understand how human activities and projected climate change may influence the emissions from both sources. Varieties of approaches used to characterize N_2O emissions generally fall into two categories: bottom-up and top-down. Bottom-up approaches estimate N_2O emissions from small scale to larger scale by extrapolating actual measurements of fluxes to larger scales or developing a models based on processes controlling N_2O fluxes and then applying it to larger scale. Top-down, also known as inverse methods use atmospheric concentration measurements or atmospheric transport models to estimate N_2O emissions at the regional or global scale.

Global production of N_2O in the ecosystems is largely attributed to microbial processes (Chap. 3). Bacteria produce N_2O through nitrification and denitrification, which are the key processes within the natural N cycle (Chap. 2). The nitrification occurs under aerobic conditions while denitrification is the dominant process under anoxic conditions. In nitrification, bacteria oxidize nitrogenous compounds (NH_3/NH_4^+) through a 2-step process oxidation of ammonia to nitrite (NO_2^-), and oxidation of nitrite to nitrate (NO_3^-). This 2-step process produces N_2O as a byproduct or as an alternate product of ammonia oxidation. In denitrification, bacteria reduce oxidized inorganic forms of N. This process may form N_2O as an intermediate byproduct, or it may consume N_2O. Therefore, denitrification can be either a source or a sink of N_2O depending on environmental conditions. Both processes can occur simultaneously in soils, water column and sediments; and the production of N_2O depends on the balance between these two processes (Conrad 1996; Butterbach-Bahl et al. 2004; Bange 2008). The rate of both processes depends on aeration and the microsite availability of the substrates. In addition, various microbiological, chemical, and physical processes in the soil influence soil N_2O emission.

The emission of N_2O from soils and aquatic systems is the result of simultaneous occurrence of production and consumption processes (Conrad 1996; Bange 2008). Hence, an adequate understanding of both production and consumption of N_2O mechanisms in soils and aquatic systems is essential to quantifying the gaseous N_2O emissions. In addition to biological processes, chemical processes also contribute a small fraction of N_2O in neutral and acidic soils (Bremner 1997). These chemical reactions are similar to biological denitrification in that they lead to formation of N_2O, NO and N_2. These processes, usually described by the term chemodenitrification involve various chemical reactions of NO_2^- ions. Under acidic conditions, (soil pH-values <4.0) the physico-chemical process of chemodenitrification is an additional process of NO production in soils.

The total emissions of N_2O from all sources are currently estimated at 18.8°Tg N_2O-N year^{-1}, which represents an increase, primarily due to human activities since the pre-industrial era. Recently, based on top-down estimates Davidson (2009) updated the global N_2O emission to 19.8 Tg N_2O-N year^{-1}. The updated global emissions are based on improved quantitative estimates of the amounts of N_2O coming from the various sources, including manure N (2%) and fertilizer N (2.5%). Future N_2O emissions will largely depend on future agricultural production and practices and climate change policies implemented. The N fertilizer use and livestock production increases are projected at 50–100% and 50–150, respectively, for the twenty-first century (Bouwman et al. 2011; van Vuuren et al. 2011).

5.2 Natural N_2O Sources

5.2.1 *Soils Under Natural Vegetation*

5.2.1.1 Factors Influencing N_2O Emissions from Soils

N_2O production by nitrifying bacteria may arise from NH_3/NH_4^+ oxidation or dissimilatory NO_2^- reduction when O_2 supply is limited. Under some conditions, other processes such as chemodenitrification or heterotrophic nitrification (Bremner 1997) can contribute to the N_2O production. The NH_4^+ oxidizing bacteria require NH_4^+, CO_2 and O_2 availability for growth and multiplication. In well-drained soils, nitrification rate and size of ammonia oxidizing bacteria limit the availability of NH_3/NH_4^+. Nitrification declines in soils that remain wetter than field capacity for extended periods, i.e., several days (Schjonning et al. 2003). During denitrification, N_2O is an intermediate product of dissimilatory nitrate and/or nitrite reduction under anoxic conditions. Denitrification requires presence of nitrogen oxides (NO_3^-, NO_2^- NO, or N_2O) in soils. Denitrifying bacteria can also consume N_2O during denitrification process. Denitrifying bacteria are ubiquitous in all normal soil environments. However, the denitrifying enzymes in soils may limit their activity and denitrification process (Smith and Parsons 1985). Denitrification thrives under anaerobic conditions. In soils, it occurs under waterlogged conditions when redox potential falls below 400 mV, because O_2 is a potent denitrification inhibitor. However, strict anaerobiosis is not a necessary precondition for denitrification because denitrifiers can operate within a range of low O_2 concentrations (Tiedje 1988). Some denitrification can even occur aerobically (Robertson and Kuenen 1984). Available C plays a significant role in determining the magnitude and products of denitrification (Firestone and Davidson 1989).

Nitrification and denitrification processes are influenced by soil temperature, pH, moisture content, texture, and labile organic C. Temperature affects microbial growth and multiplication. Optimum temperature for microbial activity in soils is between 25 and 30°C. There is a strong evidence indicating nitrifying activity under typical winter soil temperatures (2–10°C), however (Cookson et al. 2002). Microbial activity is generally optimum at neutral to slight alkaline pH range. Microbial growth rate and activity is significantly reduced outside the optimum pH range. However, adaptation and diversity of nitrifying and denitrifying bacteria have enabled nitrification and denitrification to continue throughout most of naturally occurring pH ranges in soils.

Modeling approaches is one of the tools for identifying controls of N_2O fluxes from soils. Li et al. (1992a, b) used process-based DeNitrification DeComposition (DNDC) model to simulate the factors influencing N_2O emissions from agricultural soils and observed that rainfall patterns strongly influenced N_2O emissions. In addition, the model indicated that soluble C and nitrate availability could be limiting factors for N_2O evolution during denitrification (Li et al. 1992a). On the annual basis, temperature, precipitation, organic C, clay content, and soil pH have significant influence on denitrification and N_2O emissions (Li et al. 1992b). The responses of DNDC to changes of external parameters are consistent with field and experimental results reported in the literature (Li et al. 1992b). In addition, simulation of soil moisture content and soil temperature as an indicator of N_2O fluxes for terrestrial ecosystems was successfully predicted by a daily time-step version of the Century ecosystem model (DAYCENT, Parton et al. 1998). Model simulations have an advantage of integrating various factors regulating gas emissions, and predict the overall response based on field measurements. However, the reliability of model predictions depends on availability of reliable field data.

Statistical techniques can correlate field-measured fluxes with soil physico-chemical properties. Stehfest and Bowman (2006) summarized field measurements of N_2O emissions under natural vegetation cover worldwide. On the basis of 207 studies from natural soils reviewed, it was observed that the major factors controlling N_2O emissions under wide range of environmental conditions are soil organic C content (SOC), soil pH, soil bulk density, drainage, vegetation type, length of the measurement period, and frequency of sampling. The N_2O emissions increased with increasing SOC concentration due to increased availability of substrate for microbial community in soils (Stehfest and Bouwman 2006; Kanerva et al. 2007). Bulk density and drainage affect soil hydrological conditions and gaseous exchange. Poor drainage and high bulk density both limit gaseous O_2 diffusion (Stehfest and Bouwman 2006), indicating dependency of nitrification and denitrification on oxygen availability and, thus, on the extent of anaerobic zones in the soil profile (Li et al. 2000). An increase in soil bulk density decreases total pore volume and, thus, oxygen diffusion into the soil profile. This is notable following rainfall events, when in soils with high bulk density the anaerobic volume fraction increases more rapidly as compared to those with lower bulk density (Stange et al. 2000; Kiese et al. 2005). The N_2O emissions can be the highest from wettest soils (75% WFPS), and as much as 90% of the N_2O can be produced through denitrification (Khahil and Baggs 2005). N_2O emission decreases with increasing plant species diversity, and increases in the presence of legumes (Niklaus et al. 2006). Overall, the N_2O emissions decreased with increase in soil pH, with the lowest emissions for pH >7.3. In addition, the N_2O emissions from acidic soils exceeded those reported from alkaline soils reflecting the reported higher N_2O emissions from nitrification under acidic conditions (Martikainen and de Boer 1993). The sensitivity of N_2O emissions to changes in soil reaction is due to the pH effects on N_2O production by nitrification and denitrification (Li et al. 2000; Kesik et al. 2005). This strong sensitivity of microbial N_2O production even to small changes in pH has been demonstrated in studies involving pure cultures of heterotrophic nitrifiers (Khahil and Baggs 2005), but also in field and laboratory studies (Yamulki et al. 1997; Ellis et al. 1998; Stevens et al. 1997). For the vegetation type, the hierarchy of N_2O emissions is rainforest $>$ N affected coniferous/deciduous forest $>$ savannah/tropical dry forest. In general, several microbiological, chemical, and physical properties of soil that may influence N_2O emissions are distributed throughout natural soils worldwide. Based on the statistical analysis of published data (Stehfest and Bouwman 2006), general trends of N_2O

emissions can be summarized as follows: (1) emissions of N_2O from rainforests are significantly higher than those from grasslands, savanna, and tropical dry forests, (2) emissions from grasslands are significantly lower than those from deciduous forests (Stehfest and Bouwman 2006), and (3) high N availability and high soil moisture content make soils under tropical rain forest a major source of N_2O (Hirsch et al. 2006; Keller et al. 2005).

5.2.1.2 Estimates of N_2O from Natural Soils

Fourth assessment report of the Intergovernmental Panel on Climate Change (IPCC) estimated global emissions for soils under natural vegetation at 6.6 Tg N_2O-N year^{-1} with the uncertainty range of 3.3–9.0 Tg N_2O-N year^{-1} (Denman et al. 2007).This global budget was based on data for 1990s' provided by Bouwman et al. (2002). However, some other global budgets of N_2O emissions from soils under natural vegetation based on both bottom-up and top down modeling approaches have been established (Kroeze et al. 1999; Del Grosso et al. 2005; Galloway et al. 2004; Hirsch et al. 2006). Some microbiological, chemical, physical, and environmental parameters that determine N_2O emissions create complex interactions that make extrapolating global emissions budgets difficult and uncertain. Since the publication of IPCC fourth assessment report, number of new measurements from natural soils has increased, which has enabled some improvements in N_2O budgets, especially from soils of the tropics. Statistical analysis of 207 field measurement of soils under natural vegetation enabled developing a comprehensive statistical model for estimation of global N_2O emissions (Stehfest and Bouwman 2006; Table 5.1). Due to limited number of data and lack of data for many vegetation types, global emissions for soils under natural vegetation estimated based on this statistical model is highly uncertain.

Werner et al. (2007a) used the biogeochemical model – Forest-DNDC-tropica coupled with global GIS database with high resolution (0.25° × 0.25°) to provide the relevant model initialization and driving parameters. The model estimated global strength source of soils under tropical rainforests of Africa, Asia, Oceania/ Australia, Central and South America between 1991 and 2000. Based on simulated data for 10-year-period of 1999–2000, the global mean annual N_2O emission source strength of rainforests ecosystems was estimated at 1.2 kg N_2O-N year^{-1}. This amounts to emissions of about 1.34 Tg N_2O-N year^{-1} (0.88–2.37 Tg N_2O-N year^{-1}; Table 5.1). Detailed biogeochemical models provide useful methods for global N_2O emissions estimates. However, insufficient field measurements limit application of the models to other ecosystems (Werner et al. 2007a). The data indicates that tropical rainforest soils with estimated emissions plays significant role in the global N_2O emissions (Werner et al. 2007a). Kesik et al. (2005) used Forest-DNDC model to estimate N_2O emissions from European forest soils from 1990 to 2000 at 0.07–0.087 Tg N_2O-N year^{-1}.

Although advances have been made in collecting and analyzing emissions measurements and improvements and in characterizing sources and strengths of N_2O emissions of the tropical rain forest soils (Stehfest and Bouwman 2006;

Table 5.1 Global estimates of N_2O emissions from soils under natural vegetation

Vegetation type	Land area (ha × 10^6 or Mha)	Estimated fluxes (kg N_2O-N ha^{-1} $year^{-1}$)	Estimated total emissions (Tg N_2O-N $year^{-1}$)	Factors controlling emissions	Reference
Temperate forest	230	0.64	0.15	Vegetation type, Soil organic C, soil pH, soil texture, bulk density drainage.	Stehfest and Bouwman (2006)
Open tropical forest	1,598	0.21	0.33		
Closed tropical forest	854	1.37	1.17		
Grassland/ steppe	2,765	0.56	0.40		
Total	5,447		2.05		
Global rain forests by continent					
South America	602.6	1.11 ± 0.26	0.67 ± 0.15	Soil properties i.e., soil pH, bulk density	Werner et al. (2007a)
Africa	305.5	1.13 ± 0.28	0.34 ± 0.08		
Asia	143.2	1.80 ± 0.35	0.26 ± 0.06	Climate variability	
Central America	31.0	1.63 ± 0.35	0.05 ± 0.01		
Oceania/ Australia	10.4	1.09 ± 0.27	0.01 ± 0.003		
Total tropical rain forests	1092.7	1.22 ± 0.29	1.34 ± 0.32		

Werner et al. 2007b), there are not enough field measurements. Statistical models like that of Stehfest and Bouwman (2006) rely on incomplete coverage of global vegetation cover zones and have high uncertainty in the estimated N_2O emissions. In general, statistical models such as this are useful for site-specific estimates, and are not reliable for creating global N_2O budgets. Comprehensive statistical analysis of field measurement conducted by Stehfest and Bouwman (2006) revealed lack of relevant field data from other major ecoregions. For example, data from dry tropical forests, savanna, tundra, and temperate forest ecosystems less affected by N deposition are lacking. More field measurements are also required as a crucial step for reducing model uncertainties.

Overall, the uncertainty associated with estimates of N_2O budget from soils under natural vegetation remains significantly high. Detailed biogeochemical models such as that described by Werner et al. (2007a) provide useful methods for estimation of global N_2O emissions. However, there are insufficient field measurements, including soils and vegetation data for this type of modeling to be applied to other ecosystems. Major limitations for more comprehensive estimates for N_2O emissions from natural vegetation globally are: (a) insufficient field measurement data required for models parameterization, vegetation and soil characterization from dry tropics, savannah, temperate ecosystems not affected by N deposition and tundra regions, (b) insufficient long-term soil fluxes monitoring to improve understanding of complex interactions of emission factors, (c) incomplete coverage of global vegetation zones, and (d) insufficient understanding of complex interaction of soil microbial, chemical, physical and environmental factors that regulate N_2O fluxes. The complexity of these interactions is further compounded by climate change related to global warming. Increased number of field measurements and refined global models may improve emission scenarios and reduce the existing uncertainties in global N_2O budget.

5.2.2 *Aquatic Nitrous Oxide Sources*

Human alteration of fixed N in the biosphere, particularly through production of fertilizers and cultivation of N fixing crops has more than doubled the availability of reactive N in the biosphere (Robertson and Vitousek 2009). The increased N availability is producing unintended environmental consequences, including eutrophication of aquatic ecosystems and enhanced emissions of N_2O. Microbial conversion of agriculturally derived N in soils and aquatic ecosystems is the largest source of anthropogenic N_2O to the atmosphere (IPCC 2007). Both natural and anthropogenic emissions of N_2O from streams, rivers, estuaries, lakes, and riparian zones have received much less attention compared to agricultural soils, and remains a major source of uncertainty in the global N_2O budget. Microbial denitrification is the large source of N_2O emission in aquatic ecosystems. The proportion of denitrified NO_3^- that is converted to N_2O rather than N_2 partially controls how much N_2O is produced by aquatic systems (Codispoti 2010). However, few studies provide information on the N_2O yield in streams, rivers and other aquatic ecosystems, because of the difficulty of measuring N_2 and N_2O production from these ecosystems.

5.2.2.1 Description of the Sources

The aquatic ecosystems consist of marine and fresh water sources — including oceans, estuaries, rivers, and lakes. Earth's interconnected waters form a gradient from freshwater rivers and lakes, to estuaries where freshwater and saltwater mixes, through the shallow coastal ocean on the continental shelves, and finally deep-water open oceans.

Oceans have the surface area of about 361×10^6 km^2, covering about 71% of the Earth surface. Oceans can be categorized by water column depth and distance from the land into three zones: the open oceans, coastal waters, and near-shore/inshore areas (Capone 1991).

Near-Shore Waters

These waters are shallow and generally well mixed down to the bottom sediments. In addition, biogeochemical processes in the water column and underlying sediments are often coupled. Near-shore waters can be highly influenced by the adjacent land, and commonly enriched by nutrient inputs from land runoff. As the distance from offshore increases, also the depth of water column and salinity increase, and eventually there is a tendency to develop two layers in water column separated by sharp thermal gradient, often-called thermoclines. The two layered waters — shallow and deep waters — are only loosely linked. In addition, the direct interaction of terrestrial land decreases with distance from the coast. Oceans play an important role in controlling Earth's temperature, and serve as both source and sinks for atmospheric trace gases.

Open Oceans

In the open ocean, a thin surface layer — typically 100–200 m depth, called mixed layer and containing euphotic zone, is separated from a much deeper layer containing aphotic zone (average about 4000 m deep; but differ depending on the ocean) by a permanent thermocline (Capone 1991). Sediments represent a third distinct zone of open sea in terms of N_2O budget. The euphotic zone of the open ocean is an area of active photoautotrophy and N assimilation by planktonic algae due to sufficient light penetration, which permits growth of phytoplankton. It is estimated that about 5,000 Tg year^{-1} of reactive N is assimilated and recycled globally by primary producers within euphotic zone (Capone 1991). Therefore, euphotic zone is characterized by low concentrations of inorganic N species (NO_3^-, NH_4^+). The concentration of NO_3^--N increases sharply below the thermocline and remains greater throughout aphotic zone (Capone 1991). This pattern is due to heterotrophic degradation and release of organically bound N (i.e. ammonification) and its subsequent oxidation (nitrification). In aquatic environments, both nitrification and denitrification are light inhibited processes. Therefore, a low production of N_2O occurs in the euphotic zone (Ward et al. 1989).

Coastal Shelves

Continental shelves occupy a much smaller area than the open oceans, but their emissions per unit area are much larger. Continental shelves are defined according to depth or proximity to the land. They represent an area of about 10% of that covered by ocean. Similar to the open ocean, coastal waters are a three-layered system consisting of euphotic water column, aphotic water column and underlying sediments (Capone 1991). However, the depth to the sediment is considerably less than that of the open ocean, and more materials produced in the euphotic zone may reach sediments. Likewise, the euphotic zone is substantially shallower (30–60 m);

algal densities and productivity per unit area is higher (Capone 1991). In addition, waters over the continental shelves have higher concentrations of inorganic N than the open ocean. Coastal ocean bordering the continents can be sub-divided into two systems of importance to N_2O fluxes: (1) continental shelves, and (2) upwelling regions.

The continental shelves, which constitute about 7% of the total ocean area, are commonly defined by relatively flat regions adjacent to the continents. These are the sites of great primary productivity, biological activity, and sedimentation than the deep ocean. These are commonly impacted by the anthropogenic activities of the surrounding terrestrial areas. Continental shelves receive riverine and estuarine drainage. Sometimes they are highly impacted by anthropogenic activities. About 8–40% of N inputs to rivers and estuaries are exported to coastal shelves (Boyer et al. 2006; Schaefer and Alber 2007). Continental shelves also receive N inputs through soil erosion from adjacent terrestrial lands and atmospheric deposition.

Upwelling refers to pattern of coastal and open water oceanic circulation created by persistent winds blowing across the open ocean surface. As winds move surface waters, deeper waters that are richer in nutrients and can support phytoplankton growth replace them. The upwelling regions are therefore, the areas where the nutrient–rich deep waters are brought to the surface. These regions support enhanced primary productivity, greater populations of fish and other consumers, making them some of the most productive fisheries in the world. Coastal upwelling regions which are commonly found along Eastern ocean boundaries comprise a relatively small (<10%) subset of total coastal area (Bange et al. 1996). Two factors suggest that coastal upwelling regions are particularly strong sources of N_2O to the atmosphere: (1) oceanic N_2O is produced during microbial respiration only in subsurface waters and sediments, coastal upwelling provide a pathway for venting N_2O that otherwise might remain trapped below the surface, and (2) due to surface nutrient recharge, coastal upwelling regions are characterized by high rates of primary production, which leads to large fluxes of sinking particulate OM that create a favorable conditions for enhanced subsurface microbial N_2O production (Codispoti et al. 1992). Regions of coastal upwelling include coast of Peru, Chile, the Arabian Sea, western part of South Africa, eastern New Zealand, and coast of California.

Estuaries and Rivers

Estuarine and near-shore waters represent only a small fraction of total ocean environments. However, these waters are important in terms of N_2O budget because of anthropogenic impacts. Compared to coastal waters or open ocean waters, the highest concentrations of inorganic N sources have been reported in these waters (Sharp 1983). The shallowness of near-shore and estuarine areas result in essentially two-layered systems of well-mixed water and sediments. Estuaries typically have higher organic inputs and nutrient levels than the oceans. These are relatively shallow and experience constant mixing, which can transport gases produced in the sediments into the surface where these can be released to the atmosphere. Much of the productivity of the near shore and estuarine areas depends on inputs from the anthropogenic source and rivers rather than the N recycled in the water column. The anaerobic conditions commonly observed in the sediments of these areas make them important sites for denitrification.

Rivers consists of fresh waters, while estuaries and coastal waters form a transition zone between fresh water and salt water. Traditionally, biogeochemists have regarded rivers as conduits leading from land to oceans. However, current research has shown that important biogeochemical reactions occur in river channels, which transform reactive N and organic matter during downstream transport.

Lakes

Lakes are naturally formed permanent water bodies contained on a land surface. They may consist of freshwater or salty waters. There are about 300 million natural lakes and ponds globally, with a total area of about 4,200,000 km^2, or about 3% of global surface area (Downing et al. 2006). About 90% of these lakes are <0.01 km^2 and nearly 43% of the total area is of lakes <1 km^2. Impoundments, including engineered dams cover about 250,000 km^2 (Downing et al. 2006). There is the potential for overlap between natural lakes, wetlands and rivers in some occasions, since: (1) wetland inventories include shallow lakes as wetlands, (2) the shoreline zones of many lakes has emergent vegetation and could be considered wetlands, (3) many wetlands contains small ponds, and in scaling the wetland area these ponds are included as wetlands, and (4) in tropics where remote sensing is used for quantifying areas of different land classes, a class of wetland habitat include open waters (lakes and channels), flooded forests, and aquatic macrophytes. Some fraction of this class could fall under rivers, lakes, as well as wetlands (Melack et al. 2004).

5.2.2.2 Nitrogen Transformations Under Aquatic Environments

Like in terrestrial ecosystems, N is often the primary nutrient limiting biological productivity in marine ecosystems (Howarth 1988). The N_2 gas is by far the most abundant form of N in marine environment, but nutritionally it is available only to certain prokaryotic organisms (diazotrophic organisms) with the capability of biological N-fixation. Organically bound N is the next largest pool of N, but as N_2, it is not directly available for biological productivity. Nitrates and NH_4^+, which are the reactive forms of N occur in relatively small quantities (Capone 1991). The most quantitatively important component of the marine N cycle on global basis with direct bearing on the fluxes of N_2O is the flux between inorganic (NH_4^+ and NO_3^-) and organic (living biomass and detritus) forms of N. Much of the N added

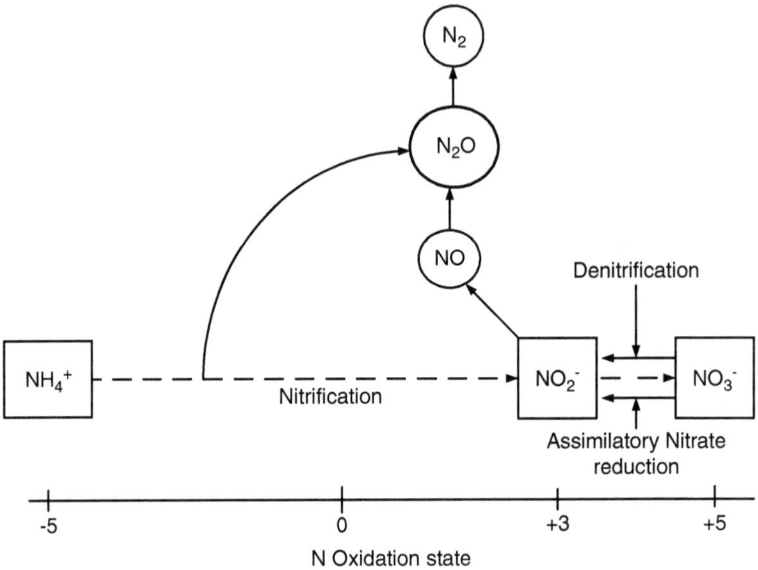

Fig. 5.1 Nitrogen transformation in marine/aquatic ecosystems involved in N_2O cycling (Modified from Capone 1991)

to the land surfaces and recycled through human and food and energy production is carried away from its original site of introduction through volatilization, leaching, and run-off. The anthropogenic changes in the N cycle have implication for aquatic emissions of N_2O because the rates of nitrification and denitrification depend on the reactive N. In general, the anthropogenic influence on N_2O emissions from the ocean decreases with increasing distance offshore, and this change affects the approaches used to estimate the fluxes from the ocean. The main biological pathways for marine N_2O cycling are through nitrification and denitrification processes. Nitrification and denitrification are processes in aquatic ecosystems which are often closely coupled, where denitrification may be limited by the rate of production of NO_3^- through nitrification (Capone 1991; Suntharalingam and Sarmiento 2000; Bange 2008).

Controversy surrounds the issue of the dominant N_2O formation pathway; water column nitrification during subsurface oxidation of organic matter is widely accepted as the main source for the majority of the open oceanic N_2O emissions. This assumption, however, has been challenged based on isotopic measurements and the evidence that N_2O cycle displays complex behavior in some oceanic environments. The main biological pathways for marine N_2O cycling are nitrification and denitrification processes (Fig. 5.1). The N_2O is produced from denitrification as an obligatory intermediate product in the multistep reduction of NO_3^- to N_2 gas, which is mediated by bacteria. However, not all strains of bacteria are capable of carrying out this process to completion and produce N_2. Many bacteria lack the necessary enzymes to metabolize all of the intermediate products. Therefore, these organisms participate in subset of steps. For example, organisms lacking N_2O–reductase enzyme cause the process to terminate at N_2O (Suntharalingam and Sarmiento 2000). Denitrification is traditionally viewed as a bacterial anaerobic respiration process, and nitrate reduction stops in the presence of O_2 because O_2 is preferentially consumed (Capone 1991).

Biological N_2O production from nitrification occurs from the oxidation of NH_4^+ to NO_2^- and NO_3^-. This process is mediated by nitrifying bacteria in water column and sediments. It is hypothesized that an intermediate compound hydroxylamine (NH_2OH) is produced (Bremner 1997). Thus, N_2O is evolved during the oxidation of NH_2OH to NO_2^-. Bacteria involved in nitrification (*Nitrobacter*, *Nitrosomonas*) require O_2 for their activity. Production of N_2O from nitrification constitutes a small fraction of marine N_2O sources (Butler et al. 1987). Nitrification is an aerobic process. However, under low O_2 conditions, N_2O yields are enhanced (de Bie et al. 2002). Using a 3-dimensional numerical modeling and analysis of surface and deep N_2O data to study the processes involved in N_2O source, Suntharalingam and Sarmiento (2000) estimated that the majority of the oceanic N_2O source (over 75%) is produced in the upper water column (above 600 m) and effluxes directly to the atmosphere in the latitude band of formation. Based on isotopic measurements of N_2O in the oligotrophic (i.e., nutrient poor, therefore low productivity) subtropical North Pacific, Dore et al. (1998) estimated that nitrification in the lower euphotic and upper aphotic zones (i.e., zones with no light – between 100 and 300 m) may supply 70–90% of the oceanic N_2O source.

The widespread and seasonally variable super saturations of N_2O gas suggest that oceans are an important atmospheric N_2O source (Schmidt 1978; Nevison et al. 1995). The N_2O production in the surface layer is small because O_2 inhibits denitrification, and light inhibits nitrifiers. Therefore, subsurface production and then transport to the surface is the major process of N_2O production in open ocean. Concentrations closer to the atmospheric equilibrium or even small under saturation have been reported in the ocean gyres — i.e., the stationary regions of open ocean with circular water movement, which tend to isolate these regions from the rest of the ocean (Nevison et al. 2003; Suntharalingam and Sarmiento 2000).

5.2.2.3 Estimates of N_2O from Aquatic Ecosystems

Earlier estimates of N_2O emissions from aquatic ecosystems were based on two principal approaches: (1) gas transfer calculations, which combine measurements of the near-surface concentrations and wind speed through gas transfer coefficients, and (2) organic matter decomposition using the N_2O fraction of nitrification and denitrification or NO_3^-. The N_2O production and consumption at low O_2 levels in aquatic systems is complex. The N_2O concentrations may be low in anoxic waters due to denitrifying organisms using the gas for respiration, but may be higher in suboxic waters or on the periphery of anoxic areas, suggesting the production in these environments. For the ecosystems impacted by anthropogenic N sources, well-developed information on anthropogenic N export to rivers has been utilized to calculate cycling through nitrification and denitrification and estimates of N_2O production. These two approaches are considered "bottom-up" techniques, which use combination of measurements, inventories, and emission coefficients to calculate N_2O fluxes and estimate global emissions.

Recent advancements in modeling techniques and expanding database on atmospheric N_2O mixing ratios have led to development of inverse modeling techniques to be utilized in estimating N_2O emissions. This third approach involves constraining the distribution and magnitude of fluxes and then estimating the actual fluxes contributing to the overall budget. This is a "top-down" budgeting approach. Most of emission estimates from aquatic environments are based on these three approaches, which provide N_2O emission estimates without measuring actual fluxes using chambers or eddy covariance techniques.

Open Ocean

It is hypothesized that terrestrial and open ocean N budgets are disconnected (Galloway et al. 2004). This hypothesis is based on the observation that most, if not all, of the anthropogenic N input to aquatic systems is consumed and recycled in rivers, estuaries and near-shore waters. Atmospheric N deposition decreases with the increase in distance from continental regions. Therefore, open ocean is not impacted by anthropogenic N sources, and all of the open oceanic N_2O emissions are considered to be of natural origin. A range of techniques have been used for estimating the global oceanic emissions, including: (1) measurement of dissolved N_2O concentrations in surface waters of the ocean and estimating N_2O gas-transfer based on wind speeds, (2) estimates of N_2O yield from nitrification and denitrification, (3) linking of N_2O production to O_2 consumption and developing relationship between dissolved N_2O and O_2, and (4) inverse modeling (i.e., chemical transport models).

Samplings from the surface waters in 1970s and 1980s demonstrated that that surface waters of the world ocean were slightly supersaturated with respect to atmospheric levels of N_2O (i.e., N_2O concentration in the water is ~4% higher than its equilibrium concentration with the air above ocean waters), and that open ocean is a net global N_2O source (Nevison et al. 1995). Much higher super saturation occurs in regions that experience coastal upwelling and affected by subsurface O_2 efficiency (Bange et al. 2001; Nevison et al. 2004a). Using 60,286 global expedition data and transfer calculations, Nevison et al. (1995) estimated global open ocean emissions at 4.0 Tg N_2O-N year^{-1}, with a range of 1.2–6.8 Tg N_2O-N year^{-1}. The large range observed for calculated emissions was attributed to: (a) uncertainties in gas-transfer coefficients, (b) seasonally biased data collection, and (c) incomplete geographic and temporal sampling coverage. Similar N_2O emissions were also calculated based on nitrification and denitrification (i.e., 4.0–5.8 Tg N year^{-1}; Table 5.2, Capone 1991; Najjar et al. 1992). Using the Global Emissions Inventory Activity (GEIA) data and Nevison et al. (1995) estimates, Bouwman et al. (1995) concluded that 45% of open ocean emissions

Table 5.2 Estimated post 1990 global N_2O emissions from open ocean

Emissions range (average in brackets) (Tg N_2O-N $year^{-1}$)	Data source	Comments
1.3	Prinn et al. (1990)	Early inverse modeling
5.8	Capone (1991)	Estimated from N_2O released during nitrification and denitrification
1.4–2.6 (2)	Houghton et al. (1992)	IPCC first assessment estimate
4	Najjar et al. (1992)	Estimated from N_2O released during nitrification
1.2–6.8 (4)	Nevison et al. (1995)	Based on surface measurements, includes upwelling
1.0–5.0 (3)	Houghton et al. (1996)	IPCC second assessment estimate
3.6	Bouwman et al. (1995)	Based on compilation of cruise data
3.5	Seitzinger et al. (2000)	Based on cruise data with continental sub-grids subtracted
2.7–8.0 (3.85)	Suntharalingam and Sarmiento (2000)	Modeled organic matter remineralization and N_2O – apparent O_2 utilization relationship
0.9–5.7 (3.0/3.6)	IPCC (2001)	IPCC third assessment estimate
3.8–7.8 (5.8)	Nevison et al. (2003)	N_2O production rate estimate; may be 1–3 Tg N_2O-N $year^{-1}$ less due to N_2O consumption by denitrifiers
3.7	Nevison et al. (2005)	Modeled using atmospheric N_2O seasonal cycle and a production as a function of O_2 consumption
4.5–6.5 (5.7)	Hirsch et al. (2006)	Inverse modeling using the global emissions inventory activity (GEIA). Includes upwelling and continental shelves emissions
1.8–5.8 (3.8)	IPCC (2007)	IPCC fourth assessment estimate
0.9–1.7 (1.2)	Rhee et al. (2009)	
0.9–8.0		Range of all data
1.0–4.0 (3.2)		Estimated open ocean only

occurs in the Southern Hemisphere (30–90°S). This estimation was based on high levels of super saturation and high winds observed in the Southern Hemisphere. It was also noted that some areas in subtropical gyres and in the North Atlantic might change seasonally from weak sinks in winter to weak sources in summer (Bouwman et al. 1995).

The relationship between N_2O production and O_2 consumption (i.e., apparent O_2 utilization (AOU)) in combination with global biogeochemical model was used to calculate global N_2O emissions from the open ocean. (AOU is the difference between measured dissolved O_2 concentration and the expected O_2 concentration when the water is at atmospheric equilibrium saturation. It is the measure of sum of the biological activity that has occurred in the water column since its last surface exposure). Based on observations of water column profiles, production of N_2O occurs at the intermediate water depths in the open ocean and nitrification is considered the dominant process. This contention is based on the observed inverse correlation between N_2O and O_2 and strong positive correlation between N_2O and NO_3^- concentrations. Water column N_2O emissions were estimated at 3.85 Tg N_2O-N year^{-1} and ranged from 2.7 to 8.0 Tg N_2O-N year^{-1} (Table 5.2; Suntharalingam and Sarmiento 2000). However, variability in the oceanic N_2O saturation and AOU relationship that may be the result of differences in the N_2O–NO_3^- yield due to the sensitivity of nitrifiers to O_2 may limit the applicability of this approach and its widespread use, and may make the estimated fluxes unreliable (Nevison et al. 2003). It has been shown that the correlation of N_2O super saturation with AOU is a poor predictor of the N_2O yield per mole O_2 consumed because the relationship is strongly affected by N_2O mixing (Nevison et al. 2003). For example, by calculating N_2O fluxes as a function of O_2 and depth, Nevison et al. (2003) estimated the oceanic N_2O emissions at 5.8 Tg N_2O-N year^{-1}. They suggested that the calculated fluxes might have been overestimated by 1.0–3.0 Tg N_2O-N year^{-1} due to denitrifier consumption of N_2O under low-O_2 environments. Contrary to earlier estimates of N_2O emissions dominance by southern latitudes, Nevison et al. (2003) observed that 70% of N_2O emissions occur in the tropics — between 30°N and 30°S of the Equator.

The apparent disagreement in the relative importance of southern latitude oceans was further examined using a 10-year atmospheric N_2O record at Cape Grim, Tasmania in Australia. Seasonal atmospheric cycles were used to calculate the transfer of N_2O from the subsurface to the atmosphere (Nevison et al. 2005). Comparison between atmospheric N_2O observations and those predicted by an ocean biogeochemical model coupled with an atmospheric transport model suggested that the Southern Ocean N_2O emissions were ~0.9 Tg N_2O-N year^{-1}, although the large corrections applied to the atmospheric N_2O data may have created significant uncertainty (Nevison et al. 2005). It was further observed that oceanic N_2O saturation is not an annual constant, as was considered earlier (Nevison et al. 1995; Suntharalingam and Sarmiento 2000). It exhibits strong seasonal variation due to thermal effects during the summer, and a larger mixing effect during the winter as the depth of the surface mixed layer deepens and deeper N_2O-enriched waters are mixed in (Nevison et al. 2005). Since the majority of surface N_2O measurements from ocean have been made during the summer, annual mean fluxes may have been overestimated, a factor that may also help to explain some of the apparent discrepancies in Southern Ocean emissions (Nevison et al. 2005).

Hirsch et al. (2006) applied inverse modeling techniques to atmospheric N_2O air samples measurements from 48 sites under Cooperative Global Air Sampling Network to estimate fluxes from open ocean between 1998 and 2001. Due to large geographical scales in which the technique was operated, data were discriminated only between "land" and "ocean," where the ocean class included all oceanic environments (i.e., open ocean, continental shelves, and upwelling areas). In addition, rivers and estuaries were included in land category because of its relatively small-scale features. Based on this technique, the oceanic fluxes are presented in three semispherical super regions: (i) southern oceans (90–30°S) at

0.4 Tg N_2O-N year^{-1}, ranging from 0 to 0.8 Tg N_2O-N year^{-1}, (ii) tropical oceans (30°S to 30°N) at 3.6 Tg N_2O-N year^{-1}, ranging from 2.6 to 4.1 Tg N_2O-N year^{-1}, and (iii) northern oceans (30–90°N) at 1.8 Tg N_2O-N year^{-1} ranging from 1.7 to 2.1 Tg N_2O-N year^{-1} (Hirsch et al. 2006). Globally, oceans are estimated to contribute an average of 5.7 Tg N_2O-N year^{-1} and ranging from 4.5 to 6.5 Tg N_2O-N year^{-1}. Consistent with other reports (Nevison et al. 2003, 2005), fluxes from Southern Ocean were lower than the earlier estimates (Bouwman et al. 1995). The data also indicated that tropical emissions are higher than GEIA estimates. This is attributed to sparse sampling in the tropical region (15°S to 15°N) which creates some difficulties in separating land and oceanic fluxes (Hirsch et al. 2006).

Accumulation of high-quality data recently has made it feasible to examine variability on shorter time scales and provide better identification of sources and sinks of N_2O. Comparison of N_2O data and a suite of chlorofluorocarbons tracers with atmospheric transport model coupled with biogeochemical models generated monthly mean oceanic N_2O fluxes, which estimated oceanic fluxes at 3.0 Tg N_2O-N year^{-1} (Nevison et al. 2007). However, uncertainties in stratospheric mixing tended to overwhelm source variability on shorter time scales (Nevison et al. 2007). Systematic study of the seasonal cycle and its longitudinal variation in N_2O data collected by National Oceanic and Atmospheric Administration-Global Monitoring Division (NOAA-GMD) revealed that peak to trough amplitude of seasonal cycles increases systematically from 0.29 ppb at the South Pole to 1.15 ppb at Alert, Nunavut, Canada (Jiang et al. 2007). The month of the seasonal minimum concentration also changes systematically, from April at the South Pole to September at Alert, Canada. The Northern Hemisphere seasonal cycle is largely driven by stratospheric influences, which control the loss of N_2O, whereas in the Southern Hemisphere, surface sources such as the oceans have greater impact on the data (Jiang et al. 2007). The data indicated that over a 3-year period from 2000 to 2002, the mean N_2O concentration in the Northern Hemisphere was 0.73 ppb greater than that in the Southern Hemisphere. This difference requires that sources in the Northern Hemisphere be 4.7 Tg N_2O-N year^{-1} higher than those in the Southern Hemisphere. The inter-hemispheric difference of 4.7 Tg N_2O-N year^{-1} observed by Jiang et al. (2007) is significantly greater than the inter-hemispheric difference of 2.657 Tg N_2O-N year^{-1} derived from GEIA inventory. However, the value is lower than the inter-hemispheric source difference of 5.5–11.9 Tg N_2O-N year^{-1} and averaged at 8.8 Tg N_2O-N year^{-1} estimated by Hirsch et al. (2006). A greater inter-hemispheric difference may be, in part, due to the increase in Northern Hemisphere anthropogenic emissions since 1995 as well as the downward revision in Southern Ocean fluxes (Jiang et al. 2007).

Despite the variety of approaches used to calculate the magnitude, the mean open ocean N_2O fluxes remain in the range of 3–6 Tg N year^{-1} (Table 5.2). A conservative estimated flux of 3.0 Tg N_2O-N year^{-1}, from open ocean means that 16% of the global total of 18.8 Tg N_2O-N year^{-1} is contributed by open ocean. Inclusion of the uncertainties for individual estimates increases this range further. The most significant revision of oceanic flux estimates in recent years after the IPCC fourth assessment report (IPCC 2007) is the reduction of the relative importance of the Southern Ocean, which once thought to make up nearly half of the oceanic source, but now is believed to contribute 7–15% of open ocean emissions.

Continental Shelves

Continental shelves, receiving waters from rivers and estuaries drainage, are strongly impacted by anthropogenic activities. Global estimates of N export to coastal regions are considered 25% of the watershed input (Galloway et al. 2004), and range from 5 to 42% of total input (Boyer et al. 2006; Schaefer and Alber 2007). The current global anthropogenic reactive N input to terrestrial ecosystem is estimated at 150 Tg N year^{-1}, of which, 34.5 Tg N year^{-1} is exported to rivers, estuaries, and coastal waters (Schlesinger 2009). The export from watersheds is affected by number of variables, including N loading (Boyer et al. 2006), temperature (Schaefer and Alber 2007), precipitation and discharge (Howarth et al. 2006). The N export from watersheds is generally lower in warmer watersheds than in cooler watersheds (Schaefer and Alber 2007). Export from estuaries depends on water residence time, extent of intertidal area and sediment O_2 consumption.

Terrestrial N exports are not among the major drivers of N_2O production on the continental shelves (Seitzinger et al. 2000). Therefore, different approaches than those used for rivers and estuaries are used to estimate the N_2O fluxes. Seitzinger and Giblin (1996) used a model correlating shelf denitrification with phytoplankton primary productivity and sediment O_2 uptake, and predicted denitrification rates varying with latitudes. Average denitrification rates were lowest in the high latitude shelves, intermediate in the tropical shelves and highest in the mid-latitude shelves (Seitzinger and Giblin 1996). This model can be used to estimate global denitrification rates and then apply the constant N_2O:N_2 fractions for river, estuaries, and continental shelves (Seitzinger and Kroeze 1998). The emission of N_2O is calculated as function of the area sediment and water surface multiplied though constant rates.

The estimated N_2O emissions from continental shelves range from 0.64 to 6.43 Tg N_2O-N year^{-1} (Table 5.3). Natural and anthropogenic sources are separated by assuming that rivers and estuaries supply 40–50% of N inputs and 50% of it is of anthropogenic origin. The remaining is assumed to originate from open ocean, and considered natural in its origin (Seitzinger and Kroeze 1998). Based on these calculations, natural sources make up 75% of the total N_2O emissions. Rates of nitrification are estimated by assuming an average shelf depth and constant nitrification rates from 25 to 50 m deep and below 50 m depth. Other estimates of continental shelf N_2O fluxes are based on surface dissolved N_2O concentration and gas-transfer calculations (Bange et al. 1996; Rhee et al. 2009), nitrification/denitrification yields, and ocean reservoir estimates (Capone 1991). Bange et al. (1996) estimated N_2O emissions from coastal waters and marginal seas ranging from 1.74 to 2.71 Tg N_2O-N year^{-1} (Table 5.3). All of these estimates have

Table 5.3 Estimated global N_2O emissions from coastal ocean and upwelling zones

Habitat	Flux range (Tg N_2O-N year^{-1})	Reference	Comment
Coast and upwelling	4.7	Capone (1991)	Estimated from N_2O released during nitrification and denitrification
Coastal ocean and marginal seas	1.74–2.71 (2.22)	Bange et al. (1996)	Mixed anthropogenic and natural sources
Continental shelves	0.64–6.43 (0.64)	Seitzinger and Kroeze (1998), Seitzinger et al (2000)	Ranges based on use of low and high emission factors
Continental shelf	0.06–1.0 (0.8)	Nevison et al. (2004b)	
Coastal ocean	0.23–0.51 (0.37)	Rhee et al. (2009)	
Upwelling	0.2–0.32 (0.26)	Bange et al. (1996)	
Upwelling	1.0	Suntharalingam et al. (2000)	Model for low O_2 zone
Upwelling	0.04–0.34	Nevison et al. (2004b)	
Upwelling	0.002–0.003 (0.003)	Rhee et al. (2009)	Extrapolation from smaller area
Upwelling and hypoxia zones	1.48–3.11	Naqvi et al. (2010)	Includes anoxic coastal zones impacted by terrestrial anthropogenic activities
Coastal ocean	0.37–4.52 (1.87)		Range of reported values
Upwelling	0.0003–1.0 (0.37)		Range of reported value (Naqvi et al. (2010) values not included)

high uncertainty (Table 5.3). Estimates by (Rhee et al. 2009), modeled based on samples from Atlantic Basin, indicate significantly lower N_2O in coastal surface waters (0.23–0.51 Tg N_2O-N year^{-1}) than those reported earlier (Bange et al. 1996; Seitzinger and Kroeze 1998; Seitzinger et al. 2000).

Areas of upwelling are also regions of enhanced primary productivity, resulting in higher fluxes of organic matter sinking into mid-waters, where it stimulates nitrification, as a result, depleting O_2 levels and creating favorable conditions for denitrifiers. The intensity of O_2 deficiency is controlled by (i) local circulation/hydrography especially stratification, (ii) upwelling intensity, (iii) shelf width, and (iv) primary production. These factors determine the balance between O_2 consumption and supply in the subsurface waters (Naqvi et al. 2010).

The N_2O is generally produced in the subsurface waters due to light inhibition. Therefore, upwelling regions provide a rapid outlet of N_2O to the surface (i.e. degassing). Low O_2 concentration enhances the production of N_2O, and N_2O has no sink in oxygenated waters. However, for the reasons not yet known, the extent to which the loss of O_2 promotes N_2O production varies from one upwelling system to another (Naqvi et al. 2010). Regional estimates suggest that upwelling contributes significantly to the global ocean N_2O sources (Law and Owens 1990; Lal and Patra 1998; Bange et al. 2001; Wittke et al. 2010). However, upwelling events takes place on a timescale of hours to days, and the episodic nature of the events make the associated air-sea gas exchange difficult to measure using traditional oceanographic methods. The associated gas exchange is also difficult to model because conventional air-sea flux calculations rely on seasonal or annual mean climatological data to estimate surface N_2O, which may not resolve upwelling signals. Global estimates of upwelling have been made using several approaches: (1) surface saturation levels and wind speed: Bange et al. (1996) estimated that upwelling areas covering only 0.2% of world oceans contributed 3% of global aquatic N_2O fluxes, (2) modeling approach and N_2O as tracer in an ocean general circulation model (Suntharalingam et al. 2000). On the contrary, some researchers have argued that the use of upwelling atmospheric flux modeling of traditional gas transfer flux calculations are likely to underestimate coastal upwelling sources by a factor of 3–8 from poorly sampled regions such as South America and Africa (Schmidt 1982; Nevison et al. 2004b). Using observation-based calculations Nevison et al. (2004b) estimated global N_2O emission from continental shelves and coastal upwelling at 0.06–1.0 and 0.04–0.06 Tg N_2O-N year^{-1}, respectively (Table 5.3).

Comparison between regional and global upwelling fluxes reveals disagreements, even though uncertainties are large. Much of the disagreements may be due to sparse global databases of N_2O emissions from these small areas and mismatches in spatial and temporal resolutions, due to episodic nature of upwelling. Based on review of global data on N_2O from major coastal O_2-deffiicient zones formed both naturally and as a result of anthropogenic activities Naqvi et al. (2010) estimated N_2O emissions from global upwelling zones at 1.48–3.11 Tg N_2O-N year^{-1} (Table 5.3). Based on substantial contribution from upwelling to global N_2O emission from areas experiencing O_2 deficiency and the sensitivity of N_2O cycling in aquatic systems to minor changes in the ambient O_2 distribution, any major alterations in O_2 distribution in coastal as well as offshore waters will have a large impact on the N_2O budget.

Estuaries and Rivers

Rivers are not large areas globally, but they are active sites for aquatic productivity and biogeochemical cycling. Despite 2 to 20-fold increase in riverine N concentration in numerous regions (Vitousek et al. 1997), N_2O fluxes from aquatic sources remains poorly characterized. In comparison with earlier estimates of river N_2O fluxes based on surface dissolved concentrations and gas transfer calculations (Bànge et al. 1996), more recent global estimates are based on estimation of N inputs and calculating the fraction released as N_2O (Seitzinger and Kroeze 1998; Kroeze and Seitzinger 1998a; Kroeze et al. 2005). The basic assumption for these calculations is that N_2O production is related to rates of nitrification and denitrification, and these processes are related to N input. Estimates of N_2O emissions from rivers and estuarine ecosystems have been made based on inventories (Seitzinger and Kroeze 1998; Kroeze and Seitzinger 1998a; Seitzinger et al. 2000), and modeling techniques

(Boyer et al. 2006; Schaefer and Alber 2007). In both approaches, N_2O emission from nitrification and denitrification is assumed to be a constant fraction of the N_2 produced — 0.3% for N_2O:N_2 for denitrification in rivers and estuaries with low N loadings (0–10 kg N ha^{-1} $year^{-1}$) and 3% for those with high loadings (<10 kg N ha^{-1} $year^{-1}$) (Seitzinger and Kroeze 1998). The same constants are used for N_2O yield from nitrification, mainly because of the scarcity of data. However, simple relationships between rates of N cycling and rates of N_2O emissions such as N_2O: NO_3^- or N_2O:N_2 does not hold across range of rivers (Beaulieu et al. 2008) due to variable contributions of nitrification and denitrification to N_2O emissions, even though the current methods to estimate global N_2O emissions from rivers are based on such assumptions.

The N_2O yield for streams and rivers has been studied mostly in lentic freshwater and marine ecosystems where N_2O:N ratio generally ranges between 0.1 and 1.0%, although yield as high as 6% have been observed (Seitzinger and Kroeze 1998). These yields are very low compared to observations from soils (Stevens et al. 1998; Schlesinger 2009; Fig. 5.2), due to relatively lower O_2 availability in the sediments of lakes and estuaries. Laboratory data and yield studies estimated ratios of N_2O:N_2 ranging between 0.1 and 0.3% with values as high as 6% for highly polluted sediments, but the database is small (Seitzinger et al. 2000). Overall, this variable is poorly known although this is a critical value upon which suites of biogeochemical variables are based (Schlesinger et al. 2006). The importance of this ratio is demonstrated by comparing global riverine and estuarine estimates made by Capone (1991) by assuming different values for nitrification (0.3%), and nitrification linked to denitrification in low O_2 waters and sediments (1%) and denitrification (5%). Even though estimates of aquatic N reservoirs are made differently, Capone (1991) estimate of near shore/estuarine fluxes of 0.74 Tg N_2O-N $year^{-1}$ are more than thrice higher than those by Seitzinger and Kroeze group—0.22 Tg N_2O-N $year^{-1}$. Denitrification is one of the most challenging biogeochemical processes, and the need for information about this process to address local, regional, and global Nr problems cannot be overemphasized. There are no conclusively robust models suitable for regulatory assessments of denitrification processes in aquatic environments. Dissolved O_2 in headwater streams is commonly near atmospheric equilibrium and the benthic algal biofilms can produce O_2 at the sediment-water interface, which creates redox gradient compatible to those in moist soils. Therefore, streams and rivers may have high and often variable N_2O yields similar to those in soils (Stevens et al. 1998). It is also known that natural aquatic inputs vary geographically, and this variation must be duly considered when anthropogenic influence to N_2O emission needs to be isolated. Part of the NO_3^- input to rivers is derived from groundwater upwelling into headwater stream, and drainage network of streams and rivers provide high ratios of bioreactive benthic surface area to water volume (Alexander et al. 2007).

Use of model suggests a different breakdown, and it incorporates regional differences in natural N sources (Boyer et al. 2006). Such models estimate export to rivers based on calculations of biological N-fixation in forests and other natural vegetation with minor inputs from N fixation by lightning. The modeling data

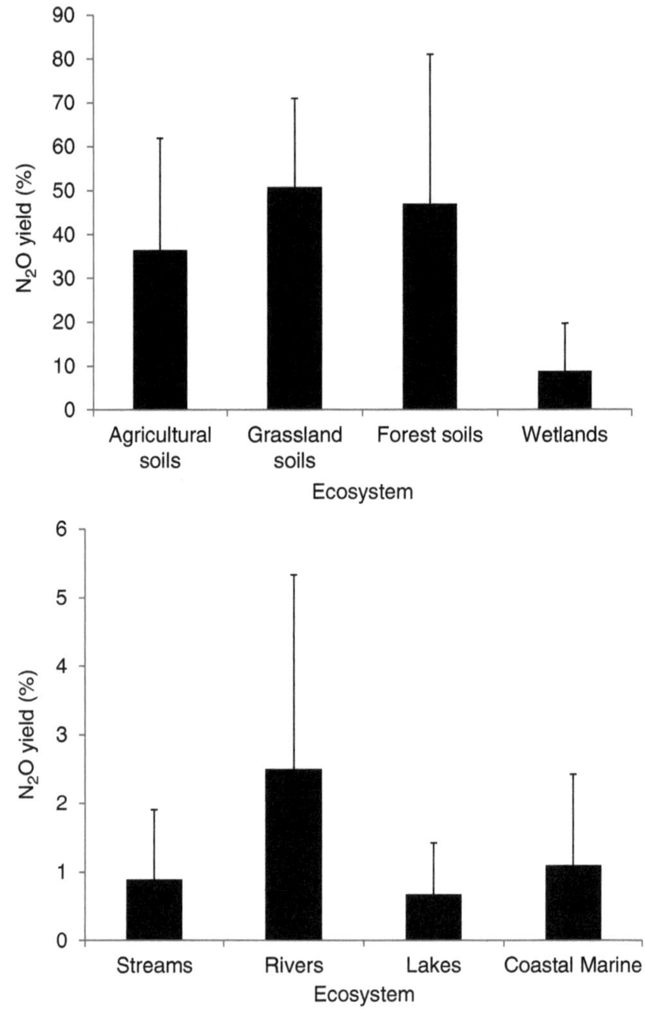

Fig. 5.2 Denitrification N_2O yields (percentage of denitrified N released as N_2O). Error bars are standard deviations. Denitrification data for soils were derived from Schlesinger 2009, denitrification data for aquatic ecosystem were derived from Beaulieu et al. 2011; Seitzinger 1988

indicate that natural N sources dominate N inputs in Africa, Latin America, and Oceania, while the anthropogenic sources are dominant N sources in Europe, North America, former Soviet Union region, and Asia. These estimates indicate approximately equal N exports from natural and anthropogenic sources, and that anthropogenic sources are concentrated in northern hemisphere. Since food and energy production patterns are changing, this geographical distribution is likely to be altered (Boyer et al. 2006). In addition, Asian sources with high N inputs and growing population and economic base will drive the overall global N budgets in the future. Using spatially explicit global model for predicting dissolved inorganic

Table 5.4 Estimated global riverine and estuarine N_2O emissions

Emission sources	N_2O emissions (Tg N_2O-N year^{-1})	Reference	Comment
Estuaries, nearshore	0.74	Capone (1991)	Estimated based on N_2O released during nitrification and denitrification
Estuaries	0.25	Kroeze et al. (2005)	Based on N export model, consisting of both natural and anthropogenic
Estuaries/ nearshore	2.34–3.63 (3.0)	Bange et al. (1996)	Consists of both natural and anthropogenic
Rivers	0.19–1.87 (1.1)	Seitzinger and Kroeze (1998), Seitzinger et al. (2000)	Calculated based on low and high emission factors
Rivers	1.26	Kroeze et al. (2005)	Based on N export model and consists of both natural and anthropogenic
Rivers	0.68	Beaulieu et al. (2011)	Estimated based on global river network model
Estuaries	0.03–0.45 (0.24)		
Rivers	0.08–0.10 (0.09)		

nitrogen (DIN) by rivers to coastal waters, Dumont et al. (2005) estimated an export of 25 Tg N year^{-1}, of which, 16 Tg N year^{-1} (64%) were from anthropogenic sources.

Beaulieu et al. (2011) used ^{15}N tracers on 72 headwater streams draining multiple land use types and demonstrated that denitrification produces N_2O at rates that increase with stream water NO_3^- concentrations. The rates of N_2O denitrified varied by land use with highest rates observed in high NO_3^- urban streams and lowest in pristine with little land conversion in their watersheds (Beaulieu et al. 2011). Stream N_2O emissions rates were also related to watershed land use, with highest emission rates from urban streams, intermediate rates in agricultural streams, and lowest rates in pristine streams. Stream NO_3^- predicted N_2O emissions only when NO_3^--N concentration exceeded 95 μg L^{-1}, but below this concentration N_2O emissions were low and unrelated to NO_3^- concentrations (Beaulieu et al. 2011). Based on the global river network model, the percentage of DIN converted to N_2O by direct denitrification of water column NO_3^- in river network across the globe ranges from 0 to 0.9%. The percentage of N input converted to N_2O decreases with increasing N loading because denitrification becomes less efficient as NO_3^- sink at higher NO_3^- concentrations (Mulholland et al. 2008). The N_2O emissions from global rivers estimated using global river network model was 0.68 Tg N_2O-N year^{-1}, and most of emission is anthropogenic (Beaulieu et al. 2011), (Table 5.4). Other model estimates indicate natural N_2O fluxes from rivers at 0.09 Tg N_2O-N year^{-1} and estuaries at 0.24 Tg N_2O-N year^{-1}

Table 5.5 Mean N_2O-N fluxes from measured from lakes

Habitat	Location	Fluxes µg N_2O-N m^{-2} h^{-1}	Reference
Freshwater lakes & reservoirs	Finland	−2.1 to 8.2	McMahon and Dennehy (1999)
Freshwater lakes-oligomesotrophic	Switzerland	0.3–24	Silvan et al. (2002)
Freshwater lakes eutrophic and aerated	Finland	9.5–19	Huttunen et al. (2002)

for a total of 0.33 Tg N_2O-N $year^{-1}$ (Table 5.4). The use more sophisticated models suggest that the IPCC Fourth Assessment classification (IPCC 2007) of all riverine and estuarine N_2O as wholly anthropogenic may be oversimplification.

5.2.2.4 Lakes

Available information on production and consumption processes of N_2O in lakes is rather limited. Generally, high rates of N_2O fluxes are associated with denitrification than nitrification (Firestone and Davidson 1989). Denitrification occurs when three conditions are satisfied: (1) nitrate is available, (2) O_2 concentration is reduced, and (3) electron donors are available (Seitzinger et al. 2006). Annual rates of denitrification are highest in lakes, followed by rivers, coastal ecosystems, and estuaries. Further, across ecosystems, rates are correlated with nitrate levels, which can explain 70% of the observed variability, and inversely correlated with O_2 levels (Pina-Ochoa and Alvarez-Cobelas 2006). Stadmark and Leonardson (2005), measuring N_2O emissions from shallow (<1.5 m) ponds constructed in southern Sweden for nitrogen retention/removal, indicated that N_2O fluxes from the ponds were below their detection limit, but that N_2O production in sediment and water incubations increased with increasing nitrate concentration. Liikanen et al. (2002) found a similar effect in incubations of shallow (4 m) and deep (8 m) sediments from a eutrophic or nutrient-rich lake in Finland. There are no global estimates of N_2O emissions from freshwater lakes. Only few N_2O emission measurements from lakes have been reported (Table 5.5). Fluxes of 0.01–0.84 µM m^2 h^{-1}, which is equivalent to 0.02–1.6 kg N_2O-N ha^{-1} $year^{-1}$ when integrated over a 250-day open-water year were measured from Swiss lakes of varying size and nutrient status (Mengis et al. 1997). In addition, N_2O emissions ranging from 0.003 to 0.025 kg N_2O-N ha were reported from boreal lakes during spring and summer and negligible during winter (Huttunen et al. 2003). Applying a mean annual flux of 0.01–0.1 kg N_2O-N ha^{-1} across 4.2 million km^2 of lakes and ponds (Downing et al. 2006) would generate an annual flux of 0.004–0.04 Tg N_2O-N $year^{-1}$. This represents much less than 1% of the global annual N_2O flux from natural sources. Assessment of the importance of lakes for N_2O fluxes requires more data on the N_2O fluxes and their regulation in different climatic settings neither of which is currently available.

5.2.3 *Wetlands*

Wetlands are ecosystems where water saturation is the main factor and it controls soil development and the plant and animal species that are present (Cowardin et al. 1979). Water saturation conditions may occur constantly, seasonally or sporadically, but it is the dominant factor which distinguishes wetlands from other ecosystems. While being distinctly different from both terrestrial and aquatic environments, wetlands depend on them both. Water may originate from precipitation, subsurface flow from ground water, or surface flow from surrounding watershed, or water body such as ocean, river, or lake. Water characteristics normally determines wetland type (e.g., saline, fresh, low or high nutrient, active water flow or stagnant). Globally, wetlands occur in all climatic regions and on every continent except Antarctica. Major factors that control production and emission of N_2O emission in wetlands include organic inputs and water level, which determine the balance between aerobic and anaerobic soil environments. The global area of wetlands is estimated at 5.2×10^6 km^2, or about 5% of Earth's surface (Matthews and Fung 1987; Prigent et al. 2007).

Emission of N_2O from wetland is influenced by process-level and ecosystem level controls. Process-level controls include soil fertility, available N (NO_3^-, NO_2^-, and NH_4^+) and O_2 (Groffman 1991). At the ecosystem level, the dominant control on N_2O emissions are the presence of water and position of the water table. Denitrification process dominates in wet soils (with more than 60% water-filled pore spaces (WFPS)), producing greater fraction of N_2O. However, water saturated soils (more than 80% WFPS) causes slow gaseous diffusion and enhance N_2O consumption. For saturated wetlands soils, therefore, N_2 becomes the dominant gas released. Under moderately wet conditions, some fraction of N_2O may diffuse to the atmosphere before consumption. In addition, it may remain in the soil and be released if soils dry (Davidson et al. 1993). Koschorreck (2005) reported high rates of N_2O flux during transition between flooded anaerobic soils and dry oxic conditions for Amazon floodplain lake. It has been demonstrated that N dynamics in seasonal wetlands may undergo annual transitions between nitrification- and denitrification-domination (Koschorreck and Darwich 2003).

Only few global estimates of N_2O emissions released from wetlands exist (Middelburg et al. 1995; Krithika et al. 2008; Allen et al. 2007). Modeling of global N_2O based on the input of organic matter, soil fertility, moisture status, temperature, and O_2 level suggests that wetlands are minor contributors to global N_2O emissions (Bouwman et al. 1993). Based on extrapolation from the measured fluxes of mangrove, the calculated global emissions from mangroves ranges from 0.004 to 0.17 Tg N_2O-N year^{-1} (Corredor et al. 1999; Allen et al. 2007). Further, coastal mangrove system release 0.076 Tg N_2O-N year^{-1}and it is concluded that these systems makes only minor contributions to global N_2O fluxes (Barnes et al. 2006). Measurement of N_2O uptake from soils suggested that low availability of NO_3^- and conditions in soils that slow diffusion, such as the water-saturation of wetlands may promote N_2O consumption (Chapuis-Lardy et al. 2007).

Fluxes of N_2O from wetlands largely remain unknown. Measurements of N_2O fluxes sampled to date indicate that fluxes are relatively low. However, seasonal changes in water level that characterize many tropical wetlands suggest that there may be a period of emission, since experimental work examining emissions over wetting and drying cycles suggests that N_2O fluxes may be enhanced by changes in moisture regime. There is a need for improved understanding of the global extent, importance of wetlands as both source and sink for N_2O emission.

5.2.4 Riparian Zones

Riparian zone soils are vegetated ecosystems along water body through which energy, materials and water pass. Because they form the interface between terrestrial and aquatic soils, they generally contain characteristics of both dry and water-saturated soils. They are important for their capacity to remove nitrates from subsurface agricultural runoff. Riparian zone soils are often permanently wet and rich in organic matter. Soil moisture content varies spatially with gradients from the hill slope down toward the water body (Hefting et al. 2006). The saturated conditions and microbially available C contribute to higher rates of production of N_2O than the associated dry upland soils.

Denitrification in riparian buffers is major a process for nitrate removal due to permanently wet conditions (Groffman et al. 1992; Pinay et al. 1993; Verchot et al. 1997). Denitrification in riparian zone generates N_2O as an intermediate product, which can be substantial. The riparian zones buffering agricultural fields tend to have high nitrate loading (Groffman et al. 1998; Hefting et al. 2003). Research over a range of riparian zones in Europe reveals accumulation of ammonium under wet soil conditions with ground water levels less than 10 cm below the soil surface, indicating that nitrification is limited under these conditions (Hefting et al. 2004). High nitrogen availability, coupled with high moisture content makes tropical soils especially likely to emit N_2O (Bouwman et al. 2002; Hirsch et al. 2006; Keller et al. 2005). However, data for N_2O emissions from the riparian zones are lacking.

5.3 Anthropogenic Sources

5.3.1 Description of the Sources

Human activities have doubled the rate of N fixation, which is reflected by increased concentrations of reactive N in terrestrial and aquatic ecosystems. Important N substances that are emitted to the atmosphere mostly because of human activities are nitric oxide and nitrogen dioxide (NO and NO_2, respectively, commonly denoted as NO_x), ammonia (NH_3) and N_2O. Many different sources are responsible for these emissions, but fossil fuel combustion, biomass burning and microbial nitrification and denitrification from agricultural soils are the most important anthropogenic source of NO_x and N_2O. The NH_3 originates from various sources, including volatilization from animal waste and synthetic fertilizers, industrial processes, biomass burning, agricultural soils, human excreta, and fossil fuel burning (Olivier et al. 1998). Once emitted, NO_x and NH_3 in the troposphere are involved in various chemical reactions, and their reaction products are transported in the atmosphere and deposited elsewhere. The NO_x are short-lived gases with an atmospheric lifetime of 1–10 days, while NH_3 has a lifetime of only few days. In addition to their role on atmospheric chemistry (see Chap. 1), deposition of their reactive products to terrestrial and aquatic ecosystems, especially in temperate forest ecosystems of Europe and North America (Butterbach-Bahl et al. 1997; van Dijk and Duyzer 1999; Zechmeister-Boltenstern et al. 2002), have contributed to enhanced emissions of N_2O.

Anthropogenic sources which contribute emissions of N_2O directly include: agriculture — synthetic fertilizers and animal manure use, increased production of biological N fixing crops and drainage and cultivation of organic soils (Galloway et al. 1995), industrial processes – adipic acid and nitric acid production, and fossil fuel burning. Estimates of global N_2O emissions based on top-down model suggest that 3–5% of reactive N input to arable soils is ultimately emitted by one pathway or another as N_2O (Crutzen et al. 2008). Emissions of N_2O from forest soils have also increased in recent decades and may probably continue to increase in the future due to anthropogenic perturbations of global N cycle (Galloway et al. 2004). Atmospheric N deposition to forest ecosystems are strongly correlated with N_2O emissions (Butterbach-Bahl et al. 1998; Zechmeister-Boltenstern et al. 2002).Therefore, high rates of atmospheric N deposition to many forest ecosystems in Europe, North America and Asia (Bowden et al. 1991) may increase N_2O emissions.

5.3.2 Agriculture

Direct agricultural sources of N_2O emissions include inorganic and organic forms of N added to soils as fertilizers, manures, and composts. Some of the inorganic N added to soils as fertilizers undergoes microbial nitrification and denitrification processes in soils and aquatic ecosystems, releasing N_2O to the atmosphere. Addition of nitrogenous fertilizers to agricultural fields results in increased levels of NH_4^+ and NO_3^- in the soil, which serve as substrates for the nitrification and denitrification (Chap. 3). The N applied to agricultural soils may also be lost from the field through surface erosion or leaching, which continues recycling in the soil-water-air system and eventually denitrified and released to atmosphere as N_2O. Other sources of N_2O fluxes include biological N fixation, manures in animal housing and storage, urine and feces deposited onto soils during animal grazing, wastewater treatment, and enhanced mineralization of SOM due to soil disturbance, hence increase mineral N, which stimulate the nitrification and denitrification processes.

5.3.2.1 Factors Influencing N_2O Emission from Agriculture

Agriculture is a major source of anthropogenic N_2O emissions (Smith and Conen 2004; Del Grosso et al. 2005; Smith 2005). Based on the statistical analysis of published measurements of N_2O emission from agricultural fields and natural vegetation, statistical models have been developed for estimating N_2O production and identifying the factors that moderate N_2O production worldwide. The N application rate, crop type, fertilizer type, SOC content, soil pH, and texture are among important parameters influencing N_2O emissions from agricultural fields (Bouwman 1996; Stehfest and Bouwman 2006). Synthesizing the published information from 1008 emission measurements for agricultural fields worldwide, Stehfest and Bouwman (2006) concluded that the N_2O emissions was significantly influenced by N application rate, crop type, soil organic C content, soil pH and texture. Studies of the impact of anthropogenic N inputs to terrestrial ecosystems on N_2O emission with focus on agricultural soils indicate that about 1.25% (0.25–2.25%) of the fertilizer N is lost from the soil as N_2O in both short and long-term basis (Bouwman 1996; Nevison et al. 1996). The production, consumption, and transport of N_2O fluxes in an agricultural soil are influenced by changes in soil structural quality and water content associated with tillage and compaction. Soil structure, soil water content and compaction determines the diffusion rate of N_2O. In most agricultural soils biogenic formation of N_2O is enhanced by increase in available mineral N, which increases nitrification and denitrification rates. Addition of mineral N fertilizers therefore, results in extra N_2O formation. The increase may arise from fertilizer N itself or enhanced mineralization of SOM due to agricultural activities. Emission of N_2O from agricultural soils is also enhanced by application of organic materials with high N content such as animal manures.

5.3.2.2 Estimates of N_2O from Agriculture

Agricultural N_2O emissions have been estimated based on three main approaches: (1) emission factor based calculations, (2) regression analyses, and (3) process-based models. Emission factor estimation assumes that a fixed proportion of the applied N is emitted as N_2O ("emission factor"), and calculates emission as ratio of applied N. The emission factor is developed from number of observations, and it represents an average value over all soil types, climate conditions, and management practices. The approach is used to quantify N_2O emissions at a relatively high aggregation levels such as national scale, rather than individual source. The IPCC guideline for national greenhouse gas inventories (IPCC 1996, 2006) is the example of emission factor based approach. It was developed with the main objective of assisting countries in reporting their greenhouse gas emissions to Climate Secretariat of U.N. Its strength is the simplicity. Generally the data needed to calculate the emissions of N_2O are easily obtainable (i.e., N fertilizer use, arable land area). The major weaknesses of the approach are: (i) large uncertainties associated with estimated emissions, due to sensitivity of N_2O emissions to soil types, climate conditions and land management factors, and (ii) the variability of emissions in time and space are not considered in this approach. Using the IPCC emission factors and FAO database for animals and fertilizer use, Mosier et al. (1998) estimated anthropogenic emissions for the year 1989. The direct emissions from agricultural soils was estimated at 2.1 Tg N_2O-N year^{-1}, direct emissions from animal production at 2.1 Tg N_2O-N year^{-1} and indirect emissions associated with agricultural input at 2.1 Tg N_2O-N year^{-1} for the annual total from agriculture of 6.3 Tg N_2O-Nyear^{-1} (Mosier et al. 1998; Table 5.6). According to IPCC methodology, about 1% of fertilizer N added to soils is directly emitted to the atmosphere as N_2O and another 1% is emitted indirectly from downwind and downstream ecosystems, which receive forms of N that are transported away from agricultural fields by wind, water, and commerce.

Regression analysis is based on the fact that agricultural emissions are controlled by a range of environmental factors, soil type and characteristics, and farm practices (Freibauer and Kaltschmitt 2003). Generally, these cannot all be investigated in detail. Regression analysis enables estimates to be made where the available data are not sufficiently detailed for the process-based modeling. This kind of analysis is widely used to predict emissions and to identify the most significant factors that control emissions. It can also be used to define inter-correlations between emissions (Smith et al. 2008). The main weakness of regression models is the tendency to oversimplify the reality, and it may lead to significant errors. Regression model for estimating global agricultural N_2O emission was developed based on 1,008 N_2O measurements made globally (Stehfest and Bouwman 2006). Based on this model, cropland emits 3.34 Tg N_2O-N year^{-1} and managed grasslands an additional 0.8 Tg N_2O-N year^{-1} (Table 5.6)

Process based models are increasingly being recognized as alternatives to emission factor and regression models. Process-based models parameterize the interlinked biogenic and abiogenic processes based on their current understanding (Li 2000). Many process-based models exist and differ in complexity. Detailed process-based models generally require a large amount of data to describe the state of the environment. Their results are generally in good agreement with the observations (e.g., Giltrap et al. 2010). However, the required input data for many models are often not available for the case study. In addition, knowledge about the biogenic and biogeochemical processes controlling the N_2O emissions from soils is often limited. Therefore, simple process based models, which require less input data are sometimes preferred. Inverse modeling has been used to estimate the global N_2O emissions using 10 years of Atmospheric Lifetime Experiment (ALE) and Global Atmospheric Gases Experiment (GAGE) from 1978 to 1988 (Prinn et al. 1990). Assuming the atmospheric lifetime of 166 years for N_2O, global N_2O emissions are estimated at 20.5 $\pm$ 1.0 Tg N_2O-N year^{-1} (Prinn et al. 1990). Stratospheric measurements have increased the understanding of atmospheric chemistry of N_2O, its atmospheric lifetime is now estimated at 122 $\pm$ 24 years, which will adjust these estimates downwards (Ehhalt et al. 2001). Global N_2O

Table 5.6 Global emissions of N_2O from anthropogenic sources

Source	Range	Mean	Comments
	Tg N_2O-N year^{-1}		
Direct agricultural soils			Estimated anthropogenic emissions for the year 1989 calculated based on IPCC (1996) revised emission factors (Mosier et al. 1998, IPCC 1992)
Synthetic fertilizers	0.18–1.6	0.9	
Animal waste	0.12–1.1	0.6	
Biological N fixation	0.02–0.2	0.1	
Crop residues	0.07–0.7	0.4	
Cultivated histosols	0.02–0.2	0.1	
Total	0.4–3.8	2.1	
Animal production			
Animal waste management	0.6–3.1	2.1	
Indirect emission from agriculture			
Sewage	0.04–2.6	0.2	
Atmospheric deposition	0.06–0.6	0.3	
N leaching and runoff	0.13–7.7	1.6	
Total	0.3–11.9	2.1	
Total anthropogenic	1.2–17.9	6.3	
Fertilizer use	0.3–2.3	1.0	Global emission inventories for the year 1990 estimated based on Emission Database for Global Research (EDGAR) (Olivier et al. 1998)
Animal waste	0.0–2.0	1.0	
Biomass burning	0.4–1.0	0.7	
Fossil fuel	0.1–0.5	0.2	
Industrial processes	0.1–0.5	0.3	
Total anthropogenic	0.9–6.3	3.2	
Agriculture	–	6.2	Global emissions for the year 1994 estimated by simple atmospheric model using IPCC (1996) revised emission factors as an input (Kroeze et al. 1999)
Energy	–	0.9	
Biomass burning	–	0.6	
Industrial processes	–	0.3	
Total anthropogenic	–	8.0	
Cropland		3.35	Statistical model developed based on 1,008 published field measurements (Stehfest and Bouwman 2006)
Managed grassland		0.81	
Agriculture	1.7–4.8	2.8	Top-down and bottom-up estimates of various sources, current burden divided atmospheric lifetime, verified by inverse modeling (Denman et al. 2007)
Rivers, estuaries, & coastal zones	0.5–2.9	1.7	
Fossil fuel & industrial processes	0.2–1.8	0.7	
Biomass & biofuel	0.2–1.0	0.7	
Atmospheric deposition	0.3–0.9	0.6	
Sewage (human excreta)	0.1–0.3	0.2	
Total		6.7	
Emissions from fertilizers	1.5–2.4	2.2	Combination of top-down and bottom up analysis with modified emission factors from fertilizer and manure and livestock management (Davidson 2009)
Manure and livestock	2.2–3.3	2.8	
Industrial and transport		0.8	
Biomass burning		0.5	
Total anthropogenic		6.3	
Estimated global budget		19.9	

Table 5.7 The 1998–2001 N_2O flux estimates based on inverse modeling for land and sea by zones units are Tg N_2O-N year^{-1} (based on Hirsch et al. 2006)

Ecosystem	Region	GEIA Data	Model estimate	Comments
Terrestrial	North Land	3.6 (1.4)	2.3–3.5	Calculated based on 122 years atmospheric lifetime
	Tropical Land	3.8 (2.9)	5.7–8.0	
	South Land	2.0 (1.9)	1.6–2.6	
Total		9.4 (3.7)	10.7–12.7	
Marine	North Ocean	1.2 (1.0)	1.7–2.1	
	Tropical Ocean	0.9 (3.3)	2.6–4.1	
	South Ocean	1.7 (2.0)	−0.1 to 0.8	
Total		3.8 (3.8)	5.4–6.5	
Global total	Global Total	13.2 (5.5)	17.4 ± 1.4	

emissions estimated based on inverse modeling using 1998–2001 data suggested a global N_2O emission total of 17.4 ± 1.4 Tg N_2O-N year^{-1} (Hirsch et al. 2006; Table 5.7).

Global N_2O inventories with 1 × 1° grid were compiled from fertilized arable land, and annual N_2O emissions have been calculated for four broad longitudinal zones covering the globe. The emission factors are assumed for each zone as follows: 30°S to 90°S – 0.02, 0° to 30°S –0.1, 0° to 30°N – 0.38 and 30°N to 90°N – 0.50. Based on N fertilizer consumption of 80 Tg N for 1990, fertilized arable land estimated at 1,435 Mha contributes 1.8 Tg N_2O-N year^{-1} (Bouwman et al. 1995). Using Emissions Database for Global Atmospheric Research (EDGAR) inventories with 1 × 1° resolution, which calculates emission of gases on the basis of activity, emission factors, and other variables by country or gridded maps, Olivier et al. (1998) estimated N_2O emissions from different sources for 1990 (Table 5.6). Fertilizer use on arable land and animal excreta contributed 0.3–2.3 and 0–2.0 Tg N_2O-N year^{-1}, respectively, out of 0.9–6.3 Tg N_2O-N year^{-1} anthropogenic emissions for year 1990 (Table 5.6; Olivier et al. 1998). Using combination of top-down analyses of historic atmospheric accumulation of N_2O constrained by bottom-up historical estimates of industrial and transportation sources to derive a time course estimated anthropogenic biological sources of N_2O, Davidson (2009) estimated that about 2% of annual manure N production and 2.5% of fertilizer N production are converted to N_2O annually. Emissions from fertilizers is estimated at 2.2 Tg N_2O-N, ranging from 1.5 to 2.4 Tg N_2O-N manure and livestock management system contributed 2.8 Tg N_2O-N, ranging from 2.2 to 3.3 Tg N_2O-N, while biomass burning and transportation/industrial processes contributed 0.5 and 0.8 Tg N_2O-N in 2005 (Davidson 2009; Table 5.6).

The global N_2O budget is restricted by strength of the sink and the atmospheric N_2O increase. National Oceanic and Atmospheric Administration/Earth System Research Laboratory (NOAA/ESRL) monitoring indicate that atmospheric N_2O concentration has increased at a relatively constant rate over the two decades ending in 2010 at 0.73 ± 0.06 ppb per year (http://www.esrl.noaa.gov/gmd/aggi), which is equal to ~3.5 Tg N_2O-N year^{-1} (Hirsch et al. 2006). The increase is largely attributed to anthropogenic activity. Isotopic measurements made on polar firn air suggest that the anthropogenic N_2O is dominated by agricultural emissions (Rockmann et al. 2003). The N_2O emissions from agriculture and fertilizer are increasing markedly, especially since 1970s, with global increase in fertilizer use (Prinn et al. 1990; Hirsch et al. 2006). During the last two centuries the world population has grown from about 1 billion in 1800s to about 7 billion in 2011 (The World Fact Book; online at http//www.cia.gov). This growth has demanded large increase in agricultural production to feed the growing population. In addition, there has also been a qualitative shift towards animal-based food products. The growth in human population and associated need to increase production of food for human and for feeding animals has led to a massive increase in the worldwide production of synthetic N fertilizers. Globally, the amount of N fertilizers used was about 10 Tg N $year^{-1}$ in the 1950s and about 80 Tg N $year^{-1}$ in 2000 (Zhu et al. 2005). Further increase in fertilizer N is projected in the future. Much of the future increase in fertilizer use will occur in the developing countries, particularly Asia, which now accounts for more than half of all N fertilizer use. Other N sources used on land (such as manures, composts and biological N fixation) are also increasing though not as rapidly. Application of N fertilizers is likely continue to increase in order to feed increasing population (projected to reach 9 billion by 2045), which may lead to accelerated N_2O emissions on a global scale, since anthropogenic emissions of N_2O are closely related to production of food for a growing population (Kreileman and Bouwman 1994; Nevison et al. 1996). Therefore, agricultural N_2O emissions might increase by 35–60% between 2010 and 2030 due to increase in use of N fertilizers and manures (FAO 2003). The U.S. EPA projects a 50% increase in N_2O emissions in USA by the year 2020 (US EPA 2010).

Other potential sources of N_2O from crop fields are the volatilization of ammonia, some of which is converted to N_2O in the atmosphere, and emissions from plants (Smart and Bloom 2001; Hakata et al. 2003; Zhu et al. 2005). In addition to soils, wheat leaves also emit significant amounts of N_2O. Further, N_2O production in the leaves occurs during photo-assimilation of nitrite ions in the chloroplast. Emissions of N_2O by plants may be as high as 12% of the global emissions of N_2O (Smart and Bloom 2001). Studies involving the use of ^{15}N labeled nitrate to assess the production and emission of N_2O from a range of plant species show that most of the species are able to convert NO_3^- or NO_2^- in their tissues and emit N_2O (Hakata et al. 2003). Indeed, there exists a negative correlation ($r = 0.72$) between N_2O emission and NO_2^- assimilation by the plants (Hakata et al. 2003). Direct N_2O fluxes from wheat plants may range from 0.3 to 3.9 mg N_2O-N m^{-2} day^{-1}, with as much as 0.23% of plant N released directly to the atmosphere as N_2O. There exists a linear relationship between plants N_2O emission and plant dark respiration, suggesting that in the absence of photosynthesis, some N_2O production in plant N assimilation is associated with plant respiration (Zou et al. 2005).

There are also indications of plant-mediated pathway in ecosystem N_2O emissions (Mosier et al. 1990; Yan et al. 2000; Chang et al. 1998; Yu et al. 1997). Studies based on comparison of N_2O fluxes in chambers with and without rice plants, have demonstrated that young rice plants facilitate the efflux of N_2O from paddy soils to the atmosphere (Mosier et al. 1990). Direct N_2O emission is predominantly (87%) through the rice plants when flooded, and mainly through soil surface in the absence of flooding (Yan et al. 2000).

The ability of plants to contribute to N_2O efflux has direct implication in developing field measurements of N_2O fluxes. Thus, there is a likelihood of underestimating the N_2O fluxes determined by chamber method if plants are excluded from the chamber. Furthermore, most of the static chamber measurements are done during the day, which may underestimate N_2O fluxes associated with dark respiration in most field measurements.

Although a large number of field measurements have been undertaken in an effort to quantify the emissions from arable lands, large uncertainties exist. More research is needed for horticulture crops, highly organic soils, and organic farming. There is also a need to improve the understanding of how the combinations of management, climate, and soils interact to control N_2O emissions within landscapes.

5.3.2.3 Indirect Nitrous Oxide Emissions

Indirect N_2O emissions constitute those arising from transfer of fertilizer N to downwind and downstream ecosystems through processes such as NH_3 volatilization, N leaching, runoff, and disposal of human sewage. In this case, N_2O production and emission occur in downstream aquatic system following leaching of fertilizer N from agricultural lands. Also included under this category is N_2O production associated with utilization of agricultural crops once they leave the field, (e.g., human sewage and livestock waste) (Kroeze and Seitzinger 1998a). This category of N_2O emission makes up about one third of all agricultural N_2O emissions. Most of this arises from leaching and runoff of N, principally in the form of NO_3^- (Kroeze and Seitzinger 1998a; Groffman et al. 2002). The indirect emissions account for the anthropogenic N_2O emitted either from riparian zones or from the aquatic systems where drainage waters enter. A combination of direct and indirect emissions accounts for all N_2O emitted from the point of fertilizer N application to N removal through nitrification and denitrification.

Using a model grid of $1° \times 1°$, global landmass can be divided into watersheds draining into major rivers and estuaries. Using this approach, estimates of N loading and N_2O emissions from rivers, estuaries and continental shelves in 1990 and projected estimates for the year 2050 are shown in Table 5.8 (Kroeze and Seitzinger 1998a). The data thus obtained suggest that N_2O emissions from rivers, estuaries and continental shelves may increase from 1.9 Tg N_2O-N in 1990 to 4.9 Tg N_2O-N in 2050 mainly due to increase in fertilizer use to feed a growing world population (Kroeze and Seitzinger 1998a). The indirect N_2O emissions are accounted for based on the ratio of N_2O:NO_3 emission factor for ground water (EF5-g), downstream (EF5-r) and estuarine (EF5-e). Current IPCC methodology uses EF5-g, EF5-r, and EF5-e values of 0.015, 0.0075, and 0.0025, respectively, for estimating the indirect N_2O emissions. Based on IPCC emission factors, rivers, estuaries, and coastal

Table 5.8 Global fluxes of N_2O emissions in rivers, estuaries, and continental shelves in 1990 and 2050 estimated based on emission factors of dissolved inorganic nitrogen (DIN) by world rivers, combined rates of pelagic nitrification and sediment denitrification in continental shelves

Source	Calculated emissions in 1990	Estimated emission in 2050
	Tg N_2O-N year^{-1}	
Rivers	0.19–1.87	0.42–4.25
Estuaries	0.07–0.69	0.16–1.56
Continental shelves	0.60–6.43	0.72–7.22

(Data from Kroeze and Seitzinger 1998b)

zones emissions are estimated at 0.5 to 2.9 Tg N_2O-N year^{-1} (Denman et al. 2007). Some researchers proposed downward revisions of EF5 to 0.012 based on 5-year Global Nitrogen Enrichment monitoring program in Europe. This revision reduces indirect emissions to 0.8 Tg N year^{-1} (Reay et al. 2005). Using the DIN model, global N_2O emissions from rivers and estuaries are estimated at 1.26 and 0.25 Tg N year^{-1}, respectively, totaling to 1.5 Tg N_2O-N year^{-1}. The modeled estimates are also in good agreement with the observed DIN concentrations (Kroeze et al. 2005).

5.3.3 Biomass Burning

Most of anthropogenic biomass burning occurs in the tropics, and nearly 87% of global emissions from biomass burning take place in the tropics (Andreae 1991). Biomass is burnt for a wide range of purposes including: (1) clearing of forest and bush land for agriculture, (2) control of brush, weed, and litter accumulation on grazing and cropland, (3) nutrient regeneration in grazing and cropland, (4) control of fuel accumulation in forest, and (5) energy production for cooking and heating (Andreae 1991). In living plants the nutrient N is concentrated in parts that most easily burn (i.e., leaves, twigs and bark). On mass basis, the nutrient element of dry plant is about 0.3–3.8% N (Andreae 1991). The chemical composition of emissions from burning biomass depends primarily on the rate of energy release (Cofer et al. 1991).

The N is present in plant biomass mostly as amino groups (R-NH_2) in amino acids of proteins (Cofer et al. 1991). It is released by pyrolitic decomposition of OM during combustion, and then partially or completely oxidized to various volatile N compounds. The NO is the single most abundant species emitted, and represents 10–20% of N initially contained in plant (Andreae 1991), and about 30% of total yield for all NO_x species. Nitrogen compounds other than NO –NO_2, N_2O, NH_3 HCN organic nitrites and nitrates account for another 10–20% of the fuel N (Andreae 1991). In comparison, direct biomass burning contributes <7% of global sources of N_2O (Cofer et al. 1991; Lobert et al. 1991).

Based on N_2O release of 0.7 ± 0.3% of N content of the material burned, Crutzen and Andreae (1990) estimated 0.1–0.3 Tg N_2O-N year^{-1}. Using emission factors of Crutzen and Andreae (1990), Bouwman et al. (1995) estimated emissions of 0.2 Tg N^2O-N for 1990 from biomass burning.

5.3.4 *Tropical Land Conversion*

Accelerated decomposition and mineralization of litter, root material, and SOM in the first few years after forest clearing may cause pulse of N_2O emissions. The N_2O emissions exceeding forest soils by a factor of 2–8 have been reported for 2–10-year old pastures established after forest clearing (Keller et al. 1993). Enhanced emissions are generally not observed in older clearing (more than 10 year. old), probably due to gradual decline of easily decomposable SOM with age. Based on tropical forest conversion rate of 2.4 Mha year^{-1} for pastures and 12.6 Mha year^{-1} for arable land, Bouwman et al. (1995) estimated 0.36 Tg N_2O-N year^{-1} for conversion of tropical forests to pasture and arable land.

5.3.5 *Fossil Fuel Burning*

Commercial energy use by stationary combustion is considered a negligible source of N_2O emissions. However, mobile combustion particularly cars equipped with catalytic converters contribute to N_2O emissions.

5.4 Conclusions

Global emission of N_2O originates from natural and anthropogenic sources. However, the atmospheric concentration increase has been attributed primarily to anthropogenic sources. Primary anthropogenic sources are agriculture and fossil fuel combustion. The natural sources include soils under natural vegetation and ocean contributes about 12.1 Tg N_2O-N year^{-1} or 64% of total global emission. Emissions of N_2O from other aquatic environments such as rivers, estuaries and coastal waters are generally considered as anthropogenic because the majority of reactive N entering these ecosystems is associated with anthropogenic activities such as agriculture. The number of N_2O emission measurements has increased steadily from both natural and agricultural soils, allowing the improvement in emission models and budgets. However, the uncertainties for both budgets and modeled results are still high. Although the dominant sources of N_2O emissions currently are natural, anthropogenic emissions contribute significant amount, and will continue to increase in the future due to expected expansion of agriculture to meet the demand for growing population. Because anthropogenic activities are the main drivers of the changes in N_2O emissions, future emissions scenarios depend on changing human activities, mainly food production. Population growth will increase the need for food for both human and animal feed, and increasing fertilizer use. In addition, this may accelerate forest clearing, especially in the tropical region, the main contributor to global N_2O budget. Soil moisture and temperature regimes are the key determinants of the microbial processes that produce N_2O. Global patterns of temperature and precipitation changes will also play a significant role in the future emissions of N_2O. The number of N_2O emission measurements has increased steadily, allowing for the improvements in emission models and budgets. However, the vast majority of studies have focused on N_2O emissions from agricultural, not natural soil sources. In addition, there are still significant model uncertainties, even for agricultural soils. Global climate change models show patterns of temperature and precipitation changes worldwide. Since soil moisture is a key determinant of the microbial processes that consume or produce N_2O, these shifting climate patterns will also determine the fluxes of N_2O into the future.

5.5 Researchable Issues

Despite many years of intensive research, there remain many gaps in knowledge. Tropical forests and grasslands, tropical soils, and soil management in the tropical environments have received relatively less attention. Additional research focused on processes that result in N_2O emissions should improve current estimates of flux and help refine future estimates. High uncertainties observed in some sources are a result of lack of basic data-flux measurements. In some ecosystems flux data are sparse from some geographic regions and some seasons. Also, in number of sources such as wetlands and lakes there are no reliable field measurements across which make it difficult to ascertain if these ecosystems are sources or sinks for N_2O. Overall, for most sources, research data from the tropical region are either lacking or sparse, increasing the uncertainty to the global estimation. Increased research linking emissions to environmental controls, long-term studies to capture the seasonality and inter-annual variability will improve model predictions. In addition, continuous monitoring will capture and enable to quantify episodic emissions and help resolve some modeling difficulties.

For soils and riparian zones the major sources contributing to emissions more field data are needed. While field data are increasing steadily, coverage of global zones and different vegetation types, crop types, agronomic practices are still sparsely distributed. In summary, future research should address the following:

1. Long-term field measurements from natural soils and arable lands to capture seasonal variation and impacts of global change for these sources.
2. Field measurements from tropical and tundra regions to minimize the uncertainties and improve models prediction.
3. Better understanding of processes controlling N_2O production will improve the mitigation strategies.

Suggested questions

1. Outline the major natural sources of N2O emission and create its global budget.
2. What are the dominant global anthropogenic sources of N2O?
3. What role does future climate may play on global N2O emissions?

References

Alexander, R.B., Boyer, E.W., Smith, R.A., Schwarz, G.E., Moore, R.B. (2007) The role of headwater streams in downstream water quality. J. Am. Water Resour. Assoc. **43**(1), 41–59. doi:10.1111/j.1752-1688.2007.00005.x

Allen, D.E., Dalal, R.C., Rennenberg, H., Meyer, R.L., Reeves, S., Schmidt, S. (2007) Spatial and temporal variation of nitrous oxide and methane flux between subtropical mangrove sediments and the atmosphere. Soil Biol. Biochem. **39**(2), 622–631. doi:10.1016/j.soilbio.2006.09.013

Andreae, M.O. (1991) Biomass burning:Its history, use and distribution and its impact on environmental quality and global climate In: Levine, J.S. (ed.) Global biomass Burning: Atmospheric, climatic and biosphere implication. MIT Press, Cambridge, MA

Bange, H.W. (2008) Gaseous nitrogen compounds (NO, N_2O, N_2 NH_3) in the ocean. In: Capone, D.G., Bronk, D.A., Muholland, M.R., Carpenter, E.J. (eds.) Nitrogen in the Marine Environment. pp. 51–94. Elsevier, Amsterdam

Bange, H.W., Andreae, M.O., Lal, S., Law, C.S., Naqvi, S.W.A., Patra, P.K., Rixen, T., Upstill-Goddard, R.C. (2001) Nitrous oxide emissions from the Arabian Sea: A synthesis. Atmos. Chem. Phys. **1**, 61–71

Bange, H.W., Rapsomanikis, S., Andreae, M.O. (1996) Nitrous oxide in coastal waters. Global Biogeochem. Cycles **10**(1), 197–207. doi:10.1029/95gb03834

Barnes, J., Ramesh, R., Purvaja, R., Rajkumar, A.N., Kumar, B.S., Krithika, K., Ravichandran, K., Uher, G., Upstill-Goddard, R. (2006) Tidal dynamics and rainfall control N_2O and CH_4 emissions from a pristine mangrove creek. Geophys. Res. Lett. **33**(15), L15405. doi:10.1029/2006gl026829

Beaulieu, J.J., Arango, C.P., Hamilton, S.K., Tank, J.L. (2008) The production and emission of nitrous oxide from headwater streams in the Midwestern United States. Global Change Biol. **14**(4), 878–894. doi:10.1111/j.1365-2486.2007.01485.x

Beaulieu, J.J., Tank, J.L., Hamilton, S.K., Wollheim, W.M., Hall, R.O., Mulholland, P.J., Peterson, B.J., Ashkenas, L.R., Cooper, L.W., Dahm, C.N., Dodds, W.K., Grimm, N.B., Johnson, S.L., McDowell, W.H., Poole, G.C., Valett, H.M., Arango, C.P., Bernot, M.J., Burgin, A.J., Crenshaw, C.L., Helton, A.M., Johnson, L.T., O'Brien, J.M., Potter, J.D., Sheibley, R.W., Sobota, D.J., Thomas, S.M. (2011) Nitrous oxide emission from denitrification in stream and river networks. Proc. Natl. Acad. Sci. USA **108**(1), 214–219. doi:10.1073/pnas.1011464108

Bouwman, A.F. (1996) Direct emission of nitrous oxide from agricultural soils. Nutr. Cycl. Agroecosys. **46**(1), 53–70

Bouwman, A.F., Boumans, L.J.M., Batjes, N.H. (2002) Modeling global annual N_2O and NO emissions from fertilized fields. Global Biogeochem. Cycles **16**(4), 1080. doi:10.1029/2001gb001812

Bouwman, A.F., Fung, I., Matthews, E., John, J. (1993) Global analysis of the potential for N_2O production in natural soils. Global Biogeochem. Cycles **7**(3), 557–597

Bouwman, A.F., Vanderhoek, K.W., Olivier, J.G.J. (1995) Uncertainties in the global source distribution of nitrous oxide. J. Geophys. Res.-Atmos. **100**(D2), 2785–2800

Bouwman, L., Goldewijk, K.K., Van Der Hoek, K.W., Beusen, A.H.W., Van Vuuren, D.P., Willems, J., Rufino, M.C., Stehfest, E. (2011) Exploring global changes in nitrogen and phosphorus cycles in agriculture induced by livestock production over the 1900–2050 period. Proceedings of the National Academy of Sciences. doi:10.1073/pnas.1012878108

Bowden, R.D., Melillo, J.M., Steudler, P.A., Aber, J.D. (1991) Effects of nitrogen additions on annual nitrous-oxide fluxes from temperate forest soils in the northeastern United-States. J. Geophys. Res.-Atmos. **96**(D5), 9321–9328

Boyer, E.W., Howarth, R.W., Galloway, J.N., Dentener, F.J., Green, P.A., Vorosmarty, C.J. (2006) Riverine nitrogen export from the continents to the coasts. Global Biogeochem. Cycles **20**(1). doi:Gb1s9110.1029/2005gb002537

Bremner, J.M. (1997) Sources of nitrous oxide in soils. Nutr. Cycl. Agroecosys. **49**(1–3), 7–16

Butler, J.H., Jones, R.D., Garber, J.H., Gordon, L.I. (1987) Seasonal distributions and turnover of reduced trace gases and hydroxylamine in Yaquina Bay, Oregon. Geochim. Cosmochim. Acta **51**(3), 697–706

Butterbach-Bahl, K., Gasche, R., Breuer, L., Papen, H. (1997) Fluxes of NO and N_2O from temperate forest soils: impact of forest type, N deposition and of liming on the NO and N_2O emissions. Nutr. Cycl. Agroecosys. **48**, 79–90

Butterbach-Bahl, K., Gasche, R., Huber, C., Kreutzer, K., Papen, H. (1998) Impact of N-input by wet deposition on N-trace gas fluxes and CH4-oxidation in spruce forest ecosystems of the temperate zone in Europe. Atmos. Environ. **32**(3), 559–564

Butterbach-Bahl, K., Kesik, M., Miehle, P., Papen, H., Li, C. (2004) Quantifying the regional source strength of N-trace gases across agricultural and forest ecosystems with process based models. Plant Soil **260**(1–2), 311–329

Capone, D.G. (1991) Aspects of the marine nitrogen cycle with relevance to the dynamics of nitrous and nitric oxide. In: Rogers, J., Whitman, W.B. (eds.) Microbial production and consumption of greenhouse gases: Methane, Nitrogen oxides, and Halomethanes. American Society of Microbiology, Washington,

Chang, C., Janzen, H.H., Cho, C.M., Nakonechny, E.M. (1998) Nitrous oxide emission through plants. Soil Sci Soc. Am. **62**(1), 35–38

Chapuis-Lardy, L., Wrage, N., Metay, A., Chotte, J.L., Bernoux, M. (2007) Soils, a sink for N_2O? A review. Global Change Biol. **13**(1), 1–17. doi:10.1111/j.1365-2486.2006.01280.x

Codispoti, L.A. (2010) Interesting Times for Marine N_2O. Science **327**(5971), 1339–1340. doi:10.1126/science.1184945

Codispoti, L.A., Elkins, L.W., Yoshinari, T., Friederich, G.E., Sakamoto, C.M., Packard, T.T. (1992) On the nitrous oxide flux from productive region that contain low oxygen water In: Desai, B.D. (ed.) Oceanography of the Indian Ocean. pp. 271–284. Oxford, New Delhi

Cofer, W.R., Levine, J.S., Winstead, E.L., Stocks, B.J. (1991) New estimates of nitrous-oxide emissions from biomass burning. Nature **349**(6311), 689–691

Conrad, R. (1996) Soil microorganisms as controllers of atmospheric trace gases (H_2, CO, CH_4, OCS, N_2O, and NO). Microbiol. Rev. **60**(4), 609–640

Cookson, W.R., Cornforth, I.S., Rowarth, J.S. (2002) Winter soil temperature (2-15°C) effects on nitrogen transformations in clover green manure amended or unamended soils; a laboratory and field study. Soil Biol. Biochem. **34**(10), 1401–1415

Corredor, J.E., Morell, J.M., Bauza, J. (1999) Atmospheric nitrous oxide fluxes from mangrove sediments. Mar. Poll. Bull. **38**(6), 473–478

Cowardin, J.E., Carter, V., Golet, F.C., LaRoe, E.T. (1979) Classification of Wetland and Deepwater Habitats of the United States. U.S. Department of Interior, Fish and Wildlife Service.

Crutzen, P.J., Andreae, M.O. (1990) Biomass burning in the tropics: Impact on atmospheric chemistry and biogeochemical cycles. Science **250**(4988), 1669–1678. doi:10.1126/science.250.4988.1669

Crutzen, P.J., Mosier, A.R., Smith, K.A., Winiwarter, W. (2008) N_2O release from agro-biofuel production negates global warming reduction by replacing fossil fuels. Atmos. Chem. Phys. **8**(2), 389–395

Davidson, E.A. (2009) The contribution of manure and fertilizer nitrogen to atmospheric nitrous oxide since 1860. Nat. Geosci. **2**(9), 659–662. doi:10.1038/ngeo608

Davidson, E.A., Matson, P.A., Vitousek, P.M., Riley, R., Dunkin, K., Garciamendez, G., Maass, J.M. (1993) Processes regulating soil emissions of NO and N_2O in a seasonally dry tropical forest. Ecology **74**(1), 130–139 de Bie, M.J.M., Middelburg, J.J., Starink, M., Laanbroek, H.J. (2002) Factors controlling nitrous oxide at the microbial community and estuarine scale. Mar. Ecol.-Prog. Ser. **240**, 1–9

Del Grosso, S.J., Mosier, A.R., Parton, W.J., Ojima, D.S. (2005) DAYCENT model analysis of past and contemporary soil N2O and net greenhouse gas flux for major crops in the USA. Soil Till. Res. **83**(1), 9–24. doi:10.1016/j.still.2005.02.007

Denman, K.L., Brasseur, G., Chidthaisong, A., Ciais, P., Cox, P.M., Dickinson, R.E., Hauglustaine, D., Heinze, C., Holland, E., Jacob, D., Lohmann, U., Ramachandran, S., Dias, P.L.d.S., Wofsy, S.C., Zhang, X. (2007) Couplings Between Changes in the Climate System and Biogeochemistry. In: Solomon, S., Qin, D., Manning, M., Chen, Z., Marquis, M., Averyt, K.B., M.Tignor, Miller, H.L. (eds.) Climate Change 2007: The Physical Science Basis Contribution of Working Group I to the Fourth Assessment Report of the Intergovernmental Panel on Climate Change. pp. 501–587. Cambridge University Press, Cambridge, United Kingdom and New York, NY, USA

Dore, J.E., Popp, B.N., Karl, D.M., Sansone, F.J. (1998) A large source of atmospheric nitrous oxide from subtropical North Pacific surface waters. Nature **396**(6706), 63–66

Downing, J.A., Prairie, Y.T., Cole, J.J., Duarte, C.M., Tranvik, L.J., Striegl, R.G., McDowell, W. H., Kortelainen, P., Caraco, N.F., Melack, J.M., Middelburg, J.J. (2006) The global abundance and size distribution of lakes, ponds, and impoundments. Limnol. Oceanogr. **51**(5), 2388–2397

Dumont, E., Harrison, J.A., Kroeze, C., Bakker, E.J., Seitzinger, S.P. (2005) Global distribution and sources of dissolved inorganic nitrogen export to the coastal zone: Results from a spatially explicit, global model. Global Biogeochem. Cycles **19**(4), GB4S02. doi:10.1029/2005GB002488

Ehhalt, D., Prather, M., Dentener, F.J., Derwent, R., Dlugokencky, E.J., Holland, E.A., Isaksen, I., Katima, J., Kirchhoff, V., Matson, P.A. (2001) Atmospheric chemistry and greenhouse gases. In: Houghton, J.T., Ding, Y., Griggs, D.J., Noguer, M., vander Linden, P.J., Dai, X., Maskell, K., Johnson, C.A. (eds.) Climate Change 2001: The Scientific Basis. Contribution of Working Group I to the Third Assessment Report of the Intergovernmental Panel on Climate Change pp. 239–287. Cambridge University Press, Cambridge, United Kingdom and New York, NY, USA

Ellis, S., Howe, M.T., Goulding, K.W.T., Mugglestone, M.A., Dendooven, L. (1998) Carbon and nitrogen dynamics in a grassland soil with varying pH: Effect of pH on the denitrification potential and dynamics of the reduction enzymes. Soil Biol. Biochem. **30**(3), 359–367

Food and Agriculture Organization of the United Nations (FAO) (2003) World Agricultural Towards 2015/2030. An FAO Perspective. pp. Food and Agricultural Organization (FAO), Rome, Italy

Firestone, M.K., Davidson, E.A. (1989) Microbial basis of NO and N_2O production and consumption in soil. In: Andreae, M.O., Schimel, D.S. (eds.) Trace Gas Exchange Between Terrestrial Ecosystem and the Atmosphere. pp. 7–21. John Wiley, Berlin, Germany

Freibauer, A., Kaltschmitt, M. (2003) Controls and models for estimating direct nitrous oxide emissions from temperate and sub-boreal agricultural mineral soils in Europe. Biogeochemistry **63**(1), 93–115

Galloway, J.N., Dentener, F.J., Capone, D.G., Boyer, E.W., Howarth, R.W., Seitzinger, S.P., Asner, G.P., Cleveland, C.C., Green, P.A., Holland, E.A., Karl, D.M., Michaels, A.F., Porter, J.H., Townsend, A.R., Vorosmarty, C.J. (2004) Nitrogen cycles: past, present, and future. Biogeochemistry **70**(2), 153–226

Galloway, J.N., Schlesinger, W.H., Levy, H., Michaels, A., Schnoor, J.L. (1995) Nitrogen fixation - anthropogenic enhancement-environmental response. Global Biogeochem. Cycles **9**(2), 235–252

Giltrap, D.L., Li, C.S., Saggar, S. (2010) DNDC: A process-based model of greenhouse gas fluxes from agricultural soils. Agric. Ecosys. Environ. **136**(3–4), 292–300. doi:10.1016/j.agee.2009.06.014

Groffman, P. (1991) Ecology of nitrification and denitrification in soil evaluated at scales relevant to atmospheric chemistry. In: Rodgers, J.E., Whitman, W.B. (eds.) Micribial Production and Consumption of Greenhouse Gases: Methane, Nitrogen Oxides and Halomethanes. pp. 201–217. American Society for Microbiology, Washington, DC

Groffman, P., Gold, A.J., Kellog, Q., Addy, K. (2002) Mechanisms, rates and assessment of N_2O in groundwater, riparian zones and rivers. In: Van Ham, J., Baede, A.P.M., Guicherit, R., Williams-Jacobse, J.G.F.M. (eds.) Proceedings of Third International Symposium of Non-CO_2 Greenhouse gases. pp. 159–166. Maasricht, The Netherlands

Groffman, P.M., Gold, A.J., Jacinthe, P.A. (1998) Nitrous oxide production in riparian zones and groundwater. Nutr. Cycl. Agroecosys. **52**(2–3), 179–186

Groffman, P.M., Gold, A.J., Simmons, R.C. (1992) Nitrate dynamics in riparian forests - microbial studies. J. Environ. Qual. **21**(4), 666–671

Hakata, M., Takahashi, M., Zumft, W., Sakamoto, A., Morikawa, H. (2003) Conversion of the nitrate nitrogen and nitrogen dioxide to nitrous oxides in plants. Acta Biotechnol. **23**(2–3), 249–257

Hefting, M., Clement, J.C., Dowrick, D., Cosandey, A.C., Bernal, S., Cimpian, C., Tatur, A., Burt, T.P., Pinay, G. (2004) Water table elevation controls on soil nitrogen cycling in riparian wetlands along a European climatic gradient. Biogeochemistry **67**(1), 113–134

Hefting, M.M., Bobbink, R., de Caluwe, H. (2003) Nitrous oxide emission and denitrification in chronically nitrate-loaded riparian buffer zones. J. Environ. Qual. **32**(4), 1194–1203

Hefting, M.M., Bobbink, R., Janssens, M.P. (2006) Spatial variation in denitrification and N_2O emission in relation to nitrate removal efficiency in a n-stressed riparian buffer zone. Ecosystems **9**(4), 550–563. doi:10.1007/s10021-006-0160-8

Hirsch, A.I., Michalak, A.M., Bruhwiler, L.M., Peters, W., Dlugokencky, E.J., Tans, P.P. (2006) Inverse modeling estimates of the global nitrous oxide surface flux from 1998–2001. Global Biogeochem. Cycles **20**(1), GB 1008. doi:10.1029/2004gb002443

Houghton, J.T., Filho, L.G.M., Callander, B.A., Harris, N., Kattenberg, A., Maskell, K. (1996) Climate Change 1995: The Science of Climate Change. Contribution of working group I to the Second Assessment report of the Intergovernmental Panel on Climate Change (IPCC). pp. 588 Cambridge University Press, Cambridge, UK

Houghton, J.T., Jenkins, G.J., Ephraums, J.J. (1992) Climate Change: The IPCC Scientific Assessment (1990). Report prepared for Intergovernmental Panel on Climate Change by Working Group I. pp. 410 Cambridge University Press, Camridge, UK

Howarth, R.W. (1988) Nutrient limitation of net primary production in marine ecosystems. Annu. Rev. Ecol. Syst. **19**, 89–110

Howarth, R.W., Swaney, D.P., Boyer, E.W., Marino, R., Jaworski, N., Goodale, C. (2006) The influence of climate on average nitrogen export from large watersheds in the Northeastern United States. Biogeochemistry **79**(1–2), 163–186. doi:10.1007/s10533-006-9010-1

Huttunen, J.T., Vaisanen, T.S., Hellsten, S.K., Heikkinen, M., Nykanen, H., Jungner, H., Niskanen, A., Virtanen, M.O., Lindqvist, O.V., Nenonen, O.S., Martikainen, P.J. (2002) Fluxes of CH_4, CO_2, and N_2O in hydroelectric reservoirs Lokka and Porttipahta in the northern boreal zone in Finland. Global Biogeochem. Cycles **16**(1). doi:10.1029/2000gb001316

Huttunen, J.T., Juutinen, S., Alm, J., Larmola, T., Hammar, T., Silvola, J., Martikainen, P.J. (2003) Nitrous oxide flux to the atmosphere from the littoral zone of a boreal lake. J. Geophys. Res. **108**(D14), 4421. doi:10.1029/2002jd002989

Intergovernmental Panel on Climate Change (IPCC) (1996) Revised 1996 IPCC guidelines for national greenhouse gas inventories. pp. IGES, Hayama, Japan

Intergovernmental Panel on Climate Change (IPCC) (2001) Climate Change 2001: The Scientific Basis. Contribution of Working Group I to the Third Assessment Report of the Intergovernmental Panel on Climate Change. Houghton, J.T., Ding, Y., Griggs, D.J., Noguer, M., vander Linden, P.J., Dai, X., Maskell, K., Johnson, C.A. (eds.). pp. 881 Cambridge University Press, Cambridge, United Kingdom and New York, NY, USA

Intergovernmental Panel on Climate Change (IPCC) (2006) 2006 IPCC guidelines for national greenhouse gas inventories. Prepared by the National Greenhouse Gas Inventories Programme. In: Eggleston, H.S., Buendia, L., Miwa, K., Ngara, T., Tanabe, K. (eds.). Institute for Global Environmental Strategies (IGES), Hayama, Japan

Intergovernmental Panel on Climate Change (IPCC) (2007) Climate Change 2007: The Physical Science Basis. Contribution of Working Group I to the Fourth Assessment Report of the Intergovernmental Panel on Climate Change. Cambridge University Press

Jiang, X., Ku, W.L., Shia, R.L., Li, Q.B., Elkins, J.W., Prinn, R.G., Yung, Y.L. (2007) Seasonal cycle of N_2O: Analysis of data. Global Biogeochem. Cycles **21**(1). doi:10.1029/2006gb002691

Kanerva, T., Regina, K., Ramo, K., Ojanpera, K., Manninen, S. (2007) Fluxes of N_2O, CH_4 and CO_2 in a meadow ecosystem exposed to elevated ozone and carbon dioxide for three years. Environ. Pollut. **145**(3), 818–828. doi:10.1016/j.envpol.2006.03.055

Keller, M., Varner, R., Dias, J.D., Silva, H., Crill, P., de Oliveira, R.C., Asner, G.P. (2005) Soil–Atmosphere Exchange of Nitrous Oxide, Nitric Oxide, Methane, and Carbon Dioxide in Logged and Undisturbed Forest in the Tapajos National Forest, Brazil. Earth Interact. **9**(23), 1–28. doi:doi:10.1175/EI125.1

Keller, M., Veldkamp, E., Weitz, A.M., Reiners, W.A. (1993) Effect of pasture age on soil trace-gas emissions from a deforested area of Costa Rica. Nature **365**(6443), 244–246. doi:10.1038/365244a0

Kesik, M., Ambus, P., Baritz, R., Bruggemann, N.B., Butterbach-Bahl, K., Damm, M., Duyzer, J., Horvath, L., Kiese, R., Kitzler, B., Leip, A., Li, C., Pihlatie, M., Pilegaard, K., Seufert, G., Simpson, D., Skiba, U., Smiatek, G., Vesala, T., Zechmeister-Boltenstern, S. (2005) Inventories of N_2O and NO emissions from European forest soils. Biogeosciences **2**(4), 353–375

Khahil, M.I., Baggs, E.M. (2005) CH_4 oxidation and N_2O emissions at varied soil water-filled pore spaces and headspace CH4 concentrations. Soil Biol. Biochem. **37**(10), 1785–1794. doi:10.1016/j.soilbio.2005.02.012

Kiese, R., Li, C.S., Hilbert, D.W., Papen, H., Butterbach-Bahl, K. (2005) Regional application of PnET-N-DNDC for estimating the N_2O source strength of tropical rainforests in the Wet Tropics of Australia. Global Change Biol. **11**(1), 128–144. doi:10.1111/j.1365-2486.2004.00873.x

Koschorreck, M. (2005) Nitrogen turnover in drying sediments of an Amazon floodplain lake. Microb. Ecol. **49**(4), 567–577. doi:10.1007/s00248-004-0087-6

Koschorreck, M., Darwich, A. (2003) Nitrogen dynamics in seasonally flooded soils in the Amazon floodplain. Wetl. Ecol. manag. **11**, 317–330

Kreileman, G.J.J., Bouwman, A.F. (1994) Computing land-use emissions of greenhouse gases. Water Air Soil Poll. **76**(1–2), 231–258

Krithika, K., Purvaja, R., Ramesh, R. (2008) Fluxes of methane and nitrous oxide from an Indian mangrove. Curr. Sci. India **94**(2), 218–224

Kroeze, C., Dumont, E., Seitzinger, S.P. (2005) New estimates of global emissions of N_2O from rivers and estuaries. Environm. Sci. **2**(2–3), 159–165. doi:10.1080/15693430500384671

Kroeze, C., Mosier, A., Bouwman, L. (1999) Closing the global N_2O budget: A retrospective analysis 1500-1994. Global Biogeochem. Cycles **13**(1), 1–8

Kroeze, C., Seitzinger, S.P. (1998a) The impact of land use on N_2O emissions from watersheds draining into the Northeastern Atlantic Ocean and European Seas. Environ. Pollut. **102**, 149–158

Kroeze, C., Seitzinger, S.P. (1998b) Nitrogen inputs to rivers, estuaries and continental shelves and related nitrous oxide emissions in 1990 and 2050: a global model. Nutr. Cycl. Agroecosys. **52**(2–3), 195–212

Lal, S., Patra, P.K. (1998) Variabilities in the fluxes and annual emissions of nitrous oxide from the Arabian sea. Global Biogeochem. Cycles **12**(2), 321–327

Law, C.S., Owens, N.J.P. (1990) Significant flux of atmospheric nitrous-oxide from the northwest Indian-Ocean. Nature **346**(6287), 826–828

Li, C.S. (2000) Modeling trace gas emissions from agricultural ecosystems. Nutr. Cycl. Agroecosys. **58**(1–3), 259–276

Li, C.S., Aber, J., Stange, F., Butterbach-Bahl, K., Papen, H. (2000) A process-oriented model of N_2O and NO emissions from forest soils: 1. Model development. J. Geophys. Res.-Atmos. **105**(D4), 4369–4384

Li, C.S., Frolking, S., Frolking, T.A. (1992a) A model of nitrous-oxide evolution from soil driven by rainfall events .2. Model applications. J. Geophys. Res.-Atmos. **97**(D9), 9777–9783

Li, C.S., Frolking, S., Frolking, T.A. (1992b) A model of nitrous-oxide evolution from soil driven by rainfall events.1. Model structure and sensitivity. J. Geophys. Res.-Atmos. **97**(D9), 9759–9776

Liikanen, A., Flojt, L., Martikainen, P. (2002) Gas dynamics in eutrophic lake sediments affected by oxygen, nitrate, and sulfate. J. Environ. Qual. **31**(1), 338–349

Lobert, J.M., Scharffe, D.H., Hao, W.M., Kuhbushsch, T.A., Seuwen, R., Warmeck, P., Crutzen, P.J. (1991) Experimental evaluation of biomass burning emission: Nitrogen and carbon-containing compounds. In: Levine, J.S. (ed.) Global biomass burning: Atmospheric, climatic, and biosphere implications. MIT Press, Cambridge, MA

Martikainen, P.J., de Boer, W. (1993) Nitrous oxide production and nitrification in acidic soil from a dutch coniferous forest. Soil Biol. Biochem. **25**(3), 343–347. doi:10.1016/0038-0717(93)90133-v

Matthews, E., Fung, I. (1987) Methane emission from natural wetlands: Global distribution, area, and environmental characteristics of sources. Global Biogeochem. Cycles **1**, 61–86

McMahon, P.B., Dennehy, K.F. (1999) N_2O emissions from a nitrogen-enriched river. Environ. Sci. Technol. **33**(1), 21–25. doi:10.1021/es980645n

Melack, J.M., Hess, L.L., Gastil, M., Forsberg, B.R., Hamilton, S.K., Lima, I.B.T., Novo, E. (2004) Regionalization of methane emissions in the Amazon Basin with microwave remote sensing. Global Change Biol. **10**(5), 530–544. doi:10.1111/j.1529-8817.2003.00763.x

Mengis, M., Auml, Chter, R.E.N., Eacute, Wehrli, B. (1997) Sources and sinks of nitrous oxide (N_2O) in deep lakes. Biogeochemistry **38**(3), 281–301

Middelburg, J.J., Klaver, G., Nieuwenhuize, J., Markusse, R.M., Vlug, T., Vandernat, F. (1995) Nitrous oxide emissions from estuarine intertidal sediments. Hydrobiologia **311**(1–3), 43–55

Mosier, A., Kroeze, C., Nevison, C., Oenema, O., Seitzinger, S., van Cleemput, O. (1998) Closing the global N_2O budget: nitrous oxide emissions through the agricultural nitrogen cycle - OECD/IPCC/IEA phase II development of IPCC guidelines for national greenhouse gas inventory methodology. Nutr. Cycl. Agroecosys. **52**(2–3), 225–248

Mosier, A.R., Mohanty, S.K., Bhadrachalam, A., Chakravorti, S.P. (1990) Evolution of dinitrogen and nitrous oxide from the soil to the atmosphere through rice plants. Biol. Fert. Soils **9**(1), 61–67

Mulholland, P.J., Helton, A.M., Poole, G.C., Hall, R.O., Hamilton, S.K., Peterson, B.J., Tank, J.L., Ashkenas, L.R., Cooper, L.W., Dahm, C.N., Dodds, W.K., Findlay, S.E.G., Gregory, S.V., Grimm, N.B., Johnson, S.L., McDowell, W.H., Meyer, J.L., Valett, H.M., Webster, J.R., Arango, C.P., Beaulieu, J.J., Bernot, M.J., Burgin, A.J., Crenshaw, C.L., Johnson, L.T., Niederlehner, B.R., O'Brien, J.M., Potter, J.D., Sheibley, R.W., Sobota, D.J., Thomas, S.M. (2008) Stream denitrification across biomes and its response to anthropogenic nitrate loading. Nature **452**(7184), 202-U246. doi:10.1038/nature06686

Najjar, R.G., Sarmiento, J.L., Toggweiler, J.R. (1992) Downward transport and fate of organic matter in the ocean: Simulations with a general circulation model. Global Biogeochem. Cycles **6**(1), 45–76. doi:10.1029/91gb02718

Naqvi, S.W.A., Bange, H.W., Farias, L., Monteiro, P.M.S., Scranton, M.I., Zhang, J. (2010) Marine hypoxia/anoxia as a source of CH_4 and N_2O. Biogeosciences **7**(7), 2159–2190. doi:10.5194/bg-7-2159-2010

Nevison, C., Butler, J.H., Elkins, J.W. (2003) Global distribution of N_2O and the ΔN_2O-AOU yield in the subsurface ocean. Global Biogeochem. Cycles **17**(4), 1119–1137. doi:10.1029/2003gb002068

Nevison, C.D., Esser, G., Holland, E.A. (1996) A global model of changing N_2O emissions from natural and perturbed soils. Climatic Change **32**(3), 327–378

Nevison, C.D., Keeling, R.F., Weiss, R.F., Popp, B.N., Jin, X., Fraser, P.J., Porter, L.W., Hess, P. G. (2005) Southern Ocean ventilation inferred from seasonal cycles of atmospheric N_2O and O_2/N_2 at Cape Grim, Tasmania. Tellus B **57**(3), 218–229

Nevison, C.D., Kinnison, D.E., Weiss, R.F. (2004a) Stratospheric influences on the tropospheric seasonal cycles of nitrous oxide and chlorofluorocarbons. Geophys. Res. Lett. **31**(20), L20103. doi:10.1029/2004gl020398

Nevison, C.D., Lueker, T.J., Weiss, R.F. (2004b) Quantifying the nitrous oxide source from coastal upwelling. Global Biogeochem. Cycles **18**(1), GB1018. doi:10.1029/2003gb002110

Nevison, C.D., Mahowald, N.M., Weiss, R.F., Prinn, R.G. (2007) Interannual and seasonal variability in atmospheric N2O. Global Biogeochem. Cycles **21**(3), GB3017. doi:10.1029/2006gb002755

Nevison, C.D., Weiss, R.F., Erickson, D.J. (1995) Global oceanic emissions of nitrous-oxide. J. Geophys. Res.-Oceans **100**(C8), 15809–15820

Niklaus, P.A., Wardle, D.A., Tate, K.R. (2006) Effects of plant species diversity and composition on nitrogen cycling and the trace gas balance of soils. Plant Soil **282**(1–2), 83–98. doi:10.1007/s11104-005-5230-8

Olivier, J.G.J., Bouwman, A.F., Van der Hoek, K.W., Berdowski, J.J.M. (1998) Global air emission inventories for anthropogenic sources of NO_x, NH_3 and N_2O in 1990. Environ. Pollut. **102**, 135–148

Parton, W.J., Hartman, M., Ojima, D.S., Schimel, D.S. (1998) DAYCENT and its land surface submodel: Description and testing. Global Planet. Change **19**(1–4), 35–48

Pina-Ochoa, E., Alvarez-Cobelas, M. (2006) Denitrification in aquatic environments: A cross-system analysis. Biogeochemistry **81**(1), 111–130. doi:10.1007/s10533-006-9033-7

Pinay, G., Roques, L., Fabre, A. (1993) Spatial and temporal patterns of denitrification in riparian forests. J. Appl. Ecol. **30**, 581–591

Prigent, C., Papa, F., Aires, F., Rossow, W.B., Matthews, E. (2007) Global inundation dynamics inferred from multiple satellite observations, 1993-2000. J. Geophys. Res.-Atmos. **112**(D12), D12107. doi:10.1029/2006jd007847

Prinn, R., Cunnold, D., Rasmussen, R., Simmonds, P., Alyea, F., Crawford, A., Fraser, P., Rosen, R. (1990) Atmospheric emissions and trends of nitrous-oxide deduced from 10 years of ALE-GAUGE data. J. Geophys. Res.-Atmos. **95**(D11), 18369–18385

Reay, D.S., Smith, K.A., Edwards, A.C., Hiscock, K.M., Dong, L.F., Nedwell, D.B. (2005) Indirect nitrous oxide emissions: Revised emission factors. Environm. Sci. **2**(2–3), 153–158

Rhee, T.S., Kettle, A.J., Andreae, M.O. (2009) Methane and nitrous oxide emissions from the ocean: A reassessment using basin-wide observations in the Atlantic. J. Geophys. Res.-Atmos. **114**, D12304. doi:10.1029/2008jd011662

Robertson, G.P., Vitousek, P.M. (2009) Nitrogen in Agriculture: Balancing the Cost of an Essential Resource. In: Annual Review of Environment and Resources, vol. 34. Annual Review of Environment and Resources, pp. 97–125

Robertson, L.A., Kuenen, J.G. (1984) Aerobic denitrification: A controversy revived. Arch. Microbiol. **139**, 351–354

Rockmann, T., Kaiser, J., Brenninkmeijer, C.A.M. (2003) The isotopic fingerprint of the pre-industrial and the anthropogenic N_2O source. Atmos. Chem. Phys. **3**, 315–323

Schaefer, S.C., Alber, M. (2007) Temperature controls a latitudinal gradient in the proportion of watershed nitrogen exported to coastal ecosystems. Biogeochemistry **85**(3), 333–346. doi:10.1007/s10533-007-9144-9

Schjonning, P., Thomsen, I.K., Moldrup, P., Christensen, B.T. (2003) Linking soil microbial activity to water- and air-phase contents and diffusivities. Soil Sci Soc. Am. **67**(1), 156–165

Schlesinger, W.H. (2009) On the fate of anthropogenic nitrogen. Proc. Natl. Acad. Sci. USA **106**(1), 203–208. doi:10.1073/pnas.0810193105

Schlesinger, W.H., Reckhow, K.H., Bernhardt, E.S. (2006) Global change: The nitrogen cycle and rivers. Water Resour. Res. **42**(3), W03S06. doi:10.1029/2005wr004300

Schmidt, E.L. (1978) Nitrifying micoorganisms and their methodology. In: Schlesinger, D. (ed.) Microbiology. pp. 282–291. American Society for Microbiology, Washington D.C.

Schmidt, E.L. (1982) Nitrification in soils. In: Stevenson, F.J. (ed.) Nitrogen in Agricultural Soils. Soil Science Society of America, Madison, Wisconsin

Seitzinger, S., Harrison, J.A., Bohlke, J.K., Bouwman, A.F., Lowrance, R., Peterson, B., Tobias, C., Van Drecht, G. (2006) Denitrification across landscapes and waterscapes: A synthesis. Ecol. Appl. **16**(6), 2064–2090

Seitzinger, S.P. (1988) Denitrification in Freshwater and Coastal Marine Ecosystems: Ecological and Geochemical Significance. Limnol. Oceanogr. **33**(4), 702–724

Seitzinger, S.P., Giblin, A.E. (1996) Estimating denitrification in North Atlantic continental shelf sediments. Biogeochemistry **35**(1), 235–260

Seitzinger, S.P., Kroeze, C. (1998) Global distribution of nitrous oxide production and N inputs in freshwater and coastal marine ecosystems. Global Biogeochem. Cycles **12**(1), 93–113

Seitzinger, S.P., Kroeze, C., Styles, R.V. (2000) Global distribution of N_2O emissions from aquatic systems: natural emissions and anthropogenic effects. Chemosphere: Global Change Sci. **2**(3–4), 267–279

Sharp, J.H. (1983) The distribution of inorganic nitrogen and dissolved particulate organic nitrogen in the sea. In: Carpenter, E.J., Capone, D.G. (eds.) Nitrogen in the marine environment. pp. 1–35. Academic Press Inc., New York

Silvan, N., Regina, K., Kitunen, V., Vasander, H., Laine, J. (2002) Gaseous nitrogen loss from a restored peatland buffer zone. Soil Biol. Biochem. **34**(5), 721–728. doi:10.1016/s0038-0717(02)00002-0

Smart, D.R., Bloom, A.J. (2001) Wheat leaves emit nitrous oxide during nitrate assimilation. Proc. Natl. Acad. Sci. USA **98**(14), 7875–7878

Smith, E., Gordon, R., Bourque, C., Campbell, A. (2008) Management strategies to simultaneously reduce ammonia, nitrous oxide and odour emissions from surface-applied swine manure. Can. J. Soil Sci. **88**(4), 571–584

Smith, K.A. (2005) The impact of agriculture and other land uses on emissions of methane and nitrous oxide. Environm. Sci. **2**(2–3), 101–108

Smith, K.A., Conen, F. (2004) Impacts of land management on fluxes of trace greenhouse gases. Soil Use Manage. **20**, 255–263. doi:10.1079/sum2004238

Smith, M.S., Parsons, L.L. (1985) Persistence of denitrifying enzyme-activity in dried soils. Appl. Environ. Microbiol. **49**(2), 316–320

Stadmark, J., Leonardson, L. (2005) Emissions of greenhouse gases from ponds constructed for nitrogen removal. Ecol. Eng. **25**(5), 542–551. doi:10.1016/j.ecoleng.2005.07.004

Stange, F., Butterbach-Bahl, K., Papen, H., Zechmeister-Boltenstern, S., Li, C.S., Aber, J. (2000) A process-oriented model of N_2O and NO emissions from forest soils 2. Sensitivity analysis and validation. J. Geophys. Res.-Atmos. **105**(D4), 4385–4398

Stehfest, E., Bouwman, L. (2006) N_2O and NO emission from agricultural fields and soils under natural vegetation: summarizing available measurement data and modeling of global annual emissions. Nutr. Cycl. Agroecosys. **74**(3), 207–228. doi:10.1007/s10705-006-9000-7

Stevens, R.J., Laughlin, R.J., Burns, L.C., Arah, J.R.M., Hood, R.C. (1997) Measuring the contributions of nitrification and denitrification to the flux of nitrous oxide from soil. Soil Biol. Biochem. **29**(2), 139–151

Stevens, R.J., Laughlin, R.J., Malone, J.P. (1998) Measuring the mole fraction and source of nitrous oxide in the field. Soil Biol. Biochem. **30**(4), 541–543. doi:10.1016/s0038-0717(97)00192-2

Suntharalingam, P., Sarmiento, J.L. (2000) Factors governing the oceanic nitrous oxide distribution: Simulations with an ocean general circulation model. Global Biogeochem. Cycles **14**(1), 429–454

Suntharalingam, P., Sarmiento, J.L., Toggweiler, J.R. (2000) Global significance of nitrous-oxide production and transport from oceanic low-oxygen zones: A modeling study. Global Biogeochem. Cycles **14**(4), 1353–1370

Tiedje, J.M. (1988) Ecology of nitrification and dissimilatory nitrate reduction to ammonium. In: Zehnder, A.J.B. (ed.) Biology of anaerobic microorganisms. John Wiley & Sons, New York

U.S. Environmental Protection Agency (US EPA) (2010) Inventory of U.S. Greenhouse gas Emissions and Sinks 1990 - 2008. U.S. Environmental Protection Agency. Washington, D.C.

van Dijk, S.M., Duyzer, J.H. (1999) Nitric oxide emissions from forest soils. J. Geophys. Res.-Atmos. **104**(D13), 15955–15961

van Vuuren, D.P., Bouwman, L.F., Smith, S.J., Dentener, F. (2011) Global projections for anthropogenic reactive nitrogen emissions to the atmosphere: an assessment of scenarios in

the scientific literature. Curr. Opin. Environ. Sustainability **3**(5), 359–369. doi:10.1016/j.cosust.2011.08.014
Verchot, L.V., Franklin, E.C., Gilliam, J.W. (1997) Nitrogen cycling in piedmont vegetated filter zones .1. Surface soil processes. J. Environ. Qual. **26**(2), 327–336
Vitousek, P.M., Aber, J.D., Howarth, R.W., Likens, G.E., Matson, P.A., Schindler, D.W., Schlesinger, W.H., Tilman, D. (1997) Human alteration of the global nitrogen cycle: Sources and consequences. Ecol. Appl. **7**(3), 737–750. doi:10.2307/2269431
Ward, B.B., Kilpatrick, K.A., Renger, E.H., Eppley, R.W. (1989) Biological nitrogen cycling in the nitraline. Limnol. Oceanogr. **34**, 493–513
Werner, C., Butterbach-Bahl, K., Haas, E., Hickler, T., Kiese, R. (2007a) A global inventory of N_2O emissions from tropical rainforest soils using a detailed biogeochemical model. Global Biogeochem. Cycles **21**(3), GB3010. doi:10.1029/2006gb002909
Werner, C., Kiese, R., Butterbach-Bahl, K. (2007b) Soil-atmosphere exchange of N_2O, CH_4, and CO_2 and controlling environmental factors for tropical rain forest sites in western Kenya. J. Geophys. Res.-Atmos. **112**(D3), D03308. doi:10.1029/2006jd007388
Wittke, F., Kock, A., Bange, H.W. (2010) Nitrous oxide emissions from the upwelling area off Mauritania (NW Africa). Geophys. Res. Lett. **37**, L12601. doi:10.1029/2010gl042442
Yamulki, S., Harrison, R.M., Goulding, K.W.T., Webster, C.P. (1997) N_2O, NO and NO_2 fluxes from a grassland: Effect of soil pH. Soil Biol. Biochem. **29**(8), 1199–1208
Yan, X., Shi, S., Du, L., Xing, G. (2000) Pathways of N_2O emission from rice paddy soil. Soil Biol. Biochem. **32**(3), 437–440
Yu, K.W., Wang, Z.P., Chen, G.X. (1997) Nitrous oxide and methane transport through rice plants. Biol. Fert. Soils **24**(3), 341–343
Zechmeister-Boltenstern, S., Hahn, M., Meger, S., Jandl, R. (2002) Nitrous oxide emissions and nitrate leaching in relation to microbial biomass dynamics in a beech forest soil. Soil Biol. Biochem. **34**(6), 823–832
Zhu, Z., Xiong, Z., Xing, G. (2005) Impacts of population growth and economic development on the nitrogen cycle in Asia. Sci China Ser. C **48**(2), 729–737. doi:10.1007/bf03187113
Zou, J.W., Huang, Y., Sun, W.J., Zheng, X.H., Wang, Y.S. (2005) Contribution of plants to N_2O emissions in soil-winter wheat ecosystem: pot and field experiments. Plant Soil **269**(1–2), 205–211. doi:10.1007/s11104-004-0484-0

Chapter 6
Land Use and Land Management Effects on Nitrous Oxide Fluxes

Contents

Abstract The terrestrial ecosystems are the major source of N_2O contributing about 65% of global emissions. Agricultural activities are the most important anthropogenic N_2O emissions, accounting for 60–80% of the anthropogenic N_2O sources, mostly as N inputs to agricultural soils. Estimates of N_2O emissions from various agricultural systems vary widely due to variations in climatic and environmental factors, soil organic carbon (SOC) concentration, soil texture, soil drainage, abundance of NO_3-N, soil pH, management practices, and crop type. Under managed pasture soils, N_2O is mainly generated from mineral N originating from urine, dung, and biologically fixed N. Fluxes of N_2O from grazed pastures are also highly variable due to patchiness of animal excreta. In grassland ecosystems,

David Ussiri and R. Lal, *Soil Emission of Nitrous Oxide and its Mitigation*,
DOI 10.1007/978-94-007-5364-8_6,

significant N_2O emissions could occur from the accumulation of mainly nitrate-N following mineralization of organic N from legume-based pastures. Tropical savannas contribute N_2O emissions from grazing animals and frequent fires. Unfertilized forestry systems may emit less but the fertilized plantations emit more N_2O than the extensive grazed pastures. Overall, there is a need to examine the emission factors used for estimating N_2O emissions. The ratio of N_2O:N_2 production depends on oxygen supply or water-filled pore space, decomposable organic carbon, N substrate supply, temperature, and pH and salinity. N_2O production from soil is sporadic both in time and space, and therefore, it is a challenge to scale up the measurements of N_2O emission from a given location and time to regional and national levels.

Keywords Water-filled pore spaces • Cropland • Fertilizer • Grasslands • Animal excreta • Forestlands

Abbreviations

SOC	Soil organic carbon
SOM	Soil organic matter
EF	Emission factors
IFA	International Fertilizer Industry Association
FAO	Food and Agriculture Organization
CT	Conventional tillage
NT	No tillage
BNF	Biological nitrogen fixation
SON	Soil organic nitrogen
OC	Organic carbon
WFPS	Water-filled pore spaces

6.1 Introduction

Land and oceans are the principal sources of N_2O with soils contributing about 65% or an estimated 9.5 Tg N_2O-N year^{-1} (Ehhalt et al. 2001; Smith and Conen 2004) and oceans about 30% of total emissions to the atmosphere (Ehhalt et al. 2001). The remaining 5% is released mainly from other aquatic ecosystems, fossil fuel burning and industrial processes. While the major greenhouse gas (GHG) for the total anthropogenic emissions is CO_2, for the agriculture, the most important is N_2O, contributing about 60% of the anthropogenic N_2O emissions (Smith et al. 2007), mainly from soils and N inputs to agricultural soil systems. Agricultural soils emit 3.5 Tg N_2O-N year^{-1}, while managed temperate grasslands emit about 1.0 Tg N_2O-N year^{-1} (Ehhalt et al. 2001). The N_2O is produced in soils mainly from the microbial activities of nitrification and denitrification (Chap. 3; Bange 2000; Wrage et al. 2004). Nitrification is the biological oxidation of ammonia (NH_3), while denitrification is the reduction of the oxides of N. The relative contribution of these two processes and the magnitude of N_2O emissions depend mostly on physico-chemical and biological properties of soils (Smith et al. 2003; Jones et al. 2005) and environmental as well as climatic factors. Both processes can occur in soils separately or simultaneously, although the rates of two processes depend on soil aeration and the microsite availability of substrates.

Nitrification occurs under aerobic conditions with up to 50% water-filled pore space (WFPS). Potentially, both nitric oxide (NO) and N_2O can be evolved during nitrification in soils. However, NO emission is much greater than N_2O (Hutchinson and Davidson 1992; Subbarao et al. 2006). Nitrification determines the form of N present in soil, and retention and utilization of N. Therefore, it has significant implications for plant productivity and environmental quality. During nitrification, the relatively immobile NH_4^+ is converted to highly mobile NO_3^-, which becomes highly susceptible to loss from the plant root zone by leaching and denitrification (Subbarao et al. 2006). Nevertheless, NO_3^- is also highly available to plants, and often the major form of N uptake. Most crops have the ability to take up and utilize both NH_4^+ and NO_3^- (Haynes and Goh 1978). However, many dry land crops show a preference for NO_3^- over NH_4^+ (Haynes and Goh 1978). Crops that are well adapted to anaerobic soil conditions (such as rice) and plants that are adapted to acidic soils generally have a preference for NH_4^+ over NO_3^- (McKane et al. 2002).

Denitrification reduces nitrogen oxides, NO_3^- and NO_2^- to the gases NO, N_2O and N_2 (Robertson and Groffman 2007). The variable portion of N is emitted as N_2O gas. Composition of the microbial population exerts a dominant control on N_2O emissions, which remains relatively constant over time (Bedard-Haughn et al. 2006). The interactions among environmental drivers and soil properties, including NO_3^- concentration, temperature, WFPS, available soil organic carbon (SOC) controls the magnitude of N_2O released from the denitrification. An important factor determining the nitrification and denitrification processes in most natural soils is the mineralization of soil organic matter (SOM), which contributes to the soil pool of available N through the release of NH_3/NH_4^+ (Bouwman et al. 2002).

Nitrification and denitrification processes in soils are controlled by physical, biological, chemical, and environmental properties of soil. The interaction of controlling factors for nitrification and denitrification are complex. For example, the amount of N_2O produced depends on the range of O_2 supply in the soil. Temperature determines the rate of nitrification and denitrification by influencing the microbial activity. At cooler temperatures, the rate of N transformation is slow, and increases to a maximum as the temperature rises. The O_2 concentration in soil is influenced by soil moisture content, decreasing drastically for saturated soils. Both N_2O and NO are by-products of the transformation from nitrite (NO_2^-) under O_2-limited conditions, when nitrifiers use NO_2^- as a terminal electron acceptor (Tiedje 1988). N_2O emissions ranging from 0.04 to 0.45% of added N originating from nitrification under fully aerobic conditions have been reported (Bremner and Blackmer 1978) in some laboratory studies, though the mechanism responsible for the N_2O production are not fully understood.

The natural processes of nitrification and denitrification in natural grasslands and tropical rainforests, and the stimulation of these processes by the large-scale addition of synthetic N fertilizers to agricultural soils and managed pastures make these ecosystems the major sources of N_2O emission in terrestrial ecosystems. The size of the emissions reflects the greater amounts of N cycled biologically in tropical regions compared to boreal and temperate forests. Deposition of N from the atmosphere into natural ecosystems, land use change, and changes in agricultural and forest management increases N_2O emissions.

The net exchange of N_2O flux between soil and the atmosphere is the result of the balance between the concurrent production and consumption mechanisms within the soil. The strength of fluxes of N_2O is affected by soil moisture, since it controls the degree of aeration and O_2 concentration, and hence determines whether nitrification or denitrification prevails (Smith et al. 2003). The N_2O fluxes from soils peak at WFPS values in the range of 50–75% (Davidson 1991) or about field capacity for most soils. Drier soil moistures inhibit nitrification, while in water-logged soils complete denitrification favors N_2 gas evolution. Soil mineral N and temperature are the other key controls on N_2O production (Skiba et al. 1998; Conen et al. 2000; Smith et al. 2003). Movement and fate of N_2O produced in the subsoil is mainly determined by the molecular diffusion, convection, ebullition, and entrapment. These processes are influenced by soil temperature, WFPS, precipitation, and other meteorological variables (Clough et al. 2005).

Land use is intricately related to both economic and development, and the ecological characteristics of the landscape. Human beings transform landscapes to obtain food, timber, fiber, shelter, and other ecosystem goods. Global impacts of land use ranges from changing atmospheric composition to extensive modification of earth's ecosystems (Vitousek et al. 1997; Matson et al. 1997; Tilman et al. 2001; Wackernagel et al. 2002). For example, anthropogenic nutrients input to the biosphere, from fertilizers and atmospheric pollutants now exceed natural sources and have widespread effects on water quality, coastal and freshwater ecosystems. Land use and land management practices, soil and climatic factors have great impact on the net emissions of N_2O into the atmosphere (Smith et al. 2000; von Arnold et al. 2005). Land use influences the production of N_2O through N inputs (Skiba et al. 1998) and management (Flechard et al. 2007).

6.2 Croplands

Anthropogenic land use activities driven by the need to provide food, fiber, water, and shelter have become a force of global significance (Foley et al. 2005). Mankind has extensively modified the Earth's land surface and altered the ecosystem structure and function. The conversion of natural ecosystems to agriculture and other human activities such as urbanization and pasturelands has strongly modified the earth's terrestrial biosphere since 1700 (Goldewijk 2001; Lambin et al. 2003). During the past 300 years, the arable land area has increased from about 3×10^6 km^2 in 1700 to ~15×10^6 km^2 at present, while the pasture land area increased from ~5×10^6 to 34×10^6 km^2 in the same period (Goldewijk et al. 2007; Ramankutty et al. 2008). Results of these alterations are enormous, including: reducing the ability of ecosystems to continue providing important resources such as food, freshwater, forest resources and important services such as good quality of air, water and soil resources. Apart from the loss of these natural ecosystems and their benefits *per se*, changes in land cover have resulted in alterations in albedo and surface roughness, leading to significant changes in regional climate (Henderson-Sellers and Gornitz 1984). Conversion of natural ecosystems to agriculture and land cover changes has also led to changes in the biogeochemical cycle of nitrogen (Chap. 2).

Agricultural activities have particularly played significant role in these ecosystem modifications (Green et al. 2005). Nearly one third of the earth's land surface is currently used for agriculture (Goldewijk et al. 2007; Ramankutty et al. 2008), and much of this land has replaced forests, savannah and grasslands (Foley et al. 2005). In addition, ~20–30% of total available surface water globally is withdrawn for irrigation (Cassman and Wood 2005). Different societies have developed various methods for improving food production, including conversion of land from forests and other ecosystems to agriculture and different agricultural intensification practices such as tilling, fertilizing the soil, use of improved seed varieties, and crop rotation (Matson et al. 1997). Probably the most dramatic improvement in food production in mankind history is the "green revolution" which resulted in many-fold increases in crop yield by chemical fertilizers application, new and improved varieties of crops, and irrigation (Borlaug 2007).

Chemical fertilizers are a major facet of the green revolution's package of yield increasing advances and techniques. The development of Haber-Bosch ammonia synthesis greatly increased the supply of N fertilizer inputs that have been harnessed around the world to increase crop productivity. The intensification of existing agricultural activity, with commensurate increases in fertilizer application, rather than cropland expansion, has been a primary driver of great growth in global agricultural production since early 1960s (Foley et al. 2005). Increase in global food production surpassed population growth between 1961 and 2000. The world grain production has doubled, exceeding ~2,000 Tg year^{-1} (i.e., 2 billion metric tons), because of green revolution technologies involving chemical fertilizers, high-yielding cultivars, mechanization and irrigation. Overall, food crop yield per unit area increased by 106%, area under irrigation increased by 97%, and nitrogen fertilizers, phosphorus fertilizers and pesticides production increased by 638, 203 and 854%, respectively (FAOSTAT 2001; Tilman et al. 2001, 2002). Global cropland and permanent pasture land area increased by only 12 and 10%, respectively, for the same period (FAOSTAT 2001; Green et al. 2005). Overall, food demand is predicted to increase by two to threefold by 2050 as population rises to eight to ten billion, and the per capita consumption increases (Tilman et al. 2002).

While the benefits of agricultural intensification to crop yields and food production cannot be overstated, it has also unleashed a litany of global environmental damages ranging from degradation of soil fertility and water quality (Galloway et al. 2008; Vitousek et al. 2009) to loss of native habitats and climate change. Nutrients applied to croplands are leaching into aquatic systems and alter ecosystem functions (Carpenter et al. 1998; Smil 2002). For example, development of "dead zones" on the continental shelves of northern Gulf of Mexico each summer is attributed, in large part, to N fertilizers use across the Mississippi River Basin (Burkart and James 1999). Similar phenomenon has been observed in other seas such as Baltic, Kattegat, Black Sea, and East China (Diaz and Rosenberg 2008). The N losses from agricultural fields are critical issue due to economic loss and its role as a source for environmental pollution. Agriculture is also partly responsible for environmental concerns such as tropical deforestation and loss of biodiversity, land use fragmentation and loss of habitats, loss of soil quality through soil erosion and salinization, decrease in quality and quantity of water and its resources, changes in regional climate, reduction of air quality and increase in infectious diseases (Foley et al. 2005).

Since 1970s emission of N_2O have increased mainly as a result of increased use of mineral fertilizers worldwide. But input of synthetic N fertilizers in Africa remains the lowest in the world, representing 3% of the world fertilizer N consumption (FAO 2011). For example, currently about 250 million people in Sub Saharan Africa are chronically malnourished (FAO 2010), and a greater N input to the soil is urgently needed to satisfy the basic food need for growing population. The increase in N inputs in Africa will likely be the largest perturbation to the N cycle in Africa in the coming decades with inevitable impact in increasing direct and indirect N_2O emission. The magnitude of increase is not known, because there are only few published studies from Africa examining the response function relating N_2O emissions to increase in N inputs. Current estimates of N_2O emissions from African Agriculture at national, regional, and continental scale are mostly based on IPCC guidelines, which do not take into account some important characteristics of African soils and climate.

The flux of N_2O has been responsible for 6% of enhanced greenhouse effect (Denman et al. 2007), and agriculture contributes 60–80% of the anthropogenic N_2O emission (Dalal et al. 2003; Smith et al. 2007). The potential sources of N_2O in agricultural soils are nitrogenous fertilizers (Maggiotto et al. 2000), manures (Brown et al. 2000; Freibauer 2003) and mineralization of SOM. Because N_2O is formed by microbial processes, its production is controlled by factors that affect the growth of microorganisms, including temperature, pH, rainfall-soil moisture content, and aeration (Sahrawat and Keeney 1986; Dalal et al. 2003). However, its production is also affected by N fertilizer rate, N mineralization, nitrate concentration, tillage practice, soil type, soil pH, salinity, O_2 concentration, availability of SOC, vegetation, land use practices, use of agricultural chemicals, irrigation practices and water holding capacity of soil (Sahrawat and Keeney 1986; Dalal et al. 2003). Fluxes of N_2O from agricultural soils are the result of complex interactions between climatic parameters and soil chemical, biological, and physical properties.

6.2.1 Nitrous Oxide Formation in Agricultural Soils

Emissions of N_2O from agricultural soils result primarily from microbially driven nitrification and denitrification processes. Nitrification is the aerobic microbial oxidation of NH_4^+ to nitrate. When O_2 is limiting, NH_4^+ oxidizers can use NO_2^- as an alternative electron acceptor and produce N_2O. It is also formed through denitrification – anaerobic microbial reduction of NO_3^- successively to NO_2^-, and gaseous NO, N_2O and N_2. Microbial production of N_2O is dependent on the presence in the soil of suitable mineral N substrates i.e. NO_3^- and NH_4^+. Therefore, mineral fertilizers and N from other sources such as animal manures, crop residues, and N_2-fixing crops are major drivers of N_2O emissions (Bockman and Olfs 1998). In addition, there is background source due to mineralization of soil organic matter (SOM). The accelerated decomposition of OM that may have accumulated under natural forest or grassland may enhance this background source, especially where soils have recently been brought into cultivation.

6.2.2 Legumes and Nitrous Oxide Emissions

Biological N fixation by leguminous crops, such as soybeans and pulses provides significant N input in many agricultural systems. Although these crops generally receive no N fertilizers, leguminous crops accentuate emissions of N_2O that are of the same level as those under fertilized non-leguminous crops. Legumes may contribute to N_2O emission in number of ways. Atmospheric N_2 fixed by legumes can be nitrified and denitrified in the same way as fertilizer N under favorable soil conditions, thus providing a source of N_2O emissions. In addition, symbiotically living *Rhizobia* in root nodules are able to denitrify and produce N_2O (O'hara and Daniel 1985). Legumes can increase N_2O emissions from pastures by a factor of 2–3 (Duxbury et al. 1982).

On the other hand, agricultural expansion and intensification has provided a crucial service to humanity by meeting the food demands for a rapidly growing global population (Cassman and Wood 2005), and therefore, necessitating a trade-off between food production and the environmental deterioration (DeFries et al. 2004; Foley et al. 2005).

6.2.3 Conversion of Forest and Grassland to Cropland

In the tropical regions where slash and burn is widely practiced, land conversion from forest or grassland to agroecosystems enhances N_2O emission that may last for several months (Weitz et al. 1998). Cultivation of forestland or grassland in temperate regions has similar effects due to accelerated mineralization of SOM caused by disturbances (Smith and Conen 2004). For example, plowing a grass sward in southeast Scotland caused a loss of 449 kg N ha^{-1}, and associated emissions of 2.0 kg N_2O-N ha^{-1} over 18 months, while plowing grass–clover sward caused a loss of 244 kg N ha^{-1} and 4.4 kg N_2O-N emissions (Davies et al. 2001). Conversion of tropical forests to crop production and pasture has a significant effect on the emissions of N_2O, due to greater amounts of N cycled biologically in this ecosystem compared to boreal or temperate forests. Emissions of N_2O can increase by a factor of 3–8 when tropical forest is clear-cut and converted to pasture (Luizao et al. 1989; Keller et al. 1993). The fluxes gradually decrease during the following 10–20 year (Keller et al. 1993). An Additional N may be transported from cleared forest soils by surface runoff through soil erosion and ground water through leaching and emitted elsewhere. It seems rational to assume that mineral N liberated by mineralization of SOM and plant remains following cultivation of forest or grassland/pasture should be regarded as a comparable potential source of N_2O to N applied to the land either as fertilizer N or manure.

6.2.4 Land Management and Nitrous Oxide Emissions

Effects of soil management practices on N_2O emissions from different climatic regions have been inconsistent mainly due to variability in weather and soil conditions—including soil water content, rate, type of fertilizer and method of application (Baggs et al. 2003; Grant et al. 2004; Venterea et al. 2005; Drury et al. 2006; Omonode et al. 2011). Type of fertilizer influences the magnitude of nitrification and denitrification and N_2O fluxes. For example, anhydrous NH_3, generally injected into the soil commonly produces higher N_2O losses than from other fertilizer types. Although N_2O production and emissions may increase with the N application rate, denitrification and fluxes of N_2O are more closely related to the amount of unused N than to the actual application rate. Therefore, the timing of fertilizer application is an important factor in controlling N losses and gaseous N emissions. Any prolongation of the period when NH_4^+- or NO_3^--based fertilizers can undergo nitrification or denitrification without competition from plant uptake is likely to increase N_2O emissions.

6.2.4.1 Fertilizers Use and Nitrous Oxide Emissions

Nitrogen input is a principal control of N_2O emissions, and generally, an increase of N input increases both nitrification and denitrification rates. Manure, fertilizer N as well as combined manure and synthetic fertilizers may lead to higher N_2O emissions directly after application. Application of fertilizers and manures generally results in increased emissions of N_2O to the atmosphere (Ehhalt et al. 2001; Denman et al. 2007). Fertilizers can be a direct as well as indirect sources of N_2O

emissions. Nitrification and denitrification of N fertilizers in farm are a direct source of increased N_2O emissions, while volatilization of NH_3 and leaching of NO_3^- from N fertilization and their nitrification and denitrification outside the farm are indirect fertilizer sources. Most of the estimates used in the IPCC reports are based on assessments from temperate areas. Assessment of N_2O fluxes from tropical regions suggests that fluxes from tropical agricultural systems may be much higher than those from the temperate regions (Mosier and Delgado 1997; Veldkamp and Keller 1997; Veldkamp et al. 1998; Passianoto et al. 2003). Fluxes of N_2O from tropical agricultural soils are generally influenced by soil conditions such as soil moisture regimes and timing of the fertilizer application rather than fertilizer type and amount (Bouwman 1998).

Despite the detrimental effects on losses of reactive N to the environment and N_2O emissions, N fertilizers remains essential to global food production. After water availability, N remains globally the most limiting plant growth factor for non-leguminous crops. The relationship between crop productivity and N input follows a diminishing return function. Highly productive agricultural systems are often associated with large N losses to the environment, including N_2O emissions. In addition to increasing direct N_2O emissions, agricultural practices can also increase NH_3 volatilization and NO_3^- leaching. Volatilized N can affect N_2O emissions through the deposition to both agricultural and non-agricultural soils and waters where it undergoes transformations and result into N_2O emissions (Del Grosso et al. 2006). A portion of NO_3^- that is leached or discharged into drainage is also denitrified and emitted as N_2O.

There is a linear relationship between fertilizer N (F; kg N ha^{-1} $year^{-1}$) applied and N_2O emission (E; kg N_2O-N ha^{-1}) (Bouwman 1996):

$$E = 1 + 0.0125F \tag{6.1}$$

Background flux of ~1 kg N_2O-N ha^{-1} $year^{-1}$ is estimated to originate from recycling of non-fertilizer N in cultivated land i.e. crop residues, SOM, atmospheric deposition and previous year fertilization. It is estimated that ~90% of all direct contributions of fertilizer N to N_2O are encompassed within the range of 0.25–2.25% of all applied N, with an average value of 1.25%. This value is the basis for the recommendations of the IPCC emission factors (EF) for calculating emissions from croplands (IPCC 1997, 2006). The experimental studies from which these values were derived were mainly conducted in USA, Canada, and North-Western Europe (Bouwman 1996). In tropical and subtropical climates, N-use efficiency tends to lower (Pilbeam and Warren 1995), and N_2O emissions rates from tropical soils are higher than their temperate counterparts (Granli and Bockman 1995). Therefore, more data on EF are required before such calculation of factors can be used with confidence for fertilized lands in warm climates (Granli and Bockman 1995).

The EF approach for estimating N_2O emissions contribution from fertilizer N has been widely scrutinized. Clayton et al. (1997) and Smith et al. (1997) showed that the general trends for EFs from agricultural sites in Scotland are somewhat

Table 6.1 Estimates of N_2O emissions from cropland in the USA, Canada and the world in 1995 (Data from FAOSTAT IFA/FAO 2001)

	Land area	Fertilizer N applied	Animal manure	N_2O-N emitted		
				Total	Fertilizer induced	
Region	Mha	(Tg N year^{-1})	N applied			%
Canada	46	1.58	0.21	0.07	0.02	24
USA	190	11.15	1.58	0.32	0.11	35
World	1,436	81.0	20.7	3.50	0.81	23

lower than those reported by Bouwman (1996). They attributed this discrepancy to lower temperatures in Scotland sites than those sites from which the relationship was developed, and concluded that there is a significant increase in N_2O emissions with increase in temperature, if other factors are not limiting (Smith et al. 1997; Smith 1997). International Fertilizer industry Association/Food and Agriculture Organization of the United Nations (IFA/FAO) estimates showed that although North American field crop agriculture accounts for ~16% of the world field crop land area and 17% of world N consumption in 1995, its N_2O emissions are only ~12% (Table 6.1; IFA/FAO 2001). This trend suggests that the N_2O emissions may be higher in other regions of the globe.

There is a wide range of factors affecting N_2O emissions from fertilized cropland (IFA/FAO 2001) including: (1) climate, SOC concentration, soil texture, drainage, abundance of NO_3-N, and soil pH; and (2) management related factors including: N application rate, fertilizer type, crop type – with major differences between grasses, legumes, annual and biannual crops. Factors affecting fertilizer-derived N_2O emissions (Eichner 1990) include: (1) management factors — fertilizer type, application rate, application technique, application timing, tillage system, use of other agricultural chemicals, crop type, irrigation, and residual N and C from previous crop and fertilizer; and (2) environmental factors – temperature, precipitation, soil moisture content, SOC concentration, soil O_2 status, soil porosity, soil pH, freezing and thawing cycles, and microorganisms abundance and activity.

Based on the above-listed factors, the fraction of applied N actually emitted as N_2O varies widely on site-specific basis. The coefficients of variation for N_2O emissions measurements typically range between 100 and 300% (Thornton and Valente 1996). The current Tier 1 method emission factor is 0.01 with an uncertainty range of 0.003–0.03 (IPCC 2006). The use of EF approach for estimating N_2O emission is appropriate only when crops are fertilized at N rates less than or equal to those required for maximum yields, because the percent of fertilizer N emitted as N_2O becomes more variable at higher N rates (McSwiney and Robertson 2005).

The top-down estimation is based on global rate of increase in atmospheric N_2O in comparison to total reactive N produced by anthropogenic activity (Crutzen et al. 2008). This approach indicates that the sum of direct and indirect emission could amount to 3–5% of N applied, a 3–5 times larger than the revised emission factor coefficients of IPCC.

6.2.4.2 Tillage Effects on Nitrous Oxide Emissions

Tillage systems may affect soil properties and influence N_2O emissions. Changes in soil biological and physical conditions resulting from tillage or its absence can also influence the N dynamics. Conservation tillage systems such as no tillage (NT) and reduced tillage (RT) are increasingly being used for crop production due to their profitability and environmental advantages over conventional tillage (CT). Studies on the effects of NT on N_2O emissions have shown variable results. Some have reported higher N_2O emissions than CT (MacKenzie et al. 1997; Ball and Ritchie 1999; Baggs et al. 2003), others have shown lower emissions from NT than CT (Kessavalou et al. 1998; Robertson et al. 2000; Ussiri et al. 2009) and still others have reported no difference among tillage practices (Choudhary et al. 2002; Elmi et al. 2003). Plowing increases aeration and increases soil evaporation. Accessibility of crop residues for soil microbes is enhanced, and it may induce pulses of N_2O emissions.

There are several soil and environmental factors that exacerbate N_2O emissions from NT soils compared to CT (Linn and Doran 1984; Weier et al. 1996; Smith et al. 2001). Important among these factors are soil compaction, reduced porosity, and increased denitrification. An example of N_2O emissions from spring-grown barley in Scotland indicated a higher N_2O fluxes under NT than CT, but the emissions from autumn-grown barley, where soil conditions were more favorable, were similar among NT and CT (Vinten et al. 2002). Synthesis of global data show that in humid ecosystems, N_2O fluxes are generally higher under NT during the first 10 years, but after 20 years, emissions under NT are lower than those under CT (Six et al. 2004). Therefore, the effects of changing from CT to NT need to be integrated over long periods to understand the net effect on N_2O emissions, which can vary among biomes, soil type, climate, and even within a specific agroecosystem.

6.2.4.3 Cropping Systems and Nitrous Oxide Emissions

Incorporation of N fixing crops in a rotation or cropping system may increase N_2O emissions in cropping systems. The apparent N input from previous year of soybean production in an NT corn-soybean rotation can drastically increase the background N_2O emissions compared to CT continuous corn, and even more strongly compared to NT continuous corn (Mosier et al. 2006). In most cases, N_2O emissions are not affected by previous spring N fertilization. Winter cover crops such as wheat (*Triticum aestivum* L) or rye (*Secale cereal* L) can prevent NO_3-leaching in the winter months on permeable soils, reduce drainage losses of NO3-N, and most likely reduce overall N_2O emissions from water sources (Feyereisen et al. 2006).

6.3 Grasslands

Grasslands covers about one quarter of earth's surface, occupying 117×10^6 km^2 of vegetated lands, and providing forage for more than 1,800 million livestock units and wildlife (WRI 2000). Globally, natural and managed grasslands contribute to the livelihood of more than 800 million people, including many smallholders. The global emissions of N_2O from grasslands are estimated at ~2.5 Tg N_2O-N year^{-1}, and comprising 18% of total N_2O emissions (Lee et al. 1997). About 22% of the terrestrial biosphere or 33×10^6 km^2 is pastoral land, used by grazing ruminants (WRI 2000; Oenema and Taminga 2005). Most of the grasslands exist under naturally–steppe, savannahs, and prairies in the semi-arid climates. However, less is known about these extensive livestock systems (Oenema and Tamminga 2005).

Managed and man-made grasslands are dominant in the more humid regions of north-west and central Europe, New Zealand, and parts of Australia and North and South America. Most of them would eventually revert to forest if not managed. In general, managed grasslands are more productive and yield more milk and beef from ruminants than natural grasslands. Three main types of managed grasslands can be distinguished based on the intensity of management: (i) extensively managed partly for agricultural production such as milk, meat and wool, and partly for the protection of wildlife and landscape, (ii) improved grass cover mainly used for meat production on large-scale farms with no or limited fertilizer input, and (iii) intensively managed, single species grass swards receiving fertilizer N, P, K, and usually involving the production of silage for use during winter season.

The intensively managed grasslands are mainly used for milk production, and the dairy herd is normally fed additional concentrates in addition to grass (Oenema and Tamminga 2005). The main feature that distinguish intensively managed grasslands from other agricultural crops are the range of species present, mainly grasses and N-fixing legumes, the wide range of conditions under which pasture species can be grown and utilized, and the uneven excretal inputs from grazing animals. Managed grasslands are highly productive, and the major goal for the pastoral farmers is to increase pasture production and raise animal productivity per unit area. The continued use and productivity of managed pastures depend on management practices, such as grazing intensity, frequency of grazing, cultivation, re-sowing, and renovation, recycling of animal wastes, and fertilizer application.

The N cycling within and from grassland ecosystems is complex and multicompartmental, with many pathways for the release of mobile forms of N into atmosphere or aquatic systems. Grassland soils differ from other managed ecosystems in the nature and extent of the internal recycled N from mineralization and excreta added to the pool of the available and potentially available N in soil (Fig. 6.1). The N transformation processes which operate in these soils are not peculiar to grasslands, but the balance of flows along the various pathways and the impact of animal production management on the competition between the processes involved have a major effect on the form of gaseous emissions released. In lands where grazing is practiced, large amount of animal excreta are directly deposited on the

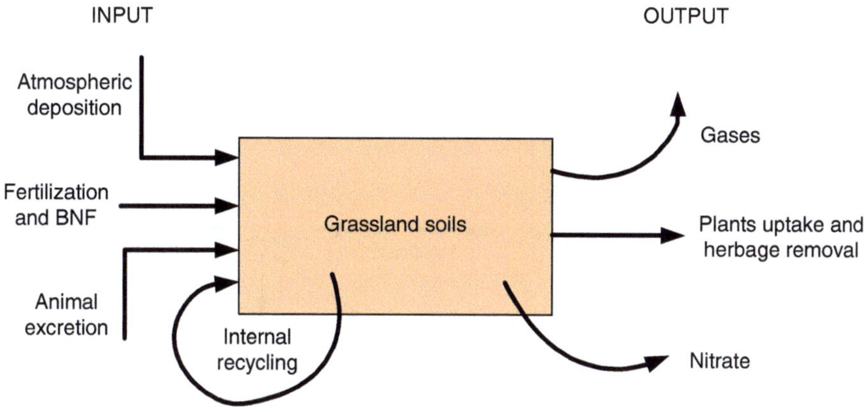

Fig. 6.1 Nitrogen flows in grassland ecosystems

land. In addition to animal excreta, N is also derived from biological N fixation (BNF), atmospheric deposition, and from manures and fertilizers (Fig. 6.1).

Temperate grasslands occupy about 9×10^6 km^2, and characterized by seasonal variations in temperature, radiation, and precipitation throughout the year. The pasture-growing season is regulated by temperature and moisture regimes. During winter, temperatures are low and the growth is limited. In spring, growth starts with temperature increase. Although in legume-based pastures most of the N could be derived from BNF, a small amount of N is traditionally added during the early spring season mainly to overcome the N deficiency caused by the slow rate of BNF and mineralization of SOM. Pasture growth may be limited by low moisture content in summer and autumn seasons. The marginal increase in pastures during the winter-to-spring transition period has high returns in milk production because of slow growth rates of pastures and high requirements by lactating cows. Managed temperate grasslands have high demands for N to maintain high productivity and long-term use. The N is often a limiting nutrient in grassland ecosystems, and mineral N fertilizers or organic manures (i.e., farmyard manures, cattle slurry) are applied to increase herbage productivity. The global demand for fertilizers has grown at an average rate of 1.7% annually since 1980s (IFA 2007). The increase represents a significant cost to the environment through increased NO_3^- leaching, enhanced NH_3 volatilization, and NO and N_2O emissions. In highly productive managed pasture soils, most of N_2O is generated by mineral N originating from applied N fertilizers, dung, urine, biologically fixed N_2, farm effluent applied primarily as a means of disposal and mineralization of soil organic N (Fig. 6.2).

The uneven depositions of animal excreta under grazed pastures are the source of large spatial and temporal variability in soil physical integrity, N availability, N and C transformations, and large uncertainty in N_2O emissions. The N_2O emissions also show strong variability in response to changes in environmental conditions such as temperature and rainfall (Smith et al. 1998; Flechard et al. 2007). The reliable measurements of N_2O fluxes from grazed pastures are therefore, very

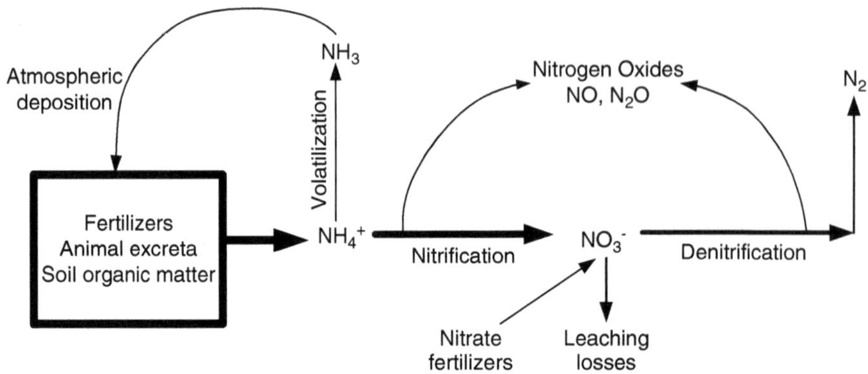

Fig. 6.2 Nitrogen emissions from grassland soils ecosystem

challenging because of strong spatial and temporal variability. The application of N inputs to grazed pastures has been increasing steadily, and is expected to continue to increase in the foreseeable future to meet the increasing demand for feeds. Increased fertilizer N inputs along with continued high intake of animal protein in developed countries and change in diet of people in developing countries are likely to exacerbate the N losses from global food production (McCarl and Schneider 2000; Mosier et al. 2001). The increasing N input in pastures has been the cause of concern regarding its impact on atmospheric, terrestrial, and aquatic environments.

Available N may temporarily exceed plant needs, and losses may occur via NO_3^- leaching to ground water and through gaseous emissions of NH_3, NO and N_2O even when timing and amounts of N-fertilization is optimized. Loss of N occurring mainly through NH_3 volatilization, biological denitrification, and NO_3^- leaching (Di and Cameron 2002a) has both economic and environmental implications. N is an important plant nutrient and its loss affects both the quality and quantity of the animal feed, leading to reduced animal productivity. In addition, environmental impacts of N loss through leaching and gaseous NO and N_2O emissions are major environmental and climate change concern. The emission of N_2O from soils is generally related directly to N inputs. Grazed pastures are identified as an important source of NH_3 and N_2O, which contributes to acid rain, ozone depletion, and global warming. The N_2O emissions from grasslands occur in short-lived bursts following the application of fertilizers (Leahy et al. 2004; Clayton et al. 1997), and as a result of animal excreta during grazing. As a result, small scale spatial variability in fluxes are very high (Ball et al. 2000), which creates technical and analytical challenges for flux measurements, and are a major source of uncertainty in N_2O fluxes at the field and annual scales (Flechard et al. 2005).

Availability of mineral N (NH_4^+ and NO_3^-), temperature, pH, and processes that reduce redox potential of soil such as changes in soil moisture, soil texture, and labile organic carbon (OC) are the major factors that affect the production of N_2O in soils. Management practices such as N fertilization, use of N inhibitors, cultivation, irrigation, and grazing intensity can have significant impacts on N_2O emissions. The N_2O production conditions under grasslands differ from those under cropland. Perennials have a longer growing season and, therefore, a prolonged N uptake as compared to annual summer crops and no period of bare soil without N uptake. This trend helps to avoid the accumulation of mineral N in soils that may happen after crop maturity and harvest under croplands.

6.3.1 *Nitrogen Inputs to Grasslands*

In most grassland ecosystems, N is often limiting and mineral and organic N fertilizers are frequently applied to increase herbage productivity and improve soil fertility. The major sources of plant available N in grasslands are N fertilizers, BNF, animal excreta, atmospheric N deposition, and mineralization of soil organic nitrogen (SON). In most pasture soils, the amount of N mineralized from SOM cannot support high pasture production, and in addition to BNF, supplemental application of N either in the form of fertilizer or manure is necessary.

6.3.1.1 Biological Nitrogen Fixation

Use of legume-based pasture is most common grazing management practice in North America, Europe, Australia, and New Zealand (Bolan et al. 2004). In such pastures, N is derived mainly from BNF by nodules of legume plants. The amount of BNF occurring in legume-based pastures depends on legume species, soil and climatic conditions, presence of microorganisms, nutrient supply, and grazing management. High levels of available phosphorus (P) in soils are essential for maintaining the presence and N_2-fixing activity of legumes in legume-based pastures. Similarly, adequate levels of other plant nutrients, including sulfur (S) and molybdenum (Mo) in particular, are required. High concentrations of inorganic N in soils tends to inhibit BNF, as legumes tend to utilize available soil N. Therefore, application of fertilizer N to pasture soils decreases BNF. Reduction of BNF ranging from 30to 70% has been reported for clover, depending on time of application and the grazing management (Ledgard et al. 1996). The BNF rates in the range of 100–300 kg N ha^{-1} $year^{-1}$ are common for grass-clover pastures (Ledgard et al. 2001; Bolan et al. 2004).

6.3.1.2 Animal Excretal Nitrogen

Globally, there are ~1,380 million cattle, 1,100 million sheep, 830 million goats, 990 pigs, and 16,800 million chicken (FAO 2006), which produce ~20,000 Tg $year^{-1}$ of fresh manure. Animal excretion (dung and urine) contributes between

80 and 130 Tg N $year^{-1}$ to the biosphere (Oenema and Tamminga 2005). The amount of N contained in animal manures is as large as, or larger than the current global annual fertilizer N consumption of ~95 Tg $year^{-1}$ by agriculture (IFA 2007). In grazed pastures, a substantial amount of N is recycled through direct deposition of animal excreta. Cattle retain up to 20% of the total N intake from fodder and feeds. The remaining intake is excreted in urine and feces. The proportion of total N intake excreted and its partition between urine and feces is dependent on type of animal, the intake of dry matter, and N concentration of the diet.

Where pastoral grazing predominates over housed animals systems, the most important source of N_2O emissions is the urine deposited by grazing ruminant animals (Di and Cameron 2002b; Saggar et al. 2005). Some N_2O is emitted as soil NH_4^+ from urine nitrifies to nitrate, but emission is mainly during denitrification of nitrate under anaerobic conditions (Saggar et al. 2004) by denitrifying soil bacteria, where there is readily available source of C (Haynes and Williams 1993). In contrast to urine, N deposited on the soil surface in feces is mainly in the organic form and undergoes relatively slow mineralization, resulting in smaller quantities of N_2O emissions. Large temporal and spatial variability of N_2O emissions from animals grazed pastures is attributed to uneven distribution of excretal N returns, soil heterogeneity, compaction caused by animal trampling and the episodic nature of N_2O emissions (van Groenigen et al. 2005; Bhandral et al. 2007; Saggar et al. 2007).

6.3.1.3 Fertilizer Nitrogen

Fertilizer N is the most effective management input for manipulating pasture production. The N fertilizers are widely used in the grass-based intensive pasture production systems of Europe and North America. Pastoral lands are receiving 14–15 Tg N $year^{-1}$ or 15–16% of global fertilizer N usage. Fertilizer application rates to pastures vary with the production objectives. Where the forages are harvested, large quantities of nutrients are needed to replace the nutrients removed, and N rates are generally greater than in grazed pastures where N is returned in the form of animal excreta. In addition, N_2O emissions differ between the cut and removal versus grazed grassland systems.

The principles of N application are more complex under pastures than under croplands because: (i) pastures have high demand for N throughout the growing season (6–9 months in temperate regions), (ii) climatic conditions vary during the growing season (iii) sward age differs, and its botanical composition is variable, and (iv) grazing management and grazing intensity differ. Pure grass pastures often responds to N linearly up to 200–400 kg N ha^{-1} $year^{-1}$. The average N application rate varies with the production objectives. Higher rates of N are needed where pastures are cut and removed, whereas lower rates can be used where the pastures are grazed.

6.4 Forest Land

Forestland covers nearly 42×10^6 km^2 in tropical, temperate, and boreal regions, or about 30% of the total land surface. Forests provides ecological, social, aesthetic and environmental services to mankind, including refuge for biodiversity, forest products, moderation of the hydrologic cycle, protection of soil resources, and recreational use. In addition, forests influence climate through exchange of energy, water, greenhouse gases, and other chemical species with atmosphere. Forests store ~45% of total terrestrial C and contribute ~50% of terrestrial net primary productivity (Sabine et al. 2004). Forests are a large sink of CO_2. Carbon uptake by forest is estimated at 2.6 Pg $year^{-1}$ (=Petagram, 1 Pg = 10^9 tons, = 10^{12} kg), or about 33% of the total anthropogenic C emissions (Denman et al. 2007). Forests have low surface albedo, which can mask high albedo of snow. In contrast, forests also sustain hydrologic cycle through evapotranspiration, which cools climate through feedback with cloud and precipitation.

The main sources of N in forest soils include the BNF of atmospheric N_2 by microbes –both free living and symbiotic, mineralization of SON, and atmospheric deposition. Similar to cropland and grassland soils, N_2O in forest soils is produced via nitrification and denitrification processes. Forest soils represent significant sources of N_2O (Schmidt et al. 1988; Skiba et al. 1994). The N_2O emissions from undisturbed forest soils are considered natural emissions. However, emissions of N_2O from forest soils have increased in recent decades, and are predicted to increase in the future as a result of the increased anthropogenic perturbations of the global N cycle (Galloway et al. 2004), and high rates of atmospheric deposition of reactive N in many forest ecosystems in Europe, North America and Asia (Kesik et al. 2005). Atmospheric N deposition to forest ecosystems is positively correlated with the N_2O emissions (Butterbach-Bahl et al. 1998; Zechmeister-Boltenstern et al. 2002).

The N_2O emissions rates from mature forest are on average higher than those from natural unfertilized and unmanaged grasslands, with deciduous forests having higher emissions than coniferous forests (Menyailo and Huwe 1999; von Arnold et al. 2005; Ambus et al. 2006). The observed difference is attributed to species- specific soil microbial community composition. Nitrogen cycling in forest soils is controlled by litter quality parameters such as C:N ratio, litter N concentration, lignin, and phenolic compounds (Venterea et al. 2004). Therefore, deciduous tree species generally have faster nutrient cycling and higher microbial activity than coniferous tree species (Smolander et al. 2005). However, fertilized grasslands and pastures typically produce higher emission rates than forests and natural ecosystems (Flechard et al. 2007; Saggar et al. 2008). The fluxes from mature undisturbed forest stands are governed by tree type and hydrologic conditions (Jungkunst et al. 2008; Ullah et al. 2008). The parameters affecting N_2O emissions differ significantly among young and mature forests, which include distribution and quality of SOM, C:N ratio, microbial community composition, and hydrological conditions.

6.4.1 Factors Affecting Nitrous Oxide Emissions from Forest Soils

The main focus of research on N_2O emissions in the past has been on agricultural ecosystems and much work remains to be done to estimate the role of forest ecosystems in N_2O emission. Production of NH_4^+ and NO_3^- in the forest floor by microbial processes of gross ammonification and gross nitrification is of critical importance for plant N supply. Nitrification and denitrification processes regulate N losses from the forest ecosystems via both hydrologic and gaseous flow paths. Generally, in the undisturbed forest ecosystems, inorganic N (NH_4^+ and NO_3^-) production is counterbalanced by microbial immobilization and plant uptake, thus resulting in moderate concentrations of inorganic N and losses along hydrological and gaseous flow paths (Vitousek and Melillo 1979). Atmospheric deposition and management interventions such as logging can alter the balance between nitrification, plant uptake, and microbial immobilization, thus leading to enhanced soil NO_3^- concentrations and N losses from the ecosystem (Di and Cameron 2002a; Borken and Matzner 2004). The N deposition may alter soil C:N ratio, and its long-term cumulated effect may control production and emission of N_2O from forest soil. Nitrification increases with decrease in C:N ratio below a threshold value of 22–25 (Lovett et al. 2004).

The emission of N_2O from forest soils is mainly the result of simultaneously occurring production and consumption processes, which are directly linked to microbial N turnover processes of N mineralization, nitrification, and denitrification (Conrad 2002). These processes vary largely on spatial and temporal scales, and number of environmental factors, such as climate and meteorological conditions, soil and even vegetation properties, influences them. Because of changes in these properties, emissions of N_2O from forest soils can vary on several orders of magnitudes among seasons, years, and even among sites within the same region (Papen and Butterbach-Bahl 1999; Butterbach-Bahl et al. 2002). Although denitrification process produces most of N_2O emissions under forest soils, nitrification processes in soils control the N_2O emissions in these soils (Ambus et al. 2006).

Tropical forests have high rates of biological productivity due to abundant solar radiation, and rapid decomposition rates of SOM due to high temperature and high rainfall. High turnover rates make these forest soils rich N sources compared to those under temperate forests. As a result, tropical rain forests emit high rates of N_2O (Davidson et al. 2000), and are an important source of global N_2O emissions. The median rate of N_2O emission from tropical forest is estimated at 1.7 kg N_2O-N ha^{-1} (Bouwman 1998).

6.4.2 Forest Clear-Cutting and Nitrous Oxide Emissions

Increased mineralization rates after clear-cutting may significantly increase N_2O emissions in short-term (Huttunen et al. 2003). Logging removes forest canopy, interrupt nutrient and hydrologic cycles, and may impact soil physical conditions.

Its impact may influence GHG fluxes. Logging also causes soil compaction and decreases air-filled pore space. The parameters affecting N_2O emissions differ significantly among young and mature forests, which include distribution and quality of SOM, C:N ratio, microbial community composition and hydrological conditions at the site (Inagaki et al. 2004; Gartzia-Bengoetxea et al. 2009; Macdonald et al. 2009).

6.5 Factors Controlling Nitrous Oxide Production in Soil

The production of N_2O from soils is dominated by biotic processes of nitrification and denitrification (Chap. 3). A prerequisite to nitrification in natural soils is the mineralization of organic N directly to NH_4^+ or via excretion of urea that is hydrolyzed to NH_4^+ by urease enzyme. Alternatively, NH_4^+ can enter the soil by addition of ammonium base fertilizers or atmospheric deposition. The prerequisite for denitrification is the presence of a source of NO_3^-. Except where NO_3^- based fertilizers are added, nitrification can be viewed as a prerequisite for denitrification. However, the different oxidative conditions required for nitrification and denitrification dictates that the processes must be separated either spatially or temporally. In addition, chemical decomposition of HNO_2 (i.e., reduction of NO_2^- by chemical reductants under limited O_2 conditions and at low pH – chemical denitrification) can produce N_2O together with NO and N_2. Chemical denitrification can occur when nitrification rates are high (e.g., after application of NH_4^+-based fertilizers). The hole-in-the-pipe concept has been used to describe the gaseous N (NO, N_2O and N_2) emissions from soils (Bouwman 1998; Davies et al. 2001). In this model, gas production and exchange with the atmosphere depends on: (i) factors controlling the amount of N flowing through the pipe (i.e., those factors which affect nitrification and denitrification rates such as N availability and temperature), and (ii) the size of the hole through which N gases leaks. Factors controlling the partition of the reacting N species to NO, N_2O or N_2 moderate the size, while the rate at which N moves through the pipe determines the importance of the leaks (Table 6.2).

The N_2O emissions from soils arise from low, relatively constant continuous emissions and generally from short emission peaks (episodic) commonly associated with denitrification (Firestone and Davidson 1989) or both denitrification and nitrification. The impact of soil factors on N_2O emissions are summarized in Table 6.2. The N_2O fluxes from soils are relatively small compared to other N fluxes, and are dependent on soil temperature, soil water content, O_2 availability, N substrate availability (NO_3^- and NH_4^+), and organic C substrate availability (Davidson 1991). These factors are influenced by climate, vegetation, soil physical and chemical properties, land use and land management factors. The N_2O production is also influenced by other interacting N processes in the plant-soil N cycle such as plant N uptake, NH_3 volatilization, and nitrate leaching. These variables contribute to high spatial and temporal variability of N_2O emissions.

Table 6.2 Factors controlling N_2O emission from soils

Factor	Effect on N_2O emission	References
Soil moisture (water filled pore spaces)	Low emissions below 40%, linear increase between 55% and 65%, exponential increase above field capacity to 90%, at waterlogged soils N_2O is reduced to N_2	Dalal et al. (2003), Granli and Bockman (1994)
Mineral N (NH_4^+ and NO_3^-)	Emissions increases with increase in mineral-N. Positive correlations with NH4+ supply, increased emissions from denitrification under anaerobic conditions. No effect of NO_3^- when available C is limiting	Granli and Bockman (1994), Dalal et al. (2003), Wrage et al. (2004)
Soil temperature	Sharp increase between 5 and 20 °C in temperate soils, and above 20 °C in sub-tropical soils. Maximum temperature is 37 °C	Dalal et al. (2003)
Soil texture	Low emissions in coarse-textured soils, high emission in fine textured soils	Saggar et al. (2004)
Soil aeration/O_2 availability/ redox potential	High water content increases emissions. Restricted diffusion in water-logged soils. Regulated within narrow redox potential range of +150 to +250	Wrage et al. (2004), Dalal et al. (2003)
Soil pH	Increasing pH increases the proportion of denitrification and reduces N_2O emissions	Godde and Conrad (2000)
Soil compaction	Emission increases with soil compaction	Granli and Brockman (1994)

Proximal soil factors – substrate availability, O_2 availability, soil moisture, and soil temperature drive the microbial processes of N_2O production and consumption at the micro scale. These proximal physical and chemical factors are themselves controlled by biological drivers, which supplies substrate – such as readily degradable SOM, and O_2 demand of the decomposing organisms and roots. For example, O_2 availability is driven by consumption through microbial and root activity as well as diffusion constraints through soil structure and soil water content. Water content is determined by water balance as a function of local factors such as precipitation, interflow, drainage rate, and evapotranspiration, which depends on climate, landscape position, soil texture, plant type, and season. Therefore, localized factors govern N_2O emission rates on a daily or weekly time scale. Land management may also contribute to N_2O emissions, but the driving forces of the microbial activity are intimately linked to climate, site properties (Smith et al. 1998) and land use history (Mosier et al. 1998).

6.5.1 *Soil Moisture and Aeration*

Soil moisture, together with temperature, controls soil processes at all levels by governing the SOM decomposition, nitrification and denitrification. Soil moisture content controls the O_2 partial pressure in the gas phase and the availability of O_2 in soils. The N_2O and NO are produced in soil in the course of two contrasting microbial processes of nitrification and denitrification. Nitrification is an aerobic process (up to 50% water filled pore spaces (WFPS)). Nevertheless, when the supply of O_2 is limited by diffusion constrains the nitrifying bacteria can use NO_2^- as an electron acceptor and reduce it to NO and N_2O (Poth and Focht 1985; Bollmann and Conrad 1998), which can be represented in Eq. 6.2:

$$\begin{array}{c} \qquad\qquad\qquad\mathrm{NO, N_2O(emission)} \\ \qquad\qquad\qquad\uparrow \\ \mathrm{NH_4^+ \rightarrow NO \rightarrow NO_2^- \rightarrow NO_3^-} \end{array} \tag{6.2}$$

Chemoautotrophic nitrifying bacteria gain energy from the oxidation process of NH_4^+ to NO_2^- and NO_3^-. Potentially, both N_2O and NO may be emitted during nitrification in soils. Normally NO emission is 10–100 times more than N_2O (Hutchinson and Davidson 1992). The N_2O emissions increase with increase in soil moisture (Hutchinson and Davidson 1992). The N lost as N2O during nitrification is $\leq 2\%$ of total nitrified N, but can be as high as 4% in some cases (Duxbury and McConnaughey 1986; Mosier et al. 1996). The N_2O and NO may also be produced from non-biological reactions of NO_2^- (HNO_2) with SOM or dissociation of NO_2^- itself. These non-biological reactions are concentration dependent, and are affected by NO_2^- concentration, soil pH and SOM concentration (Blackmer and Cerrato 1986).

Denitrification is a reduction of NO_3^- via NO_2^- and NO to N_2O and N_2 (Eq. 6.3), is favored under low O_2 supply.

$$\begin{array}{c} \mathrm{N_2O(emission)} \\ \uparrow \\ \mathrm{NO_3^+ \rightarrow NO_2^- \rightarrow NO \rightarrow \quad N_2O \quad \rightarrow N_2(emission)} \end{array} \tag{6.3}$$

The N losses as N_2O from denitrification are substantial and can occur very rapidly (i.e., within 24 h) in soils that are wet but not waterlogged; about 70% WFPS (Linn and Doran 1984) and warm enough to support rapid microbial activity (Sahrawat and Keeney 1986). Under anaerobic conditions, about 60–80% of the NO_3^--N can be lost as gases (N_2O, NO and N_2) (Mosier et al. 1996). The WFPS defined by Eq. 6.4 is a variable used for measuring soil water status in field and incubation studies. It is also used in models to predict the influence of soil water content on N_2O emissions.

$$WFPS = \frac{\theta_v}{TP} = \frac{\theta_m * \rho_b}{1 - \frac{\rho_b}{\rho_p}} \tag{6.4}$$

where θ_v is the volumetric water content (cm^3 water cm^{-3} total soil), *TP* is the total porosity (cm^3 pores cm^{-3} total soil), θ_m is the gravimetric water content (g water g^{-1} soil) ρ_b is the bulk density (g soil particles cm^{-3} total soil), and ρ_p is the particle density (g soil particles cm^{-3} soil particles).

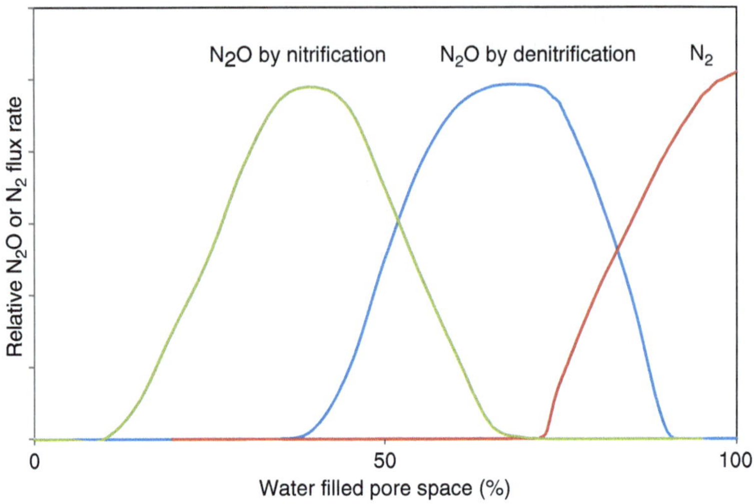

Fig. 6.3 A generalized schematic indicating relationship between water-filled pore space (WFPS) of soils and relative fluxes of N_2O and N_2. N_2O can be produced by both nitrification and denitrification, N_2 is produced by denitrification (Redrawn with from Firestone and Davidson 1989)

Nitrification and denitrification processes are controlled by several factors – particularly WFPS (Davidson 1991), which depends on the balance between the amount of water entering the soil through precipitation or irrigation and the combined effect of evapotranspiration and drainage. Soil water content influences N_2O emissions from all soil types. The N_2O emissions depend on O_2 availability for microbial processes and the ease of escape of N_2O by diffusion through soil pores. Soil water content is a controlling factor for both. Figure 6.3 shows the dependence of N_2O production by nitrification and denitrification on soil moisture in a schematic way. The N_2O emissions are favored when the soil is sufficiently wet to restrict O_2 availability. If soil becomes very wet, nitrification ceases to occur, denitrification proceeds increasingly to N_2, and gas escape from soil is hindered (Fig. 6.3). Thus, at soil water content where both nitrification and denitrification can proceed generally lead to the maximum emission of N_2O.

In general, microbial activity peaks at 30–60% WFPS. The rate of N_2O production during nitrification is normally low below 40% WFPS, but increases rapidly with increasing soil water content and peaks at 55–65% WFPS (Bouwman 1998; Davidson and Verchot 2000; Dalal et al. 2003). Above 60 –70% WFPS, an increase in soil water content hinders aeration by limiting O_2 diffusion, therefore nitrification slows down. Denitrification becomes a dominant source of N_2O between 70 and 90% WFPS. The WFPS above 90% produces undetectable N_2O emissions mainly because any N_2O produced is completely reduced to N_2. Low concentrations of O_2 in soil occur when consumption of O_2 in the soil by plant roots and soil microorganisms exceeds the rate of replenishment by diffusion from the atmosphere, and anaerobic microsites are created within the soil profile (Smith et al. 2003). The NO_3^- is the chemical species that readily acts as the electron acceptor once O_2 is exhausted. The fraction of the total gaseous products of anaerobic denitrification emitted to the atmosphere as N_2O depends on the structure and wetness of the soil. If an N_2O molecule can easily diffuse from the site of production to an oxygenated pore space, it has higher chance of being emitted to the atmosphere rather than being reduced to N_2. In contrast, N_2O molecule produced well below the surface of a saturated clod is much more likely to be reduced to N_2 than to escape. The distribution of regions in soils producing significant fluxes of N_2O are inherently very complex, and can occur as a three-dimensional mosaic of anaerobic microsites producing N_2O within an otherwise aerobic matrix (Smith et al. 2003).

The denitrification pathway to N_2O has been established as having to go via NO (Eq. 6.3). However, NO emissions are virtually absent from gaseous emissions in wet soils where anaerobic conditions occur. This trend has been attributed to a greater tendency for NO than N_2O to be consumed by the denitrifying organisms (Firestone and Davidson 1989). Thus, NO emission from denitrifying soil is usually detected only in laboratory experimental conditions when NO is swept out of a column of soil by a stream of O_2-free gas (Johansson and Galbally 1984).

The O_2, moisture status, and gas diffusion in soils depend on soil texture and drainage. Fine–textured soils have more capillary pores within the aggregates than sandy soils, and hold soil water more tightly. As a result, anaerobic conditions may be more easily reached and maintained for longer periods within aggregates in fine–textured than in coarse-textured soils. Under flooded conditions, however, any N_2O produced in soils is completely reduced to N_2. This is typically observed under paddy rice, where flooding prevents N_2O emissions. Management practices of rice paddies to control N_2O emissions is discussed in detail in Chap. 7.

The wetting of dry soils causes pulses in N mineralization, nitrification, and N_2O fluxes. The alternating drying and wetting of soils enhances the release of N_2O from the soils to the atmosphere, even though peaks in N_2O production may decline with subsequent wetting events due to N limitation. The N_2O emissions from soil also depend on tortuosity and connectivity of the pores (Rappoldt and Crawford 1999).

6.5.2 *Temperature*

It has been observed that the N_2O emission increases exponentially with increase in temperature (Brumme 1995; Flechard et al. 2007; Cantarel et al. 2011). Research on N_2O emissions from permanent grassland across Europe indicated that 48% of the temporal variability in N_2O fluxes across the sites could be explained by soil temperature and moisture with increase in N_2O emissions in warm and wet environments (Flechard et al. 2007). Formation of N_2O is reduced when soil temperature is below 10°C (Brumme 1995). At sufficient soil moisture content, the N_2O emissions depend on temperature. Decomposition of SOM and mineralization of N is stimulated by temperature increases, especially in temperate zones (Winkler et al. 1996), and the enhanced mineralization increases the emissions of N_2O. Although an increase in temperature results in increased N_2O, the increase is not straightforward because many contrasting processes are involved in the emission of N_2O. For example, denitrification produces N_2O, but it can also consume N_2O. Therefore, stimulation of denitrification by temperature increase can also result into increased consumption of N_2O.

Both nitrification and denitrification rates increase with increase in temperature until the optimum temperature for microbial growth is reached. However, there is evidence suggesting that both nitrification and denitrifying organisms can adaptation temperature (Stark and Firestone 1996; Parton et al. 2001). The temperature response to nitrification is approximately bell-shaped with an optimum between 20 and 35°C. The decline at the higher temperatures may be due to increased biological O_2 consumption. Denitrification occurs in anaerobic zones in soils. An increase in temperature leads to an increase in the size of the anaerobic zones because of increased respiration, causing larger gradients of O_2 concentrations (Smith et al. 2003). It can also lead to increased rate of denitrification per unit anaerobic volume. The overall change in the rate of N_2O production in a soil mass is the product of both temperature and WFPS. An increase in WFPS and overall decrease in O_2 diffusion rate within soil has similar effect to soil system as that induced by rise in temperature.

6.5.3 Soluble and Mineralizable Carbon

Easily mineralizable SOM content stimulates microbial activity, and thus, N_2O emissions (Chadwick et al. 2000; Azam et al. 2002). The SOC provides C and energy source for heterotrophic denitrifying organisms. Although most nitrifiers are ubiquitous autotrophs that obtain their energy from inorganic sources, heterotrophic nitrification can occur under certain conditions, and also requires a source of OC. In addition, SOC can stimulate biological O_2 demand from aerobically respiring organisms, which enhances O_2 consumption and reduce O_2 status where rates of O_2 diffusion cannot replenish the consumed O_2. This conditions leads to increase in anaerobic volume of soil and a concomitant increase in denitrification potential. Denitrification is an oxidation-reduction process that require source of energy from OC and N oxides as terminal electron acceptors in absence of preferred acceptors such as O_2. Inputs of SOC are usually from plant residues, roots exudates and other organic sources. Denitrification increases with increase in organic OC content, especially water-soluble OC (Drury et al. 1991). Incorporation of plant materials enhances the rate of denitrification (Aulakh et al. 1991). However, the ratio of N_2O/N_2 from denitrification decreases with increasing available OC (Weier et al. 1993). Addition of biomass rich in degradable OM increases N_2O production in soil containing NO_3^- or supplied with fertilizer NO_3^-.

6.5.4 Soil pH and Salinity

Soil pH is an important regulator of nitrification and denitrification rates as well as the ratio of products derived from these processes. Soil pH controls denitrification by mainly affecting the nitrification process. Soil pH may also affect other N transformation processes such as mineralization and immobilization; therefore influence N_2O emissions indirectly by altering the production and consumption of NH_4^+ and NO_3^-. The optimum pH range for nitrification and denitrification is 7–8. Although NO_3^- reduction has been detected at pH as low as 3.5, denitrification declines if soil acidity shift towards lower soil pH (Simek and Cooper 2002). The N_2O production is enhanced at pH of 5.5–6.0. As pH increases, denitrification products tend to be more or completely towards N_2 production. Soil pH has a marked effect on the products of denitrification. Denitrification rates are slower under acid than under slightly alkaline conditions, but N_2O fraction may be larger at low soil pH, particularly with an adequate NO_3^- supply. This is generally attributed to sensitivity of N_2O reductase to proton activity. Emission of N_2O decreases with increasing soil pH in acid soils. Autotrophs are the main nitrifying agents in most acid soils. The nitrification at low pH is limited by the supply of NH_3, which are the primary substrate rather than NH_4^+ (Suzuki et al. 1974).

High salinity inhibits both nitrification and denitrification (Inubushi et al. 1999). The N_2O reductase is susceptible to salt, which may result in N_2O accumulation from denitrification under saline conditions (Menyailo et al. 1997). Therefore, N_2O production from nitrification can be accentuated by increased salt concentration (Low et al. 1997).

6.5.5 Soil Nitrogen

Nitrification and denitrification rates are strongly influenced by soil N content. Therefore, N fertilizers, animal manures, crop residues, N deposition, and BNF are strong stimulants of denitrification and associated N_2O fluxes. Generally, denitrification increases with increasing NO_3^- content in soil under conditions suitable for denitrification. These include high soil moisture content, low O_2, optimum temperature, and SOC. Under most circumstances, presence of NO_3^- inhibits the reduction of N_2O to N_2, favoring N_2O emission. The magnitude of NH_3 volatilization determines the availability of N for nitrification and denitrification. Denitrification is lower when NH_3 losses are high. Under normal field conditions, nitrification is limited by release of NH_4^+ from SON mineralization. Addition of urea or NH_4^+-N to soils under aerobic conditions produces more N_2O than similar amount of N in NO_3^- form (Wang and Rees 1996).

6.5.6 Other Nutrients Limitations

Deficiency of other essential plant nutrients for plant growth limits the ability of plants to utilize NH_4^+ or NO_3^- N. For example, P-limited forest emitted more N_2O and NO than N-limited forest with similar rates of N application (Hall and Matson 1999), suggesting that P deficiency may limit N uptake by plants, favoring higher N_2O emissions.

6.6 Conclusions

Agricultural land use, consisting of croplands, permanent crops and managed grassland occupy about 40–50% of Earth' surface and accounts for 60–80% of global anthropogenic N_2O emissions or ~6 Tg N_2O-N. Current anthropogenic N_2O emissions may escalate in the future due to population growth, demand for food and changing diets, and increase in demand for fuel and bioenergy crops. The N input in the form of N fertilizers and manures into agricultural ecosystems to support increased crop and pasture production are the major sources of anthropogenic N_2O emissions. Production and emissions of N_2O in soils is mainly controlled by the factors that affect the growth of microorganisms in the environment, including temperature, moisture content, substrate availability, pH, and aeration. However, fluxes of N_2O fluxes from agricultural soils are accentuated by land management factors including land use practices, vegetation type, tillage practices, fertilizer N rates, SOC concentration, soil aeration, NO_3^- concentration, soil compaction, water-holding capacity of soil, irrigation, and use of agricultural chemicals. Therefore, fluxes of N_2O from agricultural ecosystems are the result of complex interactions of various parameters, including climate, soil physical, biological, and chemical properties, as well as land management.

Global N_2O emission from grasslands is estimated at 2.5 Tg N_2O-N $year^{-1}$. The N_2O fluxes from grazed grasslands are highly variable, which creates technical challenges for fluxes measurement and creates a significant uncertainty in global emissions estimates from grasslands. The grassland soils ecosystem differ from cropland in the nature and extent of internal recycled N from various pools. In gazed grassland, large amounts of N from animal excreta are directly deposited on the grasslands, which contribute to N_2O emissions.

Most of N_2O emissions from unmanaged forest ecosystems are of natural origin. The N_2O fluxes from mature undisturbed forest stands are controlled by tree species and hydrologic conditions, and are generally higher under deciduous than coniferous tree species. Factors influencing N_2O emission from forest ecosystem include SOM, vegetation type, C:N ratio, lignin and phenolic composition of forest litter, microbial community composition, N deposition and hydrological conditions.

The WFPS, O_2, temperature SOC, rate of SON mineralization, soil NO_3^- content, and soil pH affects the rates of nitrification and denitrification and eventually determines the rates of N_2O emissions as well as the ratio of N_2O:N_2 from soils. The magnitude of interaction of these factors controls the magnitude, spatial and temporal variability of N_2O emissions.

6.7 Research Needs

- The N losses, N fertilizer N_2O emission factors need to include the measurements of NH_3 and NO emissions and NO_3^- leaching and their contribution to N_2O emissions off site.
- Long-term monitoring of N_2O fluxes extending beyond the growing season, which encompasses crop rotation and cropping system changes, are critically needed.
- The comparison N_2O emissions among different N sources, placement, and timing to minimize the potential for confounded interpretations.
- The N_2O emission data from unmanaged grasslands and forest soils are lacking.
- Contribution of bio-energy crops to N_2O emissions, both in terms of its production and utilization for energy should receive immediate attention.

Study Questions

1. Describe how the major sources of N_2O from agricultural, grassland and forest ecosystem differ.
2. What are the major factors controlling N_2O emissions from the soils and how are they modified under each of the three major ecosystems described in this chapter?
3. Under which circumstances may human activities influence the N_2O emissions from forest ecosystems?

References

Ambus, P., Zechmeister-Boltenstern, S., Butterbach-Bahl, K. (2006) Sources of nitrous oxide emitted from European forest soils. Biogeosciences **3**(2), 135–145

Aulakh, M.S., Doran, J.W., Walters, D.T., Mosier, A.R., Francis, D.D. (1991) Crop residue type and placement effects on denitrification and mineralization. Soil Sci. Soc. Am. J. **55**(4), 1020–1025

Azam, F., Muller, C., Weiske, A., Benckiser, G., Ottow, J.C.G. (2002) Nitrification and denitrification as sources of atmospheric nitrous oxide - role of oxidizable carbon and applied nitrogen. Biol. Fertil. Soils **35**(1), 54–61. doi:10.1007/s00374-001-0441-5

Baggs, E.M., Stevenson, M., Pihlatie, M., Regar, A., Cook, H., Cadisch, G. (2003) Nitrous oxide emissions following application of residues and fertiliser under zero and conventional tillage. Plant Soil **254**(2), 361–370

Ball, B.C., Horgan, G.W., Parker, J.P. (2000) Short-range spatial variation of nitrous oxide fluxes in relation to compaction and straw residues. Eur. J. Soil Sci. **51**(4), 607–616

Ball, B.C., Ritchie, R.M. (1999) Soil and residue management effects on arable cropping conditions and nitrous oxide fluxes under controlled traffic in Scotland 1. Soil and crop responses. Soil Till. Res. **52**(3–4), 177–189

Bange, H.W. (2000) Global change - It's not a gas. Nature **408**(6810), 301–302

Bedard-Haughn, A., Matson, A.L., Pennock, D.J. (2006) Land use effects on gross nitrogen mineralization, nitrification, and N_2O emissions in ephemeral wetlands. Soil Biol. Biochem. **38**(12), 3398–3406. doi:10.1016/j.soilbio.2006.05.010

Bhandral, R., Saggar, S., Bolan, N.S., Hedley, M.J. (2007) Transformation of nitrogen and nitrous oxide emission from grassland soils as affected by compaction. Soil Till. Res. **94**(2), 482–492. doi:10.1016/j.still.2006.10.006

Blackmer, A.M., Cerrato, M.E. (1986) Soil properties affecting formation of nitric oxide by chemical reactions of nitrite. Soil Sci. Soc. Am. J. **50**(5), 1215–1218

Bockman, O.C., Olfs, H.W. (1998) Fertilizers, agronomy and N_2O. Nutr. Cycl. Agroecosys. **52**(2–3), 165–170

Bolan, N.S., Saggar, S., Luo, J., Bhandral, R., Singh, J. (2004) Gaseous Emissions of nitrogen from grazed pastures: processes, measurementsand modelling, environmental implications, and mitigation. Adv. Agron. **84**, 37–120

Bollmann, A., Conrad, R. (1998) Influence of O_2 availability on NO and N_2O release by nitrification and denitrification in soils. Global Change Biol. **4**(4), 387–396

Borken, W., Matzner, E. (2004) Nitrate leaching in forest soils: an analysis of long-term monitoring sites in Germany. J. Plant Nutr. Soil Sci. **167**(3), 277–283. doi:10.1002/jpln.200421354

Borlaug, N. (2007) Feeding a hungry world. Science **318**, 359–359. doi:10.1126/science.1151062

Bouwman, A.F. (1996) Direct emission of nitrous oxide from agricultural soils. Nutr. Cycl. Agroecosys. **46**(1), 53–70. doi:10.1007/bf00210224

Bouwman, A.F. (1998) Nitrogen oxides and tropical agriculture. Nature **392**(6679), 866–867

Bouwman, A.F., Boumans, L.J.M., Batjes, N.H. (2002) Modeling global annual N_2O and NO emissions from fertilized fields. Global Biogechem. Cycles **16**(4), 1080. doi:10.1029/2001gb001812

Bremner, J.M., Blackmer, A.M. (1978) Nitrous oxide: Emission from soils during nitrification of fertilizer nitrogen. Science **199**(4326), 295–296

Brown, H.A., Wagner-Riddle, C., Thurtell, G.W. (2000) Nitrous oxide flux from solid dairy manure in storage as affected by water content and redox potential. J. Environ. Qual. **29**(2), 630–638

Brumme, R. (1995) Mechanisms of carbon and nutrient release and retention in beech forest gaps .3. Environmental-regulation of soil respiration and nitrous-oxide emissions along a microclimatic gradient. Plant Soil **168**, 593–600

Burkart, M.R., James, D.E. (1999) Agricultural-nitrogen contributions to hypoxia in the Gulf of Mexico. J. Environ. Qual. **28**(3), 850–859

Butterbach-Bahl, K., Gasche, R., Huber, C., Kreutzer, K., Papen, H. (1998) Impact of N-input by wet deposition on N-trace gas fluxes and CH_4-oxidation in spruce forest ecosystems of the temperate zone in Europe. Atmos. Environ. **32**(3), 559–564

Butterbach-Bahl, K., Gasche, R., Willibald, G., Papen, H. (2002) Exchange of N-gases at the Hoglwald Forest - A summary. Plant Soil **240**(1), 117–123

Cantarel, A.A.M., Bloor, J.M.G., Deltroy, N., Soussana, J.F. (2011) Effects of Climate Change Drivers on Nitrous Oxide Fluxes in an Upland Temperate Grassland. Ecosystems **14**(2), 223–233. doi:10.1007/s10021-010-9405-7

Carpenter, S.R., Caraco, N.F., Correll, D.L., Howarth, R.W., Sharpley, A.N., Smith, V.H. (1998) Nonpoint pollution of surface waters with phosphorus and nitrogen. Ecol. Appl. **8**(3), 559–568

Cassman, K.G., Wood, S. (2005) Cultivated ecosystems. In: Hassan, R.M., Scholes, R., Ash, N. (eds.) Human well-being: Vol. 1. Current state and trends. pp. 745–794. Island Press, Washington,

Chadwick, D.R., Pain, B.F., Brookman, S.K.E. (2000) Nitrous oxide and methane emissions following application of animal manures to grassland. J. Environ. Qual. **29**(1), 277–287

Choudhary, M.A., Akramkhanov, A., Saggar, S. (2002) Nitrous oxide emissions from a New Zealand cropped soil: tillage effects, spatial and seasonal variability. Agric. Ecosys. Environ. **93**(1–3), 33–43

Clayton, H., McTaggart, I.P., Parker, J., Swan, L., Smith, K.A. (1997) Nitrous oxide emissions from fertilised grassland: A 2-year study of the effects of N fertiliser form and environmental conditions. Biol. Fertil. Soils **25**(3), 252–260

Clough, T.J., Sherlock, R.R., Rolston, D.E. (2005) A review of the movement and fate of N_2O in the subsoil. Nutr. Cycl. Agroecosys. **72**(1), 3–11. doi:10.1007/s10705-004-7349-z

Conen, F., Dobbie, K.E., Smith, K.A. (2000) Predicting N_2O emissions from agricultural land through related soil parameters. Global Change Biol. **6**(4), 417–426

Conrad, R. (2002) Microbiological and biochemical background of production and consumption of NO and N_2O in soil. In: Gasche, R., Papen, H., Rennenberg, H. (eds.) Trace gas exchange in forest ecosystems. pp. 3–33. Kluwer Academic Publishers, Dordrecht, Netherlands

Crutzen, P.J., Mosier, A.R., Smith, K.A., Winiwarter, W. (2008) N_2O release from agro-biofuel production negates global warming reduction by replacing fossil fuels. Atmos. Chem. Phys. **8**(2), 389–395

Dalal, R.C., Wang, W.J., Robertson, G.P., Parton, W.J. (2003) Nitrous oxide emission from Australian agricultural lands and mitigation options: a review. Austr. J. Soil Res. **41**(2), 165–195. doi:10.1071/sr02064

Davidson, E.A. (1991) Fluxes of nitrous oxide and nitric acid from terrestrial ecosystem. In: Rogers, J.E., Whitman, W.B. (eds.) Microbial production and consumption of greenhouse gases: Methane, Nitrous oxide and Halomethane. pp. 219–236. American Society of Microbiology, Washington,

Davidson, E.A., Keller, M., Erickson, H.E., Verchot, L.V., Veldkamp, E. (2000) Testing a conceptual model of soil emissions of nitrous and nitric oxides. Bioscience **50**(8), 667–680

Davidson, E.A., Verchot, L.V. (2000) Testing the Hole-in-the-Pipe Model of nitric and nitrous oxide emissions from soils using the TRAGNET Database. Global Biogechem. Cycles **14**(4), 1035–1043. doi:10.1029/1999gb001223

Davies, M.G., Smith, K.A., Vinten, A.J.A. (2001) The mineralisation and fate of nitrogen following ploughing of grass and grass-clover swards. Biol. Fertil. Soils **33**(5), 423–434

DeFries, R.S., Foley, J.A., Asner, G.P. (2004) Land-use choices: balancing human needs and ecosystem function. Front. Ecol. Environ. **2**(5), 249–257

Del Grosso, S.J., Parton, W.J., Mosier, A.R., Walsh, M.K., Ojima, D.S., Thornton, P.E. (2006) DAYCENT national-scale simulations of nitrous oxide emissions from cropped soils in the United States. J. Environ. Qual. **35**(4), 1451–1460. doi:10.2134/jeq2005.0160

Denman, K.L., Brasseur, G., Chidthaisong, A., Ciais, P., Cox, P.M., Dickinson, R.E., Hauglustaine, D., Heinze, C., Holland, E., Jacob, D., Lohmann, U., Ramachandran, S., Dias, P.L.d.S., Wofsy, S.C., Zhang, X. (2007) Couplings Between Changes in the Climate System and Biogeochemistry. In: Solomon, S., Qin, D., Manning, M., Chen, Z., Marquis, M., Averyt, K.B., M.Tignor, Miller, H.L. (eds.) Climate Change 2007: The Physical Science Basis. Contribution of Working Group I to the Fourth Assessment Report of the Intergovernmental Panel on Climate Change. pp. 499–587. Cambridge University Press, Cambridge, United Kingdom and New York, NY, USA

Di, H.J., Cameron, K.C. (2002a) Nitrate leaching in temperate agroecosystems: sources, factors and mitigating strategies. Nutr. Cycl. Agroecosys. **64**(3), 237–256

Di, H.J., Cameron, K.C. (2002b) The use of a nitrification inhibitor, dicyandiamide (DCD), to decrease nitrate leaching and nitrous oxide emissions in a simulated grazed and irrigated grassland. Soil Use Manage. **18**(4), 395–403. doi:10.1079/sum2002151

Diaz, R.J., Rosenberg, R. (2008) Spreading dead zones and consequences for marine ecosystems. Science **321**(5891), 926–929. doi:10.1126/science.1156401

Drury, C.F., McKenney, D.J., Findlay, W.I. (1991) Relationships between denitrification, microbial biomass and indigenous soil properties. Soil Biol. Biochem. **23**(8), 751–755. doi:10.1016/0038-0717(91)90145-a

Drury, C.F., Reynolds, W.D., Tan, C.S., Welacky, T.W., Calder, W., McLaughlin, N.B. (2006) Emissions of nitrous oxide and carbon dioxide: influence of tillage type and nitrogen placement depth. Soil Sci. Soc. Am. J. **70**(2), 570–581. doi:10.2136/sssaj2005.0042

Duxbury, J.M., Bouldin, D.R., Terry, R.E., Tate, R.L. (1982) Emissions of nitrous-oxide from soils. Nature **298**(5873), 462–464

Duxbury, J.M., McConnaughey, P.K. (1986) Effect of fertilizer source on denitrification and nitrous oxide emissions in a maize field. Soil Sci. Soc. Am. J. **50**(3), 644–648

Ehhalt, D., Prather, M., Dentener, F., Derwent, R., Dlugokencky, E., Holland, E., Isaksen, I., Katima, J., Kirchhoff, V., P. Matson, Midgley, P., Wang, M. (2001) Atmospheric chemistry and greenhouse gases. In: Houghton, J.T., Ding, Y., Griggs, D.J., Noguer, M., vanderLinden, P.J., Dai, X., Maskell, K., Johnson, C.A. (eds.) Climate Change 2001: The Scientific Basis. Contribution of Working Group I to the Third Assessment Report of the Intergovernmental Panel on Climate Change. p. 881. Cambridge University Press, Cambridge, United Kingdom and New York, NY, USA

Eichner, M.J. (1990) Nitrous oxide emissions from fertilized soils: Summary of available data. J. Environ. Qual. **19**(2), 272–280

Elmi, A.A., Madramootoo, C., Hamel, C., Liu, A. (2003) Denitrification and nitrous oxide to nitrous oxide plus dinitrogen ratios in the soil profile under three tillage systems. Biol. Fertil. Soils **38**(6), 340–348. doi:10.1007/s00374-003-0663-9

Food and Agriculture Organization of the United nations (FAO) (2006) World agriculture: Towards 2030/2050 - Prospects for food, nutrition, agriculture and major commodity groups. Food and Agriculture Organization of the United Nations. http://www.fao.org/docrep/009/a0607e/a0607e00.htm

Food and Agriculture Organization of the United nations (FAO) (2010) The food insecurity in the world. Food and Agriculture Organization of the United Nations.

Food and Agriculture Organization of the United nations (FAO) (2011) Food and Agriculture Organization of the United Nations. FAOSTAT: Agriculture Database. http:faostat.fao.org. (Accessed December 2011)

FAOSTAT (2001) FAO Statistical Databases. Food and Agricultural Organization of the United Nations. On line at http://fao.org (Accessed July 2011)

Feyereisen, G.W., Wilson, B.N., Sands, G.R., Strock, J.S., Porter, P.M. (2006) Potential for a rye cover crop to reduce nitrate loss in southwestern Minnesota. Agron. J. **98**(6), 1416–1426. doi:10.2134/agronj2005.0134

Firestone, M.K., Davidson, E.A. (1989) Microbiological basis of NO and N_2O production and consumption in soil. In: Andreae, M.O., Schimel, D.S. (eds.) Exchange of Trace Gases between Terrestrial Ecosystems and the Atmosphere, vol. 47. Life Sciences Research Report, pp. 7–21

Flechard, C.R., Ambus, P., Skiba, U., Rees, R.M., Hensen, A., van Amstel, A., Pol-van Dasselaar, A.V., Soussana, J.F., Jones, M., Clifton-Brown, J., Raschi, A., Horvath, L., Neftel, A., Jocher, M., Ammann, C., Leifeld, J., Fuhrer, J., Calanca, P., Thalman, E., Pilegaard, K., Di Marco, C., Campbell, C., Nemitz, E., Hargreaves, K.J., Levy, P.E., Ball, B.C., Jones, S.K., van de Bulk, W.C.M., Groot, T., Blom, M., Domingues, R., Kasper, G., Allard, V., Ceschia, E., Cellier, P., Laville, P., Henault, C., Bizouard, F., Abdalla, M., Williams, M., Baronti, S., Berretti, F., Grosz, B. (2007) Effects of climate and management intensity on nitrous oxide emissions in grassland systems across Europe. Agric. Ecosys. Environ. **121**(1–2), 135–152. doi:10.1016/j.agee.2006.12.024

Flechard, C.R., Neftel, A., Jocher, M., Ammann, C., Fuhrer, J. (2005) Bi-directional soil/atmosphere N_2O exchange over two mown grassland systems with contrasting management practices. Global Change Biol. **11**(12), 2114–2127. doi:10.1111/j.1365-2486.2005.01056.x

Foley, J.A., DeFries, R., Asner, G.P., Barford, C., Bonan, G., Carpenter, S.R., Chapin, F.S., Coe, M.T., Daily, G.C., Gibbs, H.K., Helkowski, J.H., Holloway, T., Howard, E.A., Kucharik, C.J., Monfreda, C., Patz, J.A., Prentice, I.C., Ramankutty, N., Snyder, P.K. (2005) Global consequences of land use. Science **309**(5734), 570–574. doi:10.1126/science.1111772

Freibauer, A. (2003) Regionalised inventory of biogenic greenhouse gas emissions from European agriculture. Eur. J. Agron. **19**(2), 135–160

Galloway, J.N., Dentener, F.J., Capone, D.G., Boyer, E.W., Howarth, R.W., Seitzinger, S.P., Asner, G.P., Cleveland, C.C., Green, P.A., Holland, E.A., Karl, D.M., Michaels, A.F., Porter, J.H., Townsend, A.R., Vorosmarty, C.J. (2004) Nitrogen cycles: past, present, and future. Biogeochemistry **70**(2), 153–226

Galloway, J.N., Townsend, A.R., Erisman, J.W., Bekunda, M., Cai, Z.C., Freney, J.R., Martinelli, L. A., Seitzinger, S.P., Sutton, M.A. (2008) Transformation of the nitrogen cycle: Recent trends, questions, and potential solutions. Science **320**(5878), 889–892. doi:10.1126/science.1136674

Gartzia-Bengoetxea, N., Gonzalez-Arias, A., Merino, A., de Arano, I.M. (2009) Soil organic matter in soil physical fractions in adjacent semi-natural and cultivated stands in temperate Atlantic forests. Soil Biol. Biochem. **41**(8), 1674–1683. doi:10.1016/j.soilbio.2009.05.010

Godde, M., Conrad, R. (2000) Influence of soil properties on the turnover of nitric oxide and nitrous oxide by nitrification and denitrification at constant temperature and moisture. Biol. Fertil. Soils **32**(2), 120–128. doi:10.1007/s003740000247

Goldewijk, K., Van Drecht, G.G., Bouwman, A.F. (2007) Mapping contemporary global cropland and grassland distributions on a 5 × 5 minute resolution. J. Land Use Sci. **2**(3), 167–190. doi:10.1080/17474230701622940

Goldewijk, K.K. (2001) Estimating global land use change over the past 300 years: The HYDE Database. Global Biogechem. Cycles **15**(2), 417–433. doi:10.1029/1999gb001232

Granli, T., Bockman, O.C. (1994) Nitrous oxide from agriculture. Norwegian Journal of Agricultural Sciences Supplement **12**, 1–128

Granli, T., Bockman, O.C. (1995) Nitrous oxide (N2O) emissions from soils in warm climates. Fert. Res **42**(1–3), 159–163. doi:10.1007/bf00750510

Grant, B., Smith, W.N., Desjardins, R., Lemke, R., Li, C. (2004) Estimated N_2O and CO_2 emissions as influenced by agricultural practices in Canada. Climatic Change **65**(3), 315–332

Green, R.E., Cornell, S.J., Scharlemann, J.P.W., Balmford, A. (2005) Farming and the fate of wild nature. Science **307**(5709), 550–555. doi:10.1126/science.1106049

Hall, S.J., Matson, P.A. (1999) Nitrogen oxide emissions after nitrogen additions in tropical forests. Nature **400**(6740), 152–155. doi:10.1038/22094

Haynes, R.J., Goh, K.M. (1978) Ammonium and nitrate nutrition of plants. Biol. Rev. Cambridge Phil. Soc. **53**(4), 465–510

Haynes, R.J., Williams, P.H. (1993) Nutrient cycling and soil fertility in the grazed pasture ecosystem. In: Advances in Agronomy, Vol 49, vol. 49. Advances in Agronomy, pp. 119–199.

Henderson-Sellers, A., Gornitz, V. (1984) Possible climatic impacts of land cover transformations, with particular emphasis on tropical deforestation. Climatic Change **6**(3), 231–257. doi:10.1007/bf00142475

Hutchinson, G.L., Davidson, E.A. (1992) Processes for production and consumption of gaseous nitrogen oxides in soils. In: Mosier, A.R., Duxbury, J.M., Rolston, D.E. (eds.) Agricultural systems effects on trace gases and global climate change. pp. 79–93. American Society of Agronomy, Madison, WI

Huttunen, J.T., Nykanen, H., Martikainen, P.J., Nieminen, M. (2003) Fluxes of nitrous oxide and methane from drained peatlands following forest clear-felling in southern Finland. Plant Soil **255**(2), 457–462

International Fertilizer industry Association (IFA) (2007) Sustainable management of the nitrogen cycle in agriculture and mitigation of reactive nitrogen side effects. International Fertilizer Industry Association, (IFA)

International Fertilizer industry Association /Food and Agricultural Organization of the United Nations (IFA/FAO) (2001) Global estimates of gaseous emissions of NH_3, NO, and N_2O from agricultural land. International Fertilizer Industry Association and Food and Agriculture Organization of the United Nations. On line at http://fao.org/agl/agll/docs/globest.pdfIFA/FAO

Inagaki, Y., Miura, S., Kohzu, A. (2004) Effects of forest type and stand age on litterfall quality and soil N dynamics in Shikoku district, southern Japan. For. Ecol. Manage. **202**(1–3), 107–117. doi:10.1016/j.foreco.2004.07.029

Inubushi, K., Barahona, M.A., Yamakawa, K. (1999) Effects of salts and moisture content on N_2O emission and nitrogen dynamics in Yellow soil and Andosol in model experiments. Biol. Fertil. Soils **29**(4), 401–407. doi:10.1007/s003740050571

Intergovernmental Panel on Climate Change (IPCC) (1997) Revised 1996 IPCC Guidelines for National Greenhouse Gas Inventories. Intergovernmental Panel on Climate Change

Intergovernmental Panel on Climate Change (IPCC) (2006) 2006 IPCC Guidelines for National Greenhouse Gas Inventories. Volume 4: Agriculture, Forestry and Other Land Use National Greenhouse Gas Inventories Programme

Johansson, C., Galbally, I.E. (1984) Production of nitric oxide in loam under aerobic and anaerobic conditions. Appl. Environ. Microb. **47**(6), 1284–1289

Jones, S.K., Rees, R.M., Skiba, U.M., Ball, B.C. (2005) Greenhouse gas emissions from a managed grassland. Global Planet. Change **47**(2–4), 201–211. doi:10.1016/j.gloplacha.2004.10.011

Jungkunst, H.F., Flessa, H., Scherber, C., Fiedler, S. (2008) Groundwater level controls CO_2, N_2O and CH_4 fluxes of three different hydromorphic soil types of a temperate forest ecosystem. Soil Biol. Biochem. **40**(8), 2047–2054. doi:10.1016/j.soilbio.2008.04.015

Keller, M., Veldkamp, E., Weitz, A.M., Reiners, W.A. (1993) Effect of pasture age on soil trace-gas emissions from a deforested area of Costa Rica. Nature **365**(6443), 244–246

Kesik, M., Ambus, P., Baritz, R., Bruggemann, N.B., Butterbach-Bahl, K., Damm, M., Duyzer, J., Horvath, L., Kiese, R., Kitzler, B., Leip, A., Li, C., Pihlatie, M., Pilegaard, K., Seufert, G., Simpson, D., Skiba, U., Smiatek, G., Vesala, T., Zechmeister-Boltenstern, S. (2005) Inventories of N_2O and NO emissions from European forest soils. Biogeosciences **2**(4), 353–375

Kessavalou, A., Doran, J.W., Mosier, A.R., Drijber, R.A. (1998) Greenhouse gas fluxes following tillage and wetting in a wheat-fallow cropping system. J. Environ. Qual. **27**(5), 1105–1116

Lambin, E.F., Geist, H.J., Lepers, E. (2003) Dynamics of land-use and land-cover change in tropical regions. Annu. Rev. Env. Resour. **28**, 205–241. doi:10.1146/annurev.energy.28.050302.105459

Leahy, P., Kiely, G., Scanlon, T.M. (2004) Managed grasslands: A greenhouse gas sink or source? Geophys. Res. Lett. **31**(20), L20507. doi:10.1029/2004gl021161

Ledgard, S.F., Sprosen, M.S., Penno, J.W., Rajendram, G.S. (2001) Nitrogen fixation by white clover in pastures grazed by dairy cows: Temporal variation and effects of nitrogen fertilization. Plant Soil **229**(2), 177–187

Ledgard, S.F., Sprosen, M.S., Steele, K.W. (1996) Nitrogen fixation by nine white clover cultivars in grazed pasture, as affected by nitrogen fertilization. Plant Soil **178**(2), 193–203

Lee, D.S., Bouwman, A.F., Asman, W.A.H., F.J.Dentener, derHoek, K.W., J, G.J.O. (1997) Emissions of nitric oxide, nitrous oxide and ammonia from grasslands on global scale. In: S.C., j., Pain, B.S. (eds.) Gaseous nitrogen emission from grasslands. pp. 353–371. CAB International, Wallingford,

Linn, D.M., Doran, J.W. (1984) Effect of water-filled pore-space on carbon-dioxide and nitrous-oxide production in tilled and nontilled soils. Soil Sci. Soc. Am. J. **48**(6), 1267–1272

Lovett, G.M., Weathers, K.C., Arthur, M.A., Schultz, J.C. (2004) Nitrogen cycling in a northern hardwood forest: Do species matter? Biogeochemistry **67**(3), 289–308. doi:10.1023/B:BIOG.0000015786.65466.f5

Low, A.P., Stark, J.M., Dudley, L.M. (1997) Effects of soil osmotic potential on nitrification, ammonification, N-assimilation, and nitrous oxide production. Soil Sci. **162**(1), 16–27. doi:10.1097/00010694-199701000-00004

Luizao, F., Matson, P., Livingston, G., Luizao, R., Vitousek, P.M. (1989) Nitrous oxide flux following tropical land clearing. Global Biogechem. Cycles **3**, 281–285

Macdonald, C.A., Thomas, N., Robinson, L., Tate, K.R., Ross, D.J., Dando, J., Singh, B.K. (2009) Physiological, biochemical and molecular responses of the soil microbial community after afforestation of pastures with Pinus radiata. Soil Biol. Biochem. **41**(8), 1642–1651. doi:10.1016/j.soilbio.2009.05.003

MacKenzie, A.F., Fan, M.X., Cadrin, F. (1997) Nitrous oxide emission as affected by tillage, corn-soybean-alfalfa rotations and nitrogen fertilization. Can. J, Soil Sci. **77**(2), 145–152

Maggiotto, S.R., Webb, J.A., Wagner-Riddle, C., Thurtell, G.W. (2000) Nitrous and nitrogen oxide emissions from turfgrass receiving different forms of nitrogen fertilizer. J. Environ. Qual. **29**(2), 621–630

Matson, P.A., Parton, W.J., Power, A.G., Swift, M.J. (1997) Agricultural intensification and ecosystem properties. Science **277**(5325), 504–509

McCarl, B.A., Schneider, U.A. (2000) U.S. agriculture's role in a greenhouse gas emission mitigation world: An economic perspective. Rev. Agric. Econ. **22**(1), 134–159

McKane, R.B., Johnson, L.C., Shaver, G.R., Nadelhoffer, K.J., Rastetter, E.B., Fry, B., Giblin, A.E., Kielland, K., Kwiatkowski, B.L., Laundre, J.A., Murray, G. (2002) Resource-based niches provide a basis for plant species diversity and dominance in arctic tundra. Nature **415**(6867), 68–71

McSwiney, C.P., Robertson, G.P. (2005) Nonlinear response of N_2O flux to incremental fertilizer addition in a continuous maize (*Zea mays* L.) cropping system. Global Change Biol. **11**(10), 1712–1719. doi:10.1111/j.1365-2486.2005.01040.x

Menyailo, O.V., Huwe, B. (1999) Activity of denitrification and dynamics of N_2O release in soils under six tree species and grassland in central Siberia. J. Plant Nutr. Soil Sci. **162**(5), 533–538

Menyailo, O.V., Stepanov, A.L., Umarow, M.M. (1997) The transformation of nitrous oxide by denitrifying bacteria in solonchaks. Eurasian Soil Sci. **30**(2), 178–180

Mosier, A.R., Bleken, M.A., Chaiwanakupt, P., Ellis, E.C., Freney, J.R., Howarth, R.B., Matson, P.A., Minami, K., Naylor, R., Weeks, K.N., Zhu, Z.L. (2001) Policy implications of human-accelerated nitrogen cycling. Biogeochemistry **52**(3), 281–320

Mosier, A.R., Delgado, J.A. (1997) Methane and nitrous oxide fluxes in grasslands in western Puerto Rico. Chemosphere **35**(9), 2059–2082

Mosier, A.R., Duxbury, J.M., Freney, J.R., Heinemeyer, O., Minami, K. (1996) Nitrous oxide emissions from agricultural fields: Assessment, measurement and mitigation. Plant Soil **181**(1), 95–108. doi:10.1007/bf00011296

Mosier, A.R., Duxbury, J.M., Freney, J.R., Heinemeyer, O., Minami, K. (1998) Assessing and mitigating N_2O emissions from agricultural soils. Climatic Change **40**(1), 7–38. doi:10.1023/a:1005386614431

Mosier, A.R., Halvorson, A.D., Reule, C.A., Liu, X.J.J. (2006) Net global warming potential and greenhouse gas intensity in irrigated cropping systems in northeastern Colorado. J. Environ. Qual. **35**(4), 1584–1598. doi:10.2134/jeq2005.0232

O'hara, G.W., Daniel, R.M. (1985) Rhizobial denitrification: A review. Soil Biol. Biochem. **17**(1), 1–9

Oenema, O., Tamminga, S. (2005) Nitrogen in global animal production and management options for improving nitrogen use efficiency. Sci. China Ser. C **48**, 871–887. doi:10.1360/062005-279

Omonode, R.A., Smith, D.R., Gal, A., Vyn, T.J. (2011) Soil nitrous oxide emissions in corn following three decades of tillage and rotation treatments. Soil Sci. Soc. Am. J. **75**(1), 152–163. doi:10.2136/sssaj2009.0147

Papen, H., Butterbach-Bahl, K. (1999) A 3-year continuous record of nitrogen trace gas fluxes from untreated and limed soil of a N-saturated spruce and beech forest ecosystem in Germany - 1. N2O emissions. J. Geophys. Res. -Atmos. **104**(D15), 18487–18503

Parton, W.J., Holland, E.A., Del Grosso, S.J., Hartman, M.D., Martin, R.E., Mosier, A.R., Ojima, D.S., Schimel, D.S. (2001) Generalized model for NOx and N2O emissions from soils. J. Geophys. Res. -Atmos. **106**(D15), 17403–17419. doi:10.1029/2001jd900101

Passianoto, C.C., Ahrens, T., Feigl, B.J., Steudler, P.A., do Carmo, J.B., Melillo, J.M. (2003) Emissions of CO_2, N_2O, and NO in conventional and no-till management practices in Rondonia, Brazil. Biol. Fertil. Soils **38**(4), 200–208

Pilbeam, C.J., Warren, G.P. (1995) Use of ^{15} N for fertilizer N recovery and N mineralization studies in semi-arid Kenya. Nutr. Cycl. Agroecosys. **42**(1), 123–128. doi:10.1007/bf00750506

Poth, M., Focht, D.D. (1985) ^{15} N kinetic-analysis of N_2O production by *Nitrosomonas europaea:* An examination of nitrifier denitrification. Appl. Environ. Microb. **49**(5), 1134–1141

Ramankutty, N., Evan, A.T., Monfreda, C., Foley, J.A. (2008) Farming the planet: 1. Geographic distribution of global agricultural lands in the year 2000. Global Biogechem. Cycles **22**(1), GB1003. doi:10.1029/2007gb002952

Rappoldt, C., Crawford, J.W. (1999) The distribution of anoxic volume in a fractal model of soil. Geoderma **88**(3–4), 329–347. doi:10.1016/s0016-7061(98)00112-8

Robertson, G.P., Groffman, P.M. (2007) Nitrogen transformation. In: Paul, E.A. (ed.) Soil microbiology, biochemistry and ecology. pp. 341–364. Springer, New York

Robertson, G.P., Paul, E.A., Harwood, R.R. (2000) Greenhouse gases in intensive agriculture: Contributions of individual gases to the radiative forcing of the atmosphere. Science **289** (5486), 1922–1925

Sabine, C.L., Heimann, M., Artaxo, P., Bakker, D.C.E., Chen, C.-T.A., Field, C.B., Gruber, N., Le Que´re´, C., Prinn, R.G., Richey, J.E.e.a. (2004) Current status and past trends of the global carbon cycle. In: Field, C., Raupach, M.R. (eds.) The Global Carbon Cycle: Integrating Humans, Climate, and the Natural World. pp. 17–44. Island Press, Washington, DC

Saggar, S., Andrew, R.M., Tate, K.R., Hedley, C.B., Rodda, N.J., Townsend, J.A. (2004) Modelling nitrous oxide emissions from dairy-grazed pastures. Nutr. Cycl. Agroecosys. **68**(3), 243–255

Saggar, S., Bolan, N.S., Singh, J., Blard, A. (2005) Economic and environmental impacts of increased nitrogen use in grazed pastures and role of inhibitors in mitigating nitrogen losses. N. Z. Sci. Rev. **62**, 69–74

Saggar, S., Giltrap, D.L., Li, C., Tate, K.R. (2007) Modelling nitrous oxide emissions from grazed grasslands in New Zealand. Agric. Ecosys. Environ. **119**(1–2), 205–216. doi:10.1016/j.agee.2006.07.010

Saggar, S., Tate, K.R., Giltrap, D.L., Singh, J. (2008) Soil-atmosphere exchange of nitrous oxide and methane in New Zealand terrestrial ecosystems and their mitigation options: a review. Plant Soil **309**(1–2), 25–42. doi:10.1007/s11104-007-9421-3

Sahrawat, K.L., Keeney, D.L. (1986) Nitrous oxide emissions from soils. Adv. Soil Sci. **4**, 103–148

Schmidt, J., Seiler, W., Conrad, R. (1988) Emission of nitrous oxide from temperate forest soils into the atmosphere. J. Atmos. Chem. **6**(1–2), 95–115. doi:10.1007/bf00048334

Simek, M., Cooper, J.E. (2002) The influence of soil pH on denitrification: progress towards the understanding of this interaction over the last 50 years. Eur. J. Soil Sci. **53**(3), 345–354. doi:10.1046/j.1365-2389.2002.00461.x

Six, J., Ogle, S.M., Breidt, F.J., Conant, R.T., Mosier, A.R., Paustian, K. (2004) The potential to mitigate global warming with no-tillage management is only realized when practised in the long term. Global Change Biol. **10**(2), 155–160. doi:10.1111/j.1529-8817.2003.00730.x

Skiba, U., Fowler, D., Smith, K. (1994) Emissions of no and N_2O from soils. Environ. Monit. Assess. **31**(1–2), 153–158

Skiba, U.M., Sheppard, L.J., MacDonald, J., Fowler, D. (1998) Some key environmental variables controlling nitrous oxide emissions from agricultural and semi-natural soils in Scotland. Atmos. Environ. **32**(19), 3311–3320

Smil, V. (2002) Nitrogen and food production: Proteins for human diets. Ambio **31**(2), 126–131

Smith, K.A. (1997) The potential for feedback effects induced by global warming on emissions of nitrous oxide by soils. Global Change Biol. **3**(4), 327–338. doi:10.1046/j.1365-2486.1997.00100.x

Smith, K.A., Ball, T., Conen, F., Dobbie, K.E., Massheder, J., Rey, A. (2003) Exchange of greenhouse gases between soil and atmosphere: interactions of soil physical factors and biological processes. Eur. J. Soil Sci. **54**(4), 779–791. doi:10.1046/j.1351-0754.2003.0567.x

Smith, K.A., Conen, F. (2004) Impacts of land management on fluxes of trace greenhouse gases. Soil Use Manage. **20**, 255–263. doi:10.1079/sum2004238

Smith, K.A., Dobbie, K.E., Ball, B.C., Bakken, L.R., Sitaula, B.K., Hansen, S., Brumme, R., Borken, W., Christensen, S., Prieme, A., Fowler, D., Macdonald, J.A., Skiba, U., Klemedtsson, L., Kasimir-Klemedtsson, A., Degorska, A., Orlanski, P. (2000) Oxidation of atmospheric methane in Northern European soils, comparison with other ecosystems, and uncertainties in the global terrestrial sink. Global Change Biol. **6**(7), 791–803

Smith, K.A., McTaggart, I.P., Tsuruta, H. (1997) Emissions of N_2O and NO associated with nitrogen fertilization in intensive agriculture, and the potential for mitigation. Soil Use Manage. **13**(4), 296–304. doi:10.1111/j.1475-2743.1997.tb00601.x

Smith, K.A., Thomson, P.E., Clayton, H., McTaggart, I.P., Conen, F. (1998) Effects of temperature, water content and nitrogen fertilisation on emissions of nitrous oxide by soils. Atmos. Environ. **32**(19), 3301–3309

Smith, P., Goulding, K.W., Smith, K.A., Powlson, D.S., Smith, J.U., Falloon, P., Coleman, K. (2001) Enhancing the carbon sink in European agricultural soils: including trace gas fluxes in estimates of carbon mitigation potential. Nutr. Cycl. Agroecosys. **60**(1–3), 237–252

Smith, P., Martino, D., Cai, Z., Gwary, D., Janzen, H., Kumar, P., McCarl, B., Ogle, S., O'Mara, F., Rice, C., Scholes, B., Sirotenko, O. (2007) Agriculture In: Metz, B., Davidson, O.R., Bosch, P.R., Dave, R., Meyer, L.A. (eds.) Climate Change 2007: Mitigation. Contribution of Working Group III to the Fourth Assessment Report of the Intergovernmental Panel on Climate Change. pp. 497–540. Cambridge University Press, Cambridge, United Kingdom and New York, NY, USA

Smolander, A., Loponen, J., Suominen, K., Kitunen, V. (2005) Organic matter characteristics and C and N transformations in the humus layer under two tree species, *Betula pendula* and *Picea abies*. Soil Biol. Biochem. **37**(7), 1309–1318. doi:10.1016/j.soilbio.2004.12.002

Stark, J.M., Firestone, M.K. (1996) Kinetic characteristics of ammonium-oxidizer communities in a California oak woodland-annual grassland. Soil Biol. Biochem. **28**(10–11), 1307–1317. doi:10.1016/s0038-0717(96)00133-2

Subbarao, G.V., Ito, O., Sahrawat, K.L., Berry, W.L., Nakahara, K., Ishikawa, T., Watanabe, T., Suenaga, K., Rondon, M., Rao, I.M. (2006) Scope and strategies for regulation of nitrification in agricultural systems-challenges and opportunities. Crit. Rev. Plant Sci. **25**(4), 303–335. doi:10.1080/07352680600794232

Suzuki, I., Dular, U., Kwok, S.C. (1974) Ammonia or ammonium ion as substrate for oxidation by *Nitrosomonas europaea* cells and extracts. J. Bacteriol. **120**(1), 556–558

Thornton, F.C., Valente, R.J. (1996) Soil emissions of nitric oxide and nitrous oxide from no-till corn. Soil Sci. Soc. Am. J. **60**(4), 1127–1133

Tiedje, J.M. (1988) Ecology of denitrification and the dissimilatory nitrate reduction to ammonium. In: Zehnder, J.B. (ed.) Biology of anaerobic microorganisms. pp. 179–244. Wiley, New York

Tilman, D., Cassman, K.G., Matson, P.A., Naylor, R., Polasky, S. (2002) Agricultural sustainability and intensive production practices. Nature **418**(6898), 671–677. doi:10.1038/nature01014

Tilman, D., Fargione, J., Wolff, B., D'Antonio, C., Dobson, A., Howarth, R., Schindler, D., Schlesinger, W.H., Simberloff, D., Swackhamer, D. (2001) Forecasting agriculturally driven global environmental change. Science **292**(5515), 281–284

Ullah, S., Frasier, R., King, L., Picotte-Anderson, N., Moore, T.R. (2008) Potential fluxes of N_2O and CH_4 from soils of three forest types in Eastern Canada. Soil Biol. Biochem. **40**(4), 986–994. doi:10.1016/j.soilbio.2007.11.019

Ussiri, D.A.N., Lal, R., Jarecki, M.K. (2009) Nitrous oxide and methane emissions from long-term tillage under a continuous corn cropping system in Ohio. Soil Till. Res. **104**(2), 247–255. doi:10.1016/j.still.2009.03.001

van Groenigen, J.W., Velthof, G.L., van der Bolt, F.J.E., Vos, A., Kuikman, P.J. (2005) Seasonal variation in N_2O emissions from urine patches: Effects of urine concentration, soil compaction and dung. Plant Soil **273**(1–2), 15–27. doi:10.1007/s11104-004-6261-2

Veldkamp, E., Keller, M. (1997) Nitrogen oxide emissions from a banana plantation in the humid tropics. J. Geophys. Res. -Atmos. **102**(D13), 15889–15898

Veldkamp, E., Keller, M., Nunez, M. (1998) Effects of pasture management on N_2O and NO emissions from soils in the humid tropics of Costa Rica. Global Biogechem. Cycles **12**(1), 71–79

Venterea, R.T., Burger, M., Spokas, K.A. (2005) Nitrogen oxide and methane emissions under varying tillage and fertilizer management. J. Environ. Qual. **34**(5), 1467–1477

Venterea, R.T., Groffman, P.M., Verchot, L.V., Magill, A.H., Aber, J.D. (2004) Gross nitrogen process rates in temperate forest soils exhibiting symptoms of nitrogen saturation. For. Ecol. Manage. **196**(1), 129–142. doi:10.1016/j.foreco.2004.03.016

Vinten, A.J.A., Ball, B.C., O'Sullivan, M.F., Henshall, J.K. (2002) The effects of cultivation method, fertilizer input and previous sward-type on organic C and N storage and gaseous losses under spring and-winter barley following long-term leys. J. Agric. Sci. **139**, 231–243. doi:10.1017/s0021859602002496

Vitousek, P.M., Melillo, J.M. (1979) Nitrate losses from disturbed forests: patterns and mechanisms. Forest Sci. **25**(4), 605–619

Vitousek, P.M., Mooney, H.A., Lubchenco, J., Melillo, J.M. (1997) Human domination of Earth's ecosystems. Science **277**(5325), 494–499

Vitousek, P.M., Naylor, R., Crews, T., David, M.B., Drinkwater, L.E., Holland, E., Johnes, P.J., Katzenberger, J., Martinelli, L.A., Matson, P.A., Nziguheba, G., Ojima, D., Palm, C.A., Robertson, G.P., Sanchez, P.A., Townsend, A.R., Zhang, F.S. (2009) Nutrient Imbalances in Agricultural Development. Science **324**(5934), 1519–1520. doi:10.1126/science.1170261

von Arnold, K., Nilsson, M., Hanell, B., Weslien, P., Klemedtsson, L. (2005) Fluxes of CO_2, CH_4 and N_2O from drained organic soils in deciduous forests. Soil Biol. Biochem. **37**(6), 1059–1071. doi:10.1016/j.soilbio.2004.11.004

Wackernagel, M., Schulz, N.B., Deumling, D., Linares, A.C., Jenkins, M., Kapos, V., Monfreda, C., Loh, J., Myers, N., Norgaard, R., Randers, J. (2002) Tracking the ecological overshoot of the human economy. Proc. Natl. Acad. Sci. U.S.A. **99**(14), 9266–9271. doi:10.1073/pnas.142033699

Wang, Y.P., Rees, R.M. (1996) Nitrous oxide production from different forms of nitrogen. In: van Cleemput, O., Hoffman, G. (eds.) Progress in nitrogen cycling studies: Proceedings of 8th Nitrogen Workshop, September 1994. Kluwer Academic Publishers, Dordrecht

Weier, K.L., Macrae, I.C., Myers, R.J.K. (1993) Denitrification in a clay soil under pasture and annual crop estimation of potential losses using intact soil cores. Soil Biol. Biochem. **25**(8), 991–997. doi:10.1016/0038-0717(93)90145-2

Weier, K.L., McEwan, C.W., Vallis, I., Catchpoole, V.R., Myers, R.J. (1996) Potential for biological denitrification of fertilizer nitrogen in sugarcane soils. Austr. J. Agric. Res. **47**(1), 67–79

Weitz, A.M., Veldkamp, E., Keller, M., Neff, J., Crill, P.M. (1998) Nitrous oxide, nitric oxide, and methane fluxes from soils following clearing and burning of tropical secondary forest. J. Geophys. Res. -Atmos. **103**(D21), 28047–28058

Winkler, J.P., Cherry, R.S., Schlesinger, W.H. (1996) The Q(10) relationship of microbial respiration in a temperate forest soil. Soil Biol. Biochem. **28**(8), 1067–1072

Wrage, N., Lauf, J., del Prado, A., Pinto, M., Pietrzak, S., Yamulki, S., Oenema, O., Gebauer, G. (2004) Distinguishing sources of N_2O in European grasslands by stable isotope analysis. Rapid Commun. Mass Sp. **18**(11), 1201–1207

World Resources Institute (WRI) (2000) World resources 2000–2001. World Resources Institute. Washington,

Zechmeister-Boltenstern, S., Hahn, M., Meger, S., Jandl, R. (2002) Nitrous oxide emissions and nitrate leaching in relation to microbial biomass dynamics in a beech forest soil. Soil Biol. Biochem. **34**(6), 823–832

Chapter 7
Nitrous Oxide Emissions from Rice Fields

Contents

Abstract Most of rice cultivation practices require standing water over the field for the extended periods during their lifecycle. As a result, rice soils exhibit limited O_2 supply in the plow layer during growing season. Although limited O_2 favors N_2O production during denitrification, this gas would further be reduced to N_2 under strong anaerobic conditions of flooded rice soils. Therefore, rice paddies are considered to be less important source of atmospheric N_2O emissions. However, water management has strong influence on N_2O emissions. It controls the O_2 supply of the paddy rice soils by providing suitable conditions for microbial

David Ussiri and R. Lal, *Soil Emission of Nitrous Oxide and its Mitigation*,
DOI 10.1007/978-94-007-5364-8_7,

growth and activity and restricting O_2 supply to microsites by filling soil pores and creating anaerobic conditions. Monitoring of N_2O from rice soils has not been done as extensively as CH_4. Overall amount of fertilizer N usage and area under rice cultivation is increasing rapidly, necessitating implementation of different water management strategies. Therefore, global N_2O emissions from rice cultivation might increase appreciably. Monitoring of N_2O emissions from different rice management ecosystems and estimating realistic regional and global budgets from rice ecosystems is assuming high significance due to increasing land area and changing water management regime. The N_2O emissions from rice are controlled by real time field conditions and fluctuations in cultural practices. Mitigation efforts should focus on water management without compromising crop production.

Keywords Mitigation • Paddy rice soils • Nitrogen transformations • Ammonia volatilization • Irrigation

Abbreviations

UN	United Nations
GHG	Greenhouse gas
F	Flooding
F-D-F	Flooding-drain-flooding
F-D-FM	Flooding-drainage intermittent irrigation without flooding
EF	Emission Factor
CI	Controlled irrigation
EC	Electrical conductivity
CEC	Cation exchange capacity
SOM	Soil organic matter
SON	Soil organic nitrogen

7.1 Introduction

Rice (*Oryza sativa* L.) is cultivated under a wide range of climate, soil, and water regime. Although some varieties of rice grow in upland soils, majority of rice cultivation practices require standing water over the farm for extended periods during their lifecycle. The extended water ponding saturates the plow layer and develop a reducing layer beneath an oxidized layer at the soil surface. The focus of this chapter is on the flooded rice soils, since these soils undergo different physicochemical and biological changes when flooded, and exhibit different properties compared to other upland agricultural soils. Flooding of rice fields during land preparation, as is the case with paddy rice, sets-in-motion several physicochemical and microbiological processes in soil, which are different from those under upland cultivated soils (De Datta 1995). Flooded rice soils exhibit limited O_2 supply, since the diffusivity of O_2 through water is 10,000 times less than that in the air. Thus, atmospheric O_2 does not diffuse in appreciable quantities into flooded rice soils, and anoxic conditions prevail. Partially aerobic conditions exist only in the first few millimeters of the surface soils primarily due to dissolved O_2 present in water (Ponnamperuma 1972).

Rice is the world's most important food crop, and its production has increased threefold since 1970s. Use of modern rice varieties, improved management practices, and fertilizers are the major reasons for the global increase in production, which is estimated at 2.4% per annum and 71% overall since early 1970s. Rice is food of about three billion people and is the most common staple food of the largest population on Earth (Maclean et al. 2002). In some South Asian countries, rice accounts for two thirds or more of calorific intake and approximately 60% of daily protein consumption. Even in wealthier countries, rice account for more than one-third of protein source (Maclean et al. 2002). United Nations (UN) declared the year 2004 as the "International Year of Rice," and noted that rice is the staple food for more than half of the world's population. In their declaration they also emphasized the role of rice in alleviating poverty and ensuring global food security (IRRI 2003). Therefore, rice is essential for health, wealth, and prosperity of almost half of the world's population.

Good rice crop yield depends on an adequate supply of water and nutrients. Rice is very sensitive to reduced water availability in the period around flowering as this greatly affects spikelet sterility (Ekanayake et al. 1989). Because rice is mostly grown under flooded or submerged conditions, it is one of the biggest users of the world's developed fresh water resources (Tuong and Bouman 2003). However, water is becoming increasingly scarce and major concerns exist about the sustainability of irrigated agriculture (Rijsberman 2006). Nitrogen (N) is often the yield-limiting nutrient, even though the need for inputs of other nutrients is becoming increasingly evident in some regions. In rice culture, inorganic N fertilizers are increasingly providing the required N. Applied N compensates the losses through crop removal, ammonia (NH_3) volatilization, denitrification, and nitrate (NO_3^-) leaching. Global fertilizer N use for rice production for 1995–1997 was estimated at 15 Tg N year^{-1} (Tg = teragram = 10^{12} g = 1 million ton), and is projected at 19.0 Tg N year^{-1} in 2015 or 26.6% increase (FAO 2000). The overall efficiency of N utilization in agricultural crops seems to be low at about 20%. The N lost from agriculture as nitrogen (N_2) through denitrification is waste of resource but with no environmental consequences. However, a loss in the form of N_2O, nitrogen oxides (NO and NO_2, together denoted as NO_x) and NH_3 may influence regional and global atmospheric chemistry. Additionally, N_2O is very effective at trapping heat in the atmosphere and cause global warming. NH_3 volatilization and NO_3^- leaching losses provide enhanced N inputs to other ecosystems where it can also increase the N_2O fluxes. Therefore, avoiding excessive use of N inputs in agriculture, maximizing crop N-use efficiency (NUE), and minimizing N losses to the atmosphere and aquatic systems are the important strategies to minimize N_2O emissions.

Agriculture accounts for 10–20% of the total global anthropogenic emissions of greenhouse gases (GHG), and 60% of global N_2O emissions (Smith et al. 2007). Food production is primarily a major source of the anthropogenic N_2O emissions, and these emissions are largely driven by the amount of fertilizer N applied in agriculture (van Groenigen et al. 2010). The emission of N_2O that results from anthropogenic N inputs occurs through direct emissions from added fertilizers to agricultural soils, increased leguminous crop production, and through two indirect pathways: (i) volatilization of NH_3 and NO_x, and its subsequent deposition, and (ii) through leaching and runoff. Direct emission contributes ~75% to the agricultural N_2O sources (Zheng et al. 2004).

Paddy rice ecosystems are fundamentally different from other agricultural systems, because it typically grows in flooded soils. Methane (CH_4) is the dominant GHG produced and emitted in flooded soil ecosystems, and emissions are largely controlled by water and residue management (Wassmann et al. 2000). However, rice systems also emit N_2O, and the N fertilizers rate controls the intensity of its emissions (Zou et al. 2007). Rice systems are also unique from other systems in that some of N_2O as well as majority of CH_4 are emitted through plant aerenchyma rather than soil (Yu et al. 1997). Fertilizer N use is a costly input for rice farming, accounting for up to 50% of capital outlay for each crop (Freney et al. 1990), and may become even greater fraction in the future. Despite growth in world's rice production and use, research over the past 35–55 years has shown that crop utilization of N fertilizers is generally low. Normally, crops utilizes less than 40% of N applied (De Datta et al. 1968; De Datta 1995), suggesting that large quantities of soil and applied N is lost from agricultural soils. Losses occur through volatilization, denitrification, leaching, and erosion. Recoveries of N applied to rice can be as low as 10% and rarely exceeds 50% of N fertilizer (De Datta 1981). The main causes of low NUE in flooded rice are gaseous emissions of N_2O, N_2, and/or NH_3 to the atmosphere (Galbally et al. 1987; Simpson et al. 1988), especially for traditional N fertilizers applied by traditional methods. The largest losses of N occur when fertilizer N application results into higher concentrations of NH_4^+ in floodwater (Simpson et al. 1988; Cao et al. 1984a).

Water management regime often causes sensitive changes in N_2O emissions in rice paddies (Akiyama et al. 2005). Global rice production systems practice a range of water management systems. The most common regimes include: (i) seasonal continuous flooding (F), (ii) flooding, followed by midseason drainage and re-flooding (F-D-F), (iii) flooding followed by midseason drainage with intermittent irrigation but without water logging (F-D-FM, Huang et al. 2004; Zou et al. 2005, 2007; Liu et al. 2010).

Midseason drainage for 7–10 days is a common practice for inhibiting ineffective tillers, remove toxic substances, and improve rice root activity (Zou et al. 2005). Midseason drainage in rice paddies triggers substantial N_2O emissions in contrast with continuous flooding. The N_2O emissions are also enhanced by dry-wet episodes in paddy fields (Zou et al. 2004). The potential of N_2O emissions increases when the amount of N available for microbial transformation is enhanced through N fertilizer, manure, and crop residue application (Eichner 1990; Aulakh et al. 1991), cropping of legumes, and other farm management practices.

7.2 Literature Survey on Nitrous Oxide Emissions from Rice Cultivation

Although the role of soil processes in N_2O emissions has been known for long time (i.e., Arnold 1954; Goody et al. 1954), effective monitoring of N_2O emissions from rice cultivation systems and rice soils started in late 1970s (Delariche et al. 1978; Denmead et al. 1979). Rice production occurs in all six major continents of the world, but most of the published data on N_2O emissions from rice soils are from India, Philippines, China, and Japan. The reasons behind less attention to N_2O emissions from rice fields elsewhere include: (i) low N_2O flux emissions and perceived negligible loss of added N as N_2O, (ii) unfavorable climatic conditions in most of rice growing regions, and (iii) lack of reliable N_2O flux monitoring techniques. Early studies reported negligible N_2O emissions from paddy fields (e.g., Smith et al. 1982). However, more recent studies have suggested that rice cultivation could be an important anthropogenic source of N_2O (e.g., Cai et al. 1997; Akiyama et al. 2005; Zhao et al. 2011; Linquist et al. 2012).

The N_2O emissions from rice fields originate from N fertilizers, farmyard and green manures, plant litter decomposition, and dissolved N in irrigation water. Increased use of N fertilizers and water management schemes to increase rice production under 'green revolution' in late 1960s is considered as one of the main determinants of increased N_2O emissions from agriculture and rice soils (Cai et al. 1997; Kumar et al. 2000; Majumdar et al. 2000). Research efforts to increase NUE and evaluate the impact of increased use of N fertilizers on N_2O emissions from rice production systems increased significantly since early 1980s. Loss of N applied to rice production is important since a sizeable proportion of applied N is lost from the soil, leading to wasted labor, lower yields, loss of financial resources, and environmental pollution. Research efforts to increase NUE under rice production have actually helped in mitigating N_2O emissions, since a higher NUE potentially minimizes N_2O emissions (Prasad and Power 1995).

The ^{15}N tracers can determine dominant route of fertilizer N losses from rice soils. Studies involved direct estimation of $^{15}N - [^{15}N_2 + {}^{15}N_2O]$ by isotopic mass spectroscopy and quantifying the unrecovered ^{15}N (Buresh and Austin 1988; De Datta et al. 1991). The amount of ^{15}N gas loss measured directly – $[N_2 + N_2O]$ is generally small compared to N loss estimated by a ^{15}N mass balance calculated on the soil-water-plant system. De Datta et al. (1991) observed that $[^{15}N_2 + {}^{15}N_2O]$ loss was about 0.1% of applied N for the first 8 days from a puddled rice soil in the Philippines site. Total N loss determined from the unrecovered N in ^{15}N mass balance at day 20 for three seasons were 64, 43, and 52% of N applied in the form of urea (De Datta et al. 1991). In another experiment, the measured N gas $[N_2 + N_2O]$ and total N loss for 17 days was 1.1 and 40% of ^{15}N applied, respectively, following basal application of 58 kg N ha^{-1} as urea with 2 cm of standing water (Buresh and Austin 1988). The loss decreased to 0.2 and 26% of applied N for basal application of the same rate without standing water. The $(N + N_2O)$ ^{15}N flux and total ^{15}N loss were 0.5 and 40%, respectively, following application of urea at 44 kg N ha^{-1} to

50 mm flood water at 10 day after transplanting (Buresh and Austin 1988). Losses of ^{15}N by leaching or runoff are negligible, and N_2O dissolved in the floodwater based on solubility is less than 1%. Overall, these experiments indicated that in flooded rice soils, the gaseous N loss due to nitrification and denitrification were smaller compared to total N loss. The dominant pathway of N loss is probably volatilization of NH_3.

Rice paddies are considered less important source of atmospheric N_2O emissions, because an intermediate product of denitrification, N_2O would further be reduced to N_2 under strong anaerobic conditions of flooded rice soils (Granli and Bockman 1994). However, water management has strong influence on N_2O emissions as it controls the O_2 supply of the paddy rice soils. Soil water can influence denitrification directly or indirectly through (i) providing suitable conditions for microbial growth and activity, (ii) restricting O_2 supply to microsites by filling soil pores, (iii) releasing available C and N substrates through wetting and drying cycles, and (iv) providing diffusion medium through which substrates and products are moved to and from soil microorganisms (Pathak 1999). The primary effect of flooding on N_2O emission in rice paddies is the reduction of O_2 by reducing and eventually eliminating air-water interfacial area within air-filled pores, thus creating anaerobic conditions in soils. Alternate anaerobic and aerobic cycling increases N_2O emissions considerably relative to continuous aerobic or anaerobic conditions (Smith and Patrick 1983). Data from experimental conditions to assess the effects of water-logging on N_2O emissions show that little N_2O is emitted during water-logging of well-drained soils, while N_2O emissions peak when water-logged soil is drained (Flessa and Beese 1995). Bronson et al. (1997b) monitored N_2O fluxes from flooded rice and fallow soils for two irrigated dry rice-growing seasons and one wet-rice growing season. The N_2O fluxes are generally not detectable during the growing season, except for small peaks of up to 3.5 mg N_2O-N m^{-2} day^{-2} appearing after N fertilizer applications. Midseason drainage at mid-tillering or panicle initiation increase N_2O sharply until reflooding when fluxes drop back to near zero (Bronson et al. 1997b). In addition, N_2O emission occur during fallow periods, which is attributed to nitrification of mineralized organic N in the topsoil (aerobic zone) and denitrification of mineralized N in the subsoil (anaerobic zone) (Bronson et al. 1997a). The N_2O fluxes are also the highest after rainfall and following the establishment of flooding for rice at the end of fallow period (Bronson et al. 1997a).

The relationship between applied fertilizer N rate and N_2O emissions established by Eichner (1990) and Bouwman (1996) motivated the concept of fertilizer-induced N_2O emission factor (EF) of the Intergovernmental Panel on Climate Change (IPCC 1997). It is the N_2O emission from N fertilized plots minus that from unfertilized plots (control plots) expressed as a percentage of applied fertilizer N, with all other conditions being equal to those of treated plots. The EF represents the anthropogenic N_2O emission caused by fertilizer N application, although the emission from the control plots may not be the same as natural emissions of the original vegetation in pre-agricultural soils (historical natural emissions). The emissions from control plots generally differ from historical natural emissions because of enhanced mineralization of soil organic matter (SOM), depletion of SOM, or effects of residual N fertilizer applied prior to the experiment (Mosier et al. 1998). The default EF value of the IPCC does not distinguish emissions from paddy rice fields from upland field crops (IPCC 1997, 2006).

Some studies have quantified fertilizer induced N_2O emission and the background emission from rice paddies (Yan et al. 2003; Akiyama et al. 2005). Yan et al. (2003) estimated fertilizer induced and background N_2O emission for East, Southeast, and South Asia rice growing regions. The average fertilizer induced N_2O EF from paddy fields for the region is estimated at $0.25 \pm 0.24\%$ of applied N, while background N_2O emission is estimated at 1.22 kg N_2O-N ha^{-1} (Yan et al. 2003). Based on 113 measurements from 17 sites around the globe Akiyama et al. (2005) estimated that 0.22 and 0.37% of applied fertilizer N is lost as N_2O from the flooded paddies, and flooded paddies with midseason drainage, respectively. Zou et al. (2007) estimated the fraction of fertilizer N lost as N_2O based on water management regime using the data from 71 published peer reviewed reports from mainland China. They reported that 0.02, 0.42, and 0.73% of the applied fertilizer N is emitted as N_2O from F, F-D-F, and F-D-FM water management regimes, respectively (Zou et al. 2007).

The rate of N_2O emissions from rice soils is dependent on the rate of N applied (Cai et al. 1997; Zou et al. 2007). The data from a study to estimate N_2O EF from Chinese paddy soils between 1950s and 1990s, indicated that the overall N_2O EF increased from 0.37% in 1950s to 0.46% in 1990s, mainly reflecting an overall increase in total N input from 87.5 kg N ha^{-1} in 1950s to 224.6 kg N ha^{-1} in 1990s (Zou et al. 2009). A meta-analysis of published data to assess grain yield scaled N_2O global warming potential of major cereal crops (corn, wheat and rice) globally indicated that on average continuously flooded rice emits 0.165 kg N_2O-N ha^{-1} $season^{-1}$, and overall average emissions from rice production globally is 1.38 kg N_2O-N ha^{-1} $season^{-1}$ (Linquist et al. 2012).

Water management regime imposes sensitive changes in N_2O emissions in rice paddies (Akiyama et al. 2005). Mid-season drainage in rice paddies triggers enhanced N_2O emissions from paddy soils compared to continuous flooding (Cai et al. 1997; Zheng et al. 2000). Draining paddy soils aerates the soil and accentuates nitrification. In addition, intermittent irrigation may also stimulate N_2O fluxes depending on soil water content. The N_2O fluxes may be high if dry spells occurs in-between irrigations. However, if soils remains waterlogged, N_2O produced by denitrification is further denitrified to N_2, leading to lower N_2O fluxes (Zou et al. 2005). Soil moisture content is the most sensitive factor in regulating N_2O emissions, not only from rice soils but from other soils as well. Highest N_2O emissions are observed at water-filled pore spaces (WFPS) ranging from 45 to 75%, and N_2O is mostly produced during denitrification (Hansen et al. 1993; Khahil and Baggs 2005; Ding et al. 2007). Controlled irrigation (CI) is a water saving technique designed to minimize water use under paddy rice culture, where soil remains dry 60–80% of the growing season compared to traditional flooding (Peng et al. 2011). Cumulative N_2O emissions under CI are higher, i.e., 2.5 kg N_2O-N ha^{-1} (1.0% of applied N) compared to 1.0 kg

N_2O-N ha^{-1} (0.4% of applied N) under traditional irrigation, where soils are maintained under flooding/water logging most of the growing period (Peng et al. 2011).

Fertilizer N type or formulation and application rate also plays significant role in N_2O emissions from paddy rice soils. Experiment to assess the effect of N fertilizer type demonstrated that the N_2O emission increased with increase in N application rate regardless of the fertilizer type. In addition, N_2O emission can be higher from $(NH_4)_2SO_4$ than from urea application (Cai et al. 1997). During dry season, N_2O flux can be 2.5 times higher with $(NH_4)_2SO_4$ than with urea (Bronson et al. 1997b). Also, N_2O fluxes from plots treated with $(NH_4)_2SO_4$ or urea may markedly exceed those from plots receiving the same amount of N as $Ca(NO_3)_2$ (Blackmer et al. 1980).

The N_2O emissions generally show significant spatial and temporal variability, probably reflecting the variability of the factors controlling N_2O emissions from paddy rice soils. The N_2O emissions from the same rice field were widely variable among years (Chen et al. 1997; Majumdar et al. 2000; Pathak et al. 2002). However, in some experiments, similar patterns of N_2O emission are observed every year, having two emission peaks each year, first bigger peak appearing during transplanting in summer due to production and accumulation of N_2O via nitrification and denitrification during winter or early spring. During flooding, N_2O emissions are negligible while during harvest N_2O emissions increases after water is drained (Chen et al. 1997).

7.3 Types of Rice Ecosystems

Rice is unique among the major food crops in its ability to grow in a wide range of hydrological and climatic conditions, from wettest areas of the world to dry deserts, provided the water requirement is met (Maclean et al. 2002). However, rice dominates a major portion of South, Southeast, and East Asia, a lowland area with high temperature and alternating wet and dry seasonal cycle, and contains many of world's major rivers and delta areas. The total cropland area under rice cultivation is estimated at 158.3 Mha, of which, 141.0 Mha is in Asia, 9.4 Mha is in Africa, 6.14 Mha in Latin America, 1.3 Mha is USA and the rest (<0.7 Mha) in Europe and Australia (Bouman et al. 2007, FAO 2011). Large land area under rice cultivation in this region reflects the fact that rice is the staple food in many Asian countries. Based on the hydrology of rice soils, the rice cultivation systems falls under four major categories: (i) irrigated lowland rice, (ii) rainfed lowland rice, (iii) flood-prone rice, and (iv) upland rice. The irrigated lowland rice system is the dominant production ecosystem, followed by rainfed lowland rice (Table 7.1). Each of these four major rice ecosystems affects N_2O emissions differently depending on floodwater management regime (Table 7.2).

Table 7.1 Types of global rice farming ecosystems

Ecosystem type	General characteristics	Countries practicing	Total rice area[a] (000 ha)	%
Irrigated lowland rice	Grown in bunded field, shallow water depth maintained to near maturity, highly productive system	Almost all major rice producing countries, all of Japan, USA, Europe, Australian, Pakistan rice production	86,969	55
Rainfed lowland rice	Grown in bunded fields flooded by rainfall at least part of cropping season, water depth as high as 50 cm but does not last for more than 10 consecutive days	All Asia, Africa and Latin America rice producing countries	39,575	25
Upland rice	Rice grow as a rainfed dryland crop on hillsides, undulating and steep slopes, less productive	Asia, Africa, Latin America	18,960	12
Flood-prone/ Deepwater rice	These are flood-prone lowlands where water depth as deep as 100–400 cm can be reached. Lands are salt prone	Asia, Africa	12,797	8
Total rice cultivation area			159,279	100

[a]Source: calculated based on http//faostat.fao.org, cropland area 2009 and IRRI, 1997

7.3.1 Irrigated Lowland Rice

Irrigated lowland rice fields are bunded for water retention. Farmers control water supply and try to maintain 5–10 cm floodwater on the field until maturity of the crop. Some areas are equipped only with supplemental irrigation in the wet season, and rainfall supplies the bulk of the water requirement. Irrigated lowland rice land is prepared while wet (puddling). Puddling reduces soil permeability, percolation losses, and controls weeds. Transplanted rice is most common in Asia. Elsewhere, rice is direct seeded mostly due to high labor costs needed for transplanting. In the transplanting system, seeds are pre-germinated. Seeds are seeded in wet seedbeds for a period extending from 9 to 14 days prior to transplanting. The seedlings are transplanted manually. In direct seeding, seeds are frequently pre-germinated. Seeds are hand broadcasted (in Asia), machine-drilled in puddled soil or drill-seeded into dry soil, or even spread over the water by airplane (as in the United States and Australia). It is easier to top-dress fertilizers, do intercultural operations, and monitor GHG emissions by static chamber method.

Table 7.2 The N_2O fluxes in relation to rice ecosystems

Ecosystem type	Water regime	N_2O emission processes
Irrigated rice	Continuous flooding	Nitrification inhibited due to anaerobiasis
		Enhances production by denitrification, but N_2O is converted to N_2 due to lower diffusion under submergence. Higher NH_3 volatilization may result into higher indirect N_2O emission in deposition off field
	Flooded-drained-flooded	Mid-season drainage for soil aeration enhances N_2O emissions due to nitrification of NH4 during druing
		Anaerobic-aerobic-anaerobic cycle triggers interchangeable nitrification and denitrification which enhances N_2O emissions
	Flooded- with intermittent irrigation	Aeration enhances N_2O emissions. Water content affects soil aeration and aeration status of soils affects the rates of N_2O production in soils. Intermittent irrigation increases N_2O emission
Rainfed lowland rice-	Traditional	Intercultural operations increase N_2O emissions by allowing N_2O to move up within soil more rapidly and escape to the atmosphere
Deep water or flood prone		Soil submergence enhances production of N_2O by denitrification but N_2O can be converted to N_2, reducing emissions
Upland or dry land rice		N_2O emissions from dryland rice soils are influenced by similar conditions like other terrestrial soils

Fertilization with mineral or organic fertilizers increases land productivity, especially with semi-dwarf or high-yielding varieties, as was the case in the "Green Revolution" era of 1960s. Irrigated rice ecosystem is highly productive, with up to 5 Mg ha^{-1} during the rainy season and more than 10 Mg ha^{-1} in the dry season, when adapting advanced technologies. Irrigation permits up to three rice crops per year in tropical lowlands. Irrigated rice accounts for 55% of the world rice area (Table 7.1) and about 75% of world production, and practiced in all six continents. All rice producing land in USA, Japan, Australia, and Europe is irrigated. Approximately 56% of the world's irrigated area of all crops is in Asia, where rice accounts for 46–56% of net irrigated area of all crops (Dawe 2005). Irrigated lowland rice can be subdivided into irrigated wetland rice and irrigated dry land rice. Under irrigated wetland rice, the rice soils remains waterlogged for most of the rice-growing season. Irrigated dry rice soils are waterlogged only at the appropriate times when flooding is necessary and the rest of the period, rice is maintained under semi-aerobic soils. Based on rainfall variability, irrigated rice ecosystem can be subdivided into: (i) irrigated wet season, and (ii) irrigated dry season. Irrigated wet season areas are those where irrigation water supplements rainfall. Irrigated dry season are those areas where no rice crop can grow without irrigation because rainfall is too low, cloud cover is minimal, evapotranspiration is high, and water requirement is large. In many irrigated areas, rice grows under monoculture with two crops per year. In some areas rice is also grown in rotation with a range of other crops including wheat (Dawe et al. 2004).

7.3.2 Rainfed Lowland Rice

Rainfed lowland rice based system is characterized by a monsoon season in which rice grows in the wet rainy season and various upland crops are grown in the dry season without irrigation (Tripathi et al. 1997). Rice grows in bunded fields flooded by rainfall during at least part of the cropping season. Bunds are able to retain more than 100 cm of water depth for no more than 10 days (mostly during excessive rainy periods). Rainwater or a local reception tank (filled by rainfall), passing from one paddy to another by gravity, sometimes feeds this non-irrigated rice. The risk of temporary drought and unexpected floods is the major concern for this ecosystem. Puddled field with transplanting is the conventional method of crop establishment.

Introducing this type of cultivation is difficult and it is often associated with direct seeding with some transplanting for gap filling. Fertilizer use is rare and rice yields are low. Farmers typically grow one rice crop per year, followed by minor crop if the remaining wet season permits. Some rainfed lowlands are favorable and reliably productive, but some are prone to drought, flooding, or both. This ecosystem represents 25% of the global total rice area and produces 17% of world rice production, ranking second after irrigated rice (Maclean et al. 2002). This rice farming system is located in rural regions where population density is high, often in the poorest rural and urban populations.

7.3.3 Upland or Dryland Rice Ecosystem

Upland rice grows as a rainfed dryland crop. Dry land preparation and dry seeding under nonbunded fields is common practice. Rainfed uplands are highly heterogeneous areas with climates ranging from humid to sub-humid, soils from relatively fertile to highly infertile (Piggin et al. 1998). Historically, shifting cultivation with long fallow periods (more than 15 years) was the dominant land use system in low population density areas. However, increase in population has put pressure on the system, and many upland rice areas have made transition to permanent upland cultivation systems. Currently, rotation with other crops or in shifting slash-and-burn systems with shorter fallow periods of 3–5 years sustains upland rice cultivation. Lack of humidity and normally poor soil conditions affect crops adversely and yields are often low. With only few inputs, upland rice yields are low but never- the- less critical to the household food security of some of the poor subsistence farmers. With suitable varieties, and properly managed inputs, farmers can achieve yield approaching those widely planted in rotation with pasture or soybean in Brazil, and increasingly becoming popular in China. Additionally, Africa, India, and Southeast Asia also practice this system. In some African and Latin-American countries, this system represents more than 50% of the total rice area. Upland rice represents approximately 12% of rice area in the world and 4% of global rice production. As market access remains limited, most of worlds upland rice farmers tend to be self-sufficient by producing a range of agricultural outputs.

7.3.4 Deep-Water or Flood-Prone Rice Ecosystem

The areas under this ecosystem present a range of growing conditions in both coastal and inland environments. The common identifying feature of this ecosystem is the uncontrolled flooding. Water depth ranges from 1 to 5 m and can be supplied by rivers, lakes or tides in river mouth deltas. Water depth may exceed 5 m in some parts of Bangladesh, as well as in the Mekong, Chao Phraya, and Niger deltas. Seeds are broadcasted in plowed fields, normally unbunded, in regions where the water level rises quickly after the beginning of the monsoon rains. Five major cultural conditions are widespread: (i) submergence prone-where some rice varieties tolerate being submerged for several days (1–10) days, (ii) deep water rice – where 30–300 cm of standing water may occur, (iii) very deep water rice, where 100–400 cm of standing water may occur, (iv) where plants elongates or floats to survive long inundations, and (v) tidal and dry seasons irrigation type. Coastal areas subject to tidal surges require rice varieties that tolerate high salt levels. Minerals that accumulate in waterlogged soils often render these soils infertile. Traditional varieties with long tiller and few sprouts are cultivated. The plant elongates and floats as the floodwater advances, thus its name – "the floating rice."

Deep-water rice is grown in south and Southeast Asia (Bangladesh, Thailand, Cambodia, and Sumatra), West Africa and South America. Rice productivity in this ecosystem is low, mainly due to climatic risks of extreme events such droughts and floods, and the low production potential of cultivars grown with few inputs. Nevertheless, this ecosystem meets the needs of 78–100 million people, most of them living on small family farms. The construction of dams and other hydraulic projects has led to the transformation of some parts of Bangladesh, India, Thailand, and Vietnam into irrigated rice areas. Rice yields are low and extremely variable because of problem soils and unpredictable combinations of drought and flood. This ecosystem represents about 8% of rice planted area.

7.4 Rice Soils

Majority of rice cultivation practices require standing water over the field for extended periods during their life cycle. Rice soils are saturated for at least part of the time during rice growing period but during intervals between rice crops, the soil usually dries and becomes aerated. At this time, either soil remains under fallow or upland crops are grown. Soil moisture regime may vary from near saturation to submergence or flooding, resulting in several centimeters of standing floodwater for most of the crop growing period (De Datta 1981). Flooded rice soils become anoxic, since diffusivity of O_2 through water is low. In Asia, a lowland rice soil is bunded and tilled area that has been puddled for transplanting or broadcast-seeded with rice. In United States of America, Australia, Europe, and some Asian and African countries, land is prepared dry and flooded later. Puddling for transplanting or broadcasting with pre-germinated rice is an essential ingredient of a wetland rice soil. Tillage of saturated soil, referred to as puddling, destroys soil aggregates, reduces downward water flow, and loss of nutrients by leaching, and restricts gaseous exchange between deeper soil layers and the atmosphere. In areas with a reliable irrigation supply and low percolation, puddled soils typically remain continuously saturated until drained or just before harvest. In rainfed and irrigated lowlands with inadequate or irregular water supply, soils can undergo alternate drying and rewetting during rice growing period.

Some aspects of flooded rice soils important to N transformation include: (i) retardation of gaseous exchange between flooded soil and atmosphere, (ii) decrease in soil redox potential, (iii) increase in pH of acid soils and decrease in pH of calcareous and sodic soils, (iv) increase in specific conductance (EC) and ionic strength, and (v) anaerobiosis along with anaerobic decomposition of SOM (Savant and De Datta 1982).

7.4.1 Nitrogen Transformation in Flooded Soils

Nitrogen behaves differently in flooded soils (anaerobic) compared to dry soil conditions. When aerobic soils are flooded, soil O_2 depletion occurs rapidly and soil NO_3^- becomes prone to loss by denitrification and leaching. Flooding the soil causes accumulation of NH_4^+-N, mainly due to inhibited nitrification, instability of NO_3^--N, and reduced N requirement for organic matter decomposition. Flooding also causes inefficient utilization of applied N. Under flooded soils conditions, O_2 deficiency restricts conversion of NH_4^+ to NO_3^- . Thus, mineralization stops at NH_4^+. Therefore, NH_4^+ is the form of N that accumulates. It exists in three main fractions: (i) NH_4^+ in soil solution, (ii) NH_4^+ in exchange sites, and (iii) NH_4^+ in non-exchangeable form (De Datta 1995). The NH_4^+ in solution form and at the exchange sites is easily available for uptake by rice.

The NH_4^+-N is subject to fixation by clays, loss by volatilization, leaching, runoff, seepage, and nitrification followed by loss through denitrification. Under puddled rice soils, leaching and runoff losses are not common. Therefore, the major cause of fertilizer inefficiency in flooded rice is the gaseous emission of NH_3 and N_2. The greatest losses of N are observed when fertilizer application leads to high concentration of ammoniacal N in flood water (Simpson et al. 1988), suggesting that NH_3 volatilization may be the important N loss process. Incorporation of fertilizer N into soil reduces NH_3 volatilization, but does not always reduce N loss, because when incorporated, NH_4^+ is converted to NO_3^-, denitrified and lost as N_2 and/or N_2O (Freney et al. 1990). NH_3 volatilization losses ranging from 10 to 56% of urea N broadcasted into floodwater can occur in tropical transplanted rice (Buresh and Dedatta 1990; Freney et al. 1990). Temperature, wind speed, pH, fertilizer source, cation exchange capacity (CEC) of the soil and ammoniacal $[NH_4^+ + NH_3]$-N concentration affects the pattern of NH_3 loss (Freney et al. 1990; De Datta 1995; Cai et al. 2002). During the process of volatilization, gaseous NH_3 is formed i.e.,

$$NH_4{}^+(aq) \rightarrow NH_3(g) + H^+ \tag{7.1}$$

This reaction releases H^+ ion. The pH and buffering capacity of the oxidized as well as reduced soil layers therefore, influences the volatilization process. In wetland soils, the kinetic of pH indicate that submergence causes pH values of most of the reduced acid and alkaline soils to converge between 6.5 and 7.2.

Volatilization NH_3 is negligible from direct-seeded rice in temperate regions where majority of fertilizer N is incorporated into the soil prior to flooding (Humphreys et al. 1988). Volatilized NH_4^+ can be deposited on earth by rain, leading to acidification (Kirk 2004) and unintended N inputs into natural ecosystems. Volatilized NH_3 can also become a source of indirect N_2O emissions.

In flooded rice soil, a reduced plow layer is usually sandwiched between a thin oxidized surface soil layer and partly oxidized sub-surface soil layer. Within the reduced soil layer in which most of rice roots proliferate and derive N, is a complex mosaic consisting of aerobic and anaerobic microsystem. Rice root rhizosphere is relatively oxidized compared to the rest of the soil in plow layer due to diffusion of O_2 from the rice roots. Nitrification in oxidized soil zones, rice roots rhizosphere, and floodwater converts ammoniacal N into NO_3^-, which can then move into reduced soil zones where it is denitrified to N_2 and N_2O (Reddy and Patrick 1986b). Because denitrifying bacteria needs SOM as a source of C for their activity, the nature and amount of SOM largely determines the rate of denitrification.

Wetland rice soils often undergo alternate wet and dry conditions, especially under rainfed conditions even under continuous flooding. These soils become saturated for at least part of the time during production of rice, but during intervals between rice crops, soil usually dries and become aerated. During this time, either soil remains under fallow or upland crops are grown. In aerobic dry land soil conditions, NH_4^+ from mineralization of soil organic nitrogen (SON) or N fertilizer is nitrified to NO_3^--N which can accumulate in the soil or be used by the plants. Mineralization and immobilization occur simultaneously in wetland soils and depend on soil properties and environmental factors. At the end of rice growing season, soil NO_3^- is negligible, and soil NH_4^+ is typically low because of N uptake by rice and volatilization losses (Buresh and De Datta 1991).

7.4.2 Processes Contributing to Nitrous Oxide Emission from Rice Fields

Formation and emission of N_2O from soils is a result of complex interactions of physical and biological processes of biotic (i.e. nitrification, denitrification, dissimilatory nitrate reduction and assimilatory nitrate reduction) and abiotic origin (i.e. chemodenitrification (Firestone and Davidson 1989); Chap. 3). In a fertilized flooded rice field, conditions are prevalent for denitrification, which makes it an ideal source for N_2O and N_2, but the same conditions make it possible to further reduce N_2O to N_2, the ultimate product of denitrification. Flooding of soil results in displacement of O_2 with water, and any dissolved O_2 present in the pore water is readily consumed during microbial and root respiration, thus making soil devoid of O_2. Anaerobic conditions in rice paddies inhibit nitrification, enabling accumulation of NH_4^+.

Possible mechanisms for renewing O_2 supply to flooded soils are: (i) diffusion of O_2 through the overlying floodwater, which is consumed at the soil-water interface, and (ii) transport of O_2 through the stems of rice or other wetland weeds to the roots and subsequent diffusion of O_2 into the rhizosphere (Perata and Alpi 1993; Magneschi and Perata 2009). Rice and some wetland plants have a unique feature of transporting atmospheric O_2 through stem to the roots and some of this O_2 subsequently diffuses from roots to adjacent soil layer. Some of the O_2 transported down the roots of rice is released to the surrounding soil, creating restricted aerobic zones around the roots of rice plants. Substantial differences in the oxidizing power of rice roots, among varieties have been observed, which is attributed to differences among cultivars in aerenchyma tissue differentiation resulting in different degrees of aerobic conditions in their rhizosphere (Ghosh and Kashyap 2003).

Nitrification is not a widespread process in flooded rice fields due to unfriendly water regimes, which favor anaerobic conditions, creating O_2 deficient conditions. In aerobic rice soils such as upland rice soils or intermittently flooded soils though, nitrification may be high. In intermittently flooded rice, surface soil layers become fully or partially aerobic during drainage and for sometimes during the next fresh irrigation when water brings substantial amount of dissolved O_2. In such situations with enough NH_4^+ present in the field, nitrification might produce substantial amount of N_2O from rice fields.

During drainage of water in the field, denitrification might still occur appreciably in anaerobic microcites where nitrate is present and microbial demand for O_2 exceeds supply (Arah and Smith 1989). This process might occur either at the centers of soil aggregates or in saturated regions within a structure-less soil when water impedes O_2 diffusion, or where the local O_2 demand is exceptionally high. Denitrification in soil produces and consumes N_2O, hence this bacterial process might serve both as source and sink for N_2O.

Microorganisms with assimilatory NO_3^- reduction produce N_2O with respiratory NO_3^- and with dissimilatory NO_3^- reduction to NH_4^+. All these metabolic pathways typically produce N_2O and they do not gain energy by producing N_2, and thus have been named non-respiratory N_2O producers in contrast to respiratory N_2O producing nitrifiers (Tiedje 1988). The N_2O might also be formed by chemical reaction when NO_2^- or NH_2OH are decomposed in acid soils, producing small amounts of N_2O.

7.4.3 Assessment of Special Conditions Driving Nitrous Oxide Emission in Rice Fields

The N needs for rice crop is mainly met through chemical fertilizers, and urea accounts for up to 90% of the fertilizer N share for rice production (Vlek and Byrnes 1986). While $(NH_4)_2SO_4$ is used, in many regions, NO_3^- fertilizers are seldom used because of denitrification and low use efficiency under flooded soils. The efficiency of fertilizer N is very low, generally around 30–40%, in some cases even lower (Cao et al. 1984b; Choudhury and Kennedy 2005). NH_4^+ can be nitrified in the oxidizing thin layer at the water and soil interface and form NO_3^- which moves downwards to the reducing layer where it is subsequently denitrified (De Datta 1995). Most research on denitrification in paddy fields has focused on the submerged upper plow layer. However, in the regions with high N input, NO_3^- can be leached below the plow layer to the underground saturated soil layer (subsoil) and migrate into groundwater. Denitrification in the saturated subsoil and underground waters also contributes N_2O to the atmosphere (Xing et al. 2002). For example, N_2O concentrations of up to 237.7 and 145.7 $\mu g\ L^{-1}$ were reported from observation wells of 2.5 m depth for the rice and wheat, respectively (Xing et al. 2002).

Within a thin oxidized zone present at the surface of soil, nitrifiers can transform NH_4^+ to NO_3^- via NO_2^-, and NO_3^- can be transported to lower anaerobic layers where it is transformed to N_2O and N_2 by denitrifiers (Zhu et al. 2003). Further down, where rice roots occupy much of the soil volume, a significant amount of O_2 might be present through transportation by aerenchyma (Savant and De Datta 1982). Therefore, in the predominantly anoxic zone, nitrification of NH_4^+ transported by water can also take place to produce N_2O directly or NO_3^-, which is denitrified to N_2O. Evidence for coupling of nitrification and denitrification was derived indirectly by balancing N supply with N recovery from plant and soil (Fillery et al. 1986; Reddy and Patrick 1986a), laboratory experiments conducted on the rice rhizosphere (Arth et al. 1998), using ^{15}N enriched compounds in soil (Chen et al. 1998, 1995) and field measurements using nitrification inhibitor (Carrasco et al. 2004). Rice plants also affects nitrification and denitrification indirectly by immobilizing NH_4^+ and NO_3^- from the rhizosphere and by supplying root exudates and dead root debris, which act as substrate for microorganisms. N_2O can be transported through rice plant aerenchyma system and emitted to the atmosphere (Savant and De Datta 1982).

Under rainfed, irrigated, and upland rice cultivation, rice fields may not be flooded throughout the growing season, except where permanent flooding is deliberately done. Drainage of the applied water depends on soil type and extent of puddling done before transplanting. Absence of standing water increases soil aeration, favors nitrification and incomplete denitrification. Therefore, emission of N_2 decreases and that of N_2O emission increases during the drying cycles and less intense anaerobic conditions (Chen et al. 1997; Majumdar et al. 2000). During the dry cycles, significant amount of NO_3^- accumulate in soil due to nitrification of previously accumulated NH_4^+-N. This NO_3^- is lost by denitrification once soil is flooded again, and at the end of rice season only limited amount of mineral N [NH_4^+ + NO_3^-] remains due to different loss mechanisms and plant uptake (De Datta 1995). Permanently flooded rice fields are not major source of atmospheric N_2O because it is further reduced to N_2 under the strong anaerobic conditions. Therefore, water management regime has significant influence on N_2O fluxes from paddy fields.

7.4.4 Factors Affecting Nitrous Oxide Emission from Rice Fields

N_2O emission from soils are affected by number of factors including soil water content, pH, SOM, temperature, fertilizer application, plants, and nitrification inhibitors (Pathak 1999; Chap. 6). N_2O emission primarily is a function of N source (Eichner 1990). It is not possible to determine relative contributions of crops, amount, and type of N fertilizer, management practices adopted and weather conditions on N_2O emissions (Bouwman 1996). Yet, cropping systems and soil management practices have a greater impact on N_2O emissions than N source (Mosier et al. 1996). It is the complex interplay of all these driving factors, which actually determines the rate of N_2O emission from rice paddies. The water content, aeration, N concentration, nitrification inhibitors, SOM content, soil pH, and temperature influences N_2O production. However, the primary emission control factors are water regime, inorganic N input, plant population, soil texture, and cultural practices (Table 7.3). Understanding the factors regulating emissions enables the design of management scheme to control N_2O atmospheric loading.

Table 7.3 Factors influencing N_2O production and emissions from rice fields

Factor		Impact on N_2O production in soils	Impact on N_2O emission to the atmosphere	Reference
Water regime		Controls O_2 concentrations in soils	Controls diffusion process	Granli and Bockman (1994)
	Submergence	Enhances anaerobic environment and denitrification	Flooding enhances complete denitrification to N_2 and low N_2O emission	Hou et al. (2000), Majumdar et al. (2000), Akiyama et al. (2005)
	Intermittent drainage	Aerobic-anaerobic cycling triggers interchangeable nitrification and denitrification. Increases N_2O production	Anaerobic-aerobic cycles increases N_2O emissions relative to constant aerobic or anaerobic conditions	
Fertilizer	Rate, all types	Increased N content increases nitrification and denitrification rates. N_2O formation enhanced with increase in N content	Overall, N_2O emissions increased with increase in N application rate regardless of fertilizer N type	Cai et al. (1997)
	Ammonium sulfate vs. urea fertilizers	NH_4^+ favors N_2O formation by nitrification	At the same application rate, N_2O was higher from ammonium sulfate than urea	Cai et al. (1997), Hua et al. (1997)
	Nitrate fertilizers vs. urea	NO_3^- favors N_2O formation by denitrification under flooded soils	Increased N_2O emissions than ammonium sulfate and urea	
	Farm yard manure vs. inorganic N fertilizers	FYM acts as slow N releasing via mineralization	Decreased N_2O emissions than inorganic N fertilizers	Pathak et al. (2002, 2003)
Soil texture	Heavy texture	High N_2O entrapment leading to slow diffusion	Lower N_2O emissions due to complete denitrification to N_2	Cai et al. (1999), Xu et al. (2000)
	Sandy soils	Faster diffusion of N_2O	Increased N_2O emissions due to easy mobility of N_2O	
Nitrification inhibitors		Delayed nitrification reduces N_2O formation	Decreased N_2O emissions due to controlled nitrification	Majumdar et al. (2000)

7.4.5 Sampling Techniques for Measuring Nitrous Oxide Fluxes from Rice Fields

7.4.5.1 Gas Chamber Methods

A commonly used direct sampling technique for monitoring N_2O emission from rice fields is the chamber method. Two types of chambers are static chambers and dynamic or open chambers. Chapter 4 describes the gas flux chambers sampling technique in detail. Static chambers may differ in size and shape but maintain the basic principles of operation. A floating chamber technique has been proposed for quantifying gaseous emissions from rice paddy fields (Minami and Fukushi 1984). This type of static chamber measures the gaseous emissions from soils, which pass though the floodwater layer to emit in bubbles like any other closed chamber with the only difference being the floating arrangement, which makes the chamber to float over floodwater. The larger chamber volume minimizes the variability. A closed soil cover isolates the soil surface from atmospheric pressure fluctuations, which is operative under open air soil surface. Gaseous emission measurements obtained from this method should be corrected for increased concentration in enclosed space (Focht 1974). Thus, Ryden et al. (1978) proposed using a cover coupled to external atmosphere through an air inlet vent for minimizing the differences in pressure and other conditions between the enclosed and open surfaces. Further, temperature increase inside the chambers may affect crop growth during gas measurements, especially if non-insulated transparent chambers are used. A rice yield decrease of up to 58% inside the automated chambers can occur (Bronson et al. 1997b). However, there may be no significant changes in yield of rice growing under the transparent chambers on 6 h of enclosure during sampling days (Majumdar et al. 2000). In addition, a change in temperature of up to 10 °C inside the chamber has been reported (Livingston and Hutchinson 1995), which can change the gaseous volume by 3–4%, and may also influence the plants growing in the chamber.

7.4.5.2 Micrometeorological Techniques

Micrometeorological techniques measure gaseous fluxes over a larger area than is possible with chambers. These techniques measure turbulent transfer of gases from ground surface to lower atmosphere. These techniques include eddy correlation, eddy accumulation, energy balance, aerodynamic, mass balance and flux gradient methods. Micrometeorological techniques are suitable for integrating fluxes over areas up to 1.0 km^2 depending on sampling height chosen. Detailed description and principles behind these techniques are described in more detail in Chap. 4.

7.4.5.3 Tracer Gas Method

The technique monitors simultaneous concentration ratios of target gas (N_2O) and an inert tracer released at a known rate downwind and relates their ratio fluxes. Commonly used tracer for N_2O is SF_6, which is inert, has low background concentration and easily detectable by an electron capture detector of a gas chromatograph. The tracer must be released in a way that resembles the emission of the target gas, and the tracer and target gas must be well mixed. Emission of the target gas can be obtained from Eq. 7.2:

$$Q_{N2O} = Q_{SF6} \times \frac{C_{N2O}}{C_{SF6}} \quad (7.2)$$

where Q_{N2O} is the N_2O (target) gas flux rate, Q_{SF6} is the SF_6 (tracer) gas flux rate, C_{N2O} is the measured concentration of N_2O, and C_{SF6} concentration of SF_6. This technique has been used successfully to distinguish CH_4 emissions from ruminants (Johnson et al. 1994; Pinares-Patino et al. 2011).

7.4.6 Problems Associated with Sampling and Quantification of Nitrous Oxide from Rice Fields

Flooded conditions encountered in most rice fields pose many practical problems in field gaseous sampling including:

(i) Chambers installation under flooded conditions with sanding water and puddled soils:
In experiments where very large boxes have been used, researchers have encountered difficulties in placing the collars within large fields due to standing water and puddled soils which are less stable (Majumdar et al. 2000). For the automatic chambers, due to large size of the chambers, closure time must be significantly longer to get a detectable N_2O signal inside the chambers. The longer closure times may cause changes in temperatures and pressure, which may introduce an artifact in gaseous measurements. Smaller chambers would collect detectable amount of N_2O within shorter sampling time, but chambers cannot enclose plant canopy effectively when plant height increases. Leaking of chambers during sampling can also cause significant errors.

(ii) Temporal and spatial variability:
Similar to N_2O measurements from other ecosystems, the N_2O emissions from rice ecosystem are highly variable (Majumdar et al. 2000), mainly due to variations in factors influencing N_2O emissions such as soil water content, O_2 availability, SOM content, plant population, nitrifier and denitrifier populations. Rice grown on sloping fields sometimes leads to accumulation of floodwater on the lower slope leaving thinner layer of

water on the upper-slope, which may lead to variable N_2O emissions from the two sides of the field. N_2O emissions may also vary during a 24-h cycle due to temperature differences in soil and air. Spatial variations can be minimized by increasing the number of samples from the same area, or use of micrometeorological techniques, which estimates gaseous emissions from larger areas. Water management regime also can be a source of large variations in N_2O emissions from rice soils.

(iii) N_2O dissolved in water:

A significant quantity of N_2O may remain dissolved in floodwater after emissions from soil and escape sampling. Dissolved N_2O needs to be estimated for better quantification of N_2O production from soil. Solubility of N_2O in water at different temperatures is known (Linke 1965). A solubility of 0.7 g N_2O-N L^{-1} at 25 °C can be used to estimate soluble N_2O at STP. Moraghan and Buresh (1977) proposed an equation (Eq. 7.3) for estimating dissolved N_2O entrapped in laboratory incubation system as:

$$Y = \alpha.x\left(\frac{solution\ volume}{atmosphere\ volume}\right) \tag{7.3}$$

where Y amount of dissolved N_2O (mg) in the closed chamber, α is the solubility of N_2O (cm^3) N_2O dissolved per cm^3 of water, and x is the amount of N_2O (mg) in the atmosphere, i.e. chamber headspace.

An alternative method of measuring dissolved N_2O in floodwater and released from its surface was proposed by Minami and Fukushi (1984). It involves equilibrating floodwater of known volume at 45 °C in a water-bath and stripping N_2O out with N_2 of 99.99% purity with magnetic stirring. Released N_2O is then trapped in Porapack Q packed column, cooled at −135 °C in liquid N_2. N_2O is then separated in a Porapack Q column (70 °C) and subsequently detected. Dissolved N_2O concentration ranging from 0 to 0.38 μg L^{-1} has been detected in the floodwater (Minami and Fukushi 1984).

(iv) Entrapment of N_2O in soil:

Significant quantities of N_2O remains entrapped in the soil pore spaces. The extent and quantity depend on soil texture and pressure of the standing water. During drying cycles, N_2O flux may increase due to decreased pressure. The longer the time of N_2O entrapment in the soil the higher the chance for its conversion to N_2 by denitrification. There are few experimental data for the N_2O trapped in paddy rice soils.

(v) Plant mediated N_2O emission

The rice plants can act as conduits of N_2O emission from rice soil to the atmosphere (Yan et al. 2000). Transport of soil N_2O to the atmosphere by terrestrial plants occurs through two major pathways: (i) N_2O transport in gaseous phase through plant pore spaces (aerenchyma), as commonly observed in wetland plant species including rice, and (ii) N_2O transport in liquid phase through plant fluid systems especially by transpiration process in non aerenchymous species (e.g., barley (*Hordeoum vulgare*), canola (*Brassica napis*) (Chang et al. 1998)).

As a morphological adaptation to flooding conditions, wetland plants such as rice have developed pore spaces to allow influx of O_2 through plants for root respiration (Perata and Alpi 1993). This same pathway is also used for efflux of soil gases such as N_2O, N_2 and CH_4 in the opposite direction to the atmosphere (Yu et al. 1997). As a water-soluble gas, N_2O is absorbed with water and transported to the leaves via transpiration stream. The rates of N_2O emission by transpiration-mediated mechanism largely depend on N_2O concentration in the soil solution. In general, closed chambers trap more N_2O over plant canopies than over the bare soils.

(vi) Production of N_2O by plants

Plants may produce N_2O during N assimilation processes. Hakata et al. (2003) used ^{15}N labeled NO_3^- as a tracer and provided direct evidence of N_2O emission by plants instead of microbial activities. The N_2O can be formed by enzyme activities inside plant leaves. N_2O emission is correlated with leaf NO_3^- assimilation and increases when N source is shifted from NH_4^+ to NO_3^- (Smart and Bloom 2001). Converting NO_3^- to N_2O is a common trait in many plant species. Given the large quantities of NO_3^- assimilated by plants in the terrestrial biosphere, N_2O emission during NO_{3-} photo-assimilation may be important biogenic N_2O source with global significance. Smart and Bloom (2001) estimated that enzymatic production of N_2O in plant leaves could account for 5–6% of total N_2O emission from agricultural soil-plant systems.

7.4.7 *Effects of Land Management Practices in Non-rice Growing Season on Nitrous Oxide Emissions from Rice Fields*

Monitoring of N_2O emission during non-rice growing period when the land is under fallow or rotation with other crops, which does not require flooding, can give an insight into the role of rice-based cropping systems to atmospheric loading of N_2O. Agronomic practices in non-rice growing period may influence the N_2O emissions in the next rice-growing season. In flooded soils, a significant portion of NH_4^+ ions are converted to non-exchangeable form when soil contains 2:1 type clays. The fixed NH_4^+ becomes protected from N losses via nitrification-denitrification processes during the drying and rewetting of soil. This process is controlled by the redox potential (Eh) which influences the CEC of clays. The fixed NH_4^+ becomes available to the subsequent crop from exchange sites by slow release, which is the highest in vicinity of rice roots. The release of fixed NH_4^+ is the highest in the rhizosphere of rice plants where Eh has been increased by O_2 secretion from roots and decreases with increasing distance from the root (Schneiders and Scherer 1998). The non-exchangeable NH_4^+ fixation and mobilization of NH_4^+ are strongly influenced by the redox potential in the paddy soils.

7.5 Mitigation Options of N_2O Emissions from Rice Fields

The underlying principle of reducing N_2O emission from agriculture is increasing the NUE. Management practices for minimizing N_2O emissions from rice paddy soils and increase NUE include: (i) better matching of N supply to crop demand, (ii) integrating more closely crop residue management with crop production, (iii) using improved fertilizer management techniques including controlled release fertilizers, nitrification inhibitors, fertilizer timing, and (iv) adopting efficient water management techniques.

There exists a very strong negative correlation between CH_4 and N_2O emissions from rice fields (Cai et al. 1997; Hou et al. 2000). Water management strategies for mitigating N_2O emissions from rice fields may increase CH_4 emission, and vice versa. A number of strategies specifically formulated to reduce N_2O from rice paddy fields include: (i) matching N supply with the crop demand, (ii) minimizing fallow periods to limit mineral N accumulation, (iii) optimizing split N application schemes, (iv) using controlled release fertilizers, (v) using nitrification inhibitors, and (iv) optimize tillage, irrigation and drainage (Beauchamp 1997). Mitigation strategies which reduce emissions of both CH_4 and N_2O and produce lower combined global warming potential include: (1) application of mid-season drainage which does not coincide with high NH_4^+ in soil, (2) application of urea and NH_4^+ based fertilizers in splits at critical growth stage, (3) application of foliar N spray in water-logged conditions, (4) use of nitrification inhibitors with urea and NH_4^+ fertilizers, (5) use of slow N release fertilizers, (6) incorporation or deep placement of prilled urea granules, (7) avoiding use of green manure and use of well composted organic matter, (8) use of optimum plant population, and (9) allowing aquatic weeds and algae to grow in flood water (Majumdar 2003).

7.6 Conclusions

Monitoring of N_2O from rice fields has not been done extensively worldwide mainly due to preconceived lower emissions compared with other agricultural sources and methodological difficulties encountered in paddy rice soils. Fertilizer N use in rice production is increasingly becoming major part of rice agronomy, and this may increase N_2O emissions from rice fields appreciably. Therefore, monitoring N_2O emissions from different rice ecosystems and estimating global budgets from rice fields becomes important. Modeling is one tool that has been used to predict N_2O emissions from soils, but simulation of N_2O emissions from rice fields is difficult since biochemistry of rice soils is very complex and the N_2O emissions are controlled by real-time field conditions and variations in cultural practices worldwide. Existing data are inadequate and available from a few countries – India, China, and Philippines for deployment of simulation models. In addition, wherever monitoring has been done, few studies have monitored N_2O emissions from rice fields during non-rice growing period. Existing data have suggested that paddy rice soils may be a significant source of N_2O emission during non-rice growing periods when soils become aerated. Assessment of the fate of NH_3 volatilized from rice fields may also prove to be important.

7.7 Future Research Needs

There is no extensive monitoring of N_2O emissions from paddy rice ecosystems worldwide, mainly due to preconceived low emissions of N_2O. However, fertilizer N use and acreages under rice are increasing worldwide. In addition, water conservation techniques are increasingly being employed in rice production, which favors higher N_2O fluxes from rice fields. Therefore, monitoring of emissions from different rice ecosystems such as rainfed, irrigated and deep water and estimation of realistic regional and global budget from rice needs to assume high significance. Simulation of N_2O emission from rice soils is difficult due to the complex biogeochemistry under rice production systems and the real time fluctuations of field conditions and different cultural practices worldwide. Data on N_2O emissions from rice are inadequate to support models deployment. Changing over from anaerobic rice soils to aerobic farming and vice versa needs investigation to understand the changes in biogeochemistry and its influence on N_2O emissions. In summary, future research should address the following issues:

1. Increased field data are needed on N_2O fluxes from rice agroecosystem, especially flooded rice soils to facilitate global budgets from rice production and facilitate deployment of simulation models.
2. Midseason drainage to aerate flooded rice soils might increase N_2O emission, while flooded conditions favor CH_4 emissions. Additional studies are needed to assess the overall global warming potential associated with rice production by monitoring CO_2, CH_4, and N_2O emissions for the entire growing season including the period of mid-season drainage and other water regimes.
3. Assessment of N_2O emission is needed for the entire rice based cropping system rather than for the individual crops, since pre and post-rice management practices influence N_2O emissions from rice field.

Study Questions

1. Describe the experiment needed to estimate the N_2O emissions from paddy rice field during rice growing season using a static chamber method. Describe the procedure to account for N_2O gas dissolved in the flood water?
2. Given choices, what is the best method of estimating N_2O emissions from flooded rice farm? Explain rational for the choice of a specific method.
3. What are the major limitations of modeling approaches in estimating and predicting the N_2O emissions from rice fields?

References

Akiyama, H., Yagi, K., Yan, X. (2005) Direct N_2O emissions from rice paddy fields: Summary of available data. Global Biochem. Cycles **19**(1), GB1005. doi:10.1029/2004gb002378

Arah, J.R.M., Smith, K.A. (1989) Steady-state denitrification in aggregated soils: a mathematical model. J. Soil Sci. **40**(1), 139–149

Arnold, P.W. (1954) Losses of nitrous oxide from soil. J. Soil Sci. **5**(1), 116–128

Arth, I., Frenzel, P., Conrad, R. (1998) Denitrification coupled to nitrification in the rhizosphere of rice. Soil Biol. Biochem. **30**(4), 509–515. doi:10.1016/s0038-0717(97)00143-0

Aulakh, M.S., Doran, J.W., Walters, D.T., Mosier, A.R., Francis, D.D. (1991) Crop residue type and placement effects on denitrification and mineralization. Soil Sci. Soc. Am. J. **55**(4), 1020–1025

Beauchamp, E.G. (1997) Nitrous oxide emission from agricultural soils. Can. J. Soil Sci. **77**(2), 113–123

Blackmer, A.M., Bremner, J.M., Schmidt, E.L. (1980) Production of nitrous-oxide by ammonia-oxidizing chemoautotrophic microorganisms in soil. Appl. Environ. Microbiol. **40**(6), 1060–1066

Bouman, B.A.M., Humphreys, E., Tuong, T.P., Barker, R. (2007) Rice and Water. Adv. Agron. **Volume 92**, 187–237. doi:10.1016/s0065-2113(04)92004-4

Bouwman, A.F. (1996) Direct emission of nitrous oxide from agricultural soils. Nutr. Cycling Agroecosyst. **46**(1), 53–70

Bronson, K.F., Neue, H.U., Singh, U., Abao, E.B. (1997a) Automated chamber measurements of methane and nitrous oxide flux in a flooded rice soil .1. Residue, nitrogen, and water management. Soil Sci. Soc. Am. J. **61**(3), 981–987

Bronson, K.F., Singh, U., Neue, H.U., Abao, E.B. (1997b) Automated chamber measurements of methane and nitrous oxide flux in a flooded rice soil .2. Fallow period emissions. Soil Sci. Soc. Am. J. **61**(3), 988–993

Buresh, R.J., Austin, E.R. (1988) Direct measurement of dinitrogen and nitrous-oxide flux in flooded rice fields. Soil Sci. Soc. Am. J. **52**(3), 681–688

Buresh, R.J., De Datta, S.K. (1991) Nitrogen dynamics and management in rice-legume cropping systems. Adv. Agron. **45**, 1–59

Buresh, R.J., Dedatta, S.K. (1990) Denitrification losses from puddled rice soils in the tropics. Biol. Fertil. Soils **9**(1), 1–13. doi:10.1007/bf00335854

Cai, G.X., Chen, D.L., Ding, H., Pacholski, A., Fan, X.H., Zhu, Z.L. (2002) Nitrogen losses from fertilizers applied to maize, wheat and rice in the North China Plain. Nutr. Cycling Agroecosyst. **63**(2), 187–195. doi:10.1023/a:1021198724250

Cai, Z.C., Xing, G.X., Shen, G.Y., Xu, H., Yan, X.Y., Tsuruta, H., Yagi, K., Minami, K. (1999) Measurements of CH_4 and N_2O emissions from rice paddies in Fengqiu, China. Soil Sci. Plant Nutr. **45**(1), 1–13

Cai, Z.C., Xing, G.X., Yan, X.Y., Xu, H., Tsuruta, H., Yagi, K., Minami, K. (1997) Methane and nitrous oxide emissions from rice paddy fields as affected by nitrogen fertilisers and water management. Plant Soil **196**(1), 7–14

Cao, Z.H., De Datta, S.K., Fillery, I.R.P. (1984a) Nitrogen-15 balance and residual effects of urea-N in wetland rice fields as affected by deep placement techniques. Soil Sci. Soc. Am. J. **48**(1), 203–208. doi:10.2136/sssaj1984.03615995004800010037x

Cao, Z.H., Dedatta, S.K., Fillery, I.R.P. (1984b) Effect of placement methods on floodwater properties and recovery of applied nitrogen (N-15-labeled urea) in wetland rice. Soil Sci. Soc. Am. J. **48**(1), 196–203

Carrasco, D., Fernandez-Valiente, E., Ariosa, Y., Quesada, A. (2004) Measurement of coupled nitrification-denitrification in paddy fields affected by Terrazole, a nitrification inhibitor. Biol. Fertil. Soils **39**(3), 186–192. doi:10.1007/s00374-003-0673-7

Chang, C., Janzen, H.H., Cho, C.M., Nakonechny, E.M. (1998) Nitrous oxide emission through plants. Soil Sci. Soc. Am. J. **62**(1), 35–38

Chen, D.L., Chalk, P.M., Freney, J.R., Luo, Q.X. (1998) Nitrogen transformations in a flooded soil in the presence and absence of rice plants: 1. Nitrification. Nutr. Cycling Agroecosyst. **51**(3), 259–267. doi:10.1023/a:1009736729518

Chen, D.L., Chalk, P.M., Freney, J.R., Smith, C.J., Luo, Q.X. (1995) Estimation of nitrification rates in flooded soils. Microbial Ecol. **30**(3), 269–284

Chen, G.X., Huang, G.H., Huang, B., Yu, K.W., Wu, J., Xu, H. (1997) Nitrous oxide and methane emissions from soil-plant systems. Nutr. Cycling Agroecosyst. **49**(1–3), 41–45

Choudhury, A., Kennedy, I. (2005) Nitrogen Fertilizer Losses from Rice Soils and Control of Environmental Pollution Problems. Comm. Soil Sci. Plan. **36**(11/12), 1625–1639. doi:10.1081/css-200059104

Dawe, D. (2005) Increasing water productivity in rice-based systems in Asia - Past trends, current problems, and future prospects. Plant Prod. Sci. **8**(3), 221–230. doi:10.1626/pps.8.221

Dawe, D., Frolking, S., Li, C.S. (2004) Trends in rice-wheat area in China. Field Crops Res. **87**(1), 89–95. doi:10.1016/j.fcr.2003.08.008

De Datta, S.K. (1981) Principles and practices of rice production. pp. John Wiley & Sons, New York

De Datta, S.K. (1995) Nitrogen transformations in wetland rice ecosystems. Fertil. Res. **42**(1–3), 193–203

De Datta, S.K., Buresh, R.J., Samson, M.I., Obcemea, W.N., Real, J.G. (1991) Direct measurement of ammonia and denitrification fluxes from urea applied to rice. Soil Sci. Soc. Am. J. **55**(2), 543–548

De Datta, S.K., Magnaye, C.P., Padilla, J.L., Moomaw, J.C. (1968) Efficiency of fertilizer nitrogen (^{15} N Labelled) for flooded rice In: Transactions of 9th International Soil Science Congress, vol. IV. pp. 67–76. Adelaide, Australia

Delariche, C.C., Bissell, S., Virginia, R. (1978) Soil and other sources of nitrous oxide. In: Nielsen, D.R., MacDonald, J.G. (eds.) Nitrogen in the environment: Nitrogen behaviour in the soil. pp. 459–476. Academic Press, London, UK

Denmead, O.T., Freney, J.R., Simpson, J.R. (1979) Nitrous oxide emission during denitrification in a flooded field. Soil Sci. Soc. Am. J. **43**(4), 716–718

Ding, W.X., Cai, Y., Cai, Z.C., Yagi, K., Zheng, X.H. (2007) Nitrous oxide emissions from an intensively cultivated maize-wheat rotation soil in the North China Plain. Sci. Tot. Environ. **373**(2–3), 501–511. doi:10.1016/j.scitotenv.2006.12.026

Eichner, M.J. (1990) Nitrous oxide emissions from fertilized soils: summary of available data. J. Environ. Qual. **19**(2), 272–280

Ekanayake, I.J., Dedatta, S.K., Steponkus, P.L. (1989) Spikelet sterility and flowering response of rice to water-stress at anthesis. Annal Bot. **63**(2), 257–264

Food and Agriculture Organization of the United Nations (FAO) (2000) Fertilizer requirements in 2015 and 2030. 37. Food and Agricultural Organization of the United Nations (FAO), Rome, Italy

Food and Agriculture Organization of the United Nations (FAO) (2011) FAOSTAT Land resources database. On line at http://fao.org (Accessed December 2011)

Fillery, I.R.P., Simpson, J.R., Dedatta, S.K. (1986) Contribution of ammonia volatilization to total nitrogen loss after applications of urea to wetland rice fields. Fertil. Res. **8**(3), 193–202. doi:10.1007/bf01048620

Firestone, M.K., Davidson, E.A. (1989) Microbiological basis of NO and N2O production and consumption in soils. In: Andreae, M.O., Schinel, D.S. (eds.) Exchange of trace gases between terrestrial ecosystems and atmosphere. pp. 7–21. John Wiley & Sons, New York

Flessa, H., Beese, F. (1995) Effects of sugar beet residues on soil redox potential and nitrous-oxide emission. Soil Sci. Soc. Am. J. **59**(4), 1044–1051

Focht, D.D. (1974) Effect of temperature, pH, and aeration on production of nitrous oxide and gaseous nitrogen: Zero-order kinetic-model. Soil Sci. **118**(3), 173–179

Freney, J.R., Trevitt, A.C.F., De Datta, S.K., Obcemea, W.N., Real, J.G. (1990) The interdependence of ammonia volatilization and denitrification as nitrogen loss processes in flooded rice fields in the Philippines. Biol. Fertil. Soils **9**(1), 31–36

Galbally, I.E., Freney, J.R., Muirhead, W.A., Simpson, J.R., Trevitt, A.C.F., Chalk, P.M. (1987) Emission of nitrogen-oxides (NO_x) from a flooded soil fertilized with urea: Relation to other nitrogen loss processes. J. Atmos. Chem. **5**(3), 343–365

Ghosh, P., Kashyap, A.K. (2003) Effect of rice cultivars on rate of N-mineralization, nitrification and nitrifier population size in an irrigated rice ecosystem. Appl. Soil Ecol. **24**(1), 27–41. doi:10.1016/s0929-1393(03)00068-4

Goody, R.M., Walshaw, C.D., Mason, B.J., Oddie, B.C.V., Arnold, P.W., Meiklejohn, J., Wormell, T.W., Bonacina, L.C.W. (1954) The origin of atmospheric nitrous oxide: discussion. Quarterly J. Royal Meteor. Soc. **80**(345), 462–464

Granli, T., Bockman, O.C. (1994) Nitrous oxide from agriculture. Norw. J. Agric. Sci. **12 Supplement**, 7–128

Hakata, M., Takahashi, M., Zumft, W., Sakamoto, A., Morikawa, H. (2003) Conversion of the nitrate nitrogen and nitrogen dioxide to nitrous oxides in plants. Acta Biotechnol. **23**(2–3), 249–257. doi:10.1002/abio.200390032

Hansen, S., Maehlum, J.E., Bakken, L.R. (1993) N_2O and CH_4 fluxes in soil influenced by fertilization and tractor traffic. Soil Biol. Biochem. **25**(5), 621–630

Hou, A.X., Chen, G.X., Wang, Z.P., Van Cleemput, O., Patrick, W.H. (2000) Methane and nitrous oxide emissions from a rice field in relation to soil redox and microbiological processes. Soil Sci. Soc. Am. J. **64**(6), 2180–2186

Hua, X., Guangxi, X., Cai, Z.C., Tsuruta, H. (1997) Nitrous oxide emissions from three rice paddy fields in China. Nutr. Cycling Agroecosyst. **49**(1–3), 23–28

Huang, Y., Zhang, W., Zheng, X.H., Li, J., Yu, Y.Q. (2004) Modeling methane emission from rice paddies with various agricultural practices. J. Geophys. Res. Atmos. **109**(D8). doi:10.1029/2003jd004401

Humphreys, E., Freney, J.R., Muirhead, W.A., Denmead, O.T., Simpson, J.R., Leuning, R., Trevitt, A.C.F., Obcemea, W.N., Wetselaar, R., Cai, G.X. (1988) Loss of ammonia after application of urea at different times to dry-seeded, irrigated rice. Fertil. Res. **16**(1), 47–57. doi:10.1007/bf01053314

IPCC (1997) Revised 1996 IPCC Guidelines for National Greenhouse Gas Inventories. Houghton, J.T., Filho, L.G.M., Lim, B., Treanton, K., Mamaty, I., Bonduki, Y., Griggs, D.J., Callender, B.A. (eds.). pp. UK Meteorological Office, Bracknell,

IPCC (2006) IPCC 2006, 2006 IPCC Guidelines for National Greenhouse Gas Inventories. Prepared by the National Greenhouse Gas Inventories Programme. Eggleston, H.S., Buendia, L., Miwa, K., Ngara, T., Tanabe, K. (eds.). pp. Institute for Global Environmental Strategies (IGES), Kanagawa, Japan

International Rice Research Institute (IRRI) (2003). Rice Today **2**(1), 5 Manila, Philippines

Johnson, K., Huyler, M., Westberg, H., Lamb, B., Zimmerman, P. (1994) Measurement of methane emissions from ruminant livestock using a sf6 tracer technique. Environ. Sci. Technol. **28**(2), 359–362. doi:10.1021/es00051a025

Khahil, M.I., Baggs, E.M. (2005) CH_4 oxidation and N_2O emissions at varied soil water-filled pore spaces and headspace CH_4 concentrations. Soil Biol. Biochem. **37**(10), 1785–1794. doi:10.1016/j.soilbio.2005.02.012

Kirk, G. (2004) The biochemistry of submerged soils. pp. John Wiley and Sons, Chichester, West Sussex,

Kumar, U., Jain, M.C., Pathak, H., Kumar, S., Majumdar, D. (2000) Nitrous oxide emission from different fertilizers and its mitigation by nitrification inhibitors in irrigated rice. Biol. Fertil. Soils **32**(6), 474–478

Linke, W.F. (1965) Solubilities, inorganic and metal-organic compounds, vol. II4th ed. pp. American Chemical Society, Washington,

Linquist, B., van Groenigen, K.J., Adviento-Borbe, M.A., Pittelkow, C., van Kessel, C. (2012) An agronomic assessment of greenhouse gas emissions from major cereal crops. Global Change Biol. **18**(1), 194–209. doi:10.1111/j.1365-2486.2011.02502.x

Liu, S.W., Qin, Y.M., Zou, J.W., Liu, Q.H. (2010) Effects of water regime during rice-growing season on annual direct N_2O emission in a paddy rice-winter wheat rotation system in southeast China. Sci. Tot. Environ. **408**(4), 906–913. doi:10.1016/j.scitotenv.2009.11.002

Livingston, G.P., Hutchinson, G.L. (1995) Enclosure-based measurement of trace gas exchange: Application and source of error. In: Matson, P., Harris, R.C. (eds.) Biogenic trace gases: Measuring emissions from soil and water. pp. 14–51. Blackwell Sciences, Oxford, UK

Maclean, J.L., Dawe, D.C., Hardy, B., Hettel, G.P. (2002) Rice almanac. pp. CABI Publishing, International Rice Research Institute (IRRI), Los Banos, Philippine and Willford,

Magneschi, L., Perata, P. (2009) Rice germination and seedling growth in the absence of oxygen. Annal Bot. **103**(2), 181–196. doi:10.1093/aob/mcn121

Majumdar, D. (2003) Methane and nitrous oxide emission from irrigated rice fields: Proposed mitigation strategies. Curr. Sci. **84**(10), 1317–1326

Majumdar, D., Kumar, S., Pathak, H., Jain, M.C., Kumar, U. (2000) Reducing nitrous oxide emission from an irrigated rice field of North India with nitrification inhibitors. Egric. Ecosys. Environ. **81**(3), 163–169

Minami, K., Fukushi, S. (1984) Methods for measuring N_2O flux from water-surface and N_2O dissolved in water from agricultural land. Soil Sci. Plant Nutr. **30**(4), 495–502

Moraghan, J.T., Buresh, R. (1977) Correction for dissolved nitrous oxide in nitrogen studies. Soil Sci. Soc. Am. J. **41**(6), 1201–1202

Mosier, A., Kroeze, C., Nevison, C., Oenema, O., Seitzinger, S., van Cleemput, O. (1998) Closing the global N_2O budget: nitrous oxide emissions through the agricultural nitrogen cycle - OECD/IPCC/IEA phase II development of IPCC guidelines for national greenhouse gas inventory methodology. Nutr. Cycling Agroecosyst. **52**(2–3), 225–248

Mosier, A.R., Duxbury, J.M., Freney, J.R., Heinemeyer, O., Minami, K. (1996) Nitrous oxide emissions from agricultural fields: Assessment, measurement and mitigation. Plant Soil **181**(1), 95–108

Pathak, H. (1999) Emissions of nitrous oxide from soil. Curr. Sci. **77**(3), 359–369

Pathak, H., Arti, B., Shiv, P., Shalini, S., Kumar, S., Jain, M.C., Singh, P., Bhatia, A., Prasad, S., Singh, S. (2003) Effect of DCD, FYM and moisture regime on nitrous oxide emission from an alluvial soil in rice-wheat cropping system. J. Indian Soc, Soil Sci. **51**(2), 139–145

Pathak, H., Bhatia, A., Prasad, S., Singh, S., Kumar, S., Jain, M.C., Kumar, U. (2002) Emission of nitrous oxide from rice-wheat systems of Indo-Gangetic plains of India. Environ. Monit. Asses. **77**(2), 163–178

Peng, S., Hou, H., Xu, J., Mao, Z., Abudu, S., Luo, Y. (2011) Nitrous oxide emissions from paddy fields under different water managements in southeast China. Paddy Water Environ. **9**, DOI 10.1007/s10333-10011-11275-10331

Perata, P., Alpi, A. (1993) Plant responses to anaerobiosis. Plant Sci. **93**(1–2), 1–17

Piggin, C., B.Courtois, George, T., Pandey, S., Lafitte, R., Kirk, G., Kondo, M., Leung, H., Nelson, R., Olofsdotter, M., Prot, J.C., Reversat, G., Roder, W., Schmidt, V., V.P.Singh, Trebui, G., Zeigler, R., Fehrney, K., Castella, J.C. (1998) The IRRI upland rice research program. International Rice Research Institute (IRRI)

Pinares-Patino, C.S., Lassey, K.R., Martin, R.J., Molano, G., Fernandez, M., MacLean, S., Sandoval, E., Luo, D., Clark, H. (2011) Assessment of the sulphur hexafluoride (SF(6)) tracer technique using respiration chambers for estimation of methane emissions from sheep. Anim. Feed Sci. Tech. **166–67**, 201–209. doi:10.1016/j.anifeedsci.2011.04.067

Ponnamperuma, F.N. (1972) The chemistry of submerged soils. Adv. Agron. **24**, 29–96

Prasad, R., Power, J.F. (1995) Nitrification inhibitors for agriculture, health, and the environment. Adv. Agron. **54**, 233–281

Reddy, K.R., Patrick, W.H. (1986a) Fate of fertilizer nitrogen in the rice root zone. Soil Sci. Soc. Am. J. **50**(3), 649–651

Reddy, K.R., Patrick, W.H.J. (1986b) Denitrification losses in flooded rice fields. In: De Datta, S.K., Patrick, W.H.J. (eds.) Nitrogen economy of flooded rice soils. pp. 99–116. Martinus Nijhoff Publishers, Dordrecht

Rijsberman, F.R. (2006) Water scarcity: Fact or fiction? Agric. Water Manage. **80**(1–3), 5–22. doi:10.1016/j.agwat.2005.07.001

Ryden, J.C., Lund, L.J., Focht, D.D. (1978) Direct in-field measurement of nitrous oxide flux from soils. Soil Sci. Soc. Am. J. **42**(5), 731–737

Savant, N.K., De Datta, S.K. (1982) Nitrogen transformations in wetland rice soils. Adv. Agron. **35**, 241–302

Schneiders, M., Scherer, H.W. (1998) Fixation and release of ammonium in flooded rice soils as affected by redox potential. Eur. J. Agron. **8**(3–4), 181–189

Simpson, J.R., Muirhead, W.A., Bowmer, K.H., Cai, G.X., Freney, J.R. (1988) Control of gaseous nitrogen losses from urea applied to flooded rice soils. Fertil. Res. **18**(1), 31–47

Smart, D.R., Bloom, A.J. (2001) Wheat leaves emit nitrous oxide during nitrate assimilation. Proc. Natl. Acad. Sci. U.S.A. **98**(14), 7875–7878. doi:10.1073/pnas.131572798

Smith, C.J., Brandon, M., Patrick, W.H. (1982) Nitrous oxide emission following urea-n fertilization of wetland rice. Soil Sci. Plant Nutr. **28**(2), 161–171

Smith, C.J., Patrick, W.H. (1983) Nitrous oxide emission as affected by alternate anaerobic and aerobic conditions from soil suspensions enriched with ammonium sulfate. Soil Biol. Biochem. **15**(6), 693–697

Smith, P., Martino, D., Cai, Z., Gwary, D., Janzen, H., Kumar, P., McCarl, B., Ogle, S., O'Mara, F., Rice, C., Scholes, B., Sirotenko, O. (2007) Agriculture. In: Metz, B., Davidson, O.R., Bosch, P.R., Dave, R., Meyer, L.A. (eds.) Climate Change 2007: Mitigation. Contribution of Working Group III to the Fourth Assessment Report of the Intergovernmental Panel on Climate Change. pp. 497–540. Cambridge University Press, Cambridge, United Kingdom and New York, NY, USA

Tiedje, J.M. (1988) Ecology of nitrification and dissimilatory nitrate reduction to ammonium. In: Zehnder, A.J.B. (ed.) Biology of anaerobic microorganisms. John Wiley & Sons, New York

Tripathi, B.P., Ladha, J.K., Timsina, J., Pascua, S.R. (1997) Nitrogen dynamics and balance in intensified rainfed lowland rice-based cropping systems. Soil Sci. Soc. Am. J. **61**(3), 812–821

Tuong, T.P., Bouman, B.A.M. (2003) Rice production in water scarce environments. In: Kijne, J.W., Barker, R., Molden, D. (eds.) Water productivity in Agriculture: Limitsand opportunities for improvement. pp. 53–67. CABI Publishing, Wallingford,

van Groenigen, J.W., Velthof, G.L., Oenema, O., Van Groenigen, K.J., Van Kessel, C. (2010) Towards an agronomic assessment of N_2O emissions: a case study for arable crops. Eur. J. Soil Sci. **61**(6), 903–913. doi:10.1111/j.1365-2389.2009.01217.x

Vlek, P.L.G., Byrnes, B.H. (1986) The efficacy and loss of fertilizer-N in lowland rice. Fertil. Res. **9**(1–2), 131–147. doi:10.1007/bf01048699

Wassmann, R., Lantin, R.S., Neue, H.U., Buendia, L.V., Corton, T.M., Lu, Y. (2000) Characterization of methane emissions from rice fields in Asia. III. Mitigation options and future research needs. Nutr. Cycling Agroecosyst. **58**(1–3), 23–36

Xing, G.X., Cao, Y.C., Shi, S.L., Sun, G.Q., Du, L.J., Zhu, J.G. (2002) Denitrification in underground saturated soil in a rice paddy region. Soil Biol. Biochem. **34**(11), 1593–1598

Xu, H., Xing, G., Cai, Z., Tsuruta, H., Xu, H., Xing, G.X., Cai, Z.C. (2000) Effect of soil water regime and soil texture on N_2O emission from rice paddy field. Acta Pedol. Sin. **37**(4), 499–505

Yan, X., Shi, S., Du, L., Xing, G. (2000) Pathways of N_2O emission from rice paddy soil. Soil Biol. Biochem. **32**(3), 437–440

Yan, X.Y., Akimoto, H., Ohara, T. (2003) Estimation of nitrous oxide, nitric oxide and ammonia emissions from croplands in East, Southeast and South Asia. Global Change Biol. **9**(7), 1080–1096

Yu, K.W., Wang, Z.P., Chen, G.X. (1997) Nitrous oxide and methane transport through rice plants. Biol. Fertil. Soils **24**(3), 341–343

Zhao, X., Min, J., Wang, S., Shi, W., Xing, G. (2011) Further understanding of nitrous oxide emission from paddy fields under rice/wheat rotation in south China. J. Geophys. Res.-Biogeo. **116**. doi:10.1029/2010jg001528

Zheng, X., Wang, M., Wang, Y., Shen, R., Gou, J., Li, J., Jin, J., Li, L. (2000) Impacts of soil moisture on nitrous oxide emission from croplands: a case study on the rice-based agro-ecosystem in Southeast China. Chemosphere **2**(2), 207–224

Zheng, X.H., Han, S.H., Huang, Y., Wang, Y.S., Wang, M.X. (2004) Re-quantifying the emission factors based on field measurements and estimating the direct N_2O emission from Chinese croplands. Global Biochem. Cycles **18**(2), GB2018. doi:10.1029/2003gb002167

Zhu, J.G., Liu, G., Han, Y., Zhang, Y.L., Xing, G.X. (2003) Nitrate distribution and denitrification in the saturated zone of paddy field under rice/wheat rotation. Chemosphere **50**(6), 725–732

Zou, J.W., Huang, Y., Jiang, J.Y., Zheng, X.H., Sass, R.L. (2005) A 3-year field measurement of methane and nitrous oxide emissions from rice paddies in China: Effects of water regime, crop residue, and fertilizer application. Global Biochem. Cycles **19**(2), GB2021. doi:10.1029/2004gb002401

Zou, J.W., Huang, Y., Qin, Y.M., Liu, S.W., Shen, Q.R., Pan, G.X., Lu, Y.Y., Liu, Q.H. (2009) Changes in fertilizer-induced direct N_2O emissions from paddy fields during rice-growing season in China between 1950s and 1990s. Global Change Biol. **15**(1), 229–242. doi:10.1111/j.1365-2486.2008.01775.x

Zou, J.W., Huang, Y., Zheng, X.H., Wang, Y.S. (2007) Quantifying direct N_2O emissions in paddy fields during rice growing season in mainland China: Dependence on water regime. Atmos. Environ. **41**(37), 8030–8042. doi:10.1016/j.atmosenv.2007.06.049

Zou, J.W., Huang, Y., Zong, L.G., Zheng, X.H., Wang, Y.S. (2004) Carbon dioxide, methane, and nitrous oxide emissions from a rice-wheat rotation as affected by crop residue incorporation and temperature. Adv. Atmos. Sci. **21**(5), 691–698

Chapter 8
Nitrous Oxide Sources and Mitigation Strategies

Contents

Abstract Industrial and biological fixation of N_2 has outpaced denitrification; as a result, N has become a major pollutant. Nitrogen fertilization is a substantial source of nitrogen-containing trace gases that have both regional and global consequences. Achieving synchrony between N supply and crop demand without excess or deficiency is the key to optimizing trade-offs amongst yield, profit, and environmental protection in both large-scale systems in developed countries and small-scale systems in developing countries. Improving the efficiency of N use by crops plants is crucial for meeting this challenge. Practices for better synchronization of N supply with plant needs include use of soil tests, better timing, and placement of fertilizers, cover crops during fallow periods, use of slow release and controlled release fertilizers and nitrification inhibitors. Use of available technology

David Ussiri and R. Lal, *Soil Emission of Nitrous Oxide and its Mitigation*, DOI 10.1007/978-94-007-5364-8_8, © Springer Science+Business Media Dordrecht 2013 and best management practices could reduce N_2O emission by 30–40%. Mitigation of N_2O through improved N management has additional economic benefits of decreasing fertilizer needs and minimizing production costs.

Keywords Agriculture • Mitigation • Cropland management • Nitrogen use efficiency • Best management practices

Abbreviations

Nr	Reactive nitrogen
SOC	Soil organic carbon
SOM	Soil organic matter
GHG	Greenhouse gas
CFC	Chloroflorocarbon
INM	Integrated nutrients management
TRF	Tropical rain forest
NUE	Nitrogen use efficiency
RMP	Recommended management practice
BMP	Best management practices
CT	Conventional tillage
NT	No tillage
WFPS	Water-filled pore spaces
GM	Genetic modification
GE	Genetic engineered crops

8.1 Introduction

Nitrogen plays an essential role in production of crops, both in terms of its economic and ecological aspects. There are two major pools of N with relatively little exchange between them: (1) the gaseous dinitrogen (N_2) of the atmosphere, which makes up about 99% of total N pool, and (2) the 1% of N that is chemically bound to other elements such as carbon (C), hydrogen (H), or oxygen (O) and is described as reactive nitrogen (Nr) for its tendency to react with other elements (Chap. 2) (Galloway et al. 2004). The Nr includes inorganic reduced forms such as ammonia (NH_3), ammonium (NH_4^+) inorganic oxidized forms–nitrogen oxides (NO_x), nitric acid (HNO_3), nitrous oxide (N_2O), nitrate (NO_3^-), and nitrite (NO). It also includes organic N compounds –urea, amines, proteins, and nucleic acids. The N in decomposed organic matter in soils (humus) is regarded as Nr in the long term only.

All organisms require N in order to live. In addition to being a key component of proteins, genetic material and other organic molecules of living organisms, N moderates numerous essential ecological and biogeochemical processes, including species composition, diversity, population growth and dynamics, productivity, decomposition, atmospheric chemistry, and nutrient cycling of many terrestrial, freshwater and marine ecosystems. Nitrogen is required by plants to increase quality and quantity of product. Owing to strong bond between its two N atoms, N_2 is almost inert, and therefore non-reactive. Most plants and all animals cannot use N_2 gas directly from the air because of high energy input requirement to convert N_2 into Nr forms; they must wait for N_2 gas to be 'fixed', i.e. bonded to H or O to form inorganic compounds, mainly nitrates (NO_3^-) and ammonium (NH_4^+), before they can secure their N nourishment.

Anthropogenic activities have altered the global natural N cycle by increasing biologically available N dramatically, through more than doubling the rate of transfer of N from highly abundant but biologically unavailable form–N_2 in the atmosphere to available forms—NH_4^+, NO_3^-, and nitrite (NO_2^-) (Vitousek et al. 1997; Smil 1999; Galloway et al. 2003, Chap. 2). Three major anthropogenic pathways which are responsible for increased Nr are: (i) industrial fixation of N for use as fertilizers, (ii) cultivation of soybeans, peas, beans and other crops that host symbiotic N fixing microorganisms, and (iii) mobilization from fossil N and fixation through high temperature internal combustion engine and fossil combustion for energy generation (Galloway et al. 1995, 2004). Fixation of N_2, both biological and industrial, now far outpaces historical rates of denitrification. It is the principal reason why N has become a major pollutant. Global synthetic N fertilizer consumption has increased from ~10 Tg N in 1950s to ~100 Tg N in 2008 (Robertson and Vitousek 2009). Furthermore, anthropogenic activities are also accelerating the release of N from long-term storage in soils and organic matter (OM) through land use change and biomass burning. Changes in the global N cycling have significant influence on Earth system functions both positive and negative. On positive side, increased N inputs enhance soil fertility, promote sustainable agriculture, and advance food and nutrition security. The need to feed and provide energy for growing population drives the increase in demand for fixed N. On negative side, the increase causes a number of undesired changes in biosphere associated to increased emissions of NO_x, NH_3, and N_2O to the atmosphere, leaching of NO_3^- into surface and ground water, and deposition of NO_y and NH_x to both terrestrial and aquatic ecosystems (Mosier et al. 2001). Human–induced emissions of N_2O are currently increasing by ~150 Tg N $year^{-1}$ (Mosier 2002) with the current global concentrations of N_2O in the atmosphere of about 323 ppbv versus pre-industrial concentrations of 270 ppbv (IPCC 2007; WMO 2011). Some of the changes are playing out as air and water pollution, including formation of photochemical smog, and ground level ozone, which has serious detrimental effects to human health as well as health and productivity of crops and forest. Others include acid rain, ozone damage due to enhanced N_2O and NO_x emissions from soils and aquatic systems. In addition, water enrichment with NO_3^- from leaching and runoff from cropped and fertilized fields causes eutrophication at local and global scale. N_2O has important effects on climatic system and on stratospheric ozone. It is a potent greenhouse gas (GHG) produced by nitrification and denitrification (Robertson and Groffman 2007;

Chap. 3), and its enhanced emission, together with carbon dioxide (CO_2) and methane (CH_4) are the main cause of global warming. It is the major GHG emitted by agricultural activities.

Therefore, global change in N cycling has become a source of multidimensional environmental problem, drawing interests among scientists, managers, and policy makers. The creation of Nr is closely tied to essential anthropogenic endeavors—provision of food and energy, which make the problems stemming from global increase in Nr especially difficult to solve. Use of N for agriculture, both as organic or synthetic inorganic forms is not easily substitutable, and even technological changes are unlikely to provide a replacement for N nutrition. The key to maintain adequate food supplies to meet the demand for growing global population will always be supplying plants with adequate N (Matson et al. 1997), which sets fertilizer N use and its associated environmental pollution apart from those that can be solved by technological substitution such as chlorofluorocarbon (CFC) substitution with non-ozone depleting chemicals in the industrial processes.

In addition, technological changes that increase efficiency of fuel combustion or removes NO_x from the exhaust stream could only reduce the total amount of NO_x emitted, but complete solution is closely linked to the development of non-polluting alternative energy sources. As it has been demonstrated throughout the previous chapters, that Nr compounds are highly mobile, moving easily through air, water, soils, and across geographical and ecosystem boundaries. As a result, Nr sources and sinks are often widely separated (Mosier et al. 2001), and policies to solve environmental problems associated with increased Nr in the ecosystems must be of the multinational or global in scale. Changes in Nr are interactive with other global changes, such as elevated CO_2, land use change, biological invasions, and other biogeochemical changes such as N deposition, increased tropospheric ozone, and global climate change. Only comprehensive policies to address the multiple interacting changes that are occurring globally will be able to address the impact of increasing Nr.

Although agriculture production has increased dramatically since 1960s (Borlaug and Dowswell 2008), fertilizer N use efficiency by crops remains relatively low. Crops typically take up only 40–50% of total organic and inorganic N added during each cropping season (Bleken and Bakken 1997; Olsthoorn and Fong 1998). As a result, N use in excess of crop requirements causes losses to the environment through various pathways, including volatilization, leaching, nitrification, and denitrification processes. This excessive N released into ecosystems is responsible for enhanced N_2O emissions. Improved efficiency of fertilizer N use can be attained, in part, through adopting integrated nutrients management (INM) approach, using available technology to maximize the benefits while minimizing the risks associated with the use of N in cropping and animal production (Matson et al. 1998; Mosier et al. 1998b; Luo et al. 2010; Chen et al. 2011).

8.2 Sources of Nitrous Oxide

Both natural and anthropogenic sources emit N_2O. Natural N_2O sources include varieties of biological sources in soil and water, particularly from microbial activities of nitrification and denitrification. For many years, soils have been recognized as large sources of N_2O from terrestrial ecosystems (Alexander 1961; Matson and Vitousek 1990; Bouwman et al. 1995; Kroeze et al. 1999). Estimates suggest that emissions of N_2O from the tropical soils are the largest single source worldwide. Several factors contribute to the importance of tropical soils as sources of N_2O production including: (a) high availability of N in clay-rich, weathered tropical forests, where N typically does not limit plant growth (Vitousek and Farrington 1997; Martinelli et al. 2000), (b) high temperatures, moisture and aggregation of clays in wet tropical forests create conditions where N_2O is produced by both nitrification and denitrification.

The global anthropogenic N_2O emissions are estimated at 6.0–13.0 Tg N_2O-N year^{-1} and averaged at ~7.0 Tg N_2O-N year^{-1} (Kroeze et al. 1999). Primary anthropogenic sources of N_2O are: (i) agricultural soil management, (ii) livestock production, (iii) combustion of fossil fuels, (iv) adipic acid and nitric acid production, and (v) sewage treatment. Factors that influence regional distribution of N_2O emissions are: agricultural production characteristics, combustion technologies, waste management practices, industrial characteristics, and climate and weather conditions. The only known process responsible for N_2O removal from the atmosphere is photolysis and reaction with excited singlet oxygen atoms formed by photolysis of ozone in the stratosphere. Stratospheric sink of N_2O is estimated at 12.6 Tg N_2O-N year^{-1} (Ehhalt et al. 2001; Crutzen et al. 2008). Photodissociation is a sink for ~90%, while reaction with electronically excited O_2 atom ($O(^1D)$) is the sink for the remaining 10% (Prinn and Zander 1999; Ehhalt et al. 2001). Atmospheric reaction of N_2O is one of the sources of NO that contributes to depletion of ozone that protects the biosphere from harmful ultraviolet radiation (Crutzen 1970). The concentration of N_2O in the atmosphere is increasing at the rate of 0.8 ppbv year^{-1}, which translates to an atmospheric stock increase of 4–6 Tg N_2O-N year^{-1}, and averaged at 5 Tg N_2O-N year^{-1} (Del Grosso et al. 2005). Globally, about 65% of all N_2O emissions arise from microbial processes of aerobic nitrification and anaerobic denitrification in soils (Smith and Conen 2004). The natural operations in tropical rainforests and stimulation of these processes by large-scale addition of synthetic fertilizers to agricultural soils results in these environments being two major N_2O sources.

8.2.1 Agriculture

Agriculture remains the single largest anthropogenic N_2O source (Bouwman et al. 2002; Smith and Conen 2004; Del Grosso et al. 2005), contributing up to 80% of all anthropogenic emissions (Saggar 2010). In agricultural soils, N_2O emission is mainly generated from mineral N originating from applied N fertilizers and manures, mineralization of soil organic N including crop residues, and biologically fixed N. Application of N to soils in the form of inorganic fertilizer, manure and crop residues increases direct N_2O emissions from soils (Rochette et al. 2008). Emission of N_2O is largely driven by interaction between N application and high soil moisture conditions. Where large application of fertilizers are combined with soil conditions favorable to denitrification, large amounts of nitrous oxide can be produced and emitted to the atmosphere. Similarly, the widespread and often poorly controlled use of animal waste as fertilizer can lead to substantial emissions of nitrous oxide from agricultural soils. Some additional N_2O may arise in agricultural soils through the process of biological nitrogen fixation (BNF), although the true importance of this source remains poorly defined. The N fertilizers and manure in agricultural soils enhance N_2O emissions by stimulating microbial processes of nitrification and denitrification. Ruminants cause additional N_2O emission from soils through dung and urine deposition during grazing and animal waste management and utilization.

The global fertilizer N demand has grown at an annual rate of 1.7% since 1980s to reach about 105 Tg N in 2009 (FAOSTAT 2011), and is anticipated to grow at the same annual rate during this decade. Average annual growth rate would decrease to 1.6% by 2020 and to 1.4% by 2050 if NUE does not increase (Woods et al. 2004). The increase represents significant environmental consequences due to increased NO_3^- leaching, enhanced NH_3 volatilization and N_2O emissions. The stimulatory effect of N fertilizers application on N_2O emission is generally due to increased mineral N levels in soils. The fraction of fertilizer N emitted as N_2O is site-specific and ranges from 0.1 to 5% of the applied N, depending on site conditions (Mosier et al. 1998a; Crutzen et al. 2008). Globally N_2O production has increased by 17% from 1990 to 2005, and it is expected that N_2O emissions from agricultural practices may increase by 35–60% by the year 2030 (Smith et al. 2007).

When N is applied above the crop requirement or when available N exceeds crop uptake, the risk of N_2O emissions increases. In addition, when N fixing crops are included in crop rotations, they may also contribute to post-season N_2O emissions as their plant residues decompose. In addition to increasing direct soil N_2O emissions, some agricultural practices may also increase NH_3 volatilization and NO_3^- leaching (Del Grosso et al. 2006). Volatilized NH_3 affects N_2O emission because a portion of this N may be deposited on agricultural and non-agricultural soils and water and be subjected to transformations that may result in N_2O emissions. In addition, a portion of leached NO_3^- and discharged into drainage water or streams can be denitrified and result into N_2O emissions. Considering direct and indirect N_2O emissions from applied N, Crutzen et al. (2008) estimated that 3–5% of fertilizer N could be emitted as N_2O. In contrast, Nevison et al. (2007) reported that a 2% emission factor for global anthropogenic N input was sufficient to explain N_2O increases over the last 150 years at a global scale. Estimates of the stratospheric sink of N_2O are reasonably well constrained (Ehhalt et al. 2001; Crutzen et al. 2008). Therefore, the global N_2O sources strength can be inferred from the sink and the rate N_2O accumulation in the atmosphere. Atmospheric

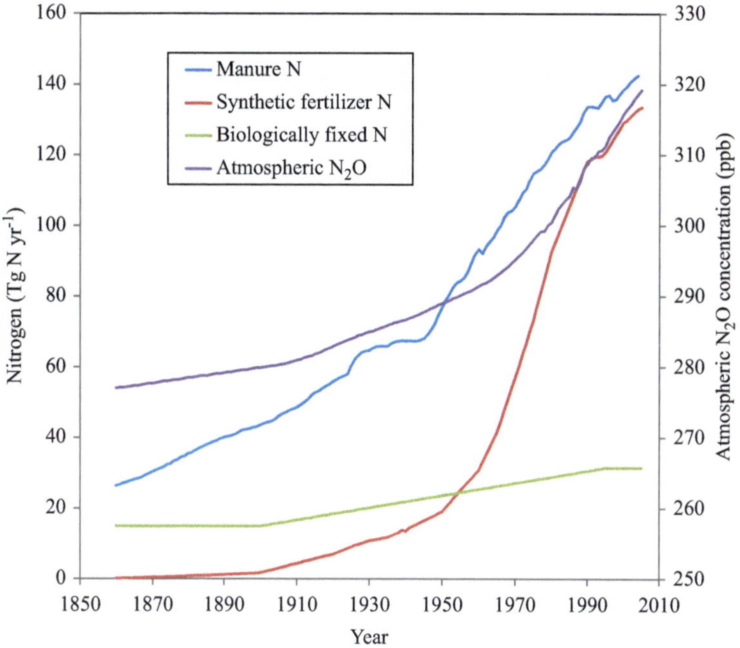

Fig. 8.1 Historical ecosystems anthropogenic N input and atmospheric N_2O concentration (Data from Davidson 2009)

record indicates that N_2O concentrations were increasing steadily from 1860 to 1960 before synthetic inorganic N fertilizers became the major anthropogenic source (Fig. 8.1). The growth rate of atmospheric N_2O concentration increased after 1960 mainly due to synthetic inorganic N use in agriculture. Using combination of top-down and bottom-up budgets of historic accumulation, Davidson (2009) suggested that 2% of the animal manure N and 2.5% of synthetic fertilizer N was converted to N_2O between 1860 and 2005. These percentages explained the patterns of atmospheric N_2O concentration over the entire period from 1860 to 2005 (Davidson 2009).

Agriculture (through soils emissions, biomass burning and animal production) contributes 65–80% of the anthropogenic N_2O emissions (Kroeze et al. 1999; Ehhalt et al. 2001; Del Grosso et al. 2005). Factors other than N concentration and N fertilization, that may impact N_2O emissions from agricultural soils include: (i) mixture of organic and inorganic fertilizers — which could emit more N_2O than inorganic or organic fertilizers alone, (ii) crop type — with major differences between grasses, legumes and annual crops, (iii) soil organic carbon (SOC) concentration — high concentration of SOC enhances microbial activity and increased N_2O emission, and (iv) soil texture and drainage — heavy-textured and poorly drained soils could emit more N_2O than light-textured and well drained soils (Laegried and Aastveit 2002; IFA/FAO 2001).

Land use change affects N_2O emissions (Neill et al. 2005). Deforestation influences N_2O emissions by altering the soil conditions of N availability and moisture content that control emission rates. Logging of tropical rain forest (TRFs) may increase N_2O emissions by 30–350%, depending on soil-specific conditions. The N_2O fluxes are generally higher under compacted clayey moist soils than sandy soils (Keller et al. 2005). The TRFs have high rates of biological productivity and rapid decomposition rates of OM due to high temperatures and heavy precipitation. Logging affects microclimate, soil physical conditions, and accelerates biogeochemical cycling.

Indirect agricultural sources of nitrous oxide remain poorly defined in most cases. Indirect N_2O emissions resulting from direct N addition to soils occur through: (i) volatilization of NH_3 and emission of NO_x from soil surfaces. These gases are then re-deposited onto soils in the form of particulate ammonium, nitric acid, and N oxides, and release N_2O in areas of deposition, and (ii) NO_3^- leached into groundwater or surface waters and subsequently denitrified to N_2O in aquatic ecosystems (Mosier et al. 1998a; Nevison 2000; Groffman et al. 2000, 2002). Indirect N_2O emissions require the knowledge of the amount of N leaching or run-off from the fields and also the variation of N_2O:N and N_2O:NOx ratios resulting from different soil, sediment and aquatic conditions (Groffman et al. 2002). The most important of these is nitrous oxide emission arising from nitrogen leaching and run-off from agricultural soils. After fertilizer application or heavy rain, large amounts of nitrogen may leach from the soil into drainage ditches, streams, rivers, and eventually estuaries. Some indirect N_2O emission may be produced from drainage waters when the leached fertilizer they contain undergoes the processes of nitrification or denitrification in aquatic and estuarine sediments. Other important indirect N_2O sources from agricultural soils include the volatilization and subsequent deposition of ammonia from fertilizer application, and the consumption of crops followed by sewage treatment.

8.2.2 Industrial Processes

Industrial processes of synthesis of adipic acid and nitric acid (HNO_3) produces N_2O as a byproduct. Adipic acid is a white crystalline solid used primarily as the main constituent of nylon (nylon-6,6), representing about half of the nylon molecule. Nylon-6,6 is used in carpet fibers, upholstery, tire reinforcements, auto parts, apparel, and other products. It is also used in the manufacturing of some low-temperature synthetic lubricants, synthetic fibers, coatings, plastics, and polyurethane resins. The U.S. is the major producer of the adipic acid, producing about 40% of world production. Other countries producing it include Brazil, Canada, China, France, Germany Italy, Japan, Singapore Ukraine and U.K. Adipic acid is a dicarboxylic acid produced by a two-stage process involving reaction of HNO_3 with cyclohexane (Eq. 8.1).

$$(CH_2)_5CO + CH_2)_5CHOH + wHNO_3 \rightarrow HOOOC(CH_2)_4COOH + xN_2O + yH_2O \tag{8.1}$$

Currently there is an effort to substitute HNO_3 with 30% aqueous hydrogen peroxide (H_2O_2) in an effort to eliminate N_2O emission (Sato et al. 1998).

The production of HNO_3 generates N_2O as a by-product of high temperature catalytic oxidation of ammonia (NH_3). The HNO_3 is primarily used to make synthetic commercial fertilizer. It is also used as a raw material for adipic acid and explosives, metal etching and processing of ferrous metals. Nitric acid production involves three distinct chemical reactions (Eqs. 8.2, 8.3, 8.4)

$$4NH_3 + 5O_2 \rightarrow 4NO + 6H_2O \tag{8.2}$$

$$2NO + O_2 \rightarrow 2NO_2 \tag{8.3}$$

$$3NO_2 + H_2O \rightarrow 2HNO_3 + NO \tag{8.4}$$

Although N_2O generation during HNO_3 production is not well documented, the steps of nitrogen oxidation are considered the potential sources of N_2O. The NO, an intermediate in the production of HNO_3 readily decomposes to N_2O and NO_2 at high pressures for temperature ranges of 30–50 °C (Cotton and Wilkenson 1998).

8.2.3 Other Sources

Other N_2O sources include sewage and wastewater treatment, and energy combustion. Sewage and wastewater treatment produces N_2O by nitrification and denitrification of N present in the form of urea, NH_4^+ and NO_3^-. Mineral N can also be generated from mineralization of proteins. Emissions from energy originate from oxidation of N in the fossil fuel as either a stationary (power plants) or mobile (traffic) sources. In addition, high temperature and pressure in internal combustion engines provide energy to convert N_2 to NO_x.

8.3 Mitigation of Nitrous Oxide Emissions from Cropland

Population pressure and economic growth have been the main drivers of growth and changes in agricultural sector since early 1960s. This growth has occurred at the expense of increased pressure on the environment and depletion of natural resources in order to keep pace with the global demand for food, feed, fiber and fuel or energy (Tilman et al. 2001; Rees 2003; Paoletti et al. 2011). The GHG emissions from agriculture will continue to increase in coming decades due to escalating demand for food and dietary shift from plants- to animal-based diet (Oenema and Tamminga 2005). Improved land and animal management practices and emerging technologies must outpace the demand for environmental and ecological services to permit a reduction in GHGs emissions per unit of food produced.

Most of the agricultural N_2O emissions arise from fertilization of soils with mineral N and animal manure. Nitrogen fertilizers are the most significant anthropogenic source of N_2O from agricultural ecosystems. Global food production needs and farmers accessibility consideration suggest that mitigation technologies should meet the following general guidelines: (i) agricultural production levels should be maintained or enhanced, especially in parts of the world where food production and population demand are in delicate balance, (ii) additional benefits such as reduced labor, reduced or more efficient use of inputs will accrue benefits to farmers, and (iii) agricultural products will be accepted by consumers (Cole et al. 1996). Successful development and implementation of mitigation strategies for agricultural sources of N_2O require an understanding of effects of land use and agricultural practices on fluxes of N_2O and on the controlling mechanisms.

Agricultural N_2O fluxes are complex and heterogeneous. Generally, the production of N_2O from soils occurs as a result of microbial processes of nitrification and denitrification (Granli and Bockman 1994). It is controlled by factors that influence the growth of microorganisms — soil O_2 content, soil temperature, mineral N content, available water content, mineralizable OM, and soil pH (Mosquera et al. 2007). In addition, under acidic soil conditions N_2O can be formed chemically in reactions involving NO_2^-, (i.e., chemodenitrification). The NO_2^- must be produced biologically first. This process is a predominant source of N_2O under specific soil conditions (Venterea and Rolston 2002). Soil management practices—land use, nutrients application via manure and N fertilizers, crop residues incorporation, tillage, reduction of soil compaction through their effect on these factors can directly and indirectly influence the N_2O fluxes from agroecosystems. Alternate wetting and drying cycles that permit nitrification to progress, and water filled pore space (WFPS) above 60% but below saturation contribute to the greatest potential for N_2O emission (Granli and Bockman 1994). The magnitude of N lost is controlled by interaction of soil moisture and N availability–principally NO_3^- (McSwiney and Robertson 2005). Due to its multiple of sources nature and multiple environmental controls, which are only partly manageable, N_2O emissions from agricultural systems are of highly stochastic nature (Oenema et al. 2005; Smith et al. 2007). Active management of the agricultural N_2O sources using current ranges of technologies offers possibilities for mitigation. Mineral N content governs the N supply to plants and denitrification. Increasing the efficiency of mineral N use to plants should result in lower amounts of mineral N available for nitrification and denitrification. The N_2O emission is often enhanced where available N exceeds plant requirements, especially under wet conditions (Smith and Conen 2004; Oenema et al. 2005).

Strategies that increase the efficiency of fertilizer N use can reduce N_2O emission (Mosier et al. 1996). Nitrogen use efficiency (NUE) is a measure of the proportion of available N taken up by a crop, and has been measured and expressed in various ways (Dobermann 2007). Several practices and technologies can be used to improve NUE while maintaining potential for increased agricultural production. These include: (i) increasing yielding potential and yield stability through genetic improvement and crop management, (ii) balanced plant nutrition to allow optimum utilization of available N, (ii) split application of fertilizer N to better match N requirements for crops through the growing season, (iv) use of more efficient fertilizer products that better synchronize N release and crop demand, such as slow- and controlled-release fertilizers, (v) use of fertilizer additives such as nitrification and urease inhibitors to reduce N losses, (vi) use of site-specific soil testing and other diagnostic tools such as plant testing to achieve precision N management rates, (vii) use computer-based models and simple field assessment tools and interpretation aids for guiding nutrient management, and (viii) genetic improvement in N recovery and N utilization for crops which have received little attention of crop breeders (IFA 2007; Lobell 2007). Strategies for mitigating N_2O emissions from N fertilizers are discussed in Chap. 10.

To reverse increasing GHG, several mitigation technologies for both crop and animal production systems of farm have been recommended. These fall into three main categories: (i) reducing emissions, (ii) enhancing removals, and (iii) avoiding or displacing emissions (Smith et al. 2008). In agricultural ecosystems, the fluxes of N_2O can be reduced by managing more efficiently the flows and cycling of N in soils. For example, delivering N nutrient more efficiently to crops and managing livestock to make most efficient use of feeds often suppresses emissions of N_2O. Recommended management practices (RMPs) which promote good land stewardship and sustain or increase net economic returns and reduce global warming potential are known as best management practices (BMPs). These are generally practical and affordable ways to minimize the environmental risks without sacrificing economic productivity of land (Jayasundara et al. 2007). The BMPs options for reducing N_2O emissions from cropland are summarized in Table 8.1.

The BMPs for reducing N_2O from cropland ensures adequate available N when required by crops and prevent N availability exceeding plant N demand. Excessive N is easily transformed among various reduced and oxidized forms and distributed by hydrologic and atmospheric processes. Excessive N can be lost from farm through soil erosion, runoff, leaching of NO_3^- or dissolved forms of organic N, gaseous emissions to the atmosphere in the form of NH_3, NO, NO_2, N_2O or N_2. Except N_2, all other avenues of N loss from farm can potentially affect one or more environmental risks. The central concept that defines the relationship between the beneficial use of N and that proportion which harms the environment is the extent to which the rate of Nr supply matches the rate of crop N demand. When the supply of Nr via fertilizer, OM mineralization and BNF is in excess of crop demand, Nr has a potential to accumulate in soils and become susceptible to loss from agroecosystems through different pathways and undergo various transformations which cause environmental risks, including increased N_2O emissions which causes

Table 8.1 Best management practices for increasing cropland nitrogen use efficiency and reducing N_2O emissions

Strategy	Practice	Activities and N_2O mitigation effect
Cropland management	Agronomy Adopt conservation tillage Residue management Plant nutrient management Water management i.e. irrigation, improving drainage Land use change (set-aside)	Use of improved crop varieties Reduced reliance on fertilizers Extending crop rotations with perennial crops Temporary vegetative cover/cover crops between successive agricultural crops to extract plant available N unused by preceding crop Increase SOM storage Reduce soil disturbance
Fertilizer management	Optimizing fertilizer N rates (right fertilizer N rate) Matching N supply to crop N demand Use the right product Right fertilizer placement to increase plant accessibility (right place)	Reduce fertilizer N use and reduce N_2O emissions Use of soil tests and plant analysis to as diagnostic tools N requirement. Increase nutrient use efficiency to reduce N_2O emissions Source of fertilizer N affects N_2O emissions Use of controlled- and slow-release N fertilizers Use of nitrification inhibitors
Organic soils/ wetland management	Avoid cultivation	Avoid cultivation of organic soils Avoid drainage of wetlands
Degraded lands	Restoration	Erosion control Organic amendments Nutrients amendment Forest restoration

global warming. In contrast, when the supply of Nr is insufficient to meet crop need there is potential for decreased yields and economic loss (Crews and Peoples 2005), mining of SOC and decline of long-term soil productivity (Jaynes and Karlen 2008). Table 8.2 summarizes the reductions in N_2O emissions from the literature survey. Strategies for N_2O abatement for croplands are discussed below.

8.3.1 Cropland Management

Prudent management of N inputs is an effective strategy to minimize N_2O emitted from croplands. For example, McSwiney and Robertson (2005) reported that N_2O fluxes were low to moderate until N input exceeded crop needs, after which the flux nearly doubled. Therefore, optimum N rates for minimizing N_2O emission are those, which do not exceed the crop demand. Practices that deliver added N

Table 8.2 Estimated N_2O mitigation potentials for different management techniques (Estimated from Smith et al. 2007, 2008)

Mitigation strategy	Emission reduction range (kg N_2O-N ha^{-1} $year^{-1}$)	Mean (kg N_2O-N ha^{-1} $year^{-1}$)
Conventional to no till	−0.9–0.7	−0.2
Conventional to conservation till	0.1–0.4	0.1
Use of winter cover crops	0.2–1.1	0.4
Crop rotation	0.0–0.3	0.1
Application of manure	−1.4–1.9	0.2
Reduce fertilizer N use	0.1–1.4	0.4
Change fertilizer N source	0.0–1.5	0.5
Use nitrification inhibitors	0.0–2.3	1.0
Improve manure management	0.4–1.3	0.9
Land conversion cropland to pasture	−0.7–4.6	1.0
Set aside land	0.4–5.2	1.4
Wetland restoration	−2.8–0	−1.4
Biochar application	0.8–3.0	1.2
Convert organic soils to natural	1.5–12.4	7.0
Change fertilizer N timing	0.0–0.5	0.4
Change fertilizer N placement	0.1–0.5	0.3

more efficiently to crops often suppress the N_2O emissions (Bouwman et al. 2001). Mode and timing of N application influence the efficiency of plant uptake, NH_3 volatilization and the availability of N for nitrification and denitrification. Generally, emissions from subsurface applied or injected N fertilizers are higher than those from broadcast mineral fertilizers and animal manure. Compared to broadcasting, subsurface applied N has resulted in higher N_2O. Use of those BMPs, which increase yields and generate higher inputs of OM residues may lead to increased N storage in soils and reduced supplemental N requirement in form of manures and/or N fertilizers.

8.3.1.1 Tillage Practice

There is no clear response for mitigation of N_2O using conservation/reduced tillage (RT) or no tillage (NT) practices compared to conventional tillage (CT). The NT and other RT practices can alter soil properties that are known to increase N_2O fluxes, including increasing soil water content and increasing soil bulk density, while other changes such as decreased soil temperature and decreased N mineralization rates may promote lower N_2O emissions (Six et al. 2002). In addition, NT and RT can alter vertical distribution of microbial population and potential enzyme activities that drive soil N_2O production (Groffman 1985). The production of N_2O depends on the coexistence of microbial enzymes with other conditions including anaerobiosis and sufficient levels of labile C and inorganic N. These contrasting effects make it difficult to predict the net impact of NT and RT on N_2O emissions (Venterea and Stanenas 2008).

The effect of RT or NT on N_2O emissions depends on soil and climatic conditions, and is not well defined (Cassman et al. 2003; Smith and Conen 2004; Li et al. 2005). In some areas, RT or NT accentuates N_2O emissions, whereas in other areas it may reduce or have no measurable influence. It appears that in some regions the benefit of RT is to increase storage of organic C and N, as soil disturbance tends to stimulate OM decomposition and soil erosion (Ogle et al. 2005), as a result, RT or NT may reduce N_2O emissions by increased soil organic N (SON) storage (Ussiri et al. 2009). In other regions, N_2O emissions increase slightly by switching from CT to NT (Ball et al. 1999; Baggs et al. 2003). Climate, soil properties, and time of application are all important factors influencing N_2O emissions from NT. High clay content, wet and humid climate, wet soils and poor aeration favor N_2O increase after implementation of NT (Rochette 2008). Increased aggregate stability and improved drainage after 4–6 years leads to reduced N_2O emissions. A review of 44 long term experiments indicated higher N_2O emissions in the initial years following transition to NT, but reduced N_2O emissions when compared to CT after NT has been practiced for 10 years or more (Six et al. 2004). The NT management have the potential of reducing agricultural GHG emissions in the Alberta Parkland region (Lemke et al. 1999). The amount of N as N_2O was higher from N-fertilized CT than N-fertilized NT in Gray Luvisol of Saskatchewan Canada (Malhi et al. 2006). N_2O emissions in temperate climate are generally event driven, with rainfall being a critical factor. Spring thaw results in large N_2O emissions due to denitrification below and in the frozen soil layer during winter. These emissions are mostly caused by anoxic conditions in the soil induced by rainfall and snowmelt water (Pattey et al. 2007). These studies have led to reduced N_2O emission estimates for land under NT in the Canadian prairie region in the 1990–2005 inventory to 0.68 ± 0.57 Tg N_2O-N year^{-1} compared to 0.95 ± 0.98 Tg N_2O-N year^{-1} for CT (Environment Canada 2007). The effects of changing from CT to NT must be integrated over time to understand its overall influence on N_2O emissions and global warming potential (GWP), since it can vary between and within agro-ecosystem (Six et al. 2004; Lee et al. 2006). In a 20-year time span, in humid climates relative decreases in N_2O under NT results in lower overall GWP compared to CT, while in drier environments NT resulted in increased N_2O and GWP (Six et al. 2004). Adoption of NT in drier climates can be combined with increase in cropping intensity to capture an excess N. Long-term use of conservation or NT cropping system generally results in reduction of N_2O emission due to increase in SOM storage, improved soil structure, enhanced aeration and reduced soil bulk density. Despite the lack of general consensus about the impact of RT on N_2O emissions (Venterea et al. 2005), research indicates that the negative impacts are generally limited to poorly aerated soils (Rochette 2008).

8.3.1.2 Crop Rotation

The N originating from biological N fixation (BNF) is generally less available for nitrification and subsequence denitrification and associated N_2O emissions, especially when legume crop is actively growing. This is because the biologically fixed

N is largely used by crop. Adoption of less intensive cropping systems which reduces the reliance on fertilizer N inputs, for example use of crop rotations with legume crops which reduce reliance on input of fertilizer N through legume-derived N can reduce N_2O emission. Higher N_2O emissions often occur from soils planted to corn than those to soybeans in a corn-soybean rotation (Parkin and Kaspar 2006). Reduced rates of fertilizer N application under crop-rotation management may also reduce fertilizer related N_2O emissions.

8.3.1.3 Catch- and Cover Crops

Agronomic practices that provide temporary vegetative cover such as catch- or cover-crops between agricultural crops (Barthes et al. 2004) also minimize N loss. These crops extract plant-available N unused by the preceding crop, thereby reducing excessive soil NO_3^- that can be denitrified and lost as N_2O emissions. Cover crops are typically grown in combination with main summer annuals such as corn, soybean, and spring cereals to control NO_3^- leaching, provide nutrients especially N as green manure, and improve soil quality. Catch crops, which may include leguminous species, are widely used in grain crop rotations to accumulate soil mineral N after harvesting of main crop. Reduction of N leaching ranging from 26 to 38% was obtained by inclusion of grass-cover catch-crop in organic crop rotation (Askegaard et al. 2005), indicating a reduction in soil N availability that would be expected to constrain N_2O emissions during autumn-winter period. Winter cover crops such as wheat (*Triticum aestivum* L.) and rye (*Scale cereal* L.) can effectively prevent NO_3-N leaching in the winter months on permeable soils and reduce drainage losses of NO_3-N (Shipley et al. 1992; Feyereisen et al. 2006) and overall reduction of N_2O emission from water resources. Winter cover crops such as wheat or rye can effectively prevent NO_3-N leaching in the winter months on permeable soils, and reduce drainage losses of NO_3-N. Reductions in NO_3-N leaching/drainage losses are likely to reduce overall N_2O emissions from water resources. Studies indicate that cover crops can significantly reduce the need for chemically derived N fertilizer, since both legumes and main crop can scavenge and recycle the mineral N that would otherwise be lost by leaching and gaseous emission (Delgado et al. 2007).

8.3.2 Fertilizer Use and Crop Yields Enhancement

The goal of fertilizer N application is to increase the crop yield by enhancing availability of inorganic soil N. Uptake of N by crop plants is time-dependent and relies on N availability in the right form and positional accessibility of nutrients to the plant roots, when demanded. Several studies have targeted optimizing the selection of N source, timing, and method of N application to minimize fertilizer N losses that may occur through leaching of NO_3^-, volatilization as NH_3, and denitrification with the potential to increase N_2O emission from fertilizer N application (Rochette et al. 2008). In general, N_2O emissions from mineral and organic N can be decreased by management practices which optimize the crops' ability to compete with processes whereby plant available N is lost from the soil system, such as NH_3 volatilization, denitrification and leaching, and directly lowering rate and duration of the loss processes. Typically, only half of N inputs are synthesized in crop biomass and the remainder is lost from the system through gaseous losses of N_2, N_2O, NO_x, or NH_3, leaching as NO_3^-, and soil erosion. The low efficiency of fertilizer N use in agricultural systems is primarily caused by the large losses of N from the agricultural systems in gaseous forms (Peoples et al. 1994) and leaching. The agricultural N_2O emissions are directly linked to these loss processes. Emissions occur both directly on agricultural lands and from N transported to non-agricultural lands through gaseous NH_3 volatilization, leaching and runoff losses from agricultural soils.

For most crops, there exists a direct relationship between soil N availability and crop yield, and the agronomic challenge is to decrease N inputs without decreasing crop yields. Improving the efficiency and effectiveness of crop N use can potentially reduce N_2O emissions by reducing the potential of elevated residual NO_3-N in the soil profile (Cassman et al. 2002; Dobermann and Cassman 2004; Dobermann 2007; Snyder and Bruulsema 2007; Snyder et al. 2009). It is estimated that improved crop N use efficiency could decrease soil-derived N_2O emissions from agriculture, largely from surplus N, by as much as 35% globally, with even greater saving in the input-intensive systems of North America and Europe (Kroeze and Mosier 2000). Better nutrients management can potentially increase crop N recovery, and minimize loss of reactive N to air and water resources (Galloway et al. 2004).

Crops do not always use N applied as fertilizers or manures efficiently (Galloway et al. 2003; Cassman et al. 2003). The surplus N is particularly susceptible to N_2O emission (McSwiney and Robertson 2005; Van Groenigen et al. 2010). Management practices that limit the available N in excess of plant requirement will limit N_2O emissions from the soil. Consequently, improving N use efficiency can reduce N_2O emissions and indirectly reduce N fertilizers need. By reducing leaching and volatilization losses, improved N use efficiency can also reduce indirect N_2O emissions offsite. Adaptation of BMPs, including those of fertilizer, can potentially increase crop N recovery and minimize its cascade to the air and water resources (Galloway et al. 2004). The fertilizer BMPs that are focused on reduction of excess N in soil will also have a possible impact on the reduction of N_2O emissions (Roberts 2007). These BMPs can be achieved through precise estimation of crop needs, use of slow- and controlled-release fertilizer forms, or the use of nitrification inhibitors. The underlying concept in limiting N_2O emission is that if N applied to increase crop yields is better utilized by crops, the amount of N needed to meet growing demand for food will be less, and N_2O emissions will decrease. Fertilizer management objectives are to: (i) increase productivity, (ii) increase profitability, (iii) maintain sustainability, and (iv) protect environment (Bruulsema et al. 2008). Practices such as the use of slow- and controlled-N-release and nitrification inhibitors, which slow the microbial processes leading to N_2O formation or can directly reduce N_2O emissions from the fields (Bronson et al. 1992; Parkin and Hatfield 2010). Testing the NO_3^- status of the soil during the early growing season of corn, when plant requirement is high may reduce the N application rate by up to 15%, and thus reduce N_2O emission (Duke 2006). Considerable economic and environmental benefits can be gained by reducing fertilizer application rates, hence significant gains in fertilizers use for high N using crops and reduced N_2O emissions.

8.3.2.1 Effect of Fertilizer Type on N_2O Emissions

Field experiments conducted to assess the N_2O production from granular fertilizer N sources indicate decreasing magnitudes from application of urea, ammonium sulfate, ammonium nitrate and calcium nitrate (Tenuta and Beauchamp 2003). Emission of N_2O under aerobic conditions is generally greater with urea than with other N fertilizers, with fewer differences at higher moisture contents. Under saturated conditions, however, both mono-ammonium and di-ammonium phosphates produce more N_2O per unit of N than other N fertilizers, including urea (Tenuta and Beauchamp 2003). The P status may also affect N_2O emissions from fertilizers. Comparison of NH_4-based and NO_3-based fertilizers indicate higher N_2O emissions from NH_4-based N fertilizers compared to NO_3-based fertilizers (Velthof et al. 2003; Tenuta and Beauchamp 2003). The higher emissions from NH_4-based fertilizers may be a result of NO_2- accumulation or N_2O production during nitrification (Venterea and Stanenas 2008).

Bouwman et al. (2002) concluded that N_2O emissions may be lower for NO_3-based fertilizers compared to NH_4-based fertilizers and organic-synthetic mixture sources. However, anhydrous NH_3 exhibits higher N_2O emissions compared with other N sources (Venterea et al. 2005),which may be related to the typical knife or coulter injection, which produces high alkaline soil zones with high NH_4^+ concentration. Stehfest and Bouwman (2006) observed that the differences among fertilizer types almost disappear after accounting for rate of application, crop type, climate, SOC concentration, soil pH and length of experimental monitoring (Table 8.3).

8.3.2.2 Fertilizer N Placement

The placement of fertilizer N into the soil and near the zone of active root uptake may both reduce surface N loss and increase plant N use resulting in reduced N_2O emissions (CAST 2004). N fertilizers can be applied by various placement methods. The banding and broadcasting are the two major modes. Banding can be drilled with the seed, side banding, mid-row banding, sub-surface banding, or nesting. Broadcasting can be performed with or without incorporation into soil. Drilling with the seed is related to high risk of seed damage when uncoated urea is used.

Table 8.3 Fertilizer induced N_2O-N from fertilized soils with various N sources (Modified from Stehfest and Bouwman 2006; Eichner 1990)

Fertilizer N source	Number of observations (n)	Balanced mean (kg N ha^{-1})	Percent of applied N emitted as N_2O
Anhydrous ammonia (NH_3)	50	1.04	0.04–19.6
Ammonium carbonate, chloride or sulfate	85	0.82	0.01–36.54
Ammonium phosphate	6	0.26	0.06–7
Ammonium nitrate	131	1.12	0–30.0
Calcium ammonium nitrate	73	1.56	0.05–11
Urea	131	0.96	0.01–46.44
Urea ammonium nitrate	40	0.78	0.03–16.03
Organic fertilizer	88	1.15	0.03–56
Organic and synthetic inorganic N fertilizer mix	48	0.81	0–31.73
Calcium nitrate, potassium nitrate, sodium nitrate	58	0.79	0–41.80

Application of urease inhibitors could reduce the seed damage (Malhi et al. 2003). Broadcasting of urea can be less effective than side-banding, and fall banded N could be inferior to spring banded N (Malhi et al. 2008). Injection of liquid urea, ammonium nitrate below the surface ~10–15 cm deep can result in 40–70% lower N_2O emissions compared to shallower (5 cm) or surface application in a fine loam soils such as those in Colorado, USA (Liu et al. 2006). In contrast, emissions of N_2O can be on average 25% higher from treatments with 10-cm deep injection of NH_4NO_3 than shallower 2-cm injection under different tillage practices on a clay loam (Drury et al. 2006). Based on these results it is difficult to generalize the benefits of fertilizer N placement for N_2O mitigation strategy.

8.3.2.3 N Rate

The NO_3^- can accumulate in soils when the N application rate exceeds crop demand and when crop recovery of the applied N is low. Such rates are known to lead to increases in NO_3^- leaching (Gehl et al. 2005) and increased N_2O fluxes (McSwiney and Robertson 2005). Numerous field studies conducted on N input in agricultural crops have shown that emissions of N_2O is correlated with fertilizer N rate (Bouwman et al. 2002; McSwiney and Robertson 2005; Drury et al. 2008; Hoben et al. 2011). Thus, this has been the basis for current Intergovernmental Panel on Climate Change (IPCC) GHG inventory calculations. Increasing amount of applied N, crop yields eventually reaches a maximum at the agronomic optimum N rate. N rates higher than agronomic optimum results in rapid increase in N_2O emissions. Use of appropriate N rates can minimize soil accumulation of NO_3-N and minimize N_2O emissions from croplands and co-benefits related to excess reactive N in the environment (Robertson and Vitousek 2009). Generalizing across

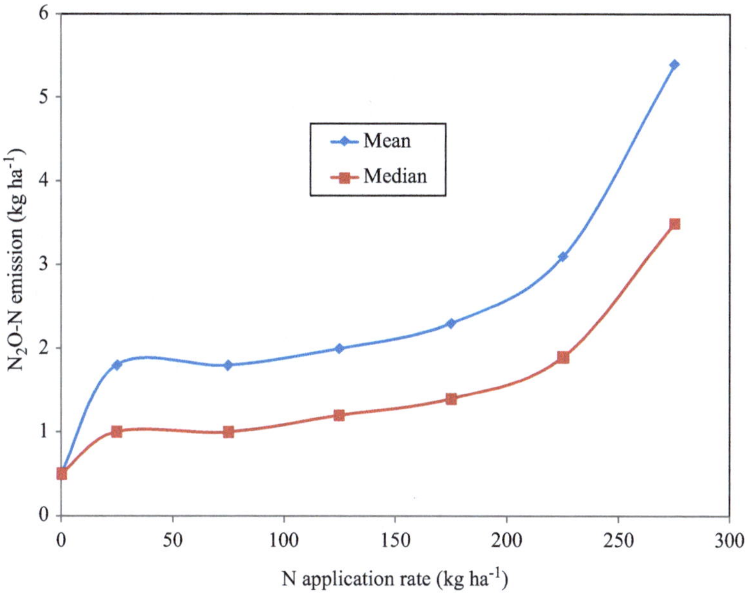

Fig. 8.2 Balance mean and median N_2O emission rates as a function of applied N (Redrawn from data presented by Bouwman et al. 2002)

multiple sites, years, sources and cropping systems, Bouwman et al. (2002) reported that N_2O emissions appears to remain relatively static across a broad range of fertilizer N rates, which is near the crop demand levels, and then tends to increase with higher rates. The data in Fig. 8.2 indicate a nonlinear N_2O emissions response with N rates (McSwiney and Robertson 2005; Ma et al. 2010), but contrasts with linear emission factor approach recommended by IPCC (2006) for Tier 1 methodology. Similar nonlinear N_2O emissions response to N rates was also reported for commercial farms in Michigan, USA (Hoben et al. 2011). The N_2O emissions increased exponentially for the N rates higher than maximum economic returns, suggesting that significant decreases in N_2O emissions may be achieved by decreasing N fertilizer inputs without affecting economic yield returns.

8.3.2.4 N Timing

Synchronizing the timing of fertilizer N with plant N demand is an important N management technique in agriculture, which determines soil N availability to crops and potentially N_2O emission from cropland. Crop N uptake capacity is generally low at the beginning of the growing season, and increases rapidly during vegetative growth and dropping sharply as the crop nears maturity. Avoiding time delays between N application and plant N uptake and placing the N more precisely into the soil to make it more accessible to crop roots ensures an adequate availability of N when needed by plants and minimize the excess N supply (Monteny et al. 2006). Application of fertilizer N in autumn prior to spring crop planting results in increased soil N due to lack of plant uptake, and therefore, exacerbates potential of the increased N_2O emission. About 30% of the area cropped to corn in USA is fertilized in autumn (CAST 2004). Thus, large emissions of N_2O could potentially be avoided with spring rather than autumn fertilization. Two studies from Canada are relevant to this discussion. In Saskatchewan, lower N_2O emissions were observed for spring compared to fall N fertilizer application (Hultgreen and Leduc 2003). In contrast, application of 100 kg N ha^{-1} to canola and wheat plots in autumn resulted in significantly greater N_2O emissions than did spring fertilizer application in Alberta (Hao et al. 2001). Prevention of excess NO_3-N accumulation in the soil especially in the presence of excess water such as in fall and over winter period is critical to reducing N losses by leaching and denitrification. Applying N according to available N reserves and matching the time of application to crop uptake, and use of catch or cover crops during inter-growing season, which may extract plant available N unused by preceding crops leads to lower residual soil NO_3^-, reduced leaching and has been suggested as an option for reducing N_2O emissions (Justes et al. 1999; Strock et al. 2004).

8.3.3 Water Management

About 18% of world's cropland receives supplementary irrigation (Smith et al. 2008). Irrigation enhances crop yields, but proper management of soil moisture and N is necessary to minimize N_2O emission. However, the effects of irrigation on N_2O emission have not been widely measured. Overall, N_2O emission reduction through irrigation improvements looks promising (Amos et al. 2005; Scheer et al. 2008a, b; Kallenbach et al. 2010). Burger et al. (2005) noted that higher N_2O emissions following each irrigation event, but when WFPS was <60% the N_2O emissions decreased significantly. Maintaining lower WFPS in surface drip irrigation compared to inundate/dry cycle of the flood irrigation limits denitrification which is tightly coupled with WFPS >60%. The WFPS is >60% only within a few cm of the drip tape in subsurface drip irrigation, with overall low WFPS of 20–30% in these systems (Kallenbach et al. 2010). Reducing irrigation intensity and irrigating cotton only when soil moisture is 65% of field capacity instead of 75% can reduce N_2O emissions by almost 50% (Scheer et al. 2008a).

Drainage of agricultural lands in humid regions can promote land productivity and suppress N_2O emissions by improving aeration (Monteny et al. 2006). Any N lost through drainage, however, may be susceptible to loss as N_2O (Reay et al. 2003). Denitrification is strongly affected by soil moisture regime and aeration. In a global review comparing 193 poorly drained soils with 460 well-drained soils, Bouwman et al. (2002) observed lower N_2O emissions by an average of 0.64 kg N_2O ha^{-1} $year^{-1}$ under the well-drained than under poorly drained soils. Drainage and irrigation management may therefore, influence soil water table and N_2O emissions. Judicious management of drainage and irrigation of croplands are desirable strategies to improve soil moisture, and affect both direct and indirect N_2O emissions.

8.3.4 *Optimization of Manure N Application*

Manure nutrients are often not fully credited as a nutrient source. There is a potential over application of N by 2–10% based on available N from fertilizer, manure, and legumes (Duke 2006). Utilization of manure nutrient value and soil N testing will reduce the need for inorganic N fertilizers in crop-livestock operations. Land application of manure in a manner that N losses are minimized will enable replacing commercial fertilizers by manure. The N_2O emissions from manure are very high during spring-thaw period (Wagner-Riddle and Thurtell 1998). This pulse could be avoided by applying manure after this period. Spring application of manure rather than fall application reduces N_2O emissions. For example, emission factors for fall and spring manure application are 1.3 and 1.8, respectively (Duke 2006). Potential losses from fall manure application are recognized as environmental risk. Surface application of liquid manure results into high NH_3 losses, which causes high indirect N_2O emission. Incorporation of manure reduces NH_3 losses from 16.9% of total N to 3.6% of total N (Rochette et al. 2001) and increase the fertilizer value of manure.

8.3.5 *Organic and Wetland Soils*

Organic soils contain high density of SOM accumulated over long periods because decomposition is suppressed by absence of O_2 under flooded conditions. To be used for agriculture, these soils are drained. Drainage and cultivation of these peat soils increase soil aeration and reverse the C flux of net CO_2 emission into the atmosphere. Cultivated organic soils are large source of CO_2 and N_2O emissions generated by the oxidation and decomposition of OM. The decomposition results in measurable subsidence of soil surface. Based on subsidence measurements, Eriksson (1991) estimated that even though the cultivated organic soils are $<10\%$ of the arable land area in Sweden, the CO_2 emissions from these soils accounts for as much as 10% of total national anthropogenic CO_2 emissions.

The N_2O is produced under sub-oxic conditions in soils as a by-product of nitrification and denitrification. The significance of cultivated organic soils is probably more important for N_2O emissions than that of CO_2. The N_2O emissions rates from cultivated peat lands exceed those from mineral agricultural soils by a factor of 2–10 (Freibauer et al. 2004). The N in the OM becomes available for N_2O-producing organisms after drainage, and organic soils drained for cultivation are a significant source of N_2O (Maljanen et al. 2003; von Arnold et al. 2005a, b). These emissions are anthropogenic, since they are induced by drainage. For example, in Finland, it is estimated that 25% (4 Tg N_2O year^{-1}) of the national anthropogenic N_2O emissions originate from cultivated organic soils (Kasimir-Klemedtsson et al. 1997). In Finland and Sweden, the total area of cultivated organic soils is estimated at 300,000 ha, representing about 10% of cropland. Most of the research on Histosols management has been conducted in Europe. However, Rochette et al. (2010) observed that organic soils in Canada exhibited similar GHG fluxes as those reported in Europe. The use of NT dramatically reduces N_2O emissions from Histosols (Elder and Lal 2008). However, the best mitigation option for these soils is to avoid drainage of the water in the first place, restoring these soils to a natural state, re-establishing a high water table where GHG emissions are still high, or conversion of arable cropping to permanent cultures as well as new crops on restored wetlands (Freibauer et al. 2004).

8.3.6 Biochar Application

Biochar have potential to mitigate N_2O by decreasing the need for fertilizer N and reducing upstream N_2O emissions (Lehmann et al. 2006) and by reducing direct N_2O emissions, possibly due to production of ethylene which inhibits microbial processes (Spokas et al. 2010). However, there are few long-term studies documenting suppression of N_2O emissions in the field, even though laboratory and short term experiments have indicated N_2O emissions reductions of 50–80% (Lehmann et al. 2006; Yanai et al. 2007; Fowles 2007). Yanai et al. (2007) observed that the impact of biochar on N_2O emissions strongly depends on soil hydrology, where emissions of N_2O varied from 89% reduction in very wet soil to a 51% increase in drier soil. Residence time of biochar in the soil may also play an important role, as increases in N_2O emissions occurs when biochar is first applied to soil with shift to N_2O emission reduction over time as sorption capacity of biochar is enhanced with aging (Singh et al. 2010).

8.3.7 Indirect Nitrous Oxide Emission from Cropland

Croplands also contribute a significant amount of N_2O by indirect emission. Indirect emission include N_2O produced in non-cropland such as aquatic or forest ecosystems originating from the N lost from cropland via leaching, runoff, and volatilization. The reliable and unambiguous method for estimating and assigning the indirect N_2O emissions does not exist. However, the IPCC approach calculates this emission by assuming the amount of N lost from agricultural ecosystems and by assuming that a fixed fraction of this N is emitted as N_2O outside agricultural ecosystem boundaries. In Canada, indirect N_2O emission accounts for approximately 22% of all agricultural N_2O produced (AAFC 2010). Indirect N_2O emission in USA accounts for nearly 25% of the cropland N_2O emissions (US EPA 2011). Factors which influence the indirect N_2O emissions include mineral N fertilizer application, rainfall, intensive swine and diary operation which produces large volumes of manure and increase the risk of ammonia volatilization and subsequent deposition in neighboring ecosystems.

8.3.8 Role of Plant Breeding and Biotechnology Advances

Yield increases are major driver for the agricultural efficiency. However, higher yields do not always correlate with reduction in agricultural land use or preclude agriculture expansion ever increasing demand (Balmford et al. 2005; Burney et al. 2010). In the context of mitigating future agricultural GHG emissions, yield increases are a key to meeting the growing global food demand, which is expected to increase by 70% by 2050 (FAO 2006). Globally, grain yields have increased significantly since 1950s, with plant breeding contributing about 50% of the increase, and the improved management contributing the remaining 50% (Duvick 2005). Although much of the discussions about increased future yield potential has centered around genetic modification (GM) or genetic engineered (GE) crops using recombinant DNA technology, some reports have indicated that these crops in the U.S. have delivered lower yields increases than the traditional breeding (Duvick 2005; Gurian-Sherman 2009). However, the impact of the use of GE crops in the US is the improvement of water and soil quality, decreases in use of insecticides, lower production costs due to higher yield returns and reduced GHG emissions (BANR 2010). The GE varieties with herbicide tolerance for crops such as wheat, canola, and soybean, and improved rooting structures in corn have reduced tillage needs and enabled better crop growth under NT management. Efforts are under way to develop new plant varieties with characteristics that improve N use efficiency and reduce irrigation requirements (Beatty et al. 2009). Researchers are interested in developing NUE crop plants that produce higher yields or maintain current yields under reduced N application rates using GE technologies. Crops that are optimized for N use can decrease fertilize N use, minimize N losses through leaching and runoff, and reduce N_2O emissions.

8.4 Conclusions

With increasing global demand for food, feed, fiber, and fuel, the appropriate strategy to mitigate N_2O emissions must involve intensive crop management practices, which enhance the N use efficiency while improving crop yields.

Fertilizer N use supports plant primary productivity. Combination of proper source selection, appropriate rate, proper timing, and the right placement method are important to optimize crop yield, increase N use efficiency, and minimize N losses. The impacts of several management practices on N_2O emissions are not known. Emissions of N_2O from croplands can be reduced by improved crop N recovery and reduced NH_3 and NO_3^- loss. Strategies with the potential for mitigating N_2O emissions are:

- Use of crop management practices that enhance efficient and effective N use such as balanced fertilization and crop rotation,
- Conservation practices that complement fertilizer N use such as crop rotations matched to specific site characteristics, effective irrigation to match crop water consumption and nutrients demand, cover- and catch crops to recover and retain residual inorganic N,
- Management of soil moisture regime through irrigation and land drainage to reduce N_2O emissions,
- Use of soil and plant analysis to identify any nutrient imbalance that may limit crop yields and lead to inefficient N use,
- Use of urease inhibitors when applying urea-containing N sources to reduce NH_3 emissions especially when surface applied to reduce potential transfer of reactive N to unintended resources,
- Use of nitrification inhibitors with ammoniacal N sources to minimize the potential for NO_3-N leaching losses and N_2O emissions, and
- Use of controlled- and slow-release N fertilizers to enhance the efficiency of N use.

8.5 Further Research Needs

Comprehensive agronomic and environmental research is needed to evaluate the potential reductions in N_2O emissions that may be achievable through crop and fertilizer BMPs. Studies are needed to evaluate the agronomic impact of controlled- and slow-release fertilizers and nitrification inhibitors. In addition, long-term monitoring of N loss and N_2O emissions beyond the period of active crop growth and encompass crop rotation and cropping system changes are needed. There is a lack of reliable data comparing effects of fertilizer source, placement, and timing on N_2O emissions and mitigation. Furthermore, choice of BMP must be based on thorough understanding of soil processes that impact/moderate the N cycling and transformation. Important among these processes are:

- Mineralization
- Nitrification
- Denitrification

Study Questions

1. Describe the best management practices (BMP) package designed to mitigate N_2O emissions from corn crop fertilized with manure and urea.
2. Describe why mitigation of N_2O from the cropland is more complex and challenging issue.
3. List principal processes that affect N_2O emissions
4. What are the soil properties that moderate nitrification and denitrification?

References

Agriculture and Aagri-Food Canada (AAFC) (2010) Environmental Sustainability of Canadian Agriculture. Agriculture and Agri-Food Canada. http://www.agr.gc.ca/env/naharp-pnarsa/index_e.php

Alexander, M. (1961) Introduction to soil microbiology. Wiley, New York

Amos, B., Arkebauer, T.J., Doran, J.W. (2005) Soil surface fluxes of greenhouse gases in an irrigated maize-based agroecosystem. Soil Sci. Soc. Amer. J. **69**(2), 387–395

Askegaard, M., Olesen, J.E., Kristensen, K. (2005) Nitrate leaching from organic arable crop rotations: effects of location, manure and catch crop. Soil Use Manage. **21**(2), 181–188. doi:10.1111/j.1475-2743.2005.tb00123.x

Baggs, E.M., Stevenson, M., Pihlatie, M., Regar, A., Cook, H., Cadisch, G. (2003) Nitrous oxide emissions following application of residues and fertiliser under zero and conventional tillage. Plant Soil **254**(2), 361–370. doi:10.1023/a:1025593121839

Ball, B.C., Scott, A., Parker, J.P. (1999) Field N_2O, CO_2 and CH_4 fluxes in relation to tillage, compaction and soil quality in Scotland. Soil Till. Res. **53**(1), 29–39. doi:10.1016/s0167-1987(99)00074-4

Balmford, A., Green, R.E., Scharlemann, J.P.W. (2005) Sparing land for nature: exploring the potential impact of changes in agricultural yield on the area needed for crop production. Global Change Biol. **11**(10), 1594–1605. doi:10.1111/j.1365-2486.2005.001035.x

Barthes, B., Azontonde, A., Blanchart, E., Girardin, C., Villenave, C., Lesaint, S., Oliver, R., Feller, C. (2004) Effect of a legume cover crop (*Mucuna pruriens* var. utilis) on soil carbon in an Ultisol under maize cultivation in southern Benin. Soil Use Manage. **20**, 231–239. doi:10.1079/sum2004235

Beatty, P.H., Shrawat, A.K., Carroll, R.T., Zhu, T., Good, A.G. (2009) Transcriptome analysis of nitrogen-efficient rice over-expressing alanine aminotransferase. Plant Biotechnol. J. **7**(6), 562–576. doi:10.1111/j.1467-7652.2009.00424.x

Bleken, M.A., Bakken, L.R. (1997) The nitrogen cost of food production: Norwegian society. Ambio **26**(3), 134–142

Board on Agriculture and Natural Resources (BANR) (2010) The impact of Genetically Engineered crops on farm sustainability in the United States., Board on Agriculture and Natural Resources (BANR), Committee on the Impact of Biotechnology on Farm Level Economics and Sustainability, Division on Earth and Life studies. National Academies Press, Washington, D.C. 250 pp

Borlaug, N.E., Dowswell, C. (2008) The second green revolution. Taeb, M., Zakri, A.H. (eds.). Agriculture, Human Security, and Peace: A Crossroad in African Development

Bouwman, A., Boumans, L.J.M., Batjes, N.H. (2001) Global estimates of gaseous emissions of NH_3, NO and N_2O from agricultural land. International Fertilizer Industry Association (IFA) & Food and Agriculture Organization of the United Nations (FAO), Rome, Italy

Bouwman, A.F., Boumans, L.J.M., Batjes, N.H. (2002) Emissions of N_2O and NO from fertilized fields: Summary of available measurement data. Global Biogeochem. Cycles **16**(4), 1058. doi:10.1029/2001gb001811

Bouwman, A.F., Vanderhoek, K.W., Olivier, J.G.J. (1995) Uncertainties in the global source distribution of nitrous oxide. J. Geophys. Res-Atmos. **100**(D2), 2785–2800. doi:10.1029/94jd02946

Bronson, K.F., Mosier, A.R., Bishnoi, S.R. (1992) Nitrous oxide emissions in irrigated corn as affected by nitrification inhibitors. Soil Sci. Soc. Amer. J. **56**(1), 161–165

Bruulsema, T.W., Chen, F., Garcia, F., Ivanova, S., Li, S., Rao, N., Witt, C. (2008) A global framework for best management practices for fertilizer use. International Plant Nutrition Institute

Burger, M., Jackson, L.E., Lundquist, E.J., Louie, D.T., Miller, R.L., Rolston, D.E., Scow, K.M. (2005) Microbial responses and nitrous oxide emissions during wetting and drying of organically and conventionally managed soil under tomatoes. Biol. Fert. Soils **42**(2), 109–118. doi:10.1007/s00374-005-0007-z

Burney, J.A., Davis, S.J., Lobell, D.B. (2010) Greenhouse gas mitigation by agricultural intensification. Proc. Nat. Acad. Sci. USA **107**(26), 12052-12057. doi:10.1073/pnas.0914216107

Cassman, K.G., Dobermann, A., Walters, D.T. (2002) Agroecosystems, nitrogen-use efficiency, and nitrogen management. Ambio **31**(2), 132–140. doi:10.1579/0044-7447-31.2.132

Cassman, K.G., Dobermann, A., Walters, D.T., Yang, H. (2003) Meeting cereal demand while protecting natural resources and improving environmental quality. Annu. Rev. Env. Resour. **28**, 315–358. doi:10.1146/annurev.energy.28.040202.122858

CAST (2004) Climate change and greenhouse gas mitigation: challenges and opportunities for agriculture. Council for Agricultural Science and Technology (CAST), Ames, IA

Chen, X.-P., Cui, Z.-L., Vitousek, P.M., Cassman, K.G., Matson, P.A., Bai, J.-S., Meng, Q.-F., Hou, P., Yue, S.-C., Roemheld, V., Zhang, F.-S. (2011) Integrated soil-crop system management for food security. Proc. Nat. Acad. Sci. USA **108**(16), 6399–6404. doi:10.1073/pnas.1101419108

Cole, V., Cerri, C., Minami, K., Mosier, A., Rosenberg, N., Sauerbeck, D. (1996) Agricultural aptions for mitigation of greenhouse gas emission. In: Watson, R.T., Zinyowera, M.C., Moss, R.H., Dokken, D.J. (eds.) Climate Change 1995. Impacts, adaptations and mitigation of climate change: Scientific technical analyses. Contribution of working group II to the second assessment report of the Intergovernmental Panel on Climate Change. pp. 745–771. Cambridge University Press, New York

Cotton, F.A., Wilkenson, G. (1998) Advanced Inorganic Chemistry. Wiley, New York

Crews, T.E., Peoples, M.B. (2005) Can the synchrony of nitrogen supply and crop demand be improved in legume and fertilizer-based agroecosystems? A review. Nutr. Cycl. Agroecosys. **72**(2), 101–120. doi:10.1007/s10705-004-6480-1

Crutzen, P.J. (1970) Influence of nitrogen oxides on atmospheric ozone content. Q. J. Roy. Meteor. Soc. **96**(408), 320–325. doi:10.1002/qj.49709640815

Crutzen, P.J., Mosier, A.R., Smith, K.A., Winiwarter, W. (2008) N_2O release from agro-biofuel production negates global warming reduction by replacing fossil fuels. Atmos. Chem. Phys. **8**(2), 389–395

Davidson, E.A. (2009) The contribution of manure and fertilizer nitrogen to atmospheric nitrous oxide since 1860. Nature Geoscience **2**(9), 659–662. doi:10.1038/ngeo608

Del Grosso, S.J., Mosier, A.R., Parton, W.J., Ojima, D.S. (2005) DAYCENT model analysis of past and contemporary soil N_2O and net greenhouse gas flux for major crops in the USA. Soil Till. Res. **83**(1), 9–24. doi:10.1016/j.still.2005.02.007

Del Grosso, S.J., Parton, W.J., Mosier, A.R., Walsh, M.K., Ojima, D.S., Thornton, P.E. (2006) DAYCENT national-scale simulations of nitrous oxide emissions from cropped soils in the United States. J. Environ. Qual. **35**(4), 1451–1460. doi:10.2134/jeq2005.0160

Delgado, J.A., Dillon, M.A., Sparks, R.T., Essah, S.Y.C. (2007) A decade of advances in cover crops. J. Soil Water Conserv. **62**(5), 110A–117A

Dobermann, A. (2007) Nutrient use efficiency – measurement and management. In: Krauss, A., Isherwood, K., Heffer, P. (eds.) Fertilizer best management practices: General principels, strategies for adoption and voluntary initiatives vs regulations. Papers presented at the IFA International Workshop, 7-9 March 2007. Brussels, Belgium. pp. 1–28. International Fertilizers Industry Association (IFA), Paris, France

Dobermann, A., Cassman, K.G. (2004) Environmental dimensions of fertilizer N: what can be done to increase nitrogen use efficiency and ensure global food security? In: Mosier, A.R., Syers, J.K., Freeney, J.R. (eds.) Agriculture and the nitrogen cycle: Assessing the impact of fertilizer use on food production and the environment. SCOPE 65. pp. 261–278. Island Press, Washington,

Drury, C.F., Reynolds, W.D., Tan, C.S., Welacky, T.W., Calder, W., McLaughlin, N.B. (2006) Emissions of nitrous oxide and carbon dioxide: Influence of tillage type and nitrogen placement depth. Soil Sci. Soc. Amer. J. **70**(2), 570–581. doi:10.2136/sssaj2005.0042

Drury, C.F., Yang, X.M., Reynolds, W.D., McLaughlin, N.B. (2008) Nitrous oxide and carbon dioxide emissions from monoculture and rotational cropping of corn, soybean and winter wheat. Can. J. Soil Sci. **88**(2), 163–174

Duke, C. (2006) The potential for agricultural greenhouse gas emission reductions in the temperate region of Canada through nutrient management planning. BIOCAP, Canada

Duvick, D.N. (2005) The contribution of breeding to yield advances in maize (*Zea mays* L.). Adv. Agron. **86**, 83–145. doi:10.1016/s0065-2113(05)86002-x

Ehhalt, D., Prather, M., Dentener, F., Derwent, R., Dlugokencky, E., Holland, E., Isaksen, I., Katima, J., Kirchhoff, V., Matson, P., Midgley, P., Wang, M. (2001) Atmospheric Chemistry and Greenhouse Gases. In: Houghton, J.T., Ding, Y., Griggs, D.J., Noguer, M., Linden, P.J.v.d., Dai, X., Maskell, K., Johnson, C.A. (eds.) Climate Change 2001: The Scientific Basis. Contribution of Working Group I to the Third Assessment Report of the Intergovernmental Panel on Climate Change, vol. . p. 881. Cambridge University Press, Cambridge, United Kingdom and New York, USA

Eichner, M.J. (1990) Nitrous oxide emissions from fertilized soils: summary of available data. J. Environ. Qual. **19**(2), 272–280

Elder, J.W., Lal, R. (2008) Tillage effects on gaseous emissions from an intensively farmed organic soil in North Central Ohio. Soil Till. Res. **98**(1), 45–55. doi:10.1016/j.still.2007.10.003

Environment Canada (2007) Greenhouse Gas Sources and Sinks in Canada, 1990–2005 National inventory report. Environment Canada, Greenhouse Gas Division. http://www.ec.gc.ca/Publications/E80FB0A6-A0C2-4E6E-A12E-005D50D041C5%5CNationalInventoryReport19902005April2007-.pdf

Eriksson, H. (1991) Sources and sinks of carbon dioxide in Sweden. Ambio **20**(3-4), 146–150

Food and Agriculture Organization of the United nations (FAO) (2006) World agriculture: Towards 2030/2050 - Prospects for food, nutrition, agriculture and major commodity groups. Food and Agriculture Organization of the United Nations. http://www.fao.org/docrep/009/a0607e/a0607e00.htm

FAOSTAT (2011) FAOSTAT fertilizers database. Food and Agriculture Organization of the United Nations. http://faostat.fao.org. Accessed November 2011

Feyereisen, G.W., Wilson, B.N., Sands, G.R., Strock, J.S., Porter, P.M. (2006) Potential for a rye cover crop to reduce nitrate loss in southwestern Minnesota. Agron. J. **98**(6), 1416–1426. doi:10.2134/agronj2005.0134

Fowles, M. (2007) Black carbon sequestration as an alternative to bioenergy. Biomass Bioen. **31**(6), 426–432. doi:10.1016/j.biombioe.2007.01.012

Freibauer, A., Rounsevell, M.D.A., Smith, P., Verhagen, J. (2004) Carbon sequestration in the agricultural soils of Europe. Geoderma **122**(1), 1–23. doi:10.1016/j.geoderma.2004.01.021

Galloway, J.N., Aber, J.D., Erisman, J.W., Seitzinger, S.P., Howarth, R.W., Cowling, E.B., Cosby, B.J. (2003) The nitrogen cascade. Bioscience **53**(4), 341–356. doi:10.1641/0006-3568(2003)053[0341:tnc]2.0.co;2

Galloway, J.N., Dentener, F.J., Capone, D.G., Boyer, E.W., Howarth, R.W., Seitzinger, S.P., Asner, G.P., Cleveland, C.C., Green, P.A., Holland, E.A., Karl, D.M., Michaels, A.F., Porter,

J.H., Townsend, A.R., Vorosmarty, C.J. (2004) Nitrogen cycles: Past, present, and future. Biogeochemistry **70**(2), 153–226. doi:10.1007/s10533-004-0370-0

Galloway, J.N., Schlesinger, W.H., Levy, H., Michaels, A., Schnoor, J.L. (1995) Nitrogen fixation: Anthropogenic enhancement-environmental response. Global Biogeochem. Cycles **9**(2), 235–252. doi:10.1029/95gb00158

Gehl, R.J., Schmidt, J.P., Maddux, L.D., Gordon, W.B. (2005) Corn yield response to nitrogen rate and timing in sandy irrigated soils. Agron. J. **97**(4), 1230–1238. doi:10.2134/agronj2004.0303

Granli, T., Bockman, O.C. (1994) Nitrous oxide from agriculture. Norw. J. Agric. Sci. Suppl. **12**, 1–128

Groffman, P.M. (1985) Nitrification and denitrification in conventional and no-tillage soils. Soil Sci. Soc. Amer. J. **49**(2), 329–334

Groffman, P.M., Gold, A.J., Addy, K. (2000) Nitrous oxide production in riparian zones and its importance to national emission inventories. Chemosphere **2**(3-4), 291–299

Groffman, P.M., Gold, A.J., Kellogg, D.Q., Addy, K. (2002) Nitrous oxide emissions derived from N leaching. In: Petersen, S.O., Olesen, J.E. (eds.) DIAS Report, Plant Production no 81, October 2002. Greenhouse Gas Inventories for Agriculture in the Nordic Countries. pp. 143–155. Danish Ministry of Food, Agriculture and Fisheries

Gurian-Sherman, D. (2009) Failure to yield. In: Union of Concerned Scientists. p. 51 pp. Cambridge, MA

Hao, X., Chang, C., Carefoot, J.M., Janzen, H.H., Ellert, B.H. (2001) Nitrous oxide emissions from an irrigated soil as affected by fertilizer and straw management. Nutr. Cycl. Agroecosys. **60**(1-3), 1–8. doi:10.1023/a:1012603732435

Hoben, J.P., Gehl, R.J., Millar, N., Grace, P.R., Robertson, G.P. (2011) Nonlinear nitrous oxide (N_2O) response to nitrogen fertilizer in on-farm corn crops of the US Midwest. Global Change Biol. **17**(2), 1140–1152. doi:10.1111/j.1365-2486.2010.02349.x

Hultgreen, G.E., Leduc, P. (2003) The effects of nitrogen fertilizer placement, formulation, timing and rate on greenhouse gas emission and agronomic performance. Saskatchewan Department of Agriculture and Food

International Fertilizer Industry Association (IFA) (2007) Sustainable management of the nitrogen cycle in agriculture and mitigation of reactive nitrogen side effects. International Fertilizer Industry Association, (IFA)

International Fertilizer industry Association/Food and Agriculture Organization of the United nations (IFA/FAO) (2001) Global estimates of gaseous emissions of NH_3, NO, and N_2O from agricultural land. International Fertilizer Industry Association and Food and Agriculture Organization of the United Nations. ftp://ftp.fao.org/agl/agll/docs/globest.pdfIFA/FAO

Intergovernmental Panel on Climate Change (IPCC) (2006) N_2O emission from managed soils and CO_2 emissions from lime and urea application Published. In: Eggleston, H.S., Buendia, L., Miwa, K., Ngara, T., Tanabe, K. (eds.) Intergovernmental Panel on Climate Change Guidelines for National Greenhouse Gas Inventories, Prepared by the National Greenhouse Gas Inventories Programme, Volume 4 Agriculture, forestry and other land uses. . IGES, Japan

Intergovernmental Panel on Climate Change (IPCC) (2007) Climate Change 2007: The Physical Science Basis. Contribution of Working Group I to the Fourth Assessment Report of the Intergovernmental Panel on Climate Change. Solomon, S., Qin, D., Manning, M., Chen, Z., Marquis, M., K.B. Averyt, Tignor, M., Miller, H.L. (eds.). Cambridge University Press, Cambridge, United Kingdom and New York, NY, USA

Jayasundara, S., Wagner-Riddle, C., Parkin, G., von Bertoldi, P., Warland, J., Kay, B., Voroney, P. (2007) Minimizing nitrogen losses from a corn-soybean-winter wheat rotation with best management practices. Nutr. Cycl. Agroecosys. **79**(2), 141–159. doi:10.1007/s10705-007-9103-9

Jaynes, D.B., Karlen, D.L. (2008) Sustaining soil resources while managing nutrients. In: Upper Mississippi River sub-basin hypoxia nutrient committee. Final Report: Gulf Hypoxia and Local Water Quality Concerns Workshop. pp. 149–158. American Society of Agricultural and Biological Engineers, St. Joseph, Michigan

Justes, E., Mary, B., Nicolardot, B. (1999) Comparing the effectiveness of radish cover crop, oilseed rape volunteers and oilseed rape residues incorporation for reducing nitrate leaching. Nutr. Cycl. Agroecosys. **55**(3), 207–220. doi:10.1023/a:1009870401779

Kallenbach, C.M., Rolston, D.E., Horwath, W.R. (2010) Cover cropping affects soil N_2O and CO_2 emissions differently depending on type of irrigation. Agric. Ecosys. Environ. **137**(3-4), 251–260. doi:10.1016/j.agee.2010.02.010

Kasimir-Klemedtsson, Å., Klemedtsson, L., Berglund, K., Martikainen, P., Silvola, J., Oenema, O. (1997) Greenhouse gas emissions from farmed organic soils: a review. Soil Use Manage. **13**, 245–250. doi:10.1111/j.1475-2743.1997.tb00595.x

Keller, M., Varner, R., Dias, J.D., Silva, H., Crill, P., de Oliveira, R.C., Jr. (2005) Soil-atmosphere exchange of nitrous oxide, nitric oxide, methane, and carbon dioxide in logged and undisturbed forest in the Tapajos National Forest, Brazil. Earth Interactions **9 Issue 23**, 1–28

Kroeze, C., Mosier, A. (2000) New estimates for emissions of nitrous oxide. In: vanHam, J.E.A., Baede, A.P.M., Meyer, L.A., Ybema, R. (eds.) Non-CO_2 Greenhouse Gases: Scientific Understanding, Control and Implementation. pp. 45–64

Kroeze, C., Mosier, A., Bouwman, L. (1999) Closing the global N_2O budget: A retrospective analysis 1500-1994. Global Biogeochem. Cycles **13**(1), 1–8. doi:10.1029/1998gb900020

Laegried, M., Aastveit, A.H. (2002) Nitrous oxide emissions from field-applied fertilizers. In: Petersen, S.O., Olesen, J.E. (eds.) DIAS Report, Plant Production no. 81, October 2002. Greenhouse Gas Inventories for Agriculture in the Nordic Countries. pp. 122–134. Danish Ministry of Food, Agriculture and Fisheries

Lee, J., Six, J., King, A.P., Van Kessel, C., Rolston, D.E. (2006) Tillage and field scale controls on greenhouse gas emissions. J. Environ. Qual. **35**(3), 714–725. doi:10.2134/jeq2005.0337

Lehmann, J., Gaunt, J., Rondon, M. (2006) Bio-char Sequestration in Terrestrial Ecosystems – A Review. Mitig. Adapt. Strateg. Glob. Change **11**(2), 395-419. doi:10.1007/s11027-005-9006-5

Lemke, R.L., Izaurralde, R.C., Nyborg, M., Solberg, E.D. (1999) Tillage and N source influence soil-emitted nitrous oxide in the Alberta Parkland region. Can. J. Soil Sci. **79**(1), 15–24. doi:10.4141/S98-013

Li, C.S., Frolking, S., Butterbach-Bahl, K. (2005) Carbon sequestration in arable soils is likely to increase nitrous oxide emissions, offsetting reductions in climate radiative forcing. Climatic Change **72**(3), 321–338. doi:10.1007/s10584-005-6791-5

Liu, X.J., Mosier, A.R., Halvorson, A.D., Zhang, F.S. (2006) The impact of nitrogen placement and tillage on NO, N_2O, CH_4 and CO_2 fluxes from a clay loam soil. Plant Soil **280**(1-2), 177–188. doi:10.1007/s11104-005-2950-8

Lobell, D.B. (2007) The cost of uncertainty for nitrogen fertilizer management: A sensitivity analysis. Field Crop Res. **100**(2-3), 210–217. doi:10.1016/j.fcr.2006.07.007

Luo, J., de Klein, C.A.M., Ledgard, S.F., Saggar, S. (2010) Management options to reduce nitrous oxide emissions from intensively grazed pastures: A review. Agric. Ecosys. Environ. **136**(3-4), 282–291. doi:10.1016/j.agee.2009.12.003

Ma, B.L., Wu, T.Y., Tremblay, N., Deen, W., Morrison, M.J., McLaughlin, N.B., Gregorich, E.G., Stewart, G. (2010) Nitrous oxide fluxes from corn fields: on-farm assessment of the amount and timing of nitrogen fertilizer. Global Change Biol. **16**(1), 156–170. doi:10.1111/j.1365-2486.2009.01932.x

Malhi, S.S., Lemke, R., Mooleki, S.P., Schoenau, J.J., Brandt, S., Lafond, G., Wang, H., Hultgreen, G.E., May, W.E. (2008) Fertilizer N management and P placement effects on yield, seed protein content and N uptake of flax under varied conditions in Saskatchewan. Can. J. Plant Sci. **88**(1), 11–33

Malhi, S.S., Lemke, R., Wang, Z.H., Chhabra, B.S. (2006) Tillage, nitrogen and crop residue effects on crop yield, nutrient uptake, soil quality, and greenhouse gas emissions. Soil Till. Res. **90**(1-2), 171–183. doi:10.1016/j.still.2005.09.001

Malhi, S.S., Oliver, E., Mayerle, G., Kruger, G., Gill, K.S. (2003) Improving effectiveness of seedrow-placed urea with urease inhibitor and polymer coating for durum wheat and canola. Comm. Soil Sci. Plant Anal. **34**(11-12), 1709–1727. doi:10.1081/css-120021307

Maljanen, M., Liikanen, A., Silvola, J., Martikainen, P.J. (2003) Nitrous oxide emissions from boreal organic soil under different land-use. Soil Biol. Biochem. **35**(5), 689–700. doi:10.1016/s0038-0717(03)00085-3

Martinelli, L.A., Almeida, S., Brown, I.F., Moreira, M.Z., Victoria, R.L., Filoso, S., Ferreira, C. A.C., Thomas, W.W. (2000) Variation in nutrient distribution and potential nutrient losses by selective logging in a humid tropical forest of Rondonia, Brazil. Biotropica **32**(4), 597–613. doi:10.1646/0006-3606(2000)032[0597:vindap]2.0.co;2

Matson, P.A., Naylor, R., Ortiz-Monasterio, I. (1998) Integration of environmental, agronomic, and economic aspects of fertilizer management. Science **280**(5360), 112–115. doi:10.1126/science.280.5360.112

Matson, P.A., Parton, W.J., Power, A.G., Swift, M.J. (1997) Agricultural intensification and ecosystem properties. Science **277**(5325), 504–509. doi:10.1126/science.277.5325.504

Matson, P.A., Vitousek, P.M. (1990) Ecosystem approach to a global nitrous oxide budget. Bioscience **40**(9), 667–671. doi:10.2307/1311434

McSwiney, C.P., Robertson, G.P. (2005) Nonlinear response of N_2O flux to incremental fertilizer addition in a continuous maize (*Zea mays* L.) cropping system. Global Change Biol. **11**(10), 1712–1719. doi:10.1111/j.1365-2486.2005.01040.x

Monteny, G.J., Bannink, A., Chadwick, D. (2006) Greenhouse gas abatement strategies for animal husbandry. Agric. Ecosys. Environ. **112**(2-3), 163–170. doi:10.1016/j.agee.2005.08.015

Mosier, A., Kroeze, C., Nevison, C., Oenema, O., Seitzinger, S., van Cleemput, O. (1998a) Closing the global N_2O budget: nitrous oxide emissions through the agricultural nitrogen cycle - OECD/IPCC/IEA phase II development of IPCC guidelines for national greenhouse gas inventory methodology. Nutr. Cycl. Agroecosys. **52**(2-3), 225–248. doi:10.1023/a:1009740530221

Mosier, A.R. (2002) Environmental challenges associated with needed increases in global nitrogen fixation. Nutr. Cycl. Agroecosys. **63**(2), 101–116. doi:10.1023/a:1021101423341

Mosier, A.R., Bleken, M.A., Chaiwanakupt, P., Ellis, E.C., Freney, J.R., Howarth, R.B., Matson, P.A., Minami, K., Naylor, R., Weeks, K.N., Zhu, Z.L. (2001) Policy implications of human-accelerated nitrogen cycling. Biogeochemistry **52**(3), 281–320. doi:10.1023/a:1006430122495

Mosier, A.R., Duxbury, J.M., Freney, J.R., Heinemeyer, O., Minami, K. (1996) Nitrous oxide emissions from agricultural fields: Assessment, measurement and mitigation. Plant Soil **181**(1), 95–108. doi:10.1007/bf00011296

Mosier, A.R., Duxbury, J.M., Freney, J.R., Heinemeyer, O., Minami, K. (1998b) Assessing and mitigating N_2O emissions from agricultural soils. Climatic Change **40**(1), 7–38. doi:10.1023/a:1005386614431

Mosquera, J., Hol, J.M.G., Rappoldt, C., Dolfing, J. (2007) Precise soil management as a tool to reduce CH_4 and N_2O emissions from agricultural soils. Report 28

Neill, C., Steudler, P.A., Garcia-Montiel, D.C., Melillo, J.M., Feigl, B.J., Piccolo, M.C., Cerri, C. C. (2005) Rates and controls of nitrous oxide and nitric oxide emissions following conversion of forest to pasture in Rondonia. Nutr. Cycl. Agroecosys. **71**(1), 1–15. doi:10.1007/s10705-004-0378-9

Nevison, C. (2000) Review of the IPCC methodology for estimating nitrous oxide emissions associated with agricultural leaching and runoff. Chemosphere **2**(3-4), 493–500

Nevison, C.D., Mahowald, N.M., Weiss, R.F., Prinn, R.G. (2007) Interannual and seasonal variability in atmospheric N_2O. Global Biogeochem. Cycles **21**(3). doi:10.1029/2006gb002755

Oenema, O., Tamminga, S. (2005) Nitrogen in global animal production and management options for improving nitrogen use efficiency. Science China C **48**, 871–887. doi:10.1360/062005-279

Oenema, O., Wrage, N., Velthof, G.L., van Groenigen, J.W., Dolfing, J., Kuikman, P.J. (2005) Trends in global nitrous oxide emissions from animal production systems. Nutr. Cycl. Agroecosys. **72**(1), 51–65. doi:10.1007/s10705-004-7354-2

Ogle, S.M., Breidt, F.J., Paustian, K. (2005) Agricultural management impacts on soil organic carbon storage under moist and dry climatic conditions of temperate and tropical regions. Biogeochemistry **72**(1), 87–121. doi:10.1007/s10533-004-0360-2

Olsthoorn, C.S.M., Fong, N.P.K. (1998) The anthropogenic nitrogen cycle in the Netherlands. Nutr. Cycl. Agroecosys. **52**(2), 269–276

Paoletti, M.G., Gomiero, T., Pimentel, D. (2011) Introduction to the Special Issue: Towards a more sustainable agriculture. Crit. Rev. Plant Sci. **30**(1-2), 2–5. doi:10.1080/07352689.2011.553148

Parkin, T.B., Hatfield, J.L. (2010) Influence of nitrapyrin on N_2O losses from soil receiving fall-applied anhydrous ammonia. Agric. Ecosys. Environ. **136**(1-2), 81–86. doi:10.1016/j.agee.2009.11.014

Parkin, T.B., Kaspar, T.C. (2006) Nitrous oxide emissions from corn-soybean systems in the Midwest. J. Environ. Qual. **35**(4), 1496–1506. doi:10.2134/jeq2005.0183

Pattey, E., Edwards, G.C., Desjardins, R.L., Pennock, D.J., Smith, W., Grant, B., MacPherson, J.I. (2007) Tools for quantifying N_2O emissions from agroecosystems. Agric. Forest Meteor. **142**(2-4), 103–119. doi:10.1016/j.agrformet.2006.05.013

Peoples, M.B., Mosier, A.R., Freney, J.R. (1994) Minimizing gaseous loss of nitrogen. In: Bacon, P.E. (ed.) Nitrogen Fertilization in the Environment. pp. 565–602. Marcel Dekker Inc, New York

Prinn, R.G., Zander, R. (1999) Long-lived ozone related compounds. In: Albritton, D.L., Aucamp, P.J., Mégie, G., Watson, R.T. (eds.) Scientific assessment of ozone depletion: 1998, Global Ozone Research and Monitoring Project - Report No. 44. pp. 1.1–54. World Meteorological Organization, Geneva, Switzerland

Reay, D.S., Smith, K.A., Edwards, A.C. (2003) Nitrous oxide emission from agricultural drainage waters. Global Change Biol. **9**(2), 195–203. doi:10.1046/j.1365-2486.2003.00584.x

Rees, W.E. (2003) A blot on the land. Nature **421**(6926), 898–898. doi:10.1038/421898a

Roberts, T.L. (2007) Right product, right rate, right time, and right place. . .The foundation of best management practices for fertilizer. In: Fertilizer best management practices: General principles, strategy for their adaptation and voluntary initiatives vs regulations: Proceedings of International Fertilizer Industry Association (IFA) Workshop, March 7-9, 2007, Brussels, Belgium. pp. 29–32. IFA, Paris, France

Robertson, G.P., Groffman, P.M. (2007) Nitrogen transformations. In: Paul, E.A. (ed.) Soil microbiology, ecology and biochemistry. pp. 341–364. Elsevier, New York

Robertson, G.P., Vitousek, P.M. (2009) Nitrogen in agriculture: Balancing the cost of an essential resource. Annu. Rev. Env. Resour. **34**, 97–125

Rochette, P. (2008) No-till only increases N_2O emissions in poorly-aerated soils. Soil Till. Res. **101**(1-2), 97–100. doi:10.1016/j.still.2008.07.011

Rochette, P., Chantigny, M.H., Angers, D.A., Bertrand, N., Cote, D. (2001) Ammonia volatilization and soil nitrogen dynamics following fall application of pig slurry on canola crop residues. Can. J. Soil Sci. **81**(3), 515–523

Rochette, P., Tremblay, N., Fallon, E., Angers, D.A., Chantigny, M.H., MacDonald, J.D., Bertrand, N., Parent, L.E. (2010) N_2O emissions from an irrigated and non-irrigated organic soil in eastern Canada as influenced by N fertilizer addition. Eur. J. Soil Sci. **61**(2), 186–196. doi:10.1111/j.1365-2389.2009.01222.x

Rochette, P., Worth, D.E., Lemke, R.L., McConkey, B.G., Pennock, D.J., Wagner-Riddle, C., Desjardins, R.L. (2008) Estimation of N_2O emissions from agricultural soils in Canada. I. Development of a country-specific methodology. Can. J. Soil Sci. **88**(5), 641–654

Saggar, S. (2010) Estimation of nitrous oxide emission from ecosystems and its mitigation technologies. Agric. Ecosys. Environ. **136**(3-4), 189–191. doi:10.1016/j.agee.2010.01.007

Sato, K., Aoki, M., Noyori, R. (1998) A "green" route to adipic acid: Direct oxidation of cyclohexenes with 30 percent hydrogen peroxide. Science **281**(5383), 1646–1647. doi:10.1126/science.281.5383.1646

Scheer, C., Wassmann, R., Kienzler, K., Ibragimov, N., Lamers, J.P.A., Martius, C. (2008a) Methane and nitrous oxide fluxes in annual and perennial land-use systems of the irrigated areas in the Aral Sea Basin. Global Change Biol. **14**(10), 2454–2468. doi:10.1111/j.1365-2486.2008.01631.x

Scheer, C., Wassmann, R., Klenzler, K., Lbragimov, N., Eschanov, R. (2008b) Nitrous oxide emissions from fertilized irrigated cotton (*Gossypium hirsutum* L.) in the Aral Sea Basin,

Uzbekistan: Influence of nitrogen applications and irrigation practices. Soil Biol. Biochem. **40**(2), 290–301. doi:10.1016/j.soilbio.2007.08.007

Shipley, P.R., Meisinger, J.J., Decker, A.M. (1992) Conserving residual corn fertilizer nitrogen with winter cover crops. Agron. J. **84**(5), 869–876

Singh, B.P., Hatton, B.J., Singh, B., Cowie, A.L., Kathuria, A. (2010) Influence of Biochars on Nitrous Oxide Emission and Nitrogen Leaching from Two Contrasting Soils. J. Environ. Qual. **39**(4), 1224–1235. doi:10.2134/jeq2009.0138

Six, J., Feller, C., Denef, K., Ogle, S.M., Sa, J.C.D., Albrecht, A. (2002) Soil organic matter, biota and aggregation in temperate and tropical soils - Effects of no-tillage. Agronomie **22**(7-8), 755–775. doi:10.1051/agro:2002043

Six, J., Ogle, S.M., Breidt, F.J., Conant, R.T., Mosier, A.R., Paustian, K. (2004) The potential to mitigate global warming with no-tillage management is only realized when practised in the long term. Global Change Biol. **10**(2), 155–160. doi:10.1111/j.1529-8817.2003.00730.x

Smil, V. (1999) Nitrogen in crop production: An account of global flows. Global Biogeochem. Cycles **13**(2), 647–662. doi:10.1029/1999gb900015

Smith, K.A., Conen, F. (2004) Impacts of land management on fluxes of trace greenhouse gases. Soil Use Manage. **20**, 255–263. doi:10.1079/sum2004238

Smith, P., Martino, D., Cai, Z., Gwary, D., Janzen, H., Kumar, P., McCarl, B., Ogle, S., O'Mara, F., Rice, C., Scholes, B., Sirotenko, O., Howden, M., McAllister, T., Pan, G., Romanenkov, V., Schneider, U., Towprayoon, S., Wattenbach, M., Smith, J. (2008) Greenhouse gas mitigation in agriculture. Pholos. T. Roy. Soc. B. **363**(1492), 789–813. doi:10.1098/rstb.2007.2184

Smith, P., Martino, D., Cai, Z., Gwary, D., Janzen, H., Kumar, P., McCarl, B., Ogle, S., O'Mara, F., Rice, C., Scholes, B., Sirotenko, O. (2007) Agriculture. In: Metz, B., Davidson, O.R., Bosch, P.R., Dave, R., Meyer, L.A. (eds.) Climate Change 2007: Mitigation. Contribution of Working Group III to the Fourth Assessment Report of the Intergovernmental Panel on Climate Change. pp. 487–540. Cambridge University Press Cambridge, United Kingdom and New York, NY, USA

Snyder, C.S., Bruulsema, T.W. (2007) Nutrient Use Efficiency and Effectiveness in North America: Indices of Agronomic and Environmental Benefit International Plant Nutrition Institute. http://www.ipni.net/ipniweb/portal.nsf/0/D58A3C2DECA9D7378525731E006066D5

Snyder, C.S., Bruulsema, T.W., Jensen, T.L., Fixen, P.E. (2009) Review of greenhouse gas emissions from crop production systems and fertilizer management effects. Agric. Ecosys. Environ. **133**(3-4), 247–266. doi:10.1016/j.agee.2009.04.021

Spokas, K.A., Baker, J.M., Reicosky, D.C. (2010) Ethylene: potential key for biochar amendment impacts. Plant Soil **333**(1-2), 443–452. doi:10.1007/s11104-010-0359-5

Stehfest, E., Bouwman, L. (2006) N_2O and NO emission from agricultural fields and soils under natural vegetation: summarizing available measurement data and modeling of global annual emissions. Nutr. Cycl. Agroecosys. **74**(3), 207–228. doi:10.1007/s10705-006-9000-7

Strock, J.S., Porter, P.M., Russelle, M.P. (2004) Cover cropping to reduce nitrate loss through subsurface drainage in the northern US Corn Belt. J. Environ. Qual. **33**(3), 1010-1016

Tenuta, M., Beauchamp, E.G. (2003) Nitrous oxide production from granular nitrogen fertilizers applied to a silt loam. Can. J. Soil Sci. **83**(5), 521–532

Tilman, D., Fargione, J., Wolff, B., D'Antonio, C., Dobson, A., Howarth, R., Schindler, D., Schlesinger, W.H., Simberloff, D., Swackhamer, D. (2001) Forecasting agriculturally driven global environmental change. Science **292**(5515), 281–284. doi:10.1126/science.1057544

U.S. Environmental Protection Agency (US EPA) (2011) Inventory of U.S. Greenhouse Gas Emissions and Sinks: 1990-2009. United States Environmental Protection Agency (US EPA)

Ussiri, D.A.N., Lal, R., Jarecki, M.K. (2009) Nitrous oxide and methane emissions from long-term tillage under a continuous corn cropping system in Ohio. Soil Till. Res. **104**(2), 247–255. doi:10.1016/j.still.2009.03.001

Van Groenigen, J.W., Velthof, G.L., Oenema, O., Van Groenigen, K.J., Van Kessel, C. (2010) Towards an agronomic assessment of N_2O emissions: a case study for arable crops. Eur. J. Soil Sci. **61**(6), 903–913. doi:10.1111/j.1365-2389.2009.01217.x

Velthof, G.L., Kuikman, P.J., Oenema, O. (2003) Nitrous oxide emission from animal manures applied to soil under controlled conditions. Biol. Fert. Soils **37**(4), 221–230. doi:10.1007/s00374-003-0589-2

Venterea, R.T., Burger, M., Spokas, K.A. (2005) Nitrogen oxide and methane emissions under varying tillage and fertilizer management. J. Environ. Qual. **34**(5), 1467–1477. doi:10.2134/jeq2005.0018

Venterea, R.T., Rolston, D.E. (2002) Nitrogen oxide trace gas transport and transformation: II. Model simulations compared with data. Soil. Sci. **167**(1), 49–61. doi:10.1097/00010694-200201000-00005

Venterea, R.T., Stanenas, A.J. (2008) Profile analysis and Modeling of reduced tillage effects on soil nitrous oxide flux. J. Environ. Qual. **37**(4), 1360–1367. doi:10.2134/jeq2007.0283

Vitousek, P.M., Aber, J., Howarth, R.W., Likens, G.E., Matson, P.A., Schindler, D.W., Schlesinger, W.H., Tilman, G.D. (1997) Human alteration of the global nitrogen cycle: Causes and consequences. Issues in Ecology **1**, 1–15

Vitousek, P.M., Farrington, H. (1997) Nutrient limitation and soil development: Experimental test of a biogeochemical theory. Biogeochemistry **37**(1), 63–75. doi:10.1023/a:1005757218475

von Arnold, K., Nilsson, M., Hanell, B., Weslien, P., Klemedtsson, L. (2005a) Fluxes of CO_2, CH_4 and N_2O from drained organic soils in deciduous forests. Soil Biol. Biochem. **37**(6), 1059–1071. doi:10.1016/j.soilbio.2004.11.004

von Arnold, K., Weslien, P., Nilsson, M., Svensson, B.H., Klemedtsson, L. (2005b) Fluxes of CO_2, CH_4 and N_2O from drained coniferous forests on organic soils. For. Ecol. Manage. **210**(1-3), 239–254. doi:10.1016/j.foreco.2005.02.031

Wagner-Riddle, C., Thurtell, G.W. (1998) Nitrous oxide emissions from agricultural fields during winter and spring thaw as affected by management practices. Nutr. Cycl. Agroecosys. **52**(2-3), 151–163. doi:10.1023/a:1009788411566

World Meteorological Organization (WMO) (2011) The State of Greenhouse Gases in the Atmosphere Based on Global Observations through 2010. World Meteorological Organization Greenhouse Gases Bulletin No. 7. World Meteorological Organization. http://www.wmo.int/gaw

Woods, S., Henao, J., Rosegrant, M. (2004) The role of nitrogen in sustaining food production and estimating future nitrogen fertilizer needs to meet food demand. In: Mosier, A.R., Syers, J.K., Freney, J.R. (eds.) Scope 65: Agriculture and the nitrogen cycle: Assessing the impacts of fertilizer use on food production and the environment. pp. 245–260. Island Press, Washington, DC, USA

Yanai, Y., Toyota, K., Okazaki, M. (2007) Effects of charcoal addition on N_2O emissions from soil resulting from rewetting air-dried soil in short-term laboratory experiments. Soil Sci. Plant Nutr. **53**(2), 181–188. doi:10.1111/j.1747-0765.2007.00123.x

Chapter 9
Mitigation Options for Livestock and Pasture Lands

Contents

Abstract Nitrous oxide emissions from grazed pastures represent a significant source of atmospheric N_2O. Wastes from animal production systems contribute as much as 30–50% to the global N_2O emissions from agriculture. Improved understanding of sources and N transformation processes, and soil and climatic conditions controlling N_2O emissions under livestock production system helps in identifying management options to reduce N_2O emissions from grazed pasture systems. The mitigation options for livestock production systems includes optimum soil and grazing land management, limiting the amount of N fertilizer or effluent applied when soil is wet, animals dietary management to decrease the amount of N excreted in animal urine through feeding low-N feed supplements as an alternative to fertilizer N boosted grass. In addition, plant and animal selection for increased N use efficiency, using N transformation inhibitors to control urea to ammonium and ammonium to nitrate conversions in soil, as well as use of stand-off/feed pads or housing systems during high-risk periods of N loss are important strategies to minimize N_2O emissions from livestock system. The use of single or multiple

David Ussiri and R. Lal, *Soil Emission of Nitrous Oxide and its Mitigation*,
DOI 10.1007/978-94-007-5364-8_9,

mitigation options always needs to be evaluated in a whole farm system context and accounting for total greenhouse gas emissions including methane and carbon dioxide. Whole-system life-cycle-based environmental analysis should also be conducted to assess overall environmental emissions associated the N_2O mitigation options. The overall focus should be ensuring tighter N cycling and increasing N use efficiency through decreasing N losses per unit of animal production.

Keywords Animal number • Animal waste • Urine and dung • Global budget • Nitrogen excretion • African rangelands

Abbreviations

BNF	Biological nitrogen fixation
SOM	Soil organic matter
GHG	Greenhouse gas
CEC	Cation exchange capacity
SON	Soil organic nitrogen
DOC	Dissolved organic carbon
OM	Organic matter
GNP	Gross national product
DCD	Dicyandiamide
UI	Urease inhibitor
NI	Nitrification inhibitors
DM	Dry matter
CP	Crude protein

9.1 Introduction

World population reached the seven billion mark in 2011, and will to increase to eight billion in 2025 (FAOSTAT 2011b). Population pressure, technological change, economic growth, and cost/price squeeze have been the major drivers of change in the agricultural sector since 1960s. With continuous decline in the availability of land area for expansion of crop production, the increase in food demand will also necessitate intensive animal production, especially in North and South Americas, Australia, New Zealand, China, and India. Thus, indigenous swards are being replaced with potentially more productive pastures (Hopkins et al. 1999) and mineral nutrient fertilizers, particularly nitrogen (N) and phosphorus (P) are used at economic optimum to increase pasture productivity (Andrews et al. 2007). For example, N application rates ranging from 100 to 689 (mean 281) kg N ha^{-1} in England and Wales are common (Jarvis 2000). Use of inorganic N fertilizers at such levels is highly inefficient especially in the intensive dairy systems where only about 20% of N input from animal feeds can be recovered from products – milk and meat (Jarvis 1993, 2000). The consequent high rates of N cycling within grassland farm through animal excreta, manures/slurries, and soil organic matter (SOM) eventually transported into aquatic ecosystems and the atmosphere with severe environmental impact. Therefore, it is necessary that livestock production is environmentally sustainable.

Grasslands worldwide cover about one quarter of earth's surface, occupying about 117 million km^2 of vegetated land. They also provide forage for more than 1,800 million livestock units and wildlife (WRI 2000). Globally, natural and managed grasslands contribute to the livelihood of over 938 million people including smallholders – about 17% of the world's population (WRI 2000; Reynolds et al. 2005). Managed pastures are generally highly productive, leading to higher animal productivity per unit area. Managed pastures produce about 27% of world milk and 23% of beef production (Sere et al. 1995). Sustaining productivity of managed pasture depends on adoption of innovative practices, including grazing intensity and frequency, fertilizer and manure application, and pesticide use. Pastures occupy about 34 million km^2 of the land surface (FAOSTAT 2011a).

Grasslands are also key sources of N_2O emissions. The global N_2O emission from grasslands is estimated at 2.7 Tg N_2O-N $year^{-1}$, comprising 18% of total anthropogenic N_2O emissions (Lee et al. 1997). Emissions of N_2O from grazed grasslands are largely a result of two microbial processes of nitrification and denitrification. The simultaneous occurrence of high soil nitrates (NO_3^-) and low soil aeration (i.e. wet or compacted soils) results in high N_2O emissions (Eckard et al. 2003). Thus, mitigation technologies are likely to be most effective if they reduce the availability of soil NO_3^- and/or improve soil aeration.

Grazing managed pastures is a major system of livestock production such as sheep, beef, and dairy cattle in many countries. Being an essential element in animal production, large quantities of N are required for growth of feed crops. Plant N, primarily in the form of protein, is an essential feed component for animal growth and development. In grazed pastures, N is derived from biological N fixation (BNF) of atmospheric N_2, mineralization of soil organic N (SON), application of N fertilizers, manures and farm effluent, and uneven deposition of animal excreta (urine and dung). Animals concentrate and excrete most of the N consumed, providing manure nutrients needed for plant growth. There has been a steady increase in N inputs to grazed pastures, and this trend will most likely continue in the future. The main problem in the cycling of N in the animal production is that large losses normally occur that contribute to the degradation of the environment. Increased fertilizer N inputs along with continued high intake of animal proteins in developed countries and changes in diets of people in developing countries are likely to exacerbate the N losses from global food production (McCarl and Schneider 2000; Mosier et al. 2001; Bleken et al. 2005). The challenge is to manage the animals, pastures and all farm components to efficiently use available N and reduce the potential loss to the environment.

Grazed grasslands are major source of N_2O emission due to large N input and rapid recycling of C and N from the animal excreta (Jarvis et al. 1995). Animal excreta (urine and dung), voided in grazed pastures and in animal housing/shelters, applied to the land is the main source of N_2O (Jarvis and Pain 1994). The major source of off-farm N_2O include ammonia (NH_3) volatilization and NO_3^- leaching from farm effluent/manure and animal excreta in grazed pastures. Oenema et al. (2005) suggested that 70% of global N_2O emissions from livestock production originate from urine and dung patches deposited during grazing. Because of all potential sites, substrates, and sources in grassland systems, the emission per ha are generally higher than those from arable land and forest soils (Oenema et al. 1998). Estimating the contribution of animal agriculture is difficult, since it requires significant data on animals and dietary composition, feed intake, manure/effluent storage and management, farm management systems, and N fertilizer usage in pasture production (Clark et al. 2005). However, N_2O emissions from N excreted as animal waste alone could be as much as 30–50% of the global agricultural N_2O emissions (Oenema et al. 2005). Thus, that the contribution of animal agriculture to total agricultural N_2O emission could be as much as 50% when including emission from fertilizer use in pasture. Field measurements indicate that high N_2O emission rates generally coincide with soil conditions that are conducive to denitrification (i.e., anaerobic environment, NO_3^-, and mineralizable C supply). This trend suggests that denitrification is the main source of N_2O supply in intensive animal production system. Nitrification is often an essential prerequisite for denitrification to convert N inputs from urine and urea or ammonium based fertilizers into NO_3^-. In animal production, the coincidence of high soil NO_3^- levels and low soil aeration such as wet and/or compacted soils, in particular, results in high N_2O emission (Eckard et al. 2003). Grazed pastures are also a significant source of NH_3 emissions and NO_3^- leaching.

Nutrient management in grassland agriculture must address multiple criteria including air, water quality, nutrient use efficiency, and animal farm economics. Effective technologies for mitigating N_2O emissions from animal agriculture can be outlined at three management levels of influence, namely: (i) manipulating soil N transformations–including management of fertilizer and manure through nitrification inhibitors (NIs), optimum N application rates, timing of N application and soil management; (ii) choice of plant type and time of grazing, (iii) animal management–feeding system management, low N supplementary feed, restricted grazing and animal diet manipulation.

9.2 Nitrogen Input to Grazed Pasture

N is often limiting in grassland ecosystems, and mineral and organic N fertilizers are frequently applied to increase herbage productivity and improve/maintain soil fertility of pasturelands. The major sources of plant-available N in grazed grasslands are animal excreta (urine and dung), BNF, N fertilizers, animal manure, atmospheric N deposition, and mineralization of soil organic N (Fig. 9.1). The contribution from atmospheric deposition is small compared to other sources. For example, deposition of N in areas of USA away from local NH_3 sources is typically

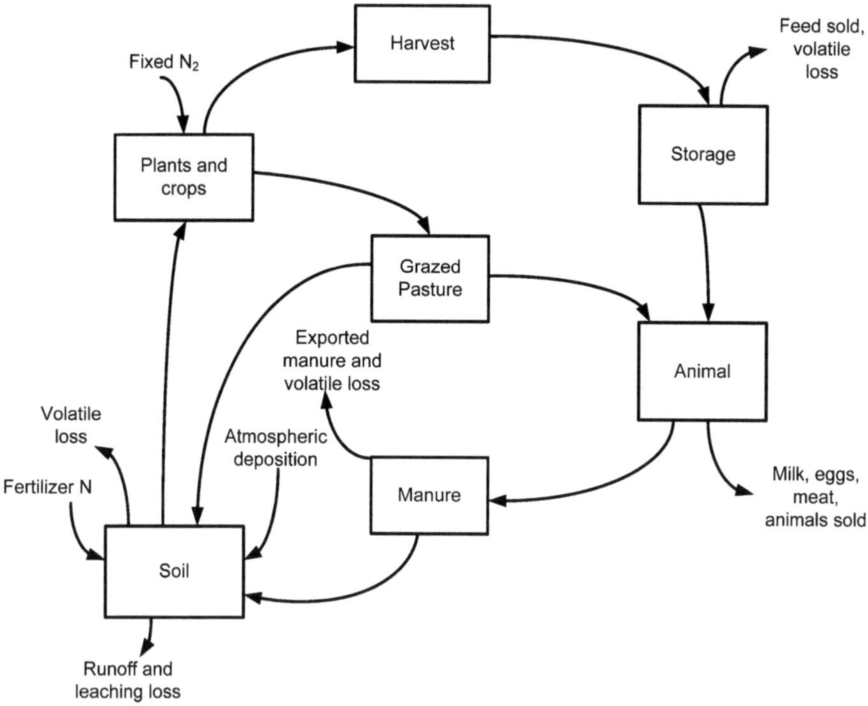

Fig. 9.1 Major N flows and cycling within and between farm and its environment

1–6 kg N ha^{-1} year^{-1}. In the animal production systems, type and number of animals, N excretion per animal, and the management of animal wastes are the major determinants of the total N_2O emissions (Oenema et al. 2005). Most of N_2O emissions from livestock production systems originate from the microbial transformations of N in the animal excreta—urine and dung of grazing animals in pastures, animal wastes of confined animals during storage, management and following application or disposal to land.

The major processes producing N_2O emissions from soils are nitrification (aerobic transformation of NH_4^+ to NO_3^-) and denitrification (anaerobic transformation of NO_3^- to N_2, Chap. 3). The poor quality of data and lack of reliable method of calculating the estimates and up-scaling results in large uncertainty about the size of N_2O emissions that are associated directly with animal production systems (Oenema et al. 2005). However, increased direct N_2O emissions from livestock production systems have been associated with increased livestock population, N fertilization of pasture to sustain increased livestock, N in animal manure, and increased production of leguminous pastures resulting into enhanced BNF. The indirect N_2O emissions from animal wastes are NH_3 and NO_x volatilization that escaped from wastes or via N leaching and runoff, and then transformed into N_2O in other locations (Oenema et al. 2005).

9.2.1 Biological Nitrogen Fixation

Use of legume-based or mixtures of legumes and sward in managed pastures is common in many areas including Europe, North America, Australia and New Zealand (Bolan et al. 2004b). In such pastures, N is mainly derived from the fixation of the atmospheric N_2 by a group of bacteria living in the root nodules of the legume plants known as "rhizobium." The quantity of N fixed by BNF depends on variety of factors such as legume species, presence of microorganisms, pasture growing conditions, soil and climatic conditions–including nutrient supply and soil pH, and grazing management (Hansen and Vinther 2001). The BNF rates in pastures ranging from 100 to 300 kg N ha^{-1} $year^{-1}$ are common for grass-clover pastures (Ledgard et al. 1990). With appropriate management and utilization of recommended cultivars, pasture production from a perennial ryegrass/white clover is similar to that receiving N at 200 kg ha^{-1} $year^{-1}$ and about 70% of that obtained with N input at 350 kg ha^{-1} $year^{-1}$ (Andrews et al. 2007). High levels of available P in soils are necessary for maintenance of N fixing activity of legumes and N inputs in these pastures. Similarly, adequate levels of other nutrients especially sulfur (S) and molybdenum (Mo) in particular, are required for effective BNF (Bolan et al. 2004b).

BNF is naturally regulated. In soils where inorganic N is high, legumes tend to utilize soil N, which results in reduced BNF activity. Therefore, addition of fertilizer N to legume pastures often decreases BNF (Ledgard et al. 1996). An annual decrease in fixed N of up to 75% have been reported (Ledgard et al. 2001). For each kg of fertilizer N applied, the BNF may decrease by between 0.3 and 0.7 kg of fixed N (Ledgard et al. 2001). Decrease in BNF due to fertilizer N application is attributed to: (1) inhibition of infection of legume roots by nodule bacteria, (2) inhibition of nitrogenase enzyme activity in the nodule due to modification of the nitrogenase protein, (3) decrease in bacterial membrane potential and the inhibition of the leghaemoglobin (i.e., an oxygen carrier and a hemoprotein found in the nitrogen-fixing root nodules of leguminous plants), (4) decrease in the supply of photosysthate to the rhizobium due to the assimilation of mineral N in the shoot, and (5) decrease in legume growth with fertilizer N application mainly due to an increased competition by the grass (Bolan et al. 2004b). In a clover-grass mixture pasture, net N mineralization can cause the grass to out-compete the clover or reduce clover nodulation. After sufficient mineral N has been removed from the soil system, the clover often re-colonizes the pasture. In addition, in intensively grazed systems, the excreta N alters soil mineral N status and may influence BNF.

The fixed N becomes available slowly over time to the grass in pastures after it is released into soil via exudates from living legume roots, by mineralization of senesced legume tissues and in excreta after consumption by grazing animals. The self-regulating nature of BNF in legume-based pastures and subsequent N transformation processes contributing to NH_3 volatilization, NO_3^- leaching, and N_2O emission suggest that BNF is less likely than fertilizer and excreted N to cause significant N losses to the environment.

9.2.2 Animal Excreta Deposition in Pasturelands

Globally, about 80–130 Tg N year^{-1} are produced as manures from cattle, sheep, goats, pigs and chicken (Oenema and Tamminga 2005; Oenema 2006). This amount is similar in magnitude to that of current global inorganic N fertilizer production (Canfield et al. 2010). Cattle account for 56% of manure N, swine for 11%, and poultry for 9% (Oenema 2006). Generally, pasture plants require significantly higher N content for optimal growth than that needed for protein synthesis of grazing ruminants. Ruminants utilize relatively less N in feed, and can excrete between 75 and 90% of the ingested N (Whitehead 1995). The low N utilization reflects the relatively high concentrations of N in pasture plants required for metabolic functions and optimum growth compared to that needed by grazing ruminant for amino acid and protein synthesis (Haynes and Williams 1993). Increasing the N concentration of the animal diet generally increases the excretion of urinary N in both absolute terms and as a percentage of total N excreted by the animal.

Manures in animal housing and storage, and urine and feces deposited onto the soils during grazing are significant source of N_2O emissions under managed grasslands. The proportion of total N intake excreted and its partition between urine and feces depends on animal type, dry matter intake, and N concentration of the diet. For example, cattle retain up to 20% of the total N intake via fodder and animal feeds in animal products such as milk and meat. The remaining N intake is excreted in urine and feces. Sheep and dairy cattle excrete 70–75% and 60–65% of N in urine, respectively, when grazing N-rich grass/legume pastures (Oenema et al. 1997). However, the concentration of N in urine may vary from 1 to 20 g N L^{-1} depending on N content of the diet and volume of water consumption. The proportion of N in urine present as urea increases with an increase in N intake. Over 70% of N in urine is urea, which is rapidly hydrolyzed in soil to NH_4^+ within 3–5 days, and subsequently nitrified to NO_3^-, the rest consists of amino acids, and peptides (Haynes and Williams 1993; Bolan et al. 2004b). However, the bulk of N in feces occurs in organic form. About 20–25% of fecal N is water-soluble, 15–25% is undigested dietary N and the remaining 50–65% is in the form of bacterial cells (Oenema et al. 1997). This organic N is not as readily available for N_2O production as urea N of urine, because the rate of mineralization to NH_4^+ is generally much slower than urea hydrolysis. Animal urine patches are the dominant substrate for N_2O production in grazed systems due to highly localized concentrations of readily available N in these patches.

Grazing animals deposit excreta unevenly across pasturelands. The loading rate under a urine patch can be as high as 1,000 kg N ha^{-1}, which is well in excess of that which can be taken up by the pasture in a growing season (Haynes and Williams 1993). The large surplus N and return of N in localized patches of excreta at high N input rates under intensively grazed pasture systems increases the risk of N loss to waterways and atmosphere. Depending on stocking rate, cattle urine patches covers between 20 and 30% of the grazed pasture area per year. Within the patches, the availability of NH_4^+ and NO_3^- from excreta N, and the factors that alter the redox potential of soil, such as changes in soil moisture, soil texture, and readily oxidizable organic C plays major role in emissions of N_2O from grazed pastures. The excretal returns, particularly urinary N from grazing ruminants is typically the major source of N lost from pastures.

9.2.3 Animal Manures and Effluent Application in Pasturelands

Increased production of livestock and poultry products for human consumption leads to increased manure by-products from these industries, which need to be dealt with to abide with the environmental regulations – including safe disposal onto land. Confined animal production – beef cattle, poultry, and swine are the major sources of manure by-products in many countries. For example, about 45.4 Tg of dairy and beef manure, and 27 Tg of poultry and swine manure are produced in USA annually, which generates 7.5 Tg N $year^{-1}$ (Bolan et al. 2004a). Manure by-products can be used on agricultural land to alter soil properties, such as nutrient availability, soil reaction, soil organic matter (SOM) content, cation exchange capacity (CEC), and water holding capacity. Application of manures to pasture and crop lands increases dry matter yield and enhances nutrient status of soils (Cameron et al. 1997).

Environmental concerns associated with the land application of manures from confined animal industry encompass all aspects of non-point source pollution, including leaching losses of N to subsurface drainage and ground water contamination, reduced air quality by odors and emission of volatile organic compounds, increased toxic metals input such as As, Cu, Zn and Ni, and emission of greenhouse gases (GHG). The optimum use of these animal by-products for land application requires knowledge of their composition for both beneficial nutrients such as N and P and the environmental implications. Maintaining the environmental quality is a major consideration when developing management practices to use manure by-products effectively as a nutrient source and soil conditioner in agricultural production system (Sharpley et al. 1998).

9.2.4 Inorganic Nitrogen Fertilizer Application in Pasturelands

Grasslands are inherently deficient of nutrients, the amount of N mineralized from organic matter cannot support potential pasture production, and fertilizer application is generally used to increase grassland productivity for pasture. The high-yielding improved pasture species used in managed ecosystems are generally adapted to high-fertility soil conditions and may not perform well in less fertile soils. N fertilizers are widely used in grass-based intensive managed pasture production as the effective management manipulation to increase productivity in North America, Europe, New Zealand, and Australia. Increased use of N fertilizers in grazed pastures has been attributed to: (i) need for extra feed throughout the year to support the increased stocking rate, achieve early calving, extend lactation later into autumn and make high quality silage to feed later in the lactation, (ii) feed obtained from N fertilizer application used to replace more expensive protein feed supplements, and (iii) the productivity and the profitability of the farm increased by fertilizer N application (Bolan et al. 2004b). Pure grass pastures often respond linearly up to 200–400 kg N ha^{-1} $year^{-1}$, and application rates in these ranges are rather common (Whitehead 1995), although the average application rates vary with the production objectives. Where pastures are cut and removed for off-site feeding, large quantities of nutrients are removed and the optimum N rates can exceed those under grazed swards, where N is returned to pasture in the form of animal excreta. In addition, the N_2O fluxes differ between the cut-and-carry and grazed grassland systems.

In legume-based pastures, a small amount of fertilizer N is added mainly during winter/spring period to increase production during this period of low growth and high requirement of pastures by lactating cows. During this period, BNF by legumes and rate of organic matter (OM) mineralization is very low, and levels of mineral N are not adequate to meet N demand for plant uptake, hence pasture respond to N fertilizer application (Bolan et al. 2004b). The principles of N use and N_2O emissions in grasslands are more complex than those for the croplands because of: (a) high N demand throughout the 6–9 months growing season, (b) differing sward age and botanical composition (i.e. annuals, short- and long-term, permanent pasture, pure grass, mixtures or legume based), (c) varying soil and climatic conditions, and (d) differing grazing management–rotational, set stocked, different habits of grazing animals–dairy, beef cattle, sheep, sheep and beef mixture. All these factors affect pasture management, including fertilizer N requirement, and tend to influence the N_2O emissions from grasslands.

Managed temperate grasslands have a high demand for N for plant growth and mixed swards of perennial ryegrass and white clover are widely used in temperate grasslands. N is the major nutrient element that mostly regulates pasture production. N is also a major contributor to environment degradation. The major challenge facing farmers is to maximize farm productivity while limiting the release of reactive N into the environment. In highly productive managed grassland soils, most of N_2O is generated from mineral N originating from applied fertilizer N, urine, dung, farm effluents applied primarily as a means of disposal, biologically fixed N and mineralization of SON. In pastures where open grazing is practiced, the uneven deposition of animal excreta in these lands causes a large spatial and temporal variability in soil physical integrity, N availability, and N and C transformations, creating large uncertainties in N_2O emissions.

9.3 Nitrogen Transformations and Loss from Pasture Soils

The primary N transformation reactions controlling N cycling in soils include: (i) mineralization, (ii) nitrification, (iii) denitrification, (iv) immobilization, (v) NH_3 volatilization, (vi) NH_4^+ fixation, and (vii) NO_3^- leaching. The first four

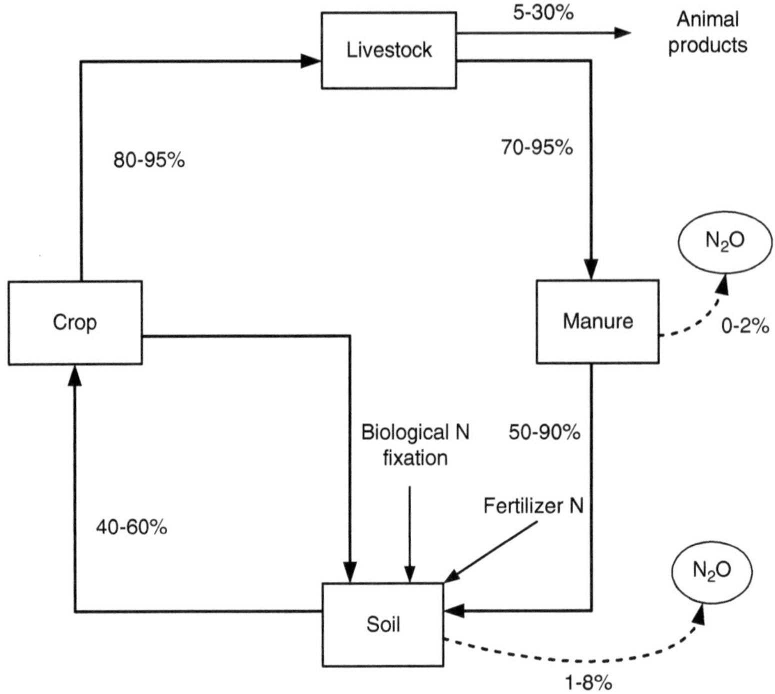

Fig. 9.2 Nitrogen cycling in animal production system (Modified from Oenema et al. 2005)

are the biotic processes involving soil microorganisms, while the last three are the abiotic physico-chemical processes. The only acceptable N loss from animal farm is via products such as milk, meat, eggs, wool (Fig. 9.2). Retention of consumed feed nutrients in ruminant animals is low, ranging from 5 to 30%. Lower retention rates occur in animal tissues and higher rates in milk.

The bulk of N in urine and feces is in organic forms, and it must undergo microbial transformation before it is released as mineral N. The mineralization process involves the conversion of plant unavailable organic N forms into inorganic forms by soil microorganisms. The process involves ammonification, where organic N, including undigested ingested herbage N, urine N, amines, and amino acids are converted by microbial processes to NH_4^+ ions. The mineralization of N from feces is slower than that from urine. The NH_4^+ ions formed through ammonification is subjected to several fates in the soil, including (i) uptake by plants, (ii) nitrification, i.e. conversion to NO_2^- and NO_3^- (Chap. 3), (iii) utilization by microorganisms, i.e. immobilization, (iv) retention onto soil, i.e. NH_4^+ fixation, (v) loss through NH_3 volatilization. Higher dissolved organic carbon (DOC) availability and possible nitrate formation are two conditions, which support denitrification in urine patches (Chantigny 2003).

Nitrates formed through nitrification or added as nitrate fertilizer is subjected to various processes including: (i) plant uptake, (ii) leaching losses, (iii) immobilization, and (iv) denitrification. Immobilization is a biological process in which plant available NH_4^+ and NO_3^- ions are converted to plant unavailable organic N. In arable soils, the addition of carbon-rich substances such as maize stubble and cereal straw promote the immobilization and reduce N availability to plants. Similarly, in grazed pastures the addition of silage promotes immobilization of N. Soil microbes play an important role in decomposition of plant residues, and the rate of decomposition and release of plant nutrients depends on quality of the residues. The quantity of C relative to N (i.e., C:N ratio) of the decomposing OM is the indication of the likelihood of N shortage and competition between microbes and higher plants for the available N in the soil. Microbes tend to maintain a low C:N ratio of about eight in their body, and during the decomposition of OM they need one unit of N for every eight units of C assimilated. Therefore, when plant residues with high C:N ratios are added to the soils, the lack of N inhibits the decomposition, or the microbes compete for available N from soil, resulting in a deficiency of N for the subsequent crop. In contrast, plant residues with low C:N ratio induce the mineralization of N from plant residues, thereby increasing the available N in the soil.

Ammonium fixation occur when NH_4^+ ions are retained on negatively charged mineral soil surfaces and organic particles by cation exchange reactions or fixed in the interlayers of 2:1 phylosilicate clay minerals such as mica, vermiculite and illite. Being a cation, NH_4^+ is strongly retained into soil matrix, whereas NO_3^- is weakly adsorbed and easily moves with water and subsequently lost from rooting zone. Leaching losses of NO_3^- to waterways and emissions of NH_3 and N_2O to the atmosphere from grazed pastures have significant environmental implications (Di and Cameron 2002). The NO_3^- leaching from grasslands increases with N input, whether that input is from BNF, animal excreta or N fertilizer (Ledgard 2001). Sward uptake plays a significant role in reducing soil solution NO_3^- concentration. Therefore, losses increase during slow growth periods such as during winter and after dry periods.

The amount of nutrient loss often increases with production intensity, because of inherent limits to the efficiency of production. In addition, when N loss by one pathway is reduced, losses by one or more remaining pathways may increase, because N transformations and losses are concentration-dependent (Brink et al. 2001). Once production is maximized in systems near equilibrium with respect to soil N, all further N inputs are lost to the environment. The major soil N reservoir is OM. Under productive grassland, this pool of N mineralizes faster than under cropped land. Livestock accelerate nutrient cycling directly through decomposition and excretion of plant-derived nutrients and indirectly through the effects of grazing and excreta on soil biota (Bardgett and Wardle 2003).

9.4 Rangelands

Rangelands are lands dominated by self-propagating vegetation comprised predominantly of grasses, grass-like species (i.e., forbes) with or without scattered woody plants (i.e., shrubs), occupying between 18 and 23% of world land area excluding

Antarctica (Blench and Sommer 1999). Several terms are normally used to describe world's rangelands including African savanna, Eurasian steppe, South American savanna, North American prairies, Indian savanna, and Australian grasslands (Groomsbridge 1992; Moore 1970; Solbrig 1996). Rangelands are broader term than grasslands, and include regions where woody vegetation may dominate. They are lands generally characterized by low and/or erratic precipitation, poor drainage, and often-low soil fertility. Rainfall is the main environmental factor determining the plant survival (Ludwig et al. 1997) and productivity of the rangelands in arid and semi-arid climates (Snyman and du Preez 2005).

Rangelands are home to significant concentrations of large mammals and plants with high value to human populations that historically are excluded and marginalized, including pastoralists, and hunter-gatherers (Blench and Sommer 1999). Rangelands provide fodder for ~360 million cattle, >600 million sheep and goats, and are source of 9% of global beef and 30% of sheep and goat meat (de Haan et al. 1997). They provide livelihood for ~100 million people in arid regions of the world where livestock keeping is the only possible source of food (de Haan et al. 1997). The majority of world's rangelands are anthropic creations, particularly where dominant subsistence strategy is pastoralism. Thus, they do not have a natural biodiversity, making problematic the argument that they should either be preserved as they are, or returned to their original state. Where large mammals are involved as part of biodiversity protection, the emotion has frequently triumphed over science in terms of management and investment strategies. Similarly, where powerful economic interests of large-scale ranching predominates the rangeland management, biodiversity generally becomes less important or completely ignored.

Typical inhabitants of anthropic rangelands are pastoralists, hunter-gatherers, and increasingly subsistence farmers depending on uncertain rainfed crops or irrigating semi-arid land from non-rechargeable water sources (Blench and Sommer 1999). Due to increasing population pressure and demand for natural resources, these groups compete for a shrinking land resource. Globally subsistence farmers are increasingly displacing the more marginal groups – the hunter-gatherers and pastoralists. Fire, rainfall (plant available moisture), soil type, plant available nutrients, and grazing animals are the driving forces determining plant species composition, distribution, and productivity of these land areas. The combination of these factors prevents the establishment and growth of trees and other woody plants in high densities (Babier et al. 1994; Solbrig 1996) although their significance varies in different parts of the world. The South African savannah, North and South American rangelands are good examples of climate and soil favoring production of grass and herbaceous species rather than trees. In some rangelands some wild grazing animal species co-evolved with the rangelands, for example, savannahs of Africa (antelopes and zebras), steppes of Asia and Eastern Europe (gazelles, goats, camel, bison and wild horses); and prairies of North America (deer and bison).

Currently livestock production dominates the world's rangelands, and all other forms of land uses are of minor importance (Solbrig 1993). In North America,

Europe, and Australia, livestock production has been intensified through the application of new technologies and improved practices, including the use of fertilizers, the seeding of high yielding grass and legume species, and modification of natural water regime through irrigation during dry season and heavy grazing through high stocking rates (Solbrig 1996). These practices are transforming rangeland ecosystems by focusing biomass production towards the need for particular species. The management of rangelands varies from nomadic pastoralism, subsistence farming, commercial ranching, and wildlife. Non-sustainable land use practices such as inappropriate plowing, overgrazing and excessive firewood use are the main causes of degradation of rangeland ecosystems (Ojima et al. 1993, 1994). Overgrazing is the most important cause of rangeland degradation. Sustained grazing may also alter the botanical composition of semi-arid rangelands from long-lived perennials to annuals or short-lived perennials, with a concomitant decrease in production (Ludwig et al. 1997; Oconnor 1994; Wiegand et al. 2004) and an increase in its variability over time (O'Connor et al. 2001). Severe grazing may reduce both aerial and basal cover of rangeland (Wiegand et al. 2004). The ecologically sensitive semi-arid areas are increasingly subject to severe grazing pressure (Oesterheld et al. 2001), which cause their rapid deterioration (Oconnor 1994).

9.4.1 *African Rangelands*

African rangelands (i.e., grasslands, savannas, and woodlands which contains both grasses and woody plants) cover ~2.1×10^7 km^2, and support livestock population of about 184 million cattle, 3.72 million small ruminants (sheep and goats), and 17 million camels which extract ~80% of its nutrition from these rangelands (IPCC 1996). Africa keeps about 14 and 21% of world's cattle and small ruminants, respectively, on a land base comprising of 25% of world's total area of rangeland. The number of people engaged in extensive livestock production in Africa is higher than any other region in the world. About 30–40 million pastoralists rely extensively on livestock production for most of their income and subsistence needs, and more than half reside in Africa. Although commonly used for nomadic and transhumant pastoralism, African rangelands contains by far the widest variety of extant large and medium sized herbivores. In addition to livestock, African rangelands support a vibrant tourist industry, one of the leading contributors to gross national product (GNP) in many African countries. Because of growing population (2.6% annually), African rangelands have been the area of intense human and animal conflict.

From land use perspective, there are differences between West Africa and East Africa in the way rangelands are used. In West Africa, rangelands are arid and semi-arid with a unimodal (i.e., one peak) annual rainfall ranging between 5 and 600 mm per annum. These areas are normally cultivated with millet during rainy season, which lasts for 3–4 months, then the fields remain fallow for the rest of the dry season where livestock eat all crop residues. In this region, cultivation dominates the land use, with the livestock playing subsidiary role in village economy (Ellis and Galvin 1994). In contrast, rainfall is bimodal (i.e., two peaks per year) in East Africa, and areas with up to 600 mm rainfall are normally inhabited by pastoral nomads, completely dependent on livestock for food (Galvin 1992; Thornton et al. 2003). Cultivation is uncommon, and occurs mainly where irrigation or water conservation for cropping is possible (Boone et al. 2002; Ellis and Galvin 1994). This rainfall pattern has important implication for natural vegetation and rain-fed agriculture. The best years for livestock production are generally not those with greatest rainfall, rather, those in which moderate rainfall extends over several months resulting in longer period of forage and milk production.

Overgrazing is a common cause of rangeland degradation in Africa. When the productivity of rangeland is overestimated, the resultant overgrazing causes decrease of palatable perennial plants in favor of less palatable undesirable vegetation (Snyman 2004). Such changes usually lead to soil compaction, reduced soil aggregates stability, decreased soil fertility and SOM content. The N_2O emission is part of N cycling and is subject to increase with alteration of N cycle through land management and land use changes (Ojima et al. 1994). There are no reliable estimates for N_2O emissions from African rangelands. Anthropogenic activities which increase N_2O emissions from African rangelands include: (i) wild fires, (ii) animal excreta.

In Africa, N input from synthetic N fertilizers is among the lowest in the world, representing only 3% of the world fertilizer consumption (FAO 2011). The African continent and Sub Saharan Africa in particular, makes small contribution to global greenhouse emission from agriculture. Africa has been estimated to produce only 15% of global agriculture soil N_2O emissions and 3% of emissions from animal production. The largest source of agricultural N_2O emissions is livestock, where combined emissions from paddocks, ranges and pastures (0.5 Tg N_2O-N $year^{-1}$) accounts for 74% of all African agricultural emissions excluding savannah and grassland burning which contributes additional 0.46 Tg N_2O-N $year^{-1}$ (Hickman et al. 2011). Manure application rates in African agriculture is quite low because cattle graze widely, making manure collection challenging compared to more intensive livestock operations. Also, farmers in some regions rarely use manure as fertilizer (Batterbury and Warren 2001; Rufino et al. 2006). Savannah and grassland fires are substantial source of N_2O emissions in Africa. Much of burning can be attributed to land preparation. Changes in human diets towards more meat consumption will play significant role in future N_2O emissions from Africa. Per capita animal protein consumption is increasing by a factor of 2–3 (Bouwman et al. 2009).

9.5 Nitrous Oxide Emissions from Livestock Management Systems

The primary pathways of N loss are volatile emissions into atmosphere and leaching and runoff losses to ground and surface waters (de Vries et al. 2001). Animal production systems have relatively large share in the emission of NH_3,

Table 9.1 Nitrous oxide emission factors (N_2O-N as % of N applied) after application of different N sources

Soil conditions	Urea	Urine	Dairy farm effluent
<Field capacity	0–0.10	0.02–0.17	0.01–0.03
Field capacity	0.40–0.59	0.72–0.75	–
>Field capacity	1.27–1.57	0.64–1.59	1.97–2.49
>Field capacity + grazing[a]	–	–	4.93
References	Luo et al. (2007)	Luo et al. (2008b)	Luo et al. (2008c)

[a]Effluent-N applied immediately after grazing

N_2O, and CH_4 into the atmosphere (Oenema et al. 2001, 2005). The main source of the gaseous N emissions is the urine and dung excreted by the animals either in pastures or in confinements–stables, barns, sheds, and corrals. Four direct sources of N_2O from animal production can be distinguished globally: (i) urine and dung from grazing animals in pasture; (ii) wastes from temporarily confined animals during storage and handling; (iii) animal wastes from confined animals following application to land; and (iv) dung collected from pastures and paddocks for use as biofuel. In addition, the indirect source is associated with N lost from animal wastes (i.e. NO_3^- leaching, NH_3 volatilization) that enters other ecosystems and is subjected to N_2O producing processes away from the source (Mosier et al. 1998a).

Animal agriculture emits 22–32 Tg NH_3-N year^{-1} into the atmosphere, which is about 50–75% of the total anthropogenic NH_3 emissions (Bouwman et al. 1997). The N_2O emissions from animal production system and animal waste are ~2.7 Tg N_2O-N year^{-1} with a range of 0.7–4.2 Tg N_2O-N year^{-1}, which is equivalent to 30–50% of total N_2O emission from agriculture (Mosier et al. 1998a; Oenema et al. 2005). These estimates can be improved, as the magnitude of gas flux from the agricultural sector still has large knowledge gaps. Although the magnitude of global N production in animal waste is similar to that of fertilizer N produced annually (i.e., 90–100 Tg N year^{-1}; Fixen and West 2002; Canfield et al. 2010), and the N_2O emission per unit of N from animal waste is as large as that of N from fertilizer (Table 9.1; Mosier et al. 1998a), there is less emphasis on the role of animal waste N environmental impacts than that of fertilizer N. As a result, there is much larger knowledge gap and higher uncertainty on N_2O emissions from animal waste compared to that caused by N fertilizers.

Animal production systems transform animal feed–carbohydrates and proteins – into milk, meat, and eggs. However, only a small fraction of the ingested N in animal feed (about 5–30%) is retained in animal products depending on animal type and management. The bulk part, 70–95%, is voided via urine and dung. Animal excreta and N fertilizers are the main sources of N_2O within the soil-plant-animal system (Clark et al. 2005). In areas where per hectare animal productivity is an important goal for pastureland management, input of resources (including fertilizer N to the managed pasture) is substantial and results in a large soil N surplus. The ability of soils to conserve this surplus N is limited, thus the majority of the surplus N is lost through leaching as NO_3^- or emitted as gaseous N – NH_3, NO, N_2O, and N_2 causing both economic loss and environmental impacts. For example, N surplus of 150–250 kg ha^{-1} $year^{-1}$ occur in highly productive dairy farm systems in Germany and the Netherlands (Rotz et al. 2005). This surplus result in increasing NO_3^- contamination of shallow groundwater, because soils lack the mechanisms for long-term storage of inorganic N. Nitrogen surplus is a main indicator for the pressure of agricultural N use on the environment. In animal production systems, N use efficiency depends on animal category and management and manure management. Under intensive animal production systems, N use efficiency is low and N losses are high (Powell et al. 2010). Losses are governed by fertilizer type (animal manure or chemical fertilizer), climate, soil hydrological conditions, and N loss mitigation measures (Hatfield and Follett 2009). Emission losses from animal production system are of major concern because of negative impacts of reactive N gases to the atmospheric chemistry.

9.5.1 Approaches to Reduce Gaseous Nitrogen Emission from Animal Production Systems

Managing livestock production for high productivity and profitability is the main goal that farmers seek to achieve. Intensification of production to meet farmers' goals has increased incomes with very high environmental cost and risks. Animals have very low N use efficiency. They concentrate and excrete large quantities of plant-surplus N in urine and feces. Thus, intensification of livestock results in further losses of N and enhanced N_2O emissions. The main strategy for reducing N_2O emission from animal agriculture is to reduce gaseous reactive N emitted from animal excreta. A simple measure to achieve abatement of these gases is to reduce the size of livestock and therefore the amount of excreta. Thus, reduction of animal products in industrial world may result in reduction of animal stocks and the gaseous emission. However, the stocks reduction in developed world may be offset by increasing demand for animal products in developing countries. As shown in Fig. 9.3, various approaches can be used to minimize N losses and reduce N_2O emissions. Manipulation of N-use efficiency of the animal reduces excreted N. Excreta N may be reduced by feeding pastures grown with moderate fertilizer application and corn-based energy supplement formulated to provide low degradable protein (Kebreab et al. 2001).

The emissions of N_2O from pastures is influenced by rainfall, quality and frequency of N input from both fertilizer and stocking rate and SOC content (Saggar et al. 2007). In general, N_2O emission increases with increased N inputs (Gregorich et al. 2005). Reducing N loss from farm must begin with proper animal feeding and management to reduce N excretion. Manure and farm byproduct management should focus on retaining N that is contained in manure until it is applied to the land, and applying the appropriate amount of manure in timely manner to enhance plant N uptake. Manure slurry ponds are typically a significant source of CH_4, and N_2O (Berg et al. 2006). Cattle over wintering areas can be a source of significant N_2O emission because of a large accumulation of excreta, high

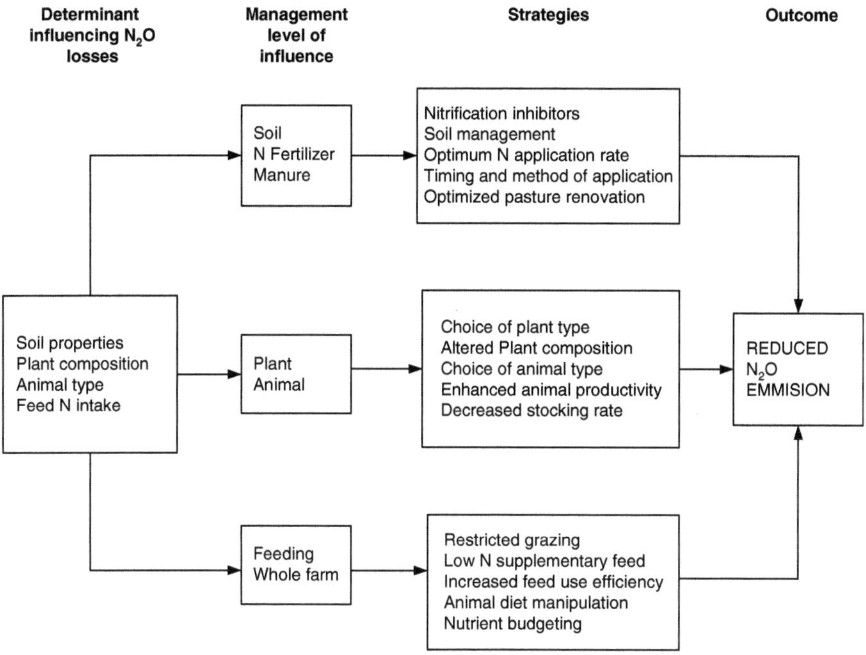

Fig. 9.3 Management strategies for reducing N_2O losses from livestock agriculture (Modified from Luo et al. 2010)

NO_3^- and wet soils (Hynst et al. 2007). Whether volatile loss of N from stored manure is in the form of NH_3 or N_2O depends on how manure is stored and subsequently applied to the land (Thorman et al. 2006). Covering manure with plastic sheet during storage and rapid incorporation after land spreading are some of successful techniques to reduce NH_3 and N_2O emission during storage and land application steps (Thorman et al. 2006).

Mitigation options aim at increasing efficiency of fertilizer N use and on reducing the amount of N cycling through pasture production system. The primary consideration for mitigating gaseous N emissions from pastoral land is to match the supply of mineral N from fertilizer, manure, and BNF to its spatial and temporal needs. It is possible to achieve uniform application of fertilizer N, but it is difficult to control uneven excretal distribution in the annually grazed pastures. Management practices can optimize the pasture's natural ability to compete with processes such as leaching, denitrification, and NH_3 volatilization, all of which lead to loss of the plant-available N from the soil-plant system. The NH_3 and NO_3^- lost from animal excreta are precursors for N_2O (Clemens and Ahlgrimm 2001), and abatement of these losses results in decreasing indirect N_2O emission caused by transformation of NH_3 and NO_3^- off-site. These management practices include optimum N supply to pasture crops, proper animal residue management, controlled-release fertilizer use, NIs and proper water management (Table 9.2).

Table 9.2 Estimated reduction of targeted N_2O abatement strategies from animal agriculture

	Reduction potential (%)			
Management	N_2O	Urinary N	Comment	References
Management intervention				
Fertilizer management	13		N fertilizer sources	Eckard et al. (2003)
	2–7		Increasing N rate on denitrification	Whitehead (1995)
Effluent management	50		Animal housing system	Oenema et al. (2005)
Wet season grazing management	7–11			de Klein et al. (2006), Luo et al. (2008a)
Soil water management	57–59		Flood irrigation	Phillips et al. (2007)
Nitrification inhibitors	61–86		Urine patches on pasture soil (autumn and spring)	Clough et al. (2007), Di and Cameron (2006), Di et al. (2007)
Feed intervention				
Improve protein to energy ratio		10–45		Miller et al. (2001), Misselbrook et al. (2005)
Condensed tannin (CT)		9	1.5–2.45% CT in diet	Carulla et al. (2005)
		45–59	3.5 CT in diet	de Klein and Eckard (2008)
		25		Misselbrook et al. (2005)
Animal intervention				
Nitrification inhibitor bolus	30–60		No direct measurements, likely to have similar effect as land-applied NI	Ledgard et al. (2008)
Animal breeding		3	Increased N use efficiency	Coffey (1996)
Diuretic (e.g. salt)		50	Reduction in urine N concentration, increased urine spread	Ledgard et al. (2007)

9.5.1.1 Soil Management Interventions

Grazing Lands Management

The N recycled through animal excreta, especially in urine is the main source of N_2O and NH_3 emission from grazed pastures (Bolan et al. 2004b). The N_2O emissions vary with soil type. For example, poorly drainage soils have higher denitrification and N_2O losses and lower NO_3^- leaching (de Klein et al. 2003), this is also true for clay-textured soil. In grazed pasture areas of greatest soil compaction and high levels of excreta, the ratio of N:N_2O can be five times higher than that in areas with the least animal impact. Further, the highest N_2O emissions are often observed under pastures with limited compaction (Simek et al. 2006). The O_2 availability is one of the most important factors affecting denitrification and N_2O production in soils. Increased denitrification rate is associated with restricted soil aeration at high water contents in pasture systems (de Klein and van Logtestijn 1994; Ledgard et al. 1999; Luo et al. 2000). Soil water content, compaction, and soil texture affects soil aeration and O_2 supply. Soil aggregate size is also an important factor when considering the effect of compaction on N_2O emissions. Soils with larger aggregate sizes can withstand significantly higher compaction pressure (Uchida et al. 2008). In well-aerated free-draining soils, heavy rainfall or irrigation events can result in short sharp bursts of N_2O emissions (Phillips et al. 2007). Low O_2 content can result from soil compaction and plugging through animal treading. Compacted soils of high water content could increase denitrification and N_2O emissions (Luo et al. 1999), and the use of machinery tends to exacerbate compaction (de Klein and Eckard 2008).

Grazed winter forage plants may be source of high N_2O emissions due to soil compaction. As soil moisture content is generally elevated during winter, forage crops grazed in winter are especially vulnerable to compaction and consequently produce high N_2O emissions. The type of tillage used to establish winter pasture crops affect the extent of subsequent soil drainage and compaction due to gazing animals. For example, in New Zealand, the N_2O emissions were six times higher when intensive tillage practice was used in combination with grazing of winter forage crop at high soil moisture content (Thomas et al. 2008). The results indicate that soil compaction from grazing may be the result of intensive tillage practices and it could have a significant effect on N_2O emissions. The N_2O emissions from two dairy cows grazed pastures in New Zealand, one on well-drained and the other on poorly drained soils over a year following grazing events were nearly 2% of the excretal and fertilizer N inputs (Saggar et al. 2004b), which are twice those measured from pasture plots with animal exclusion (de Klein et al. 2003).

Management practices that avoid prolonged soil wetness or minimize soil compaction are likely to reduce N_2O emissions. Reductions in N_2O losses can be achieved by altering soil conditions such as improving drainage, liming (for acidic soils), and avoiding soil compaction. However, the general applicability of these methods is limited. In the laboratory conditions, liming can increase N_2 emission and lower N_2O:N ratio (Zaman et al. 2007). On one hand, improved drainage can reduce direct N_2O emissions, but it also can increase NO_3^- leaching, and increase indirect N_2O emission from aquatic ecosystems. Restricted grazing when soil is wet can also reduce the risks of increasing N_2O emissions. In addition, strategic immobilization of excess N prior to high N loss periods, such as by carbon addition (Luo et al. 2010; Hoogendoorn et al. 2011), or by controlling N transformation processes in soil through inhibiting nitrification (Ledgard et al. 2008) may reduce N_2O emissions. Use of direct drilling methods to establish forage crops and restricted grazing when soils are wet minimizes risks of high N_2O emissions.

Application of Manure and Farm Effluent

About 40–60% of animal excreta are deposited in confined systems – stables, barns, sheds, corals and animal housing, where they are collected, and stored for subsequent application to the land (Oenema and Tamminga 2005). About 30% of the global animal excreta N collected in housing system are applied to pasture as slurry or effluent manure (Oenema et al. 2005). Effluent irrigation can lead to increased CH_4 and N_2O emissions depending on farming system and effluent composition. Storage conditions and treatment of the animal manures greatly affect the fate and environmental burden of N contained in manure (Oenema et al. 2005). In land based animal systems, most of the N in animal manure is returned to the land that produced animal feed, while in intensive animal production systems, animal manure is disposed of elsewhere, since the land base is missing. The manures are transported to land of other farmers nearby or processed, composted, treated and discharged or dumped. Smil (1999) estimated that 24% of total N in the manures is applied to the land, while the rest is either denitrified to N_2 or left as the environmental burden.

Manures and farm effluent application during winter or spring season under high soil moisture conditions enhance N_2O emissions and also increase direct leaching of applied N (van der Meer 2008). Significant amount of N can be lost as NH_3, NO_3^-, or N_2O after land application of effluent and/or manure. Estimates of up to 50% of mineral N are not uncommon (Mosier et al. 1998b). Most of gaseous losses are in the form of NH_3, causing air quality problems and N deposition in natural ecosystems. Leaching losses of NO_3^- reduces water quality. Because the N loss pathways are interconnected, controlling direct N_2O emissions from effluent and manure application can lower NH_3 and NO_3^- losses, and can significantly reduce indirect N_2O emissions. For example, the USA national N_2O emissions estimates from managed manures ranges from 8.4 to 98.7 Gg N_2O $year^{-1}$ (US EPA 2010; Del Grosso et al. 2008). The N_2O emission rates are highly variable depending on manure storage and management, time elapsed since manure applications, type of manure applied, climatic conditions, and amount of water available in the soil and manure (Saggar et al. 2004a). Other factors that improve the efficiency of N use by plants are climate and soil parameters, rainfall and soil drainage, and seasonal patterns of plant growth at the time of application. The use of slurry manure effluent increases N_2O emissions when applied to wet soils (Saggar et al. 2004a). The proportion of N denitrified and lost as N_2O is often the largest after liquid manure application.

Strategic timing of application of dairy farm effluent to pastures to coincide with low soil moisture status (i.e., drier soil) and low temperatures reduces N_2O emissions by up to 96% (Luo et al. 2008a, b). Delayed application of effluent after a grazing event could also minimize N_2O emissions by reducing the level of surplus mineral N in soil. In addition, NIs (Saggar et al. 2004a), and using anaerobic instead of aerobic storage reduces N_2O emissions both in storage and upon field application (Mosier et al. 1998b). The N_2O emissions from manure are also influenced by type of manure slurry, and the C:N ratio (Bhandral et al. 2007).

The best time for applying animal slurry manure for minimizing N_2O emissions is during the dry period of summer and autumn. Incorporation and injection of manure and slurry can also minimize NH_3 volatilization during and after field application. The most promising approach to minimizing N_2O emission is to adjust inorganic N fertilizer application rate to account for N addition in manure. Almost 40% of farmers in USA do not make the adjustments (USDA 2009). Incorporation and injection techniques reduce NH_3 volatilization and minimize N_2O emissions, but may increase CO_2 emissions because more fuel is required for application of manure or effluent. In addition to lower fertilizer cost, incorporation and injection also reduce N_2O emissions. Trade-off between N_2O emissions and pasture production must be considered in reducing manure rates.

Application of N Fertilizers

The ability of soil to conserve surplus N is limited. Therefore, most of surplus N is lost through leaching as NO_3^-, N_2O, and N_2. The N_2O emissions following fertilizer N application are highest in wet soils (Luo et al. 2007). Limiting the amount of fertilizer N applied during late autumn/winter or early spring when pasture growth is slow and soil is wet decreases N_2O emissions from grazed pastures. This type of management also reduces a potential for N leaching and indirect N_2O emissions off site. In addition, reducing N application rate on grazed pastures to take into account the deposition of animal excreta during grazing can minimize N_2O emissions and NO_3^- leaching. An optimum level of N fertilizers use depends on the particular production system desired. In systems that apply farm effluent and/or manure, growing plants absorb N. Hence, the fertilizer N application need to be adjusted to take into account the loading of applied manure effluent.

Nitrogen Transformation Inhibitors

The N transformation inhibitors are viable alternatives to reducing NO_3^- leaching and N_2O emissions from grazed pasture soils and increase N availability for plants. Only a limited number of chemicals are available for use as N transformation inhibitors in grazed pastures, and many of these have limitations to their use (Saggar et al. 2005). The general theory behind the use of NIs is that they slow N turnover by reducing the oxidation of NO_3^- retaining N in the form of NH_4^+. Because most soils have negative charges, the positively charged NH_4^+ ions are held onto negatively charged soil exchange surface, providing a greater opportunity to be taken up by plants. NH_4^+ can also be immobilized into SOM or fixed into certain 2:1 type clay mineral interlayers and prevent them from leaching losses. One of the compounds, which have been tested for grazed pastures, is Dicyandiamide (DCD). It inhibits the first stage of nitrification, i.e. the oxidation of NH_4^+ to NO_2^- by rendering the ammonia monooxygenase enzyme in *Nitrosomonas* bacteria ineffective (Amberger 1989). It is biodegradable, and decomposes in soils into NH_4^+ and CO_2.

Urease inhibitors (UIs) have also been used to reduce N losses from grazed pastures to the environment, particularly NH_3 volatilization to the atmosphere (Watson et al. 1998). The UIs function by inactivating the urease enzyme in soil, thus slowing down the rate of hydrolysis of urea to NH_4^+, inhibiting NH_4^+ formation and minimizing NH_3 volatilization and nitrification. The UIs and NIs are used as to control N loses from urine, dung and N fertilizers by acting on the N transformation processes of urea hydrolysis and nitrification, respectively. However, NIs and UIs cannot inhibit nitrification or urea hydrolysis indefinitely.

Quantitative knowledge of the interactions between N_2O, NH_3 emissions and NO_3^- and NH_4^+ leaching is key to understand how the pasture systems behave and respond to N transformation inhibition, and to determine the effective land management strategies to reduce overall N losses. Both NO_3^- leaching and N_2O emissions from urine patches can potentially be reduced by up to 70% with land application of NIs onto pastures (Di and Cameron 2003, 2004b, 2005, 2006; Moir et al. 2007). Di and Cameron (2004b, 2005) observed 68–76% and 70% reduction in NO_3^- leaching and N_2O emissions from cow urine-N patches, respectively, following application of DCD. An additional benefit with the use of DCD in grazed pastures is the reduction of cation leaching losses associated with NO3- leaching (Di and Cameron 2004a), and also increased pasture production. Pasture responses of >20% were recorded with the use of DCD (Di and Cameron 2004a, 2005; Moir et al. 2007). By applying NIs during autumn and winter when soil temperatures are low, the longevity, and effectiveness of nitrification inhibitor is prolonged (Kelliher et al. 2008).

The effectiveness of NI in mitigating N_2O emissions depends on their rate of degradation and persistence in soils, and key soil and environmental factors influencing NI are poorly understood. In practice, most of NO_3^- leaching occurs from urine patches deposited over a series of grazing during autumn, winter, and early spring (Monaghan et al. 2007a). Therefore, the benefits of DCD are influenced by timing of application relative to grazing. Strategic application of DCD to target this critical period of late autumn to early spring increases the efficiency of DCD (Di and Cameron 2004b, 2005). The N_2O emissions from grazed pasture plots treated with DCD can be 30–90% lower than those that are not treated (Smith et al. 2008a, b). Alternatively, the reduction in N_2O emissions could be achieved by strategically targeting urine patches through delivering NIs to the animals using slow release bolus. These are excreted intact in the urine resulting in inhibition of nitrification of urine N deposition to soil (see below).

Combined effect of urease and NIs on changes in N losses (NH_3 volatilization, NO_3^- leaching and N_2O emissions) and plant production from applied urea or animal urine indicate that N losses can be reduced when both UIs and NIs are applied together to urea or urine in pasture soils. Zaman et al. (2009) observed that combined N-(n-butyl) thiophosphoric triamide and DCD significantly reduced NH_3 volatilization, N_2O while increasing pasture dry matter and N uptake. However, the combined effects of inhibitors (UIs + NIs) are not necessarily greater than the additive effects of UIs and NIs. In another study, Zaman et al. (2008) indicated that the reduction in NH_3 volatilization due to combined inhibitors (UIs + NIs) are often not as high as reductions caused by UIs alone. Similarly, the reductions in N_2O emissions due to combined inhibitors may also not be as high as reductions associated with NIs alone. Overall, these results suggest that a combination of UIs and NIs may reduce the effectiveness of UIs on urea hydrolysis or the effectiveness of NIs in inhibiting nitrification. It is thus hypothesized that the combined effects of UIs and NIs may have prolonged the residence time of urea and NH_4^+ in the soil, possibly making them prone to leaching (Menneer et al. 2008). A fairly new groups of plant-based inhibitors popularly known as biological nitrification inhibitors (BNI), have been identified (White 1994). Some of these metabolites include monotarpenes (White 1994; Paavolainen et al. 1998). The ideal inhibitor for wide use in agriculture should: (i) specifically block an enzymatic reaction, (ii) remain in close contact with N compounds, (iii) not adversely affect other beneficial soil microorganisms and higher plants, (iv) remain effective for several weeks after fertilizer N application and excretal deposition, (v) not be toxic to animals and humans in the amounts used to effectively inhibit nitrification, and (vi) be cost effective (Saggar et al. 2005, 2008; Singh et al. 2008).

9.5.1.2 Plant and Animal Management

Plant Selection

The physical and chemical characteristics of pasture plants can influence the potential for NO_3^- leaching losses and N_2O emissions in grazed pasture systems. Plant characteristics which minimize N_2O emissions include: high root density, deep-rooting depth, active winter growth, high tannin content, high water soluble C (sugar) content, reduced N concentration, and improved residue quality (Monaghan et al. 2007b). For example, some strains of ryegrass have deeper rooting depth which increases their ability to remove N from sub-soil and thus increase N recovery from soil (Crush and Easton 2004; Crush et al. 2007), potentially reducing the risk of N losses. In addition, grass species with active winter growth have greater ability to absorb N from soil during this high leaching risk period than winter-dormant species. When plants with high tannin levels are consumed by animals, less N is excreted in urine relative to dung (Misselbrook et al. 2005). Overall, the N_2O losses are the highest from urine than from dung, because urine N is easily transformed compared to that contained in dung. Feeding beef cattle with grass silage containing elevated concentrations of water-soluble carbohydrates increases N-use efficiency for microbial growth in rumen by 46–68% (Merry et al. 2003). In addition, feeding dairy cows with high sugar variety of perennial ryegrass can reduce excreted N by 18% and urine N by 29% (Miller et al. 2001), thereby increasing efficiency of N cycling. Therefore, manipulating plant composition of animal feed offers some potential to reduce N excretion in urine, and reduce the subsequent N_2O emissions from animal urine. Plants also have the potential of altering soil N cycling through the quality of their residues. Unfertilized grass residues with high C:N ratio can immobilize inorganic N and control N_2O emissions from N sources such as animal urine, N fertilizer, and effluent/manure.

Animal Selection

Animal type influences the efficiency of N cycling by altering the spatial distribution of animal excreta or the fate of the N excreted. The NO_3^- leaching per hectare is twice as much for beef cattle compared to sheep and deer with the same level of N intake (Hoogendoorn et al. 2008). Such a differential leaching of NO_3 is attributed to spread of urine N from sheep and deer to larger area compared to beef cattle's fewer patches at higher N concentration, resulting into lower plant N recovery, and reduced efficiency of N cycling. Potentially, this beneficial effect of increased area of urine spread at lower N concentration can be achieved by dietary supplementation of diuretic compounds (e.g., salt). Salt supplementation of dairy cows feed reduces the N concentration in their urine by increasing the water intake of the animals (Luo et al. 2010). The animals urinates more frequently which, in combination with lower N concentration in the urine, avoids patches of high N concentrations due to animal urine and results in a more uniform distribution across the grazed pastures. Since the total amount of N in urine is not reduced, reduced N concentration in urine can only result in reduced N_2O emission if the N_2O emission factor is also reduced.

Breeding and selection of grazing animals for increased pasture productivity is an option that can be explored to reduce N losses. Increasing the efficiency of milk production in dairy cattle can partition more N to milk formation relative to maintenance and reduce the amount of N that ends up in excreta. In addition, growing meat production animals to the designed harvest weight more quickly reduces the associated maintenance requirements thereby reducing total feed intake and N excretion, CH_4 and N_2O emission and their GHG footprint.

The simplest approach to mitigate N_2O emission in grazed pasture systems is to reduce livestock, hence the amount of excreta N deposition. However, the demand by human population for animal protein continues to increase, and reducing the number of animals is not feasible unless the productivity per animal can also be substantially increased. Strategic selection of animals, choice of animal type and animal breeding offer a large potential to reduce N_2O emission.

Reduction of Animal Numbers

Reduction of animal stock and hence reducing the amount of animal excreta N deposition is the simplest approach to mitigate anthropogenic N_2O emissions in managed grasslands systems. The average annual per capita consumption of meat and milk in the developed countries is about 80–200 kg compared to 30–45 kg year^{-1} for the developing countries (FAO 2012). Consumption of meat is growing at 4%

and that of milk and dairy products at 2–3% per annum in the developing countries of Asia where the bulk of world population increase has occurred (FAO 2012). Reduction of meat and milk consumption in the developed world would thus reduce animal stocks and associated N_2O emissions. However, this fundamental shift in food production and consumption patterns is unlikely because the average per capita consumption of animal protein in developing countries is still in most cases far below the desire levels (Delgado 2005). There is a rapid rise in the combined consumption of meat and milk as billions of people in the developing countries diversify their diets from primarily starch-based to supplementation by milk and meat. As the economies of developing countries continue to improve, this trend will increase at a greater rate than that of the population increase. Therefore, reducing the N_2O emissions by reducing the number of animals is not a viable option in any foreseeable future.

9.5.1.3 Feeding System Management

Animal Diet Manipulation

Pastures typically require significantly higher N content for optimal growth than is needed for protein synthesis in animals. Grazing ruminants utilize relatively little of the N in feed and can excrete between 75 and 90% of ingested N (Whitehead 1995). This N is excreted either in animal housing system and applied to land as effluent or manure, or deposited directly by grazing animals as urine and dung. The animal excreta deposited in animal housing system, including both emissions from the housing system itself and those following land application of the collected excreta, contribute ~30% of global N_2O emissions from excreta (Oenema et al. 2005). On average, about one third of N feed is transformed into protein of animal products, and approximately two thirds of N intake is excreted in urine and feces of different livestock (Kirchgessner et al. 1994). About a quarter of this excreted N is emitted directly as ammonia during manure storage. Emissions of NH_3 and N_2O largely depend on the amount of N excreted by animals.

Dairy cows releases dietary N via milk (30%), urine (20–40%) and dung (30–40%). The amount of N excreted depends closely on the feed intake and, therefore, on targeted milk yield of cows (Clemens and Ahlgrimm 2001). Increase in milk also requires increase in crude proteins content of the diet, and high-producing dairy cows require a proper balance of rumen non-degradable proteins to meet their requirements for metabolizable protein. Animals on maintenance diet require ~7% of their dry matter (DM) intake as crude protein (CP). Pregnant and lactating animals require 10–12 and 15–20% of dry matter intake as CP, respectively (Thompson and Poppi 1990). Partitioning of N between urine and dung in ruminants depends on N content of the diet, with dung excretion remaining largely constant per unit of DM consumed and intake in excess of N requirement increasingly excreted via urine (Whitehead 1995). Balancing the protein-to-energy ratios

Table 9.3 Amount of animal feed needed for production of desired animal products

	Feed conversion ratio (kg kg^{-1})		
Animal product	Feed into live weight	Fee edible animal product	Feed into edible protein
Chicken meat	2.5	4.5	5
Egg	n.a.	2.3	3.3
Pork	5.0	9.4	10
Beef	10.0	25	25
Milk	n.a.	0.7	2.5

in ruminant diet is important strategy to minimize excessive urinary N and reduce N_2O emissions. Dairy cows fed on a 14% CP diet can excrete 29% less N compared to those fed with 19% CP (Misselbrook et al. 2005). Overall, the amount of N excreted in urine may decrease by 45% while that of dung increased by 4%.

Generally, the nutrition performances of animals have a significant influence on the amount of N excretion. Adjusting feeding composition to decrease the amount of N excreted is a suitable method of reducing ammonia emissions and N losses to water and air (e.g., NO_3^- leaching and N_2O emissions). Since a lower N content of fodder reduces the N excretion per animal, and consequently NH_3 and N_2O emissions from livestock (Velthof et al. 1998), modification of animal feeding strategies and feed low N diet may have significant impact on both direct and indirect N_2O emissions if livestock population remains constant. This can be achieved by: (i) reduction of N applied to grassland or substitution of grass by silage for cattle, (ii) tuning of compound feed to nutrients needs of the animals for pigs and poultry, (ii) change of raw material for pigs and poultry, (iv) pelleting of feeds to improve feeding efficiency through increased energy and protein digestibility for pigs and poultry, (v) supplementing diets with synthetic amino acids for pigs and poultry, and (vi) replacement of grass and grass silage with corn for cattle. The amount of feed needed for different desired products are presented in Table 9.3.

For dairy cattle, decreasing the daily surplus N intake per cow may decrease urinary N concentration (Petersen et al. 1998). A lower proportion of N is excreted in urine when dairy cows grazing perennial ryegrass pasture are fed supplements containing a low protein concentration and high fermentable organic matter (Mulligan et al. 2004). A 20% reduction in N excretion can be achieved in urine and feces (Nielsen et al. 2003).

For pigs, better feeding efficiency is the most efficient strategy for reducing N excretion. This can be achieved through proper processing to increase digestibility. Matching the available nutrients to animals' requirement is also a key for reducing excreta N from pigs, since animals may excrete all nutrients they are unable to use for maintenance and growth. However, the protein demand of the animals changes considerably during the course of the production cycle (e.g., pregnancy, lactation, start, or end of fattening). Therefore dietary protein composition should be adjusted depending on the production cycle.

Diet Additives

A lower N content of the pasture can reduce N excretion by animals and consequently gaseous N_2O emissions. Therefore, grass grown with moderate fertilizer N application and substitution of silage and corn-based energy supplements for grass can also reduce excretal-N. thus, food supplements can be formulated to provide low contents of degradable protein and with N intakes of less than 400 g day^{-1} for average yielding cows (Bolan et al. 2004b).

Other dietary manipulations include lowering the dietary crude protein (Misselbrook et al. 2005), increasing carbohydrate content to reduce microbial protein synthesis and ammonia loss from rumen (Kebreab et al. 2001). Use of feeding supplements such as low protein corn silage to replace fertilizer N boosted grass also lowers the amount of N excreted in urine and increases the overall efficiency of nutrient conversion into milk, thereby resulting in reduced N_2O emissions.

Nitrification inhibitor can also be delivered to animals using a slow-release bolus where unaltered inhibitor is excreted in the urine resulting in inhibition of nitrification of urine N on deposition to soil (Ledgard et al. 2008). In addition, animal diet manipulation to increase hippuric acid ($C_6H_5CONHCH_2COOH$) content of the urine has a potential for N_2O abatement (Kool et al. 2006). Increasing hippuric acid concentration in urine reduces N_2O emissions by up to 50%. Hippuric acid inhibits denitrification in soils when present in concentrations above 3.9 mmol kg^{-1} soil (van Groenigen et al. 2006). Increasing hippuric acid concentrations from 0.4 to 5.6 mmol kg^{-1} soil can result in N_2O flux reduction of 54%. Hippuric acid breaks down into benzoic acid, and evidence of decreased denitrification and N_2O reduction with benzoic acid has been observed (van Groenigen et al. 2006). Further research on effects of different diet manipulations on N_2O emissions is required.

Condensed tannins (CT) have demonstrated a potential for decreasing excreta N. They complex with protein in the rumen and protect it from microbial digestion, resulting in more efficient digestion in the lower intestine or tannin-protein complex being excreted in the dung. Sheep, fed with CT extract from *Acacia mearnesii* increases their partitioning of N from urine to feces, reducing urinary N by as much as 9.3% as a proportion of total N excreted (Carulla et al. 2005). Similarly, dairy cows on a 3.5% CT diet can excrete 25% less urine N, 60% more dung N and 8% more N overall compared with cows on a 1% CT diet (Misselbrook et al. 2005). The CT diet additive may reduce urinary N, increase fecal N, and improve N retention in animal. By reducing urinary N the emissions of N_2O from the readily available N is reduced.

9.5.1.4 Restricted or Non-grazing System

Livestock grazing of pastures generally occur all year round in temperate environment with relatively wet but mild winters. Practices involving the use of stand-off or feeding pads or housing system that are adopted to reduced soil physical damage due to grazing on wet soils can also reduce N_2O emissions and NO_3^- leaching (Luo et al. 2006). With this practice, animals are kept off grazing paddocks so excreta deposition is reduced at a time when it leads to the greatest N losses (e.g., late autumn/winter). This practice provides opportunity for controlling N losses as animal excreta is collected and can be applied evenly to pasture at targeted rates and optimum time when the risk for N losses is minimal (van der Meer 2008). Restricted grazing management, where animals are removed from pasture to a feeding pad from late autumn until early spring (4 months) with collection of effluent and reapplication during late spring/summer reduced NO_3^- leaching by 60% and potentially N_2O emissions (Monaghan et al. 2007b). Similar reductions when beef cattle are wintered off have been reported for beef cattle (Betteridge et al. 2005). In the restricted grazing system, manure/effluent management becomes critical for N use efficiency, because there are many processes through which N compounds can escape into the environment.

9.5.1.5 Animal Housing

Animal housing systems determine whether the excreta are slurry or straw and hence N_2O emissions and NH_3 volatilization. Addition of straw influences microbial activity in farm yard manure (FYM) by improving C:N ratio and reducing NH_3 emissions (Andersson 1996; Bussink and Oenema 1998). However, deep and wet litter straw system for pig housing may increase N_2O emissions due to denitrification, especially during summer.

9.6 Conclusions

About one seventh of global fertilizer N is applied to grassland ecosystems, but they are major contributors to global N_2O emissions, and overall animal agriculture contribute up to 50% of anthropogenic N_2O emissions. Livestock farming system uses nutrients inefficiently. Animals excrete between 70 and 95% of the ingested N, and more N (14–21 g) is needed to produce a unit (1 g) of animal protein than the equal amount of plant protein (approximately 3 g of N). A large proportion of the N consumed by grazing animals is excreted in urine and dung, which represent the main source of N_2O emissions and other losses from animal agriculture. There are ranges of possible management options that can reduce the N_2O emissions at the farm level in grazing systems, including:

- Optimizing soil management and N inputs including manure, effluent and fertilizer N,
- Optimizing pasture renovation,
- Manipulating soil N cycling processes though N transformation inhibitors such as nitrification and urease inhibitors,

- Selection for plants that maximize N utilization (e.g., high root density and deep rooted plants) to increase N uptake,
- Selection and breeding for animals that maximize N utilization,
- Altering grazing and feeding management to minimize N losses and improve animals N-use efficiency, respectively.

Some options can be more effective by targeting the source of N loss or by timing the period of greatest N loss risk. Use of stand-off or feed pads during high risk periods of N loss can eliminate excreta deposition in the pastures at these times. Combining this strategy with proper storage and utilization of manures collected from these pads minimizes risk of high N_2O emissions. The advantage of feed pads is that they provide an opportunity to optimize dietary N levels to ensure maximum feed use efficiency.

Further research employing plant breeding and genetic engineering hold a promise for breeding forages that utilize more N from grater soil volume and grow longer in colder or drier seasons leaving less NO_3^- in soil available for leaching and denitrification losses. Grass species with deeper and more adventitious rooting systems could improve N uptake by plants and thus potentially reduce N_2O emission. Animal breeding may also provide improvement in N conversion efficiency leading to lower excreta N concentration. Furthermore, genetic engineering can engineer denitrifying bacteria to utilize alternate electro acceptor other than NO_3^- and NO_2^- in soil thereby eliminating denitrification. Release of natural nitrification inhibitors from roots of *Brachiaria humidicola* can be high. Possibilities of finding other plants with a genetic trait of producing natural nitrification inhibitors, which could be exploited in crops and grasses as potential N_2O abatement strategy should be explored.

Study Questions

1. Outline the cost effective N_2O emissions abatement for dairy farm.
2. Due to invasive species, a farmer decides to rehabilitate a pasture land by applying herbicide to kill all weeds, and then plow grassland. The average tillage depth was 20 cm, within 1 year soil organic carbon of the top 20 cm decreased from 1.5 to 0.75%. The average annual N_2O fluxes increased from 2 μg m^{-2} day^{-1} to 10 ug m^{-2} day^{-1}, and CH_4 fluxes changed from −0.5 to 5 ug m^{-2} day^{-1}.

   a. Calculate the impact of pasture rehabilitation to N_2O emission
   b. Estimate 100-year global warming potential for this disturbance.
   c. What are the options for minimizing N_2O emissions from this operation?

References

Amberger, A. (1989) Research on dicyandiamide as a nitrification inhibitor and future outlook. Comm. Soil Sci. Plant Anal. **20**(19–20), 1933–1955. doi:10.1080/00103628909368195

Andersson, M. (1996) Performance of bedding materials in affecting ammonia emissions from pig manure. J. Agr. Eng. Res. **65**(3), 213–222. doi:10.1006/jaer.1996.0094

Andrews, M., Scholefield, D., Abberton, M.T., McKenzie, B.A., Hodge, S., Raven, J.A. (2007) Use of white clover as an alternative to nitrogen fertiliser for dairy pastures in nitrate vulnerable zones in the UK: productivity, environmental impact and economic considerations. Ann. Appl. Biol. **151**(1), 11–23. doi:10.1111/j.1744-7348.2007.00137.x

Babier, E.B., Burgess, J.C., Folke, C. (1994) Paradise lost? The ecological economics of biodiversity. pp. Earthscan, London, UK

Bardgett, R.D., Wardle, D.A. (2003) Herbivore-mediated linkages between aboveground and belowground communities. Ecology **84**(9), 2258–2268. doi:10.1890/02-0274

Batterbury, S., Warren, A. (2001) The African Sahel 25 years after the great drought: assessing progress and moving towards new agendas and approaches. Global Environ. Chang. 11(1), 1–8. doi:10.1016/s0959-3780(00)00040-6

Berg, W., Brunsch, R., Pazsiczki, I. (2006) Greenhouse gas emissions from covered slurry compared with uncovered during storage. Agric. Ecosys. Environ. **112**(2–3), 129–134. doi:10.1016/j.agee.2005.08.031

Betteridge, K., Ledgard, S.L., Lambert, M.G., Thorrold, B., Costall, D.A., Theobald, P.W., Hoogendoorn, C.J., Park, Z.A. (2005) Reduced nitrate leaching from livestock in a large lake catchment in New Zealand. Cox, S. (ed.). Precision Livestock Farming '05. 49–56

Bhandral, R., Bolan, N.S., Saggar, S., Hedley, M.J. (2007) Nitrogen transformation and nitrous oxide emissions from various types of farm effluents. Nutr. Cycl. Agroecosys. **79**(2), 193–208. doi:10.1007/s10705-007-9107-5

Bleken, M.A., Steinshamn, H., Hansen, S. (2005) High nitrogen costs of dairy production in Europe: Worsened by intensification. Ambio **34**(8), 598–606. doi:10.1639/0044-7447034 [0598:hncodp]2.0.co;2

Blench, R., Sommer, F. (1999) Understanding rangeland biodiversity. Working Paper No. 121. Overseas Development Institute (ODI)

Bolan, N.S., Adriano, D.C., Mahimairaja, S. (2004a) Distribution and bioavailability of trace elements in livestock and poultry manure by-products. Crit. Rev. Environ. Sci. Technol. **34**(3), 291–338. doi:10.1080/10643380490434128

Bolan, N.S., Saggar, S., Luo, J.F., Bhandral, R., Singh, J. (2004b) Gaseous emissions of nitrogen from grazed pastures: Processes, measurements and modelling, environmental implications, and mitigation. Adv. Agron. **84**, 37–120. doi:10.1016/s0065-2113(04)84002-1

Boone, R.B., Coughenour, M.B., Galvin, K.A., Ellis, J.E. (2002) Addressing management questions for Ngorongoro Conservation Area, Tanzania, using the Savanna modelling system. Afr. J. Ecol. **40**(2), 138–150. doi:10.1046/j.1365-2028.2002.00357.x

Bouwman, A.F., Beusen, A.H.W., Billen, G. (2009) Human alteration of the global nitrogen and phosphorus soil balances for the period 1970-2050. Global Biogeochem. Cycles **23**, GB0A04. doi:10.1029/2009gb003576

Bouwman, A.F., Lee, D.S., Asman, W.A.H., Dentener, F.J., VanderHoek, K.W., Olivier, J.G.J. (1997) A global high-resolution emission inventory for ammonia. Global Biogeochem. Cycles **11**(4), 561–587. doi:10.1029/97gb02266

Brink, C., Kroeze, C., Klimont, Z. (2001) Ammonia abatement and its impact on emissions of nitrous oxide and methane in Europe - Part 1: Method. Atmos. Environ. **35**(36), 6299–6312. doi:10.1016/s1352-2310(01)00434-4

Bussink, D.W., Oenema, O. (1998) Ammonia volatilization from dairy farming systems in temperate areas: A review. Nutr. Cycl. Agroecosys. **51**(1), 19–33. doi:10.1023/a:1009747109538

Cameron, K.C., Di, H.J., McLaren, R.G. (1997) Is soil an appropriate dumping ground for our wastes? Austr. J. Soil Res. **35**(5), 995–1035. doi:10.1071/s96099

Canfield, D.E., Glazer, A.N., Falkowski, P.G. (2010) The evolution and future of Earth's nitrogen cycle. Science **330**(6001), 192–196. doi:10.1126/science.1186120

Carulla, J.E., Kreuzer, M., Machmuller, A., Hess, H.D. (2005) Supplementation of *Acacia mearnsii* tannins decreases methanogenesis and urinary nitrogen in forage-fed sheep. Aus. J. Agric. Res. **56**(9), 961–970. doi:10.1071/ar05022

Chantigny, M.H. (2003) Dissolved and water-extractable organic matter in soils: a review on the influence of land use and management practices. Geoderma **113**(3–4), 357–380. doi:10.1016/s0016-7061(02)00370-1

Clark, H., Pinares, C., de Klein, C. (2005) Methane and nitrous oxide emissions from grazed grasslands. In: McGilloway, D.A. (ed.) Grassland: a global resource. pp. 279–294. Wageningen Academic Publishers, Wageningen, The Netherlands

Clemens, J., Ahlgrimm, H.-J. (2001) Greenhouse gases from animal husbandry: mitigation options. Nutr. Cycl. Agroecosys. **60**(1), 287–300. doi:10.1023/a:1012712532720

Clough, T.J., Di, H.J., Cameron, K.C., Sherlock, R.R., Metherell, A.K., Clark, H., Rys, G. (2007) Accounting for the utilization of a N_2O mitigation tool in the IPCC inventory methodology for agricultural soils. Nutr. Cycl. Agroecosys. **78**(1), 1–14. doi:10.1007/s10705-006-9069-z

Coffey, M.T. (1996) Environmental challenges as related to animal agriculture - swine. In: Kornegay, E.T. (ed.) Nutrient management of food animals to enhance and protect the environment. pp. 29–39. CRC, Boca Raton, FL

Crush, J.R., Easton, H.S. (2004) Screening for variation in root distribution in ryegrass (*Lolium perenne* L.). Acta Physiol. Plant **26**(3), 71–71

Crush, J.R., Easton, H.S., Waller, J.E., Hume, D.E., Faville, M.J. (2007) Genotypic variation in patterns of root distribution, nitrate interception and response to moisture stress of a perennial ryegrass (*Lolium perenne* L.) mapping population. Grass Forage Sci. **62**(3), 265–273. doi:10.1111/j.1365-2494.2007.00583.x

de Haan, C., Steinfield, H., Blackburn, H. (1997) Livestock and the environment. Finding a balance. Report of Study by the Commission of the European Communities, the World Bank and the governments of Denmark, France, Germany, The Netherlands, United Kingdom and United States of America

de Klein, C.A.M., Barton, L., Sherlock, R.R., Li, Z., Littlejohn, R.P. (2003) Estimating a nitrous oxide emission factor for animal urine from some New Zealand pastoral soils. Soil Res. **41**(3), 381–399. doi:10.1071/SR02128

de Klein, C.A.M., Eckard, R.J. (2008) Targeted technologies for nitrous oxide abatement from animal agriculture. Aus. J. Exper. Agric. **48**(2), 14–20. doi:10.1071/EA07217

de Klein, C.A.M., Smith, L.C., Monaghan, R.M. (2006) Restricted autumn grazing to reduce nitrous oxide emissions from dairy pastures in Southland, New Zealand. Agric. Ecosys. Environ. **112**(2–3), 192–199. doi:10.1016/j.agee.2005.08.019

de Klein, C.A.M., van Logtestijn, R.S.P. (1994) Denitrification and N_2O emission from urine-affected grassland soil. Plant Soil **163**(2), 235–241

de Vries, W., Kros, H., Oenema, O. (2001) Modeled impacts of farming practices and structural agricultural changes on nitrogen fluxes in the Netherlands. The Scientific World J. **1 Suppl 2**, 664–672

Del Grosso, S.J., Ogle, S., Wirth, J., Skiles, S. (2008) US Agriculture and foerstry greenhouse gas inventory: 1990–2005. US Department of Agriculture

Delgado, C.L. (2005) Rising demand for meat and milk in developing countries: Implications for grassland-based livestock production. In: McGilloway, D.A. (ed.) Grassland: A global resources. pp. 29–39. Academic Publishers, Wageningen, The Netherlands

Di, H.J., Cameron, K.C. (2002) The use of a nitrification inhibitor, dicyandiamide (DCD), to decrease nitrate leaching and nitrous oxide emissions in a simulated grazed and irrigated grassland. Soil Use Manage. **18**(4), 395–403. doi:10.1111/j.1475-2743.2002.tb00258.x

Di, H.J., Cameron, K.C. (2003) Mitigation of nitrous oxide emissions in spray-irrigated grazed grassland by treating the soil with dicyandiamide, a nitrification inhibitor. Soil Use Manage. **19**(4), 284–290. doi:10.1111/j.1475-2743.2003.tb00317.x

Di, H.J., Cameron, K.C. (2004a) Effects of the nitrification inhibitor dicyandiamide on potassium, magnesium and calcium leaching in grazed grassland. Soil Use Manage. **20**(1), 2–7. doi:10.1079/sum2003205

Di, H.J., Cameron, K.C. (2004b) Treating grazed pasture soil with a nitrification inhibitor, eco-n (TM), to decrease nitrate leaching in a deep sandy soil under spray irrigation - A lysimeter study. New Zeal. J Agric. Res. **47**(3), 351–361

Di, H.J., Cameron, K.C. (2005) Reducing environmental impacts of agriculture by using a fine particle suspension nitrification inhibitor to decrease nitrate leaching from grazed pastures. Agric. Ecosys. Environ. **109**(3–4), 202–212. doi:10.1016/j.agee.2005.03.006

Di, H.J., Cameron, K.C. (2006) Nitrous oxide emissions from two dairy pasture soils as affected by different rates of a fine particle suspension nitrification inhibitor, dicyandiamide. Biol. Fert. Soils **42**(6), 472–480. doi:10.1007/s00374-005-0038-5

Di, H.J., Cameron, K.C., Sherlock, R.R. (2007) Comparison of the effectiveness of a nitrification inhibitor, dicyandiamide, in reducing nitrous oxide emissions in four different soils under different climatic and management conditions. Soil Use Manage. **23**(1), 1–9. doi:10.1111/j.1475-2743.2006.00057.x

Eckard, R.J., Chen, D., White, R.E., Chapman, D.F. (2003) Gaseous nitrogen loss from temperate perennial grass and clover dairy pastures in south-eastern Australia. Aus. J. Agric. Res. **54**(6), 561–570. doi:10.1071/ar02100

Ellis, J., Galvin, K.A. (1994) Climate patterns and land-use practices in the dry zones of Africa. Bioscience **44**(5), 340–349

Food and Agricultural Organization of the United Nations (FAO) (2011) Food and Agricultural Organization of the United Nations. FAOSTAT Agricultural database. On line at: http://faostat.fao.org. (Accessed April 2011)

Food and Agricultural Organization of the United Nations (FAO) (2012) FAO statistical yearbook 2012: World and Agriculture. Rome, Italy. On line at http://fao.org/docrep/ (Accessed January 2012)

FAOSTAT (2011a) FAOSTAT Land Resource Data. Food and Agriculture Organization of the United Nations. On line at: htpp.//faostat.fao.org. (Accessed November 2011)

FAOSTAT (2011b) FAOSTAT Population Estimate and Projection. Food and Agriculture Organization of the United Nations. On line at: http://faostat.fao.org. (Accessed December 2011)

Fixen, P.E., West, F.B. (2002) Nitrogen fertilizers: Meeting contemporary challenges. Ambio **31** (2), 169–176. doi:10.1639/0044-7447(2002)031[0169:nfmcc]2.0.co;2

Galvin, K.A. (1992) Nutritional ecology of pastoralists in dry Tropical Africa. Am. J. Hum. Biol. **4** (2), 209–221. doi:10.1002/ajhb.1310040206

Gregorich, E.G., Rochette, P., VandenBygaart, A.J., Angers, D.A. (2005) Greenhouse gas contributions of agricultural soils and potential mitigation practices in Eastern Canada. Soil Till. Res. **83**(1), 53–72. doi:10.1016/j.still.2005.02.009

Groomsbridge, B. (1992) Global biodiversity. Status of the earth's living resources. pp. Chapman & Hall, London,

Hansen, J.P., Vinther, F.P. (2001) Spatial variability of symbiotic N_2 fixation in grass-white clover pastures estimated by the ^{15} N isotope dilution method and the natural ^{15} N abundance method. Plant Soil **230**(2), 257–266. doi:10.1023/a:1010390901845

Hatfield, J.L., Follett, R.F. (eds.): Nitrogen in the environment: Source, problems, and management, 2nd ed. Elsevier, Amsterdam (2009)

Haynes, R.J., Williams, P.H. (1993) Nutrient cycling and soil fertility in the grazed pasture ecosystem. Adv. Agron. **49**, 119–199. doi:10.1016/s0065-2113(08)60794-4

Hickman, J.E., Havlikova, M., Kroeze, C., Palm, C.A. (2011) Current and future nitrous oxide emissions from African agriculture. Current Opinion in Environmental Sustainability **3**(5), 370–378. doi:10.1016/j.cosust.2011.08.001

Hoogendoorn, C.J., Betteridge, K., Ledgard, S.F., Costall, D.A., Park, Z.A., Theobald, P.W. (2011) Nitrogen leaching from sheep-, cattle- and deer-grazed pastures in the Lake Taupo catchment in New Zealand. Anim. Prod. Sci. **51**(5), 416–425. doi:10.1071/AN10179

Hoogendoorn, C.J., de Klein, C.A.M., Rutherford, A.J., Letica, S., Devantier, B.P. (2008) The effect of increasing rates of nitrogen fertilizer and a nitrification inhibitor on nitrous oxide emissions from urine patches on sheep grazed hill country pasture. Aus. J. Exper. Agric. **48** (1–2), 147–151. doi:10.1071/ea07238

Hopkins, A., Pywell, R.F., Peel, S., Johnson, R.H., Bowling, P.J. (1999) Enhancement of botanical diversity of permanent grassland and impact on hay production in environmentally sensitive areas in the UK. Grass Forage Sci. **54**(2), 163–173

Hynst, J., Simek, M., Brucek, P., Petersen, S.O. (2007) High fluxes but different patterns of nitrous oxide and carbon dioxide emissions from soil in a cattle overwintering area. Agric. Ecosys. Environ. **120**(2–4), 269–279. doi:10.1016/j.agee.2006.10.003

IPCC (1996) Climate Change 1995: Impacts, Adaptations and Mitigation of Climate Change: Scientific-Technical Analysis. Contribution of Working Group II to the Second Assessment Report of the Intergovernmental Panel on Climate Change. Watson, R.T., M.C. Zinyowera, and R.H. Moss (ed.). pp. Cambridge University Press, New York, NY, USA, and Cambridge, UK

Jarvis, S.C. (1993) Nitrogen cycling and losses from dairy farms. Soil Use Manage. **9**(3), 99–105. doi:10.1111/j.1475-2743.1993.tb00937.x

Jarvis, S.C. (2000) Progress in studies of nitrate leaching from grassland soils. Soil Use Manage. **16**, 152–156

Jarvis, S.C., Pain, B.F. (1994) Greenhouse gas emissions from intensive livestock systems: their estimation and technologies for reduction. Climatic Change **27**(1), 27–38. doi:10.1007/bf01098471

Jarvis, S.C., Scholefield, D., Pain, B.F. (1995) Nitrogen cycling in grazing systems In: Bacon, P.E. (ed.) Nitrogen fertilization in the environment. pp. 381–419. Marcel Dekker, New York

Kebreab, E., France, J., Beever, D.E., Castillo, A.R. (2001) Nitrogen pollution by dairy cows and its mitigation by dietary manipulation. Nutr. Cycl. Agroecosys. **60**(1–3), 275–285. doi:10.1023/a:1012668109662

Kelliher, F.M., Clough, T.J., Clark, H., Rys, G., Sedcole, J.R. (2008) The temperature dependence of dicyandiamide (DCD) degradation in soils: A data synthesis. Soil Biol. Biochem. **40**(7), 1878–1882. doi:10.1016/j.soilbio.2008.03.013

Kirchgessner, M., Windisch, W., Roth, F.X. (1994) The efficiency of nitrogen conversion in animal nutrition. In: Nova Acta Leopoldina; The terrestrial nitrogen cycle as influenced by man, vol. 70. Nova Acta Leopoldina, pp. 393–412.

Kool, D.M., Hoffland, E., Hummelink, E.W.J., van Groenigen, J.W. (2006) Increased hippuric acid content of urine can reduce soil N_2O fluxes. Soil Biol. Biochem. **38**(5), 1021–1027. doi:10.1016/j.soilbio.2005.08.017

Ledgard, S.F. (2001) Nitrogen cycling in low input legume-based agriculture, with emphasis on legume/grass pastures. Plant Soil **228**(1), 43–59. doi:10.1023/a:1004810620983

Ledgard, S.F., Brier, G.J., Upsdell, M.P. (1990) Effect of clover cultivar on production and nitrogen-fixation in clover-ryegrass swards under dairy cow grazing. New Zeal. J Agric. Res. **33**(2), 243–249

Ledgard, S.F., Menneer, J.C., Dexter, M.M., Kear, M.J., Lindsey, S., Peters, J.S., Pacheco, D. (2008) A novel concept to reduce nitrogen losses from grazed pastures by administering soil nitrogen process inhibitors to ruminant animals: A study with sheep. Agric. Ecosys. Environ. **125**(1–4), 148–158. doi:10.1016/j.agee.2007.12.006

Ledgard, S.F., Penno, J.W., Sprosen, M.S. (1999) Nitrogen inputs and losses from clover/grass pastures grazed by dairy cows, as affected by nitrogen fertilizer application. J. Agric. Sci. **132**, 215–225. doi:10.1017/s002185969800625x

Ledgard, S.F., Sprosen, M.S., Brier, G.J., Nemaia, E.K.K., Clark, D.A. (1996) Nitrogen inputs and losses from New Zealand dairy farmlets, as affected by nitrogen fertilizer application: Year one. Plant Soil **181**(1), 65–69. doi:10.1007/bf00011293

Ledgard, S.F., Sprosen, M.S., Penno, J.W., Rajendram, G.S. (2001) Nitrogen fixation by white clover in pastures grazed by dairy cows: Temporal variation and effects of nitrogen fertilization. Plant Soil **229**(2), 177–187. doi:10.1023/a:1004833804002

Ledgard, S.F., Welten, B.G., Menneer, J.C., Betteridge, K., Crush, J.R., Barton, M.D. (2007) New nitrogen mitigation technologies for evaluation in the Lake Taupo catchment. Proceedings New Zeal. Grass Ass. **69**, 117–121

Lee, D.S., Bouwman, A.F., Asman, W.A.H., Dentener, F.J., van der Hook, K.W. (1997) Emissions of nitrous oxide, nitric oxide and ammonia from grasslands on a global scale. In: Jarvis, S.C.,

Pain, B.F. (eds.) Gaseous nitrogen emissions from grasslands. pp. 353–371. CAB International, Willingford,
Ludwig, J., Tongway, D., Freudenberger, D., Noble, J., Hodgkinson, K. (1997) Landscape Ecology. CSIRO, Collinwood
Luo, J., de Klein, C.A.M., Ledgard, S.F., Saggar, S. (2010) Management options to reduce nitrous oxide emissions from intensively grazed pastures: A review. Agric. Ecosys. Environ. **136**(3–4), 282–291. doi:10.1016/j.agee.2009.12.003
Luo, J., Donnison, A., Ross, C., Ledgard, S., Longhurst, B. (2006) Control of pollutants using stand-off pads containing different natural materials. Proceedings New Zeal. Grass Ass. **68**, 315–320
Luo, J., Ledgard, S.F., Lindsey, S.B. (2007) Nitrous oxide emissions from application of urea on New Zealand pasture. New Zeal. J Agric. Res. **50**(1), 1–11
Luo, J., Ledgard, S.F., Lindsey, S.B. (2008a) A test of a winter farm management option for mitigating nitrous oxide emissions from a dairy farm. Soil Use Manage. **24**(2), 121–130. doi:10.1111/j.1475-2743.2007.00140.x
Luo, J., Lindsey, S.B., Ledgard, S.F. (2008b) Nitrous oxide emissions from animal urine application on a New Zealand pasture. Biol. Fert. Soils **44**(3), 463–470. doi:10.1007/s00374-007-0228-4
Luo, J., Saggar, S., Bhandral, R., Bolan, N., Ledgard, S., Lindsey, S., Sun, W. (2008c) Effects of irrigating dairy-grazed grassland with farm dairy effluent on nitrous oxide emissions. Plant Soil **309**(1), 119–130. doi:10.1007/s11104-008-9550-3
Luo, J., Tillman, R.W., Ball, P.R. (1999) Factors regulating denitrification in a soil under pasture. Soil Biol. Biochem. **31**(6), 913–927. doi:10.1016/s0038-0717(99)00013-9
Luo, J., Tillman, R.W., Ball, P.R. (2000) Nitrogen loss through denitrification in a soil under pasture in New Zealand. Soil Biol. Biochem. **32**(4), 497–509. doi:10.1016/s0038-0717(99)00179-0
McCarl, B.A., Schneider, U.A. (2000) U.S. Agriculture's role in a greenhouse gas emission mitigation world: An economic perspective. Rev. Agric. Econ. **22**(1), 134
Menneer, J.C., Ledgard, S., Sprosen, M. (2008) Soil N process inhibitors alter nitrogen leaching dynamics in a pumice soil. Austr. J. Soil Res. **46**(4), 323–331. doi:10.1071/sr06180
Merry, R.J., Lee, M.R.F., Davies, D.R., Moorby, J.M., Dewhurst, R.J., Scollan, N.D. (2003) Nitrogen and energy use efficiency in the rumen of cattle fed high sugar grass and/or red clover silage. Aspects Appl. Biol.(70), 87–92
Miller, L.A., Moorby, J.M., Davies, D.R., Humphreys, M.O., Scollan, N.D., MacRae, J.C., Theodorou, M.K. (2001) Increased concentration of water-soluble carbohydrate in perennial ryegrass (*Lolium perenne* L.): milk production from late-lactation dairy cows. Grass Forage Sci. **56**(4), 383–394. doi:10.1046/j.1365-2494.2001.00288.x
Misselbrook, T.H., Powell, J.M., Broderick, G.A., Grabber, J.H. (2005) Dietary manipulation in dairy cattle: Laboratory experiments to assess the influence on ammonia emissions. J. Dairy Sci. **88**(5), 1765–1777
Moir, J.L., Cameron, K.C., Di, H.J. (2007) Effects of the nitrification inhibitor dicyandiamide on soil mineral N, pasture yield, nutrient uptake and pasture quality in a grazed pasture system. Soil Use Manage. **23**(2), 111–120. doi:10.1111/j.1475-2743.2006.00078.x
Monaghan, R.M., Hedley, M.J., Di, H.J., McDowell, R.W., Cameron, K.C., Ledgard, S.F. (2007a) Nutrient management in New Zealand pastures— recent developments and future issues. New Zeal. J Agric. Res. **50**(2), 181–201. doi:10.1080/00288230709510290
Monaghan, R.M., Hedley, M.J., Di, H.J., McDowelli, R.W., Cameron, K.C., Ledgard, S.F. (2007b) Nutrient management in New Zealand pastures— recent developments and future issues. New Zeal. J Agric. Res. **50**(2), 181–201. doi:10.1080/00288230709510290
Moore, R.M. (1970) Australian grasslands. pp. Alexander Bros, Melbourne, Australia
Mosier, A., Kroeze, C., Nevison, C., Oenema, O., Seitzinger, S., van Cleemput, O. (1998a) Closing the global N_2O budget: nitrous oxide emissions through the agricultural nitrogen cycle - OECD/IPCC/IEA phase II development of IPCC guidelines for national greenhouse gas inventory methodology. Nutr. Cycl. Agroecosys. **52**(2–3), 225–248. doi:10.1023/a:1009740530221

Mosier, A.R., Bleken, M.A., Chaiwanakupt, P., Ellis, E.C., Freney, J.R., Howarth, R.B., Matson, P.A., Minami, K., Naylor, R., Weeks, K.N., Zhu, Z.L. (2001) Policy implications of human-accelerated nitrogen cycling. Biogeochemistry **52**(3), 281–320. doi:10.1023/a:1006430122495

Mosier, A.R., Duxbury, J.M., Freney, J.R., Heinemeyer, O., Minami, K., Johnson, D.E. (1998b) Mitigating agricultural emissions of methane. Climatic Change **40**(1), 39–80. doi:10.1023/a:1005338731269

Mulligan, F.J., Dillon, P., Callan, J.J., Rath, M., O'Mara, F.P. (2004) Supplementary concentrate type affects nitrogen excretion of grazing dairy cows. J. Dairy Sci. **87**(10), 3451–3460

Nielsen, N.M., Kristensen, T., Norgaard, P., Hansen, H. (2003) The effect of low protein supplementation to dairy cows grazing clover grass during half of the day. Livest. Produ. Sci. **81** (2–3), 293–306. doi:10.1016/s0301-6226(02)00229-4

O'Connor, T.G., Haines, L.M., Snyman, H.A. (2001) Influence of precipitation and species composition on phytomass of a semi-arid African grassland. J. Ecol. **89**(5), 850–860. doi:10.1046/j.0022-0477.2001.00605.x

Oconnor, T.G. (1994) Composition and population responses of an African savanna grassland to rainfall and grazing. J. Appl. Ecol. **31**(1), 155–171. doi:10.2307/2404608

Oenema, J., Bannink, A., Sommer, S.G., Velthof, G.L. (2001) Gaseous nitrogen emissions from animal production systems. In: Follett, R.F., Hattfield, J.L. (eds.) Nitrogen in the environment: Sources, problems, and management. pp. 255–289. Elsevier Sciences, Amsterdam, The Netherland

Oenema, O. (2006) Nitrogen budgets and losses in livestock systems. Int. Congr. Ser. **1293**(0), 262–271. doi:10.1016/j.ics.2006.02.040

Oenema, O., Gebauer, G., Rodriguez, M., Sapek, A., Jarvis, S.C., Corré, W.J., Yamulki, S. (1998) Controlling nitrous oxide emissions from grassland livestock production systems. Nutr. Cycl. Agroecosys. **52**(2), 141–149. doi:10.1023/a:1009732226587

Oenema, O., Tamminga, S. (2005) Nitrogen in global animal production and management options for improving nitrogen use efficiency. Science China C **48**, 871–887. doi:10.1360/062005-279

Oenema, O., Velthof, G.L., Yamulki, S., Jarvis, S.C. (1997) Nitrous oxide emissions from grazed grassland. Soil Use Manage. **13**(4), 288–295. doi:10.1111/j.1475-2743.1997.tb00600.x

Oenema, O., Wrage, N., Velthof, G.L., van Groenigen, J.W., Dolfing, J., Kuikman, P.J. (2005) Trends in global nitrous oxide emissions from animal production systems. Nutr. Cycl. Agroecosys. **72**(1), 51–65. doi:10.1007/s10705-004-7354-2

Oesterheld, M., Loreti, J., Semmartin, M., Sala, O.E. (2001) Inter-annual variation in primary production of a semi-arid grassland related to previous-year production. Journal of Vegetation Science **12**(1), 137–142. doi:10.2307/3236681

Ojima, D.S., Dirks, B.O.M., Glenn, E.P., Owensby, C.E., Scurlock, J.O. (1993) Assessment of C budget for grasslands and drylands of the world. Water Air Soil Polut. **70**(1–4), 95–109. doi:10.1007/bf01104990

Ojima, D.S., Galvin, K.A., Turner, B.L. (1994) The global impact of land-use change. Bioscience **44**(5), 300–304. doi:10.2307/1312379

Paavolainen, L., Kitunen, V., Smolander, A. (1998) Inhibition of nitrification in forest soil by monoterpenes. Plant Soil **205**(2), 147–154. doi:10.1023/a:1004335419358

Petersen, S.O., Sommer, S.G., Aaes, O., Soegaard, K. (1998) Ammonia losses from urine and dung of grazing cattle: Effect of N intake. Atmos. Environ. **32**(3), 295–300. doi:10.1016/s1352-2310(97)00043-5

Phillips, F.A., Leuning, R., Baigenta, R., Kelly, K.B., Denmead, O.T. (2007) Nitrous oxide flux measurements from an intensively managed irrigated pasture using micrometeorological techniques. Agric. Forest Meteor. **143**(1–2), 92–105. doi:10.1016/j.agrformet.2006.11.011

Powell, J.M., Gourley, C.J.P., Rotz, C.A., Weaver, D.M. (2010) Nitrogen use efficiency: A potential performance indicator and policy tool for dairy farms. Environ. Sci. Policy **13**(3), 217–228. doi:10.1016/j.envsci.2010.03.007

Reynolds, S.G., Bartello, C., Baas, S., Mack, S. (2005) Grassland and forage to improve livestock and reduce poverty. In: McGilloway, D.A. (ed.) Grassland: A global resource. pp. 323–338. Academic Publishers, Wageningen, The Netherlands

Rotz, C.A., Taube, F., Russelle, M.P., Oenema, J., Sanderson, M.A., Wachendorf, M. (2005) Whole-farm perspectives of nutrient flows in grassland agriculture. Crop Sci. **45**(6), 2139–2159. doi:10.2135/cropsci2004.0523

Rufino, M.C., Rowe, E.C., Delve, R.J., Giller, K.E. (2006) Nitrogen cycling efficiencies through resource-poor African crop-livestock systems. Agric. Ecosys. Environ. **112**(4), 261–282. doi:10.1016/j.agee.2005.08.028

Saggar, S., Bolan, N., Singh, J., Blard, A. (2005) Economic and environmental impacts of increased nitrogen use in grazed pastures and role of inhibitors in mitigating nitrogen losses New Zeal. Sci. Review **62**, 62–67

Saggar, S., Bolan, N.S., Bhandral, R., Hedley, C.B., Luo, J. (2004a) A review of emissions of methane, ammonia, and nitrous oxide from animal excreta deposition and farm effluent application in grazed pastures. New Zeal. J Agric. Res. **47**(4), 513–544

Saggar, S., Andrew, R.M., Tate, K.R., Hedley, C.B., Rodda, N.J., Townsend, J.A. (2004b) Modelling nitrous oxide emissions from dairy-grazed pastures. Nutr. Cycl. Agroecosys. **68**(3), 243–255. doi:10.1023/B:FRES.0000019463.92440.a3

Saggar, S., Giltrap, D.L., Li, C., Tate, K.R. (2007) Modelling nitrous oxide emissions from grazed grasslands in New Zealand. Agric. Ecosys. Environ. **119**(1–2), 205–216. doi:10.1016/j.agee.2006.07.010

Saggar, S., Tate, K.R., Giltrap, D.L., Singh, J. (2008) Soil-atmosphere exchange of nitrous oxide and methane in New Zealand terrestrial ecosystems and their mitigation options: A review. Plant Soil **309**(1–2), 25–42. doi:10.1007/s11104-007-9421-3

Sere, C., Stenifeld, H., Groenewold, J. (1995) World livestock production systems: Current status, issues, and trends. pp. Food and Agricultural Organization of the United Nations, Rome, Italy

Sharpley, A., Meisinger, J.J., Breeuwsma, A., Sims, J.T., Danie, T.C., Schepers, J.S. (1998) Impact of animal manure management on ground and surface water quality. In: Hatfield, J.L., Stewart, B.A. (eds.) Utilization: Efffective use of manure as a soil resource. pp. 173–242. Ann Arbor Press, Chelsea, MI

Simek, M., Brucek, P., Hynst, J., Uhlirova, E., Petersen, S.O. (2006) Effects of excretal returns and soil compaction on nitrous oxide emissions from a cattle overwintering area. Agric. Ecosys. Environ. **112**(2–3), 186–191. doi:10.1016/j.agee.2005.08.018

Singh, J., Bolan, N.S., Saggar, S., Zaman, M. (2008) The role of inhibitors in controlling the bioavailability and losses of nitrogen. In: Naidu, N.S., Bolan, N.S., Megharaj, A., Juhasz, S., Gupta, B., Clothier, B., Schulin, R. (eds.) Chemical bioavailability in terrestrial environment. pp. 329–362. Elsevier, Amsterdam, The Netherlands

Smil, V. (1999) Nitrogen in crop production: An account of global flows. Global Biogeochem. Cycles **13**(2), 647–662. doi:10.1029/1999gb900015

Smith, L.C., de Klein, C.A.M., Catto, W.D. (2008a) Effect of dicyandiamide applied in a granular form on nitrous oxide emissions from a grazed dairy pasture in Southland, New Zealand. New Zeal. J Agric. Res. **51**(4), 387–396

Smith, L.C., de Klein, C.A.M., Monaghan, R.M., Catto, W.D. (2008b) The effectiveness of dicyandiamide in reducing nitrous oxide emissions from a cattle-grazed, winter forage crop in Southland, New Zealand. Aus. J. Exper. Agric. **48**(1–2), 160–164. doi:10.1071/ea07262

Snyman, H.A. (2004) Soil seed bank evaluation and seedling establishment along a degradation gradient in a semi-arid rangeland Afr. J. Range For. Sci. **21**, 37–47

Snyman, H.A., du Preez, C.C. (2005) Rangeland degradation in a semi-arid South Africa - II: influence on soil quality. J. Arid Environ. **60**(3), 483–507. doi:10.1016/j.jaridenv.2004.06.005

Solbrig, O. (1996) The diversity of savanna ecosystems. In: Solbrig, O.T., Medina, E., Silva, J.F. (eds.) Biodiversity and savanna ecosystem processes. pp. 1–30. Springer, Berlin, Germany

Solbrig, O.T. (1993) Ecological constraints of savanna land use. In: Young, M.D., Solbrig, O.T. (eds.) The world's savannas. Economic driving forces, ecological constraints and policy option for sustainable land use. pp. 21–48. UNESCO and The Parthenon Publishing Group, Paris, France

Thomas, S.M., Beare, M.H., Francis, G.S., Barlow, H.E., Hedderley, D.I. (2008) Effects of tillage, simulated cattle grazing and soil moisture on N_2O emissions from a winter forage crop. Plant Soil **309**(1–2), 131–145. doi:10.1007/s11104-008-9586-4

Thompson, K.F., Poppi, D.P. (1990) Livestock production from pasture. In: Langer, R.H.M. (ed.) Pastures: Their ecology and management. pp. 263–283. Oxford University Press, Oxford

Thorman, R.E., Chadwick, D.R., Boyles, L.O., Matthews, R., Sagoo, E., Harrison, R. (2006) Nitrous oxide emissions during storage of broiler litter and following application to arable land. Int. Congr. Ser. **1293**(0), 355–358. doi:10.1016/j.ics.2006.04.005

Thornton, P.K., Galvin, K.A., Boone, R.B. (2003) An agro-pastoral household model for the rangelands of East Africa. Agr. Syst. **76**(2), 601–622. doi:10.1016/s0308-521x(01)00094-4

Uchida, Y., Clough, T.J., Kelliher, F.M., Sherlock, R.R. (2008) Effects of aggregate size, soil compaction, and bovine urine on N_2O emissions from a pasture soil. Soil Biol. Biochem. **40**(4), 924–931. doi:10.1016/j.soilbio.2007.11.007

United States Department of Agriculture (USDA) (2009) Manure use for fertilizer and for energy: Report to Congress. US Department of Agriculture, Economic Research Service. On line at http://ers.usda.gov/publications/ap/ap037/ (Accessed in March 2012)

U.S. Environmental Protection Agency (US EPA). (2010) Inventory of US greenhouse gas emissions and sinks: 1990–2008. U.S. Environmental Protection Agency. Online at: http// www.epa.gov/climatechange/emissions/usinventoryreport.html. (Accessed November 2011)

van der Meer, H.G. (2008) Optimising manure management for GHG outcomes. Aus. J. Exper. Agric. **48**(1–2), 38–45. doi:10.1071/ea07214

van Groenigen, J.W., Palermo, V., Kool, D.M., Kuikman, P.J. (2006) Inhibition of denitrification and N_2O emission by urine-derived benzoic and hippuric acid. Soil Biol. Biochem. **38**(8), 2499–2502. doi:10.1016/j.soilbio.2006.02.023

Velthof, G.L., van Beusichem, M.L., Oenema, O. (1998) Mitigation of nitrous oxide emission from dairy farming systems. Environ. Pollut. **102**, 173–178. doi:10.1016/s0269-7491(98)80030-4

Watson, C.J., Poland, P., Allen, M.B.D. (1998) The efficacy of repeated applications of the urease inhibitor N-(n-butyl) thiophosphoric triamide for improving the efficiency of urea fertilizer utilization on temperate grassland. Grass Forage Sci. **53**(2), 137–145. doi:10.1046/j.1365-2494.1998.5320137.x

White, C.S. (1994) Monoterpenes: Their effects on ecosystem nutrient cycling. J. Chem. Ecol. **20**(6), 1381–1406. doi:10.1007/bf02059813

Whitehead, D.C. (1995) Grassland nitrogen. pp. CAB International, Wallingford, UK

Wiegand, T., Snyman, H.A., Kellner, K., Paruelo, J.M. (2004) Do grasslands have a memory: Modeling phytomass production of a semiarid South African grassland. Ecosystems **7**(3), 243–258. doi:10.1007/s10021-003-0235-8

WRI (2000) A guide to World Resources 2000–2001: People and Ecosystems. pp. World Resources Institute (WRI), Washington, DC

Zaman, M., Nguyen, M.L., Blennerhassett, J.D., Quin, B.F. (2008) Reducing NH_3, N_2O and NO_3^--N losses from a pasture soil with urease or nitrification inhibitors and elemental S-amended nitrogenous fertilizers. Biol. Fert. Soils **44**(5), 693–705. doi:10.1007/s00374-007-0252-4

Zaman, M., Nguyen, M.L., Matheson, F., Blennerhassett, J.D., Quin, B.F. (2007) Can soil amendments (zeolite or lime) shift the balance between nitrous oxide and dinitrogen emissions from pasture and wetland soils receiving urine or urea-N? Austr. J. Soil Res. **45**(7), 543–553. doi:10.1071/sr07034

Zaman, M., Saggar, S., Blennerhassett, J.D., Singh, J. (2009) Effect of urease and nitrification inhibitors on N transformation, gaseous emissions of ammonia and nitrous oxide, pasture yield and N uptake in grazed pasture system. Soil Biol. Biochem. **41**(6), 1270–1280. doi:10.1016/j.soilbio.2009.03.011

Chapter 10
The Role of Fertilizer Management in Mitigating Nitrous Oxide Emissions

Contents

Abstract Fertilizer N use in agriculture is the major source of direct and indirect N_2O emissions. The loss of N not only increases emissions, but also decreases water quality and decrease N use efficiency (NUE) and cause economic loss. Improving NUE is desirable for improving crop yields, reducing cost of production, and maintaining environmental quality. Cereals, the largest consumer of fertilizer N has an average NUE of 33%. To improve N efficiency in agriculture, integrated N management strategies that take into consideration the use of improved fertilizers, such as enhanced efficiency fertilizers and stabilized fertilizers, along with soil and crop management practices are necessary. Incorporating livestock production within cropping system offers one of the best opportunities to improve NUE. Synchrony of N supply with crop demand is essential in order to ensure adequate quantity of uptake and utilization and optimum yield. This chapter discusses N dynamics in soil-plant systems, and outlines management options for enhancing N use by crops.

Keywords Fertilizer management practices • Controlled-release fertilizer • Nitrification inhibitors • Polymer-coated fertilizers • Slow-release fertilizer • Urease inhibitors

David Ussiri and R. Lal, *Soil Emission of Nitrous Oxide and its Mitigation*,
DOI 10.1007/978-94-007-5364-8_10,

Abbreviations

FBMP	Fertilizers best management practices
GHG	Greenhouse gas
NUE	Nitrogen use efficiency
UF	Urea formaldehyde
IBDU	Isobutylene diurea/crotonylidene diurea
CDU	Urea-acetaldehyde/cyclo diurea
SON	Soil organic nitrogen
AAPFCO	Association of American Plant Control Officials
EEF	Enhanced efficiency fertilizers
SRF	Slow-release fertilizers
CRF	Controlled-release fertilizers
DCD	Dicyandiamide
DMPP	Dimethyl pyrazole phosphate

10.1 Introduction

Mineral fertilizers have sustained world agriculture and thus a global population and wealth growth for more than 100 years (Smil 2001; Stewart et al. 2005). Their contribution to increasing crop yields per unit of land area has spared millions of hectares of natural ecosystems that otherwise would have been cleared and converted to agriculture (Balmford et al. 2005), and saved hundreds of millions from starvation or malnutrition. Moreover, the productivity gains since 1950s resulted from development of farming systems that rely heavily on external inputs of energy and chemicals to replace management of on-farm resources (Oberle 1994). Imbalanced and inappropriate or excessive use of fertilizer nutrients in agricultural systems in some regions of the world, and lack of fertilizer nutrients application in others remains a major concern. For example, African agricultural land receives an average of 8 kg N ha^{-1} compared to 100 kg N ha^{-1} in the United States, and 220 kg ha^{-1} in China (Vitousek et al. 2009). Due to extremely low fertilizer inputs in African agriculture, N removal through crop harvest and erosion over the last several decades has led to negative nutrient balances, declining grain yields, and consequently to insufficient food intake to meet dietary protein and energy requirements (Sanchez 2002). Nutrient mining is a major cause for low crop yields in parts of the developing world, particularly Africa (Smaling et al. 1997; Vanlauwe et al. 2010). In other situations, high quantities of inorganic fertilizers, particularly N used to increase world food production, has contributed to a large increase in mineral N surpluses in soils (Follett 2001; Ramirez and Reheul 2009). Nitrogen use efficiency (NUE) of a cropping system is the proportion of all N inputs that are removed in harvested biomass, contained in recycled crop residues, and incorporated into soil organic matter and inorganic N pools. Nitrogen not recovered in these sinks is lost from the cropping system and contributes to the reactive N (Nr) loading into environment.

The surplus N moves beyond the agricultural fields and pollute the environment, because the management practices used fail to achieve good balance between nutrient supply and crop nutrient demand (van Noordwijk and Cadisch 2002). The loss of N not only decreases NUE and hence causes the economic loss, but just as important, it also influences water quality and contributes to greenhouse gas (GHG) emissions. If left unchecked, such losses may bear significant costs to the society. For example, about 1.7 Tg N_2O-N year^{-1} emission is directly attributed to the use of synthetic N fertilizers (Prather et al. 1995; Cole et al. 1996; Smith et al. 2007). This quantity increased because of the increased use of synthetic N fertilizers since early 1890s, and projected to increase by 35–60% by 2030 (Smith et al. 2007). Fertilizers can also be an indirect source of N_2O emissions, via deposition of volatilized ammonia on natural ecosystems and denitrification of leached nitrate in subsoil, surface and groundwater and sediments (Cole et al. 1996). Estimations based on top-down analysis have suggested that for every kg of N fertilizer applied, 30–50 g of N is lost as N_2O through direct and indirect emissions (Crutzen et al. 2008). Therefore, increasing N use efficiency (NUE) continues to be a major challenge for the global agricultural production.

Application of chemical fertilizers at high rates has increased crop yields and resulted in pollution of soil, water, and air. As the intensity of agricultural production increases to keep pace with population growth and increasing demand for food, animal feed, fiber, and bioenergy, use of N fertilizer in agricultural production continues to increase. However, due to an inefficient use of N applied, crops seldom assimilates >50% of added N by crops (Newbould 1989). For most crops and pastures, plant N uptake ranges from 6 to 59% of the applied N. For example, the mean recovery of applied N to cereals worldwide is estimated at 33% (Raun and Johnson 1999). The low recovery of N in annual crops is mainly attributed to its loss through ammonia (NH_3) volatilization, nitrification, denitrification, leaching, erosion, and runoff. The relative importance of these processes varies widely depending on the agroecosystem, fertilizer formulation, and method/mode of its application. As much as 92% of the N applied can be lost from the plant-soil system (Fageria and Baligar 2005).

As a nutrient, N occupies a unique position among the essential plant nutrients because of its large demand by plants and heavy losses from the soil-plant systems. Most of N fertilizer which is not absorbed by plants is transported either into the ground- and surface waters and/or emitted into the atmosphere, and thus, contributes to degradation of soil and water quality, and to an overall environmental degradation. The N not assimilated by plants represents a large economic loss to farmers, and has significant adverse impact on the environment and human health. The NO_3^- leaching and NH_3 volatilization produced from N fertilizers causes: (i) eutrophication of lakes, rivers, estuaries, coastal and marine ecosystems, (ii) decrease biodiversity in terrestrial and aquatic ecosystems, (iii) increases invasion of N-loving weeds, (iv) causes acidification of soils and water resources, and (v) may lead to increased N_2O emissions (Krupa 2003; Loubet et al. 2010). Gaseous emissions (i.e., NH_3, NO_x and N_2O) also affect atmospheric visibility (i.e. smog or haze), reduce the stratospheric ozone layer that protects the earth from harmful ultraviolet radiation, increased ozone concentrations in troposphere which causes injury to crops and forest and possible health side-effects to humans, and contribute to global warming (Peoples et al. 2004; Barton and Colmer 2006). Furthermore, Nitrate leaching is linked to agricultural practices, mainly land cultivation and N fertilization (Jalali 2005). For example, 60% of coastal waters, rivers, and bays in the US are moderately or severely degraded by nutrient pollution, especially N (Howarth et al. 2002). Contamination of drinking water with NO_3^- has become a serious problem in many parts of the world (Newbould 1989; Fueleky and Benedek 2011). Drinking water containing more than 10 mg NO_3^--N L^{-1} is unfit for human consumption (US EPA 2009). An increased level of NO_3^- and NO_2^- in plants can also adversely affect the health of humans and animals (Shaviv and Mikkelsen 1993).

Fertilizers best management practices (FBMP) that synchronize N uptake with N availability increases NUE and minimize N losses to the environment (Delgado et al. 1998, 2001, 2011; Lal et al. 2011). The NUE is defined as the ability of agroecosystem to convert N inputs into outputs (Fageria and Baligar 2005). Higher NUE by plants reduces N fertilizer input costs, enhances crop yields, and reduces N_2O emissions (Baligar et al. 2001). Overall, NUE in plants is a function of the capacity of soil to supply adequate N, ability of plant to acquire N and transport it to roots and shoots, and synthesize it into other components. Therefore, NUE is influenced by genetic, morphological, and physical traits of plants and their interactions with external factors such as soil moisture, temperature, solar radiation, management practices, soil biological properties, and fertilizer material (Baligar et al. 2001). The objectives of this chapter are (i) to outline the approaches for increase NUE, (ii) describe FBMP to minimize N_2O emissions, and (iii) describe the role of enhanced efficiency fertilizers (EEF) in improving NUE and minimizing N_2O emissions.

10.2 Nitrogen Use Efficiency

The main goal of applying fertilizers to crops is to provide plant nutrients to increase or sustain optimal crop yields. Hence, improving NUE through increasing nutrient uptake and crop yield is the main priority of fertilizer producers and users. However, N fertilizers, either in the inorganic or organic form can also pollute the environments. Researchers and fertilizer producers are searching for ways to achieve higher defined goal of N fertilizer uptake efficiency to ensure that proper use of N fertilizers is beneficial to both crop production and the environment. Fertilizer uptake efficiency is defined as the percentage of fertilizer N recovered in aboveground biomass during the growing season, (i.e., N fertilizer recovery efficiency – NRE) by improving NUE and minimizing the environmental impacts. Efficiency is generally estimated as the ratio of nutrients output to input in agricultural cropping system. An agricultural cropping system consists of combinations of components, including: soil matrix, soil solution, soil microbes, roots, plants, and their management in the context of soil and regional specific parameters. Inorganic N fertilizers, biological N fixation by legumes and other N fixing organisms (BNF), atmospheric deposition, animal manures and crop residues accounts for 46, 20, 12, 11, and 7% respectively, of the total N input to global cropland (Cassman et al. 2002). Because N fertilizers are the largest source of N input and losses from cropland, improving NRE has substantial impact on the amount of Nr that escapes from cropland.

Nitrogen is the most limiting nutrient for crop production in most of world's agricultural areas, and its efficient use is important to the economic sustainability of cropping system. However, the NUE is usually low. Based on data from more than 800 experiments, it is estimated that on average, only 51% of the N applied is recovered by the aboveground parts of cereals (Dobermann 2005). Similar results were reported using ^{15}N isotope labeling for sugar cane (*Saccharum offinarum*) in Brazil (Ambrosano et al. 2011). However, N recovery may be even lower, especially under inappropriate soil and crop management. For example, the average N fertilizer recovery in cereals in China is 30–35% (Fan et al. 2004). Applied N that is not taken up by the plants can be immobilized in soil organic N (SON) pool and microbial biomass or lost to the environment (Cassman et al. 2002). Loss of N to the environment usually occurs when high concentrations of soluble N are present in the soil solution in excess of the amount that plants can absorb, or in the periods or locations in the soil profile where there are no plants or plant roots to absorb and synthesize it (Chien et al. 2009). The root system of most arable crops only explores 20–25% of the available soil volume for nutrients during a growing season. The nutrient availability in the soil-plant system is determined by complex interactions between plant roots, soil microorganisms, chemical reactions, and pathways for loss. Thus, the efficiency of N assimilation by crop also depends on these complex interactions among root system, soil microorganisms, chemical reactions taking place in soil and the processes leading to elimination of N from the soil. Generally, good N fertilizer management practices which relies on principles of using the right source of N, selecting appropriate rates of application compatible with plant needs, placing fertilizer within the root zone where plants can access the nutrients, and at the right time of application (Roberts 2007), may minimize N losses the and associated environmental impacts. Therefore, splitting N application into two or more doses may help achieve N FBMPs.

The N is absorbed from soil by plant root system as NH_4^+ and NO_3^- mainly through the process of mass transfer. Plant-available pool of soil N represent a very small fraction of soil N, and both indigenous soil resources and applied N input contribute to this pool. In order to make N assimilation more effective, the fertilizer should be delivered within 3–4 cm from the root system of the plant. The concentration-dependence of most processes that nutrients undergo in soil implies that when the immediate supply of N exceeds the demand, ranges of processes that lead to reduction in its concentration are set in motion. The extent by which these processes compete with plant uptake can thus affect both NUE and the environment. The overall NUE of cropping system can therefore, be increased by achieving greater uptake efficiency from applied N input, and reducing the amount of N lost from both organic and inorganic N pools (Cassman et al. 2002). Management of N fertilizers is further complicated by climatic conditions such as drought, excess rain and inundation, early frost, and other soil physical and chemical factors, which may limit plant growth and N uptake. These factors can lead to further increases in amount of residual soil N at the end of the growing season.

Increasing the NUE is crucial to increasing crop yields and quality, environmental safety, and the farm profit margin (Fageria and Baligar 2005). The NUE can be defined in many ways depending on the specific perspective. However, most of the definitions denote the ability of agroecosystem to convert N fertilizer inputs into economic outputs. Definitions of nutrient use efficiencies are sub-grouped into indices of nutrient recovery, nutrient utilization, physiological, agronomic, and partial factor productivity of applied N fertilizer, among others (Table 10.1; Cassman et al. 2002; Dobermann 2007). The NUE is generally defined as the maximum economic yield produced by a unit of N applied, absorbed, or utilized by plant to produce grain and straw (Fageria and Baligar 2005). Detailed studies on the fate of N in agroecosystems often involve isotope tracers that are particularly useful for understanding loss, fixation, immobilization, and release mechanisms (Cleemput et al. 2008). Use of ^{15}N-labeled fertilizers is especially useful for estimating crop uptake and soil recovery of applied N (Dobermann 2007).

The determination of NUE in crop plants is important to evaluating the fate of applied chemical fertilizers and their role in improving crop yields. In field studies, the NUE is either calculated based on differences in crop yields and/or nutrient uptake between fertilized plots and unfertilized control plots, or by using isotope labeled fertilizers to estimate crop and soil recovery of applied N (Dobermann 2007). The time scale for measurement is normally one cropping season, and spatial scale for measurement can range from plot scale to field or farm scale. For the same soil and cropping conditions, the NUE generally decreases with increase in amount of N added. Crop yield and plant uptake/accumulation typically increase with increase in N addition and gradually approaches the plateau, which is determined by climatic and genetic yield potential. At low levels of N supply, rates of increase of crop yield and nutrient uptake are large because N is the primary factor limiting plant growth. As N supply increases, incremental yield/uptake gains become smaller because yield determinants other than N become more limiting as the yield potential is approached.

Increasing the NUE continues to be a major challenge for the world agriculture. The global food demand will double by 2050 (Glenn et al. 2011), which is equivalent to maintaining a proportional increase in food production at the rate of about 2.4% per year over the next 40 years. This is a very large magnitude of increase compared to the historical cereal production trends, which have been increasing linearly at an average rate of 1.2–1.3% per annum for about half a century (FAO 2011). Meeting such a demand is a great challenge and requires close cooperation and understanding among disciplines across geographical regions and between public and private sectors. Increasing societal demand (for food, animal feed, fiber, and fuel), intense financial stress, growing concerns over impacts

Table 10.1 Indices of nutrient use efficiency, method of calculation and their interpretation

Index	Calculation	Interpretation	Nitrogen in cereals
Apparent crop recovery efficiency (kg increase in N uptake per kg N applied)	$RE = (U-U_0)/F$	Depends on the pant nutrient demand and release from fertilizer Affected by method of application, amount, timing, and N form Affected by factors that determine size and crop nutrient sink – genotype, climate, plant density, and biotic/abiotic stress	0.3–0.5 kg/kg, for a well-managed systems 0.5–0.8 kg/kg at low levels of N use, or at low soil N supply
Physiological efficiency of applied N (kg increase in N uptake from fertilizer)	$PE = (Y-Y_0)/(U-U_0)$	Ability of plant to transform nutrients acquired from fertilizer into economic yield i.e. grain Depends on genotype, environment and management Low PE suggests suboptimal growth i.e., nutrient deficiencies, drought stress, pests or toxicity	40–60 kg/kg, well managed system, at low levels of N use or at low soil N supply
Internal utilization efficiency of a nutrient (kg yield per kg nutrient uptake)	$IE = Y/U$	Ability of plant to transform nutrients acquired from all sources (soil, fertilizer) into economic yield i.e. grain Depends on genotype environment and management A very high IE suggests deficiency of nutrient Low IE suggests poor internal nutrient conversion due to other stresses i.e. nutrient deficiencies drought stress mineral toxicity, pest	30–90 kg/kg 55–65 kg/kg is the optimal range for balanced nutrition at high yield levels
Agronomic efficiency of applied nutrient (kg yield increase per nutrient applied)	$AE = (Y-Y_0)/F$, also $AE = RE*PE$	Product of nutrient recovery from mineral or organic fertilizer (RE) and the efficiency with which the plant uses each additional unit of nutrient (PE) AE depends on management practices that affect RE and PE	10–30 kg/kg; >25 kg/kg in well-managed systems, at low levels of N use, or at low soil N supply
Partial factor productivity of applied nutrient (kg harvested product per kg nutrient applied)	$PFP = Y/F$ or $PFP = (Yo/F) + AE$	Most important for farmers because it integrates the use efficiency of both indigenous and applied nutrients High indigenous soil nutrient supply (Yo) and high AE are equally important for PFP	40–80 kg/kg; >60 kg/kg in well-managed systems, at low levels of N use, or at low soil N supply

on water and air quality, simultaneous improvement of productivity and resource use efficiency including NUE are among the essential goals of sustainable agriculture (Fixen 2009).

Total nutrient requirement, specific periods of peak demand and preferred chemical formulations are specific to crop type and variety. Mineral constituent uptake by plants in their vegetation cycle follows a sigmoidal response (Shaviv and Mikkelsen 1993). Thus, matching or synchronizing nutrient supply to plant demand is likely to both reduce losses by processes competing with uptake and provide optimal nutrition for plant growth (Hauck 1985). An application of controlled-release fertilizers (CRF) which release their nutrients in a way better fitting to plants' requirements ensures an improved effectiveness of fertilizing through minimizing the losses between application and absorption (Shaviv 2005a). In addition, CRF reduces the negative influence of N fertilizers to the environment by minimizing surplus N in soil. For the conventional fertilizers, N release lasts between 30 and 60 days, which, given a 100–120 days of crop growth cycle means that N fertilizers must be applied two to three times to meet N requirements during the growing season. The CRF release their nutrients slowly and gradually during the whole vegetation season. Consequently, they need to be applied only once, which greatly reduces labor and energy consumption (Trenkel 2010).

10.2.1 Approaches to Improve Nitrogen Use Efficiency in Agriculture

Adopting appropriate N management strategies is an important step towards improving the NUE, and efficient crop production. These involve manipulating soil, plant, climatic and fertilizer variables to create synergism among crop production variables. Soil acidity adversely affects physiological, morphological and biochemical processes in plants and consequently N uptake and use efficiency (Baligar et al. 1997). Liming is the most common and effective practice of reducing soil acidity and the related problems. Other manageable factors include fertilizer type, amount, method, time of application, moisture status of soil (controlled by irrigation and drainage soil moisture conservation), and compaction (managed by tillage and controlling traffic). Generally, N loss can be decreased by management practices that increase crops' ability to compete for nutrients uptake against loss processes by erosion, leaching, or volatilization (Minami 1997).

Principal approaches to improving the efficiency of N fertilizer include: (i) use of soil and plant testing to make best use of indigenous soil N (Johnkutty and Palaniappan 1996; Dobermann and Cassman 2004), (ii) use of right N fertilizer formulation, rate, method and time of application (Smith et al. 1997; Roberts 2007), (iii) incorporation or deep placement of fertilizer (Cai et al. 2002; Roy and Hammond 2004), (iv) using split applications, i.e. breaking N application rates into several applications instead of single large rate to ensure the supply of N throughout the growing season (Chen et al. 2008), (v) minimizing application of N during the wet season to reduce leaching and denitrification losses (McTaggart et al. 1994), (vi) integrating fertilizer application with crop demand, i.e. delaying N application until substantial canopy has established (Humphreys et al. 1988), and (vii) using inter-seasonal cover crops to minimize the accumulation of nitrate during fallow periods (Wagner-Riddle and Thurtell 1998).

10.2.1.1 Nitrogen Sources and Placement

The concept of FBMPs is based on the principle of matching the nutrient supply with crop requirements while minimizing nutrient losses from the field. In a good fertilizer program, appropriate sources, adequate rates, efficient methods, and application timing should be determined according to crop demand. These factors are vital to improving the NUE, maintaining sustainable crop production, and reducing NH_3 and N_2O emissions. The N sources and method of application significantly influence N uptake efficiency by plants. Readily soluble N fertilizers such as urea, anhydrous ammonia, urea-ammonium nitrate, and ammonium sulfate are the most commonly used fertilizer formulations. However, the important considerations for selecting sources of N by farmers are availability, economics, convenience in storage and handling, and effectiveness of the N carrier. For example, in the US agriculture, anhydrous ammonia (NH_3) has been the important source of N fertilizer, accounting for 35% of the total use, while urea accounts for 24% of all N fertilizers consumed in 2008 (ERS 2011). Ammonia is injected into the soil to reduce loss through volatilization. Once injected, the NH_3 gas protonates to NH_4^+ ions, which are retained in cation exchange sites of the soil. The major advantages of anhydrous NH_3 are its high N analysis (82% N) and low cost of transportation and handling. However, specific equipment is required for storage, handling, and application. Therefore, anhydrous NH_3 is not a popular N carrier in the developing countries. In these countries, urea and ammonium sulfate are the preferred carriers of N fertilizers (FAOSTAT 2011).

The choice of N source is influenced by economics and logistical requirements of minimizing risks of N loss. The N source selection can affect the proper rate, timing, and placement method. Applying N as NH_4^+ rather than as NO_3^- is likely to minimize total N_2O emissions. However, fertilizing with NH_4^+ fertilizers increases the potential for NH_3 volatilization (Harrison and Webb 2001). The NO_3^- form is more vulnerable to leaching and denitrification, which results in N_2O and N_2 gaseous losses (Harrison and Webb 2001). The NH_4^+-based fertilizer is also preferred over NO_3^- sources when N is applied early in the season before the crop root system is well established (Harrison and Webb 2001). However, the source of N used as topdressing is less critical where NH_4^+ and NO_3^- forms appear to be equally effective. This is because when crop is fully established the NO_3^- form of N is rapidly absorbed by the crop before it can be leached into subsoil layers or is denitrified. The sources of N containing NO_3^- should not be used for soils that are typically inundated or waterlogged early in the season. Water logging during warm conditions (i.e., in late spring through summer) can exacerbate N_2O emissions (Snyder 2008). The use of NO_3^- fertilizers should be avoided where denitrification is likely to be the main process responsible for N loss. Matching fertilizer type with rainfall and soil moisture regimes can significantly influence the N loss (McTaggart et al. 1994). For example, trickle or drip irrigation system, which allows delivery of N to the area of maximum crop uptake enables the application rate to be matched with plant requirement, and with careful operation trickle systems reduces deep percolation, runoff, and denitrification. Thus, judicious water management reduces denitrification by ensuring that water-filled pore spaces (WFPS) remain <60% at all times (Smith et al. 1997; Mosier et al. 2002).

The urea-containing sources of N should only be used when incorporation in soil to 8–15 cm is possible and planned/timed so that there is either rainfall or irrigation within 24–48 h after application. This practice is recommended where the environmental conditions are conducive to NH_3 volatilization losses (Trenkel 1997a). There is growing evidence that enhanced efficiency fertilizers can enhance the crop N recovery and minimize N losses to the environment (Chen et al. 2008). Emission of N_2O after application of anhydrous NH_3 is two to four times higher than surface applied urea, ammonium nitrate or broadcasted urea (Venterea et al. 2005). Use of urease inhibitors to delay the hydrolysis of urea and minimize NH_3 volatilization is discussed in this chapter. Placement of any N sources in direct contact with plant residues may increase the risk of N_2O emissions (Parkin and Kaspar 2006).

10.2.1.2 Nitrogen Rate

Use of appropriate N rates and balancing these with other essential nutrients increases the NUE, minimize N losses, optimizes crop yields, and protects the environment. The management of N rate is aimed at minimizing residual NO_3^--N and reducing the risk of N_2O emissions (McSwiney and Robertson 2005; Halvorson et al. 2008). A number of studies have shown that the emissions of N_2O are well correlated with fertilizer N rate (Bouwman et al. 2002; McSwiney and Robertson 2005; Hoben et al. 2011). These studies indicates that increasing the amount of N added results in increased N_2O emissions. This is the foundation for the current IPCC (2006) N_2O inventory calculations. Management of N that considers the soil supply of N and incorporates all N sources (i.e., SON mineralization, biological N fixation, manures, irrigation water, and atmospheric deposition) minimizes surplus N that may be lost via N_2O emission and NO_3^- leaching. The N management can be refined and fine-tuned by the knowledge of partial factor productivity, partial nutrient balance, agronomic efficiency and fertilizer N recovery efficiency (Snyder and Bruulsema 2007; Snyder 2008).

10.2.1.3 Timing of Nitrogen Fertilizer Application

Synchronizing timing of N application with plant demand is an important factor in determining soil N availability and N_2O emissions. Crop N uptake capacity at the beginning of growing season is generally low, increasing rapidly during vegetative growth and dropping sharply as the crop nears maturity. Timing of N application affects crop N uptake and potential for elevated NO_3^-, which increases the risk of NO_3^- leaching and N_2O emission. Fall application of NH_4^+ sources for spring planted crops should only be done when soil temperature at 10–15 cm depth remains below 10 °C throughout the winter period to minimize NO_3^- loss and N_2O emissions (Snyder et al. 2001). Fall N application should be avoided for medium-textured and well-drained soils in the humid regions. The application of N may be timed to coincide with N uptake demand so that crop uptake is maximized while the excess residual NO_3^- is minimized to reduce N losses. Use of split-application also reduces N loss, and maximizes crop uptake. Avoiding N application to wet or waterlogged soils, except for flooded rice culture, minimizes N loss through NO_3^- leaching, NH_3 volatilization, and N_2O emissions. Use nitrification and urease inhibitors with NH_4^+ and urea N fertilizers, respectively, (this chapter) are also important to reducing losses.

10.2.1.4 Other Farm Management Practices

Incorporating plant residues with high C:N ratio into soils may help immobilize mineral N and reduce N losses. The immobilized N can then become available later when residues are mineralized (Aulakh et al. 1991). Uptake of N by plants is also optimized by the management of total crop nutrients requirement. The supply of one nutrient affects the absorption, distribution, or function of another. For example, insufficient amounts of plant available P, K, or S reduces the NUE. The N recovery is 70% when P is adequate, and decreases to <40% for P deficient corn (Oberle and Keeney 1990). The NUE is also increased by the adoption of those crop management practices which optimize N effectiveness and minimize N losses (i.e., selection of high yielding crop varieties or hybrids with superior genetics, use of conservation tillage, appropriate pests, diseases, and weed control) also increases NUE.

Optimum N rates vary spatially and seasonally. Therefore, diagnostic tools to assess soil and crop N status and make decisions on the amount of N to be applied during growing season can drastically improve the NUE (Schroder et al. 2000). The NUE can also be improved by integration of biophysical variables and interacting processes and management practices using decision support system based on comprehensive and process-based agro-ecosystem models for optimum N fertilizer management can improve the NUE. The FBMPs can be identified by simulating the combination of different management practices and their interaction of N application rate, time, and trade-off between economic and environmental interests

(Chen et al. 2006). In addition to improvements in fertilizers, use of plant breeding and genetic engineering can also lead to substantial improvements in farming efficiency by developing modern fertilizer-efficient cultivars for most crops (Meredith et al. 1997; Cassman et al. 2002; Weisler et al. 2001; Ronald 2011).

10.3 Enhanced Efficiency Fertilizers

In addition to the production of conventional N fertilizer types, the science and fertilizer industry has aimed at developing special types of N fertilizers, which avoids or substantially reduces N losses while maintaining N availability to plants and increasing the NUE. The Association of American Plant Food Control Officials (AAPFCO) have adopted the term enhanced efficiency fertilizers (EEF) to characterize the products which can minimize the potential of nutrients loss to the environment compared to reference soluble conventional fertilizers (Hall 2005). The fertilizers that fall under this category include slow-release fertilizers (SRF), controlled-release fertilizers (CRF) and stabilized fertilizers. There is considerable interest in using these newer forms of N fertilizers which releases N at a reduced or controlled rate and render it less transferable into the environment. These fertilizers are grouped into two general categories: (i) slow- and controlled-release fertilizers, and (ii) stabilized fertilizers – with nitrification or urease inhibitors (Weiske 2006).

The ideal fertilizer should have three characteristics (Shoji and Gandeza 1992): (i) requires a single application throughout the entire growing season to supply the necessary amount of nutrients for optimum plant growth (ii) have a high maximum percentage recovery and a higher return to the production input, and (iii) have a minimal detrimental effect to soil, water, and atmospheric environment. Thus, SRF, CRF, as well as stabilized N fertilizers meet some of the requirements for ideal fertilizers to a considerable extent (Weiske 2006), because these formulations significantly reduce N_2O emissions (Akiyama et al. 2010; Halvorson et al. 2010).

In addition to competition between soil and plant roots for available nutrients in the soil plant system, there are other complex interactions between plant roots and soil microorganisms, chemical and physical reactions on and within soil particles, and soil conditions conducive to losses of plant nutrients, which may affect nutrient availability to plants. Most of the transformations that nutrients undergo in the soil (i.e., soil solution interface) are concentration-dependent (Shaviv 2005a). Any surplus plant nutrient in the soil that is not taken up by plants may undergo the following reactions that decreases their availability to plants: (i) microbial transformation – nitrification, denitrification, or immobilization, (ii) chemical – exchange, fixation, precipitation or hydrolysis, and (iii) physical – leaching, runoff or volatilization (Trenkel 2010). Use of CRF, SRF, and stabilized fertilizers reduces N losses by these three processes (Chen et al. 2008). Thus, FBMPs are aimed at avoiding surplus N in the soil. While the management techniques described in Sect. 10.2.1 have potential to increase NUE and reduce N_2O emissions, considerable N losses still occurs. For example, under flooded rice (*Oryza sativa*), the time of application has a large effect on the agronomic efficiency of N fertilizer and NH_3 volatilization. However, even with the adoption of FBMP devised, ~40% of applied N can be lost through leaching, NH_3 volatilization, or denitrification (Humphreys et al. 1988). The application of conventional N fertilizers, especially with single dose application, results in too large amount of available N in the early stages and too little at later stages of crop growth.

Numerous studies have evaluated EEF and reported variable results (Smith et al. 1997; Trenkel 1997b). The high variability in efficiency is mainly due to lack of understanding of the interaction of these chemicals with soil and environmental variables (Mosier et al. 2002). For example, dicyandiamide ($H_4C_2N_4$), (DCD) is an effective nitrification inhibitor under cold climatic conditions. In warm and humid climates, however, it is less effective due to its rapid decomposition (Di and Cameron 2004; Hatch et al. 2005). Several studies have evaluated the effects of EEFs on biomass and/or grain yield (Trenkel 1997b; Guertal 2009) and gaseous N losses (Smith et al. 1997; Akiyama et al. 2010), but only few have evaluated their effectiveness on NO_3^- leaching (Diez et al. 1994, 1996).

10.3.1 Slow and Controlled Release Fertilizers

The SRF or CRF are defined as a fertilizer containing a plant nutrient in a form, which delays its availability to the plant significantly longer than a reference fertilizer (Trenkel 2010). Such a delay of initial availability or extended time of continued availability of nutrients is attributed to different mechanisms, including: (a) controlled water solubility of materials such as semi-permeable coating, occlusion of nutrient, or by inherent water insolubility of polymers, natural nitrogenous organics, protein materials, or other chemical forms, (b) low solubility of inorganic materials such as metal-ammonium phosphates, (c) presence of relatively soluble materials which gradually decompose in soil, and (d) slow hydrolysis of water soluble low molecular weight compounds or by other unidentified mechanism (Hauck 1985; Weiske 2006). The CRF are fertilizers in which the factors determining the rate, pattern and duration of release are well known and controllable during fertilizer manufacturing process. In contrast, SRF are fertilizers which release the nutrient at a slower rate than is usual, but the rate, pattern and duration of release are not well controlled (Trenkel 2010).

The CRF and SRF are schematically classified into three main types (Shaviv 2005a, b): (1) organic low solubility compounds e.g. urea formaldehyde (UF), isobutylene diurea/crotonylidene diurea (IBDU), and urea-acetaldehyde/cyclo diurea (CDU) (i.e., condensation product of urea and acetaldehyde); (2) inorganic low solubility compounds e.g. magnesium ammonium phosphate, and (3) coated materials in which physical barrier control the release, e.g., granules coated with hydrophobic polymers or matrices in which the soluble fertilizer is dispersed so dissolution of fertilizer is restricted.

The coated N fertilizers can be further grouped into: (i) sulfur-coated urea, (ii) sulfur and polymer-coated urea – including wax polymeric materials, and (iii) polymeric/polyolefin material coated fertilizers (Guertal 2009; Trenkel 2010). Sulfur coatings provide highly variable N release patterns depending on coating damage that may occur. As much as one-third of N can be released instantaneously. To improve its performance in terms of the N release patterns, a polymer coating is added to sulfur-coated urea. Polymer – sulfur coated urea reduces N_2O emissions and leaching losses to as low as 1.2% of the applied N in turf lawn in a temperate climate (Guillard and Kopp 2004), and improves turf quality (Miltner et al. 2004).

Nutrient release from polymer coating is controlled by the thickness of coating (Trenkel 2010; Shaviv 2005a). The most accurate controlled release of N is provided by polyolefin thermoplastic-coated fertilizers (Chen et al. 2008). Fertilizer release is controlled by polyolefin coating such as polyethylene and polypropylene, which have low water permeability and ethylene vinyl acetate, which has high water permeability. The pattern of nutrient release from coated fertilizers can be parabolic, linear or sigmoidal and long-term or short-term (Shaviv 2005a).

Encapsulated fertilizers have the granules encapsulated or coated with an inert layer. After application, water penetrates through a hydrophobic membrane into the inside of a granule and dissolves nutrients, and the increase in osmotic pressure leads to a partial tearing of membrane or its expansion, which allows ions transportation through coating into the soil (Kochba et al. 1990). The rate of nutrient release is controlled by the coating diffusion coefficient. Examples of encapsulated fertilizers includes sulfur-coated urea (Jarrell and Boersma 1979), polysulfide coated urea (Tomaszewska and Jarosiewicz 2002); other coating materials include gum, waxes, paraffin, ester copolymers, and polyolefines. Other categories of CRFs are those with no physical barrier in the form of polymer material, but the release of nutrient N is determined by the solubility a given fertilizer. Commonly used low-solubility fertilizers include a group of inorganic a compounds having a general formula of $MeNH_4PO_4 \times H_2O$, where *Me* stands for divalent cation such as Mg^{2+}, Fe^{2+}, Zn^{2+}, or Mn^{2+}. In some cases, mixtures of K^+ and NH_4^+ salts are used to provide slow-release N-P-K fertilizer (Shaviv 2005a).

10.3.1.1 Effects of Controlled- and Slow-Release Fertilizers on Crop Yields

Both CRF and SRF release N into the soil environment in a slower manner, which more or less matches the plant demand. The supply of N by a single application satisfies plant requirement, maintains low concentrations of mineral N in soil throughout the growing season, and can substitute for split application. Thus, they reduce the requirement for multiple field operations, reduces labor requirement, and minimizes fuel costs. As a result, they offer an effective way to improve NUE and reduce the environmental impact by decreasing the magnitude of N loss (Hauck 1985; Shaviv 2001), and improving crop yields. However, the largest proportion of these fertilizers are currently consumed in non-agricultural markets (e.g., lawn care, golf courses, and landscaping) and only ~10% is used in agriculture

(Shaviv 2001), mainly in high-value crops (e.g., vegetable crops and *Citrus* spp.) (Obreza and Rouse 2006; Morgan et al. 2009).

The N uptake of seasonal and perennial plant species is generally sigmoidal (Shoji et al. 2001; Shaviv 2005a). The pattern of nutrient N release from coated fertilizers can be sigmoidal, parabolic or linear, and short- or long-term and good synchronization between N release and plant demand can be achieved (Shaviv 2005a). Reduction in soil microbial populations by fumigants, used for reduction of weeds and soil-borne insects and diseases, is the greatest impediment to the use of some CRF and SRF products. Reduction in soil microbial population can delay fertilizer release by 6–8 weeks (Shoji and Kanno 1994), limiting applicability of some products to relatively short-term vegetable crops for which soil fumigation is an acceptable practice.

Use of polyolefin-coated urea instead of uncoated fertilizer increases yields and NUE in a range of crops (Shoji et al. 2001; Fashola et al. 2002; Shoji 2005). Improved growth and yield with CRF compared with soluble fertilizers also have been documented in potato (*Solanum tuberosum*) (Pack et al. 2006). Ammonium-based fertilizers have been coated with polyolefin for use in vegetable production to prevent the build-up of NO_3^-, which affects quality of vegetable and constitutes a health risk (Shoji 2005; Wang et al. 2005). Addition of nitrification inhibitor DCD to the fertilizer before coating the mixture with polyolefin can result in further decrease in NO_3^- uptake by vegetables. However, the use of CRF may result in N in excess of the crop's requirements, and the excess N remaining in the soil after the crop harvest. This excess N may then be lost into the environment, as does the uncoated fertilizer (Delgado and Mosier 1996). Thus, coated urea usually performs better than conventional fertilizers by increasing grain yield and N uptake such as for rice in Spain (Carreres et al. 2003), winter wheat (*Triticum aestivum*) in China (Fan et al. 2004), peanut (*Arachis hypogea*) in Japan (Wen et al. 2001), potatoes in USA (Munoz et al. 2005), and maize (*Zea mays*) in Japan (Shoji et al. 2001).

10.3.1.2 Effects on Nitrate Leaching Gaseous Losses of Nitrous Oxide

Application of SRF and CRF formulations are effective in reduction of NO_3^- leaching losses (Wang and Alva 1996; Owens et al. 1999; Pack et al. 2006). Wang and Alva (1996) observed large reduction in NO_3^- leaching in sandy soils amended with IBDU and polyolefin resin-coated urea as compared to application of NH_4NO_3 fertilizer. Overall, the benefits of using SRF and CRF to reduce NO_3^- leaching depends on soil properties and rainfall regime. Decrease in N_2O emissions have also been observed with the application of CRF and SRF compared to conventional fertilizers such as urea under diverse crops and soil types (McTaggart and Tsuruta 2003; Shoji et al. 2001; Vallejo et al. 2001). However, reduction of soil N_2O emissions by use of SRF and CRF has not always been consistent from year-to-year and season-to-season. These variable results are indicative of the complex soil, climatic and management factors that affect soil N_2O emissions and high spatial and temporal variability in emissions of N_2O under field conditions.

The CRFs are useful to the reduction of N_2O emission from fertilized soils. The reduction in N_2O emission occurs by affecting the timing of N release from fertilizer, which has a potential to reduce losses by leaching of NO_3^-, volatilization of N as NH_3, and N_2O as gaseous emission (Shaviv 2001). Reduction in losses may improve the NUE and enhance the fertilizer N performance. Large reductions in N_2O have been achieved by using polyolefin coated ammonium nitrate, ammonium sulfate and urea compared to uncoated fertilizers (Smith et al. 1997; Shoji et al. 2001).

Application of a slow-release polymer-coated urea fertilizer reduces initial N_2O emissions compared to that of urea with or without nitrification inhibitor dicyandiamide to barley (*Hordeum vulgare*) on clayey soil (Delgado and Mosier 1996). However, N_2O emissions can continue after the first 60–80 days following fertilization, such that total N_2O emission from plots treated with polymer-coated urea (PCU) can be higher than from plots treated with urea alone. A meta-analysis of published field experiment data across 113 datasets from 35 field studies revealed that application of polymer-coated enhanced efficiency fertilizers (PCF) significantly reduced N_2O emissions by 35% (range 14–58%) compared to conventional fertilizers (Akiyama et al. 2010). However, the effect of PCF on N_2O emissions produced contrasting results among land uses and soil types. The PCF are very effective on imperfectly drained Gleysol grassland, but not so effective in well-drained Andosol upland fields (Akiyama et al. 2010). Soil drainage strongly affects N_2O emissions, and emissions are lower from well drained than those from poorly drained soils (Akiyama et al. 2006). Furthermore, PCF are more effective in reducing N_2O when the emissions are relatively large.

10.3.2 Stabilized Fertilizers

The other group of the EEF consists of stabilized N fertilizers. These are the fertilizers that are treated with inhibitors such as nitrification or urease inhibitors, which would avoid rapid transformation of N into forms that are less stable in soil environments (Chien et al. 2009). Stabilizing compounds are added to extend the time the N component of the fertilizer remains in the soil in the form of urea N or ammoniacal N. Nitrification inhibitors are substances which inhibit the biological oxidation of NH_4^+-N to NO_3^--N. Similarly, urease inhibitors are substances that inhibit hydrolytic action on urea by enzyme urease.

Soil NH_4^+ is derived from several sources including (Fig. 10.1): (i) mineralization of organic N, (ii) addition of NH_4^+-containing fertilizers (iii) as a result of hydrolysis of urine and urea. Hydrolysis of urine and urea is usually very rapid (few days) and is facilitated by a ubiquitous soil microbial enzyme "urease." Volatilization of NH_3 to the atmosphere only occurs at high pH (pH >7.0), and thus, mostly in all alkaline soils. However, in soils where urea has been applied or in urine patches of pasture soil, the process of hydrolysis produces alkaline conditions in the immediate vicinity of contact with soil.

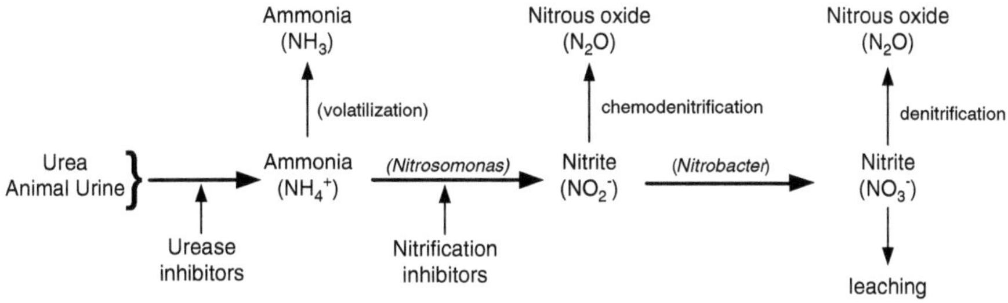

Fig. 10.1 Some important biochemical reactions relevant to the use of urease and nitrification inhibitors

$$(NH_2)_2CO + H^+ + H_2O \rightarrow 2NH_4^+ + HCO_3^- \quad (10.1)$$

$$HCO_3^- + H^+ \rightarrow CO_2 + H_2O \quad (10.2)$$

Under typical conditions NH_4^+ is oxidized first to nitrite (NO_2^-) by *Nitrosomonas* bacteria, and then to NO_3^- by *Nitrobacter*. The relative speed of these reactions is such that, under normal soil conditions ammonia and NO_2^- concentrations in soils remains low relative to that of the NO_3^-. However, plants can utilize both NH_4^+ and NO_3^- as sources of N. The NO_3^- is soluble and is not held by soil, and is subject to leaching under appropriate conditions. Furthermore, NO_3^- can be reduced by denitrifying bacteria to N_2O and N_2 gases through the denitrification process under anaerobic conditions. These gases can also be produced from NH_3 and NO_2^- under aerobic conditions by the chemodenitrification process. In contrast, NH_4^+ is retained by soil cation exchange capacity (CEC). Controlling nitrification and/or urease hydrolysis potentially restricts NO_3^- leaching, GHG emissions and NH_3 volatilization from soils. As a result, N cycling is more efficient and the NUE is increased.

10.3.2.1 Nitrification Inhibitors

Maintaining N in the NH_4^+ form in the soil can prevent its loss by leaching, nitrification and denitrification, and it can be achieved by adding a nitrification inhibitor with the N fertilizer which slows the microbial conversion of NH_4^+ to NO_3^-. Therefore, it prevents NO_3^- leaching and production of NO and N_2O by both nitrification and denitrification processes. Even though this process does not always increase crop yields, it does provide a tool for reducing NO_3^- leaching and N_2O production. Surveys of USA farmers have indicated that about 9% of the national corn production area is treated with nitrification inhibitors (NI), and this proportion has remained unchanged in recent years (Christensen 2002).

Decreasing denitrification of NO_3^--N to N_2O and NO can also benefit the environment, as N_2O and NO are greenhouse gases, with a high radiative forcing (Snyder et al. 2009). Three commonly used NIs are Nitrapin – 2-chloro-6-(trichloromethyl), dicyandiamine (DCD, $H_4C_2N_4$), and 3,4-dimethylpyrazole phosphate (DMPP) (Zerulla et al. 2000). In addition, a wide variety of plant extracts inhibit culturable nitrifiers, and few of these, such as neem oil extracted from Indian neem tree (*Azadirachta indica*) have shown to slow nitrification in situ (Kumar et al. 2008). The majority of the research indicates that NIs, when applied to soils in conjunction with N fertilizers, reduces NO_3^- leaching and N_2O emissions and increases plant growth by increasing the NUE (McTaggart et al. 1997; Merino et al. 2002). However, effects of NIs on N losses and plant yields may be none or variable (Prasad and Power 1995; Merino et al. 2001). In general, DCD increases NH_3 volatilization by increasing NH_4^+ concentration and hence N_2O emissions (Davies and Williams 1995; Nastri et al. 2000; Gioacchini et al. 2002). The use of DCD may have priming effect on net mineralization of SON resulting in greater N losses in the long-term (Gioachini et al. 2002). Some nitrification inhibitors, including DCD may have toxicity effect on some plants (Prasad and Power 1995; Macadam et al. 2003). The variable results are associated with the fact that the effectiveness of NIs decreases with increase in time since application to soils. The effectiveness of NIs are also influenced by soil temperature, soil moisture, soil pH, and SOM content. The effects of NI are more difficult to predict under the field conditions. They tend to have greater effects on those soils that are rich in N and where N losses due to leaching and denitrification are large. The expression of these effects to plant growth depends on the N status of a soil. Limiting N losses in soils that are rich in N may have little effect on plant growth. Thus, N status and other soil factors (i.e., texture, moisture content, SOM content, pH, temperature), climatic factors (i.e., temperature, rainfall intensity and frequency) which determines the size of these losses also influence the effectiveness of NIs. Large beneficial effects of NIs on leaching are more likely on free draining soils under high rainfall than on slowly draining soils. The soil N status determines the yield benefits. With high soil N levels, conserving N may have little effects on crop yield. Further, NIs may have little effect on N leaching from heavy clays with impeded drainage. In clayey soils with low permeability, NIs may increase NH_3 volatilization by increasing the soil NH_4^+ concentration given appropriate conditions.

The NIs are the most widely tested mitigation option for N_2O emission from agricultural soils. Because nitrification and denitrification are the important pathways for N_2O production (Conrad 1996), the NIs inhibit ammonium monooxygenase and block the first reaction of NH_4^+ to NO_2^- (Subbarao et al. 2006). By minimizing the rate of nitrification until the crop is in its lag phase of growth, NIs enhances absorption of NO_3^-, reduces denitrification and leaching losses, increasing NUE and potentially reducing N_2O emissions. The meta-analysis of 85 dataset from literature indicated that application of NIs reduced N_2O emissions by an average of 38% (range 31–44%, Akiyama et al. 2010). The NIs included in the meta-analysis are: DCD (42), nitrapyrin (10) and 3, 4-dimethyl pyrazole phosphate (DMPP) (12), calcium carbide (8) neem (*Azadirachta indica*) (8) and thiosulfate (4). This analysis indicated that the effectiveness of NIs increased with increasing emission of N_2O (Akiyama et al. 2010).

10.3.2.2 Urease Inhibitors

Urea is widely used form of N worldwide for crop production because of its high N content (46% N), which makes it less expensive and easier to transport (FAOSTAT 2011). Globally, urea consumption has stabilized at 30 Tg yr^{-1} for the developed countries, while the consumption is still increasing dramatically in the developing countries, and reached about at 55 Tg in 2002 (IFA 2006). However, high losses of N can occurs if urea is not incorporated into the soil soon after application. The NH_3 volatilization also occurs in neutral, alkaline, and flooded soils during the early stages of plant growth. In addition, hydrolysis of urea $[(NH_2)_2CO]$ to NH_4^+ HCO_3 produces high pH that induces NH_3 volatilization under conditions of high wind, moistened soil surface, low plant canopy and high temperatures. The soil pH in the vicinity of urea granules increases due to of hydrolysis (Eqs. 10.1 and 10.2), exacerbating the volatilization of NH_3 into the atmosphere. Typical losses can range from 5 to 20% of the total N applied, but results are highly variable and can be as high as 50% in extreme conditions (Harrison and Webb 2001; Cai et al. 2002). Similar losses also occur from urine-affected pastures (Ledgard 2001). The loss occurs after urea is converted to NH_3 at soil surface by reaction with the enzyme urease. The most important factors determining NH_3 volatilization are soil pH, temperature, moisture, and rainfall.

The effects of treating urea with urease inhibitors (UIs) are also variable depending on the same variables that control NH_3 volatilization. The UIs can slow the rate of hydrolysis, and reduce the effects of high NH_4^+ concentrations, which can inhibit germination and damage the seedlings (Watson 2000; Malhi et al. 2003).

The use of UIs to reduce NH_3 volatilization from urea hydrolysis has thus been considered an effective strategy to increase NUE of urea-based fertilizers. Many chemical compounds with differing characteristics have been tested for their ability to inhibit urease activity (Watson 2005). Some inhibit the enzyme by reacting with active sites on the enzyme or a key functional group elsewhere in the molecule, or by changing the conformation of the active site. Many organic and inorganic compounds and metal ions inhibit urease by reacting with the sulfhydryl groups in the enzyme, and others by complexing with Ni in the active site, and by reacting with carboxylic acid group. The most effective compounds for inhibition of urease activity are phosphoryl amides (e.g., N-(n-butyl) thiophosphoric triamide (nBTPT) and cyclohexylophosphoric triamide), and these are the most commonly used UIs (Chen et al. 2008).

The interest in urease inhibitors is well justified because the risks of N losses via NH_3 volatilization significantly contribute to low NUE. However, the research data show that UIs cannot fully control NH_3 loss when urea is surface-applied to soils because the inhibitory effect depends on soil conditions. The UIs available so far can prevent urea hydrolysis for about 1–2 weeks, during this period the fertilizer should ideally be incorporated into the soil by rain, irrigation, or mechanical methods. However, such incorporation may not always happen. The short-lived effect of UIs is also a constraint when urea is surface applied to a flooded rice field. Nevertheless, despite having only partial capacity for reducing NH_3 losses, UIs is an option that cannot be disregarded due to growing presence of urea in the fertilizer market.

The nBTPT at concentration of 0.01% of the applied urea N can reduce NH_3 volatilization by about 50% in temperate grasslands (Watson et al. 1994a, b). In addition, nBTPT can significantly delay and reduce NH_3 and N_2O emissions from soil after application of urea, urine (Grant et al. 1996; Wang and Douglas 1996) and urea ammonium nitrate and produce significant improvement in the NUE of corn following application of urea ammonium nitrate (Fox and Piekielek 1993). Both nBTPT and CPT have also been used successfully to control NH_3 emission from animal wastes, prevent environmental degradation and to produce a more balanced N and P levels in manure (Varel 1997; Varel et al. 1999).

The UIs slows the conversion of urea to NH_4^+ and hence reduce the concentration of NH_4^+ present in the soil solution and the potential NH_3 volatilization (Watson 2005). Together with uptake by plants, a lower NH_4^+ concentration in the soil makes less N available for nitrification and denitrification. However, a meta-analysis of the published data has indicated that UIs may not reduce N_2O emissions relative to conventional fertilizers (Akiyama et al. 2010), probably because hydrolysis of urea is not directly related to N_2O emission as is the case for nitrification. If the uptake of NH_4^+ does not increase significantly, a similar amount of NH_4^+ may eventually undergo nitrification and denitrification as compared to application of conventional urea.

10.3.3 Advantages and Disadvantages of Enhanced Efficiency Fertilizers

The Third Assessment Report of the International Panel on Climate Change (IPCC) stated that fertilizer management practices such as the use of SRF, NIs, and organic manure might globally reduce N_2O emissions from fertilizer use by 30% on global basis (Moomaw et al. 2001). The IPCC Fourth Assessment Report also considered nutrient management, including enhanced efficiency fertilizers as one of the mitigation options, with average mitigation potential of 0.07 Mg CO_2 eq. ha^{-1} $year^{-1}$ (Smith et al. 2007). However, NIs are the most widely tested mitigation option for N_2O emissions from agricultural soils. The effectiveness of nutrient supply control in increasing NUE and reducing environmental impacts depends mainly on two factors: (i) matching nutrient supply with plant demand, and (ii) maintaining nutrient availability (Chen et al. 2008; Trenkel 2010). The N requirement, specific periods of peak demand and preferred chemical formulations are specific to each crop and variety. The time pattern of N uptake by seasonal crops is generally sigmoidal (Shoji and Kanno 1994). Therefore, synchronizing N uptake with plant demand using a sigmoidal pattern of supply can provide an optimal nutrition for plant growth and reduce losses by the processes competing with nutrient uptake.

Advantages:
The use of SRF and CRF offer the following advantages over conventional fertilizers:

1. They can meet the crop N demand for the entire season through single application, generating savings in fertilizer application energy and labor costs. The lag in nutrient release enables fertilizers to be applied prior to 'annual spring rush' or when fields are less trafficable such as during fall application for winter- or spring-planted crops (Randall et al. 1985). It also reduces the demand for labor for top-dressing that is required during critical periods (Shaviv 2001),
2. They have lower toxicity to seedlings, which is caused by quick dissolution of conventional soluble fertilizers. Excessive nutrient supply commonly resulting from an application of conventional soluble fertilizers may result in high concentration of soluble salts in the root zone. This can cause toxicity to seeds and seedlings because of a high ionic concentration resulting from quick dissolution of conventional soluble fertilizers at high rates. Thus, use of SRF/CRF reduces damage caused by high ionic concentrations resulting from the quick dissolution of N carriers such as ammonia after application of urea, contributing to improved agronomic safety (Trenkel 1997b; Shaviv 2001).
3. They reduce toxicity and the salt content of substrates, permitting the application of substantially larger fertilizer dressings thereby reducing the application frequency as compared to conventional soluble fertilizers. This results in significant savings in labor, time, and energy, as well as in making the use of the fertilizer more convenient (Trenkel 1997b).
4. They significantly reduce possible losses of nutrients, particularly those of NO_3-N, between applications and uptake by the plants through gradual nutrient release. They also reduce evaporation losses of ammonia, which substantially decreases the risk of environmental pollution (Shaviv and Mikkelsen 1993; Mikkelsen et al. 1994).
5. They significantly reduce NO_3^- leaching and movement of NO_3 into water sources. They also reduce NH_3 volatilization, particularly from top-dressed agricultural systems where incorporation is not an option (e.g., pastures).
6. They reduce emissions of N_2O and NO_x (Bockman and Olfs 1998).
7. They contribute to advanced fertilizer management programs and to innovative farming systems such as no tillage farming with single co-situs fertilizer application (i.e., the placement with one application of a sufficient amount of controlled-release fertilizer for an entire growing season at any site (Trenkel 1997b).

Disadvantages:
The use of SRF and CRF can lead to the following disadvantages (Trenkel 1997b):

1. There are no standardized methods for reliable determination of nutrient release patterns. Most of existing data have been determined under laboratory conditions, which do not correlate well with the actual performance of fertilizers under field conditions. In addition, there is not sufficient data comparing enhanced efficiency fertilizers (i.e., CRF and SRF) with conventional fertilizers under FBMP.

2. Application of polymer-coated or encapsulated fertilizers may leave the undesired residues of the synthetic material in the field. Some of the synthetic material currently in use decompose extremely slowly or do not decompose at all in the soil, resulting in accumulation of undesired residues in the field.
3. Some chemical reaction products such as urea formaldehyde fertilizers release the contained fertilizers extremely slow, which makes it difficult to predict the desired application rate.
4. Some products (i.e., sulfur-coated fertilizers) release nutrient too rapid initially which can cause damage to the crop. In addition, sulfur may react in soils to produce acid, which degrades soil by increasing acidity.
5. The cost of enhanced efficiency fertilizers (i.e., CRF) is considerably higher compared to that of conventional fertilizers.
6. Depending on the specific NI, the activity of soil bacteria may not only be interrupted for a certain period but the soil bacteria may actually be killed. This is an undesirable interference in soil's natural processes.

10.4 Precision Fertilization

Increasing societal demand for food, fiber and fuel, global financial stress and growing pollution impact on quality of soil, water, and air requires simultaneous improvement of productivity and resource use efficiency to be an essential goal for agriculture. Sustainable nutrient management must support cropping systems and contribute to economic, social, and environmental quality improvement. Precision farming, also called the soil-specific farming, is the management approach that promotes environmental monitoring and control in agriculture (Robert 2002; Gebbers and Adamchuk 2010). This system is technology and information-based management that promotes controlled agricultural practices (Haneklaus and Schnug 2000). Principal goals of precision fertilization are to: (i) enhance understanding of the spatial distribution of factors affecting the crop growth, (ii) manage spatial variability by applying a variable rate of treatment of agricultural inputs within the field according to site conditions, and (iii) maximize profit and minimize the environmental impacts. The main technologies available to farmers for precision farming include: global positioning systems (GPS), field sensors, variable rate applicators for nutrients and agrochemicals, yield monitors for harvesting; microcomputer systems and user–friendly software for data collection, storage and feedback control systems, remote sensing devices and systems, soil sampling, and the geographic information systems (GIS). Initially, precision agriculture was used to adapt fertilizer application under varying soil conditions across the field. Since then, additional technologies have evolved such as automatic guidance of agricultural vehicles and implements, autonomous machinery and processes, product traceability, on farm research and software for overall management of agricultural production system (Gebbers and Adamchuk 2010). Benefits of precision farming include: being profitable, productive, sustainable, maintains crop quality,

increases food safety, protects environment, maintains on-farm quality of life, and enables rural development (Robert 2002).

It has been observed that N_2O emissions from two high yielding areas or regions of a farm receiving different amounts of fertilizer may not significantly differ, because the NO_3^- content in the soil may not limit N_2O emission. However, in low yielding soil or areas of the field, reduction of the amount of fertilizer can drastically reduce N_2O emissions, often by as much as 35% (Shey et al. 2001).

Principal technical challenges for the precision agriculture include the lack of technologies for effective monitoring and the appropriate "on the go" diagnostic tools for analysis of key parameters such as soil NO_3^- contents and plant health. Thus, there is an immediate need for development of real time automatic sensors to generate data that can be readily integrated into decision-making components of precision farming.

10.5 Conclusions

The rapid increase in the anthropogenic production of N fertilizers has been a major factor, accounting for the growth of food production. Despite the overall benefits experienced from the use of reactive N in agriculture, major environmental problems have emerged due to the presence of excessive reactive nitrogen in the environment. Considering the growing global demand for food, animal feed, fiber, and fuel, the appropriate management strategies involving intensive fertilizer and crop management practices that enhance nitrogen-use efficiency need to be developed and utilized to manage greenhouse gas emissions and minimize the agricultural footprint in the environment. The possible fertilizer management options for increasing crop nitrogen-use efficiency minimize agricultural footprint in the environment and minimize nitrous oxide emissions includes:

- Use of soil and plant testing to make best use of indigenous soil nitrogen,
- Use of right fertilizer form, rate, method and time of application,
- Synchronizing fertilizer application with the crop demand,
- Incorporation of inter-seasonal cover- and catch-crops to minimize nitrogen loss during off season and fallow periods,
- Use of enhanced efficiency fertilizers to increase nitrogen-use efficiency and minimize losses,
- Management of total plant nutrients to increase the efficiency of nitrogen utilization by crops, and
- Adoption of precision agriculture or soil-specific farming to maximize the farm profitability and minimize agriculture impact to the environment.

Study Questions

1. Outline the consequences of improper fertilizer use to the environment.
2. Describe the fertilizer best management practices.
3. Explain soil processes affecting N_2O emissions.

4. How can the leaching losses of NO_3 be minimized.
5. Can the precision farming be used by small size landholders? Explain.
6. What farming practices can reduce the loss of reactive N into the environment?
7. What are the mechanisms of action for CRF, SRF, NI, and UI?
8. What role do tillage and residue management play in enhancing the NUE?
9. How does soil structure and reaction affect N_2O emission?
10. Why does no-till farming increase N_2O emission?
11. Tabulate pros and cons of CRFs, SRFs and UIs
12. Can integrated nutrient management reduce dependence on fertilizers and meet the demands of growing population?

References

Akiyama, H., Yan, X., Yagi, K. (2010) Evaluation of effectiveness of enhanced-efficiency fertilizers as mitigation options for N_2O and NO emissions from agricultural soils: Meta-analysis. Global Change Biol. **16**(6), 1837–1846. doi:10.1111/j.1365-2486.2009.02031.x

Akiyama, H., Yan, X.Y., Yagi, K. (2006) Estimations of emission factors for fertilizer-induced direct N_2O emissions from agricultural soils in Japan: Summary of available data. Soil Sci. Plant Nutr. **52**(6), 774–787. doi:10.1111/j.1747-0765.2006.00097.x

Ambrosano, E.J., Trivelin, P.C.O., Cantarella, H., Ambrosano, G.M.B., Schammass, E.A., Muraoka, T., Rossi, F. (2011) ^{15}N labeled nitrogen from green manure and ammonium sulfate utilization by the sugarcane ratoon. Scientia Agricola **68**(3), 361–368

Aulakh, M.S., Doran, J.W., Walters, D.T., Mosier, A.R., Francis, D.D. (1991) Crop residue type and placement effects on denitrification and mineralization. Soil Sci. Soc. Amer. J. **55**(4), 1020–1025

Baligar, V., Fageria, N., He, Z. (2001) Nutrient use efficiency in plants. Comm. Soil Sci. Plant Anal. **32**(7/8), 921–950

Baligar, V.C., Pitta, G.V.E., Gama, E.E.G., Schaffert, R.E., Bahia, A.F.D., Clark, R.B. (1997) Soil acidity effects on nutrient use efficiency in exotic maize genotypes. Plant Soil **192**(1), 9–13. doi:10.1023/a:1004260222681

Balmford, A., Green, R.E., Scharlemann, J.P.W. (2005) Sparing land for nature: exploring the potential impact of changes in agricultural yield on the area needed for crop production. Global Change Biol. **11**(10), 1594–1605. doi:10.1111/j.1365-2486.2005.001035.x

Barton, L., Colmer, T.D. (2006) Irrigation and fertiliser strategies for minimising nitrogen leaching from turfgrass. Agric. Water Manage. **80**(1–3), 160–175. doi:10.1016/j.agwat.2005.07.011

Bockman, O.C., Olfs, H.W. (1998) Fertilizers, agronomy and N_2O. Nutr. Cycl. Agroecosys. **52** (2–3), 165–170. doi:10.1023/a:1009736327495

Bouwman, A.F., Boumans, L.J.M., Batjes, N.H. (2002) Emissions of N_2O and NO from fertilized fields: Summary of available measurement data. Global Biogeochem. Cycles **16**(4), 1058. doi:10.1029/2001gb001811

Cai, G., Chen, D., White, R.E., Fan, X.H., Pacholski, A., Zhu, Z.L., Ding, H. (2002) Gaseous nitrogen losses from urea applied to maize on a calcareous fluvo-aquic soil in the North China Plain. Austr. J. Soil Res. **40**(5), 737–748. doi:10.1071/sr01011

Carreres, R., Sendra, J., Ballesteros, R., Valiente, E.F., Quesada, A., Carrasco, D., Leganes, F., de la Cuadra, J.G. (2003) Assessment of slow release fertilizers and nitrification inhibitors in flooded rice. Biol. Fert. Soils **39**(2), 80–87. doi:10.1007/s00374-003-0684-4

Cassman, K.G., Dobermann, A., Walters, D.T. (2002) Agroecosystems, nitrogen-use efficiency, and nitrogen management. Ambio **31**(2), 132–140. doi:10.1579/0044-7447-31.2.132

Chen, D., Suter, H., Islam, A., Edis, R., Freney, J.R., Walker, C.N. (2008) Prospects of improving efficiency of fertiliser nitrogen in Australian agriculture: A review of enhanced efficiency fertilisers. Soil Res. **46**(4), 289–301. doi:10.1071/SR07197

Chen, D., White, R.E., Li, Y., Zhang, J.B., Li, B.G., Zhang, Y.M., Edis, R., Huang, Y.F., Cai, G.X., Wei, Y.P., Zhu, A.N., Hu, K.L., Li, G.T., Zhu, Z.L. (2006) Conservation management of water and nitrogen in the North China Plain In: Willett, I.R., Gao, Z. (eds.) Agricultural Water Management in China. ACIAR Proceedings No. 123, ACIAR. pp. 26–38. Canberra

Chien, S.H., Prochnow, L.I., Cantarella, H. (2009) Recent developments of fertilizer production and use to improve nutrient efficiency and minimize environmental impacts. Adv. Agron. **102**, 267–322. doi:10.1016/s0065-2113(09)01008-6

Christensen, L.A. (2002) Soil, nutrient, and water management systems used in US corn production. Economic Research Service, United States Department of Agriculture.

Cleemput, O.v., Zapata, F., Vanlauwe, B. (2008) Use of tracer technology in mineral fertilizer management. In: Guidelines on Nitrogen Management in Agricultural Systems. pp. 19–126. International Atomic Energy Agency (IAEA), Vienna, Austria

Cole, V., Cerri, C., Minami, K., Mosier, A., Rosenberg, N., Sauerbeck, D. (1996) Agricultural options for mitigation of greenhouse gas emissions In: Watson, R.T., Zinyowera, M.C., Moss, R.H., Dokken, D.J. (eds.) Climate change 1995. Impacts, adaptations and mitigation of climate change: Scientific technical analyses. pp. 745–771. Cambridge University Press, Cambridge

Conrad, R. (1996) Soil microorganisms as controllers of atmospheric trace gases (H_2, CO, CH_4, OCS, N_2O, and NO). Microbiol. Reviews **60**(4), 609–640

Crutzen, P.J., Mosier, A.R., Smith, K.A., Winiwarter, W. (2008) N_2O release from agro-biofuel production negates global warming reduction by replacing fossil fuels. Atmos. Chem. Phys. **8** (2), 389–395

Davies, D.M., Williams, P.J. (1995) The effect of the nitrification inhibitor dicyandiamide on nitrate leaching and ammonia volatilization - a UK nitrate sensitive areas perspective. J. Environ. Manage. **45**(3), 263–272. doi:10.1006/jema.1995.0074

Delgado, J.A., Follett, R.F., Sharkoff, J.L., Brodahl, M.K., Shaffer, M.J. (1998) NLEAP facts about nitrogen management. J. Soil Water Conserv. **53**(4), 332–337

Delgado, J.A., Groffman, P.M., Nearing, M.A., Goddard, T., Reic!sky, D., Lal, R., Kitchen, N.R., Rice, C.W., Towery, D., Salon, P. (2011) Conservation practices to mitigate and adapt to climate change. J. Soil Water Conserv. **66**(4), 118A-129A. doi:10.2489/jswc.66.4.118A

Delgado, J.A., Mosier, A.R. (1996) Mitigation alternatives to decrease nitrous oxides emissions and urea-nitrogen loss and their effect on methane flux. J. Environ. Qual. **25**(5), 1105–1111

Delgado, J.A., Riggenbach, R.R., Sparks, R.T., Dillon, M.A., Kawanabe, L.M., Ristau, R.J. (2001) Evaluation of nitrate-nitrogen transport in a potato-barley rotation. Soil Sci. Soc. Amer. J. **65**(3), 878–883

Di, H.J., Cameron, K.C. (2004) Effects of temperature and application rate of a nitrification inhibitor, dicyandiamide (DCD), on nitrification rate and microbial biomass in a grazed pasture soil. Austr. J. Soil Res. **42**(8), 927–932. doi:10.1071/sr04050

Diez, J.A., Caballero, R., Bustos, A., Roman, R., Cartagena, M.C., Vallejo, A. (1996) Control of nitrate pollution by application of controlled release fertilizer (CRF), compost and an optimized irrigation system. Fert. Res. **43**(1–3), 191–195. doi:10.1007/bf00747701

Diez, J.A., Roman, R., Cartagena, M.C., Vallejo, A., Bustos, A., Caballero, R. (1994) Controlling nitrate pollution of aquifers by using different nitrogenous controlled-release fertilizers in maize crop. Agric. Ecosys. Environ. **48**(1), 49–56. doi:10.1016/0167-8809(94)90074-4

Dobermann, A. (2005) Nitrogen use efficiency - State of the art. In: Proceedings of the Interational Workshop on Enhanced-Efficiency Fertilizers, June 28–30, 2005. Frankfurt, Germany. pp. 1–18. International Fertilizer Industry Association Paris, France

Dobermann, A. (2007) Nutrient use efficiency – measurement and management. In: Krauss, A., Isherwood, K., Heffer, P. (eds.) Fertilizer best management practices: General principles, strategies for adoption and voluntary initiatives vs regulations. Papers presented at the IFA International Workshop, 7–9 March 2007. Brussels, Belgium. pp. 1–28. International Fertilizers Industry Association (IFA), Paris, France

Dobermann, A., Cassman, K.G. (2004) Environmental dimensions of fertilizer nitrogen: What can be done to increase nitrogen use efficiency and ensure global food security. In: Mosier, A.K., Syers, J.K., Freeney, J.R. (eds.) Agriculture and the nitrogen cycle: Assessing the impact of fertilizer use on food production and the environment. pp. 261–278. Island Press, Washington, D.C.

Economic Research Service (ERS) (2011) Fertilizer use and price. United States Department of Agriculture (USDA) Economic Research Service Datasets Washington, D.C. On line at http// ers.usda.gov/data/fertilizeuse/ (Acessed December 2011)

Fageria, N.K., Baligar, V.C. (2005) Enhancing nitrogen use efficiency in crop plants. Adv. Agron. **88**, 97–185. doi:10.1016/s0065-2113(05)88004-6

Fan, X.L., Li, F.M., Liu, F., Kumar, D. (2004) Fertilization with a new type of coated urea: Evaluation for nitrogen efficiency and yield in winter wheat. J. Plant Nutr. **27**(5), 853–865. doi:10.1081/pln-120030675

Food and Agriculture Organization of the United Nations (FAO) (2011) FAOSTAT. FAO Statistics Division. Food and Agricultural Organization of the United Nations. On line at http:// faostat.fao.org. (Accessed March 2012)

FAOSTAT (2011) FAOSTAT fertilizers database. Food and Agriculture Organization of the United Nations. On line at: http://faostat.fao.org. (Accessed November 2011)

Fashola, O.O., Hayashi, K., Wakatsuki, T. (2002) Effect of water management and polyolefin-coated urea on growth and nitrogen uptake of indica rice. J. Plant Nutr. **25**(10), 2173–2190. doi:10.1081/pln-120014069

Fixen, P.E. (2009) Maximizing productivity and efficiency in contemporary agriculture. In: The Proceedings of the International Plant Nutrition Colloquium XVI, Department of Plant Sciences, UC Davis, UC Davis. California

Follett, R. (2001) Innovative ^{15}N microplot research techniques to study nitrogen use efficiency under different ecosystems. Comm. Soil Sci. Plant Anal. **32**(7/8), 951

Fox, R.H., Piekielek, W.P. (1993) Management and urease inhibitor effects on nitrogen use efficiency in no-till corn. J. Prod. Agric. **6**(2), 195–200

Fueleky, G., Benedek, S. (2011) Ecological Fertilization. In: Lichtfouse, E. (ed.) Sustainable Agriculture Reviews. Volume 6: Alternative Faming Systems, Biotechnology, Drought Stress and Ecological Fertilisation, vol. 6. pp. 215–243. Springer Science, New York

Gebbers, R., Adamchuk, V.I. (2010) Precision Agriculture and Food Security. Science **327**(5967), 828–831. doi:10.1126/science.1183899

Gioacchini, P., Nastri, A., Marzadori, C., Giovannini, C., Antisari, L.V., Gessa, C. (2002) Influence of urease and nitrification inhibitors on N losses from soils fertilized with urea. Biol. Fert. Soils **36**(2), 129–135. doi:10.1007/s00374-002-0521-1

Glenn, J.C., Gordon, T.J., Florescu, E. (2011) The milleniumproject: State of the future. World Federation of UN Associations.

Grant, C.A., Jia, S., Brown, K.R., Bailey, L.D. (1996) Volatile losses of NH_3 from surface-applied urea and urea ammonium nitrate with and without the urease inhibitors NBPT or ammonium thiosulphate. Can. J. Soil Sci. **76**(3), 417–419

Guertal, E.A. (2009) Slow-release Nitrogen Fertilizers in Vegetable Production: A Review. Horttechnology **19**(1), 16–19

Guillard, K., Kopp, K.L. (2004) Nitrogen fertilizer form and associated nitrate leaching from cool-season lawn turf. J. Environ. Qual. **33**(5), 1822–1827

Hall, A. (2005) Benefits of enhanced-efficiency fertilizer for the environment. In: International Fertilizer Association International workshop on enhanced-efficiency fertilizers. Frankfurt, Germany

Halvorson, A.D., Del Grosso, S.J., Alluvione, F. (2010) Nitrogen source effects on nitrous oxide emissions from irrigated no-till corn. J. Environ. Qual. **39**(5), 1554–1562. doi:10.2134/ jeq2010.0041

Halvorson, A.D., Del Grosso, S.J., Reule, C.A. (2008) Nitrogen, tillage, and crop rotation effects on nitrous oxide emissions from irrigated cropping systems. J. Environ. Qual. **37**(4), 1337–1344. doi:10.2134/jeq2007.0268

Haneklaus, S., Schnug, E. (2000) Decision-making strategies for the variable-rate application of compound fertilizers. Comm. Soil Sci. Plant Anal. **31**(11–14), 1863–1873. doi:10.1080/00103620009370543

Harrison, R., Webb, J. (2001) A review of the effect of N fertilizer type on gaseous emissions. Adv. Agron. **73**, 65–108. doi:10.1016/s0065-2113(01)73005-2

Hatch, D., Trindade, H., Cardenas, L., Carneiro, J., Hawkins, J., Scholefield, D., Chadwick, D. (2005) Laboratory study of the effects of two nitrification inhibitors on greenhouse gas emissions from a slurry-treated arable soil: impact of diurnal temperature cycle. Biol. Fert. Soils **41**(4), 225–232. doi:10.1007/s00374-005-0836-9

Hauck, R.D. (1985) Slow release and bio-inhibitor-amended nitrogen fertilizers. In: Engelstad, O.P. (ed.) Fertilizer technology and use. pp. 293–322. Soil Science Society of America, Madison, WI

Hoben, J.P., Gehl, R.J., Millar, N., Grace, P.R., Robertson, G.P. (2011) Nonlinear nitrous oxide (N(2)O) response to nitrogen fertilizer in on-farm corn crops of the US Midwest. Global Change Biol. **17**(2), 1140–1152. doi:10.1111/j.1365-2486.2010.02349.x

Howarth, R.W., Sharpley, A., Walker, D. (2002) Sources of nutrient pollution to coastal waters in the United States: Implications for achieving coastal water quality goals. Estuaries **25**(4B), 656–676. doi:10.1007/bf02804898

Humphreys, E., Freney, J.R., Muirhead, W.A., Denmead, O.T., Simpson, J.R., Leuning, R., Trevitt, A.C.F., Obcemea, W.N., Wetselaar, R., Cai, G.X. (1988) Loss of ammonia after application of urea at different times to dry-seeded, irrigated rice. Fert. Res. **16**(1), 47–57. doi:10.1007/bf01053314

Intergovernmental Panel on Climate Change (IPCC) (2006) N_2O emission from managed soils and CO_2 emissions from lime and urea application. In: Eggleston, H.S., Buendia, L., Miwa, K., Ngara, T., Tanabe, K. (eds.) Intergovernmental Panel on Climate Change Guidelines for National Greenhouse Gas Inventories, Prepared by the National Greenhouse Gas Inventories Programme, Volume 4 Agriculture, forestry and other land uses IGES, Japan

International Fertilizer industry Association (IFA) (2006) Annual Report 2005. International Fertilizer Industry Association. Paris, France

Jalali, M. (2005) Nitrates leaching from agricultural land in Hamadan, western Iran. Agric. Ecosys. Environ. **110**(3–4), 210–218. doi:10.1016/j.agee.2005.04.011

Jarrell, W.M., Boersma, L. (1979) Model for the release of urea by granules of sulfur-coated urea applied to soil. Soil Sci. Soc. Amer. J. **43**(5), 1044–1050

Johnkutty, I., Palaniappan, S.P. (1996) Use of chlorophyll meter for nitrogen management in lowland rice. Fert. Res. **45**(1), 21–24

Kochba, M., Gambash, S., Avnimelech, Y. (1990) Studies on slow release fertilizers .1. Effects of temperature, soil-moisture, and water-vapor pressure. Soil. Sci. **149**(6), 339–343. doi:10.1097/00010694-199006000-00004

Krupa, S.V. (2003) Effects of atmospheric ammonia (NH_3) on terrestrial vegetation: a review. Environ. Pollut. **124**(2), 179–221. doi:10.1016/s0269-7491(02)00434-7

Kumar, R., Devakumar, C., Kumar, D., Panneerselvam, P., Kakkar, G., Arivalagan, T. (2008) Influence of Edaphic Factors on the Mineralization of Neem Oil Coated Urea in Four Indian Soils. J. Agric. Food Chem. **56**(21), 10183–10191. doi:10.1021/jf802139q

Lal, R., Delgado, J.A., Groffman, P.M., Millar, N., Dell, C., Rotz, A. (2011) Management to mitigate and adapt to climate change. J. Soil Water Conserv. **66**(4), 276–285. doi:10.2489/jswc.66.4.276

Ledgard, S.F. (2001) Nitrogen cycling in low input legume-based agriculture, with emphasis on legume/grass pastures. Plant Soil **228**(1), 43–59. doi:10.1023/a:1004810620983

Loubet, B., Genermont, S., Ferrara, R., Bedos, G., Decuq, G., Personne, E., Fanucci, O., Durand, B., Rana, G., Cellier, P. (2010) An inverse model to estimate ammonia emissions from fields. Eur. J. Soil Sci. **61**(5), 793–805. doi:10.1111/j.1365-2389.2010.01268.x

Macadam, X.M.B., del Prado, A., Merino, P., Estavillo, J.M., Pinto, M., Gonzalez-Murua, C. (2003) Dicyandiamide and 3,4-dimethyl pyrazole phosphate decrease N_2O emissions from grassland but dicyandiamide produces deleterious effects in clover. J. Plant Physiol. **160**(12), 1517–1523. doi:10.1078/0176-1617-01006

Malhi, S.S., Oliver, E., Mayerle, G., Kruger, G., Gill, K.S. (2003) Improving effectiveness of seedrow-placed urea with urease inhibitor and polymer coating for durum wheat and canola. Comm. Soil Sci. Plant Anal. **34**(11–12), 1709–1727. doi:10.1081/css-120021307

McSwiney, C.P., Robertson, G.P. (2005) Nonlinear response of N_2O flux to incremental fertilizer addition in a continuous maize (*Zea mays* L.) cropping system. Global Change Biol. **11**(10), 1712–1719. doi:10.1111/j.1365-2486.2005.01040.x

McTaggart, I.P., Clayton, H., Parker, J., Swan, L., Smith, K.A. (1997) Nitrous oxide emissions from grassland and spring barley, following N fertiliser application with and without nitrification inhibitors. Biol. Fert. Soils **25**(3), 261–268. doi:10.1007/s003740050312

McTaggart, I.P., Clayton, H., Smith, K.A. (1994) Nitrous oxide flux from fertilized grassland: Strategies for reducing emission. In: van Ham, J., Jansen, I.J., Swart, R.J. (eds.) Non-CO_2 greenhouse gases: Why and how to control. pp. 421–426. Kluwer Academic Publishers, Dordrecht, The Netherlands

McTaggart, I.P., Tsuruta, H. (2003) The influence of controlled release fertilisers and the form of applied fertiliser nitrogen on nitrous oxide emissions from an andosol. Nutr. Cycl. Agroecosys. **67**(1), 47–54. doi:10.1023/a:1025108911676

Meredith, W.R., Heitholt, J.J., Pettigrew, W.T., Rayburn, S.T. (1997) Comparison of obsolete and modern cotton cultivars at two nitrogen levels. Crop Sci. **37**(5), 1453–1457

Merino, P., Estavillo, J.M., Besga, G., Pinto, M., Gonzalez-Murua, C. (2001) Nitrification and denitrification derived N_2O production from a grassland soil under application of DCD and Actilith F2. Nutr. Cycl. Agroecosys. **60**(1–3), 9–14. doi:10.1023/a:1012623515882

Merino, P., Estavillo, J.M., Graciolli, L.A., Pinto, M., Lacuesta, M., Munoz-Rueda, A., Gonzalez-Murua, C. (2002) Mitigation of N_2O emissions from grassland by nitrification inhibitor and Actilith F2 applied with fertilizer and cattle slurry. Soil Use Manage. **18**(2), 135–141. doi:10.1079/sum2002120

Mikkelsen, R.L., Williams, H.M., Behel, A.D. (1994) Nitrogen leaching and plant uptake from controlled-release fertilizers. Fert. Res. **37**(1), 43–50. doi:10.1007/bf00750672

Miltner, E.D., Stahnke, G.K., Johnston, W.J., Golob, C.T. (2004) Late fall and winter nitrogen fertilization of turfgrass in two Pacific Northwest climates. Hortscience **39**(7), 1745–1749

Minami, K. (1997) Atmospheric methane and nitrous oxide: sources, sinks and strategies for reducing agricultural emissions. Nutr. Cycl. Agroecosys. **49**(1–3), 203–211. doi:10.1023/a:1009730618454

Moomaw, W.R., Moreira, J., Blok, K., Greene, D.L., Gregory, K., Jaszay, T., Kashiwagi, T., Levine, M., McFarland, M., Prasad, N.S., Price, L., Rogner, H.-H., Sims, R., Zhou, F., Zhou, P. (2001) Technological and economic potential of greenhouse gas emissions reduction In: B., M., O., D., R., S., J., P. (eds.) Climate Change 2001: Mitigation. Contribution of Working Group III to the Third Assessment Report of the Intergovernmental Panel on Climate Change (IPCC). pp. 171–299. Cambridge University Press, Cambridge, UK and New York, NY, USA

Morgan, K.T., Cushman, K.E., Sato, S. (2009) Release Mechanisms for Slow- and Controlled-release Fertilizers and Strategies for Their Use in Vegetable Production. Horttechnology **19**(1), 10–12

Mosier, A.R., Doran, J.W., Freney, J.R. (2002) Managing soil denitrification. J. Soil Water Conserv. **57**(6), 505–513

Munoz, F., Mylavarapu, R.S., Hutchinson, C.M. (2005) Environmentally responsible potato production systems: A review. J. Plant Nutr. **28**(8), 1287–1309. doi:10.1081/pln-200067434

Nastri, A., Toderi, G., Bernati, E., Govi, G. (2000) Ammonia volatilization and yield response from urea applied to wheat with urease (NBPT) and nitrification (DCD) inhibitors. Agrochimica **44**(5–6), 231–239

Newbould, P. (1989) The use of nitrogen fertilizer in agriculture. Where do we go practically and ecologically. Plant Soil **115**(2), 297–311. doi:10.1007/bf02202596

Oberle, S. (1994) Farming systems options for united-states agriculture - an agroecological perspective. J. Prod. Agric. **7**(1), 119–123

Oberle, S.L., Keeney, D.R. (1990) Factors influencing corn fertilizer n requirements in the northern united-states corn belt. J. Prod. Agric. **3**(4), 527–534

Obreza, T.A., Rouse, R.E. (2006) Long-term response of 'Hamlin' orange trees to controlled-release nitrogen fertilizers. Hortscience **41**(2), 423–426

Owens, L.B., Edwards, W.M., Van Keuren, R.W. (1999) Nitrate leaching from grassed lysimeters treated with ammonium nitrate or slow-release nitrogen fertilizer. J. Environ. Qual. **28**(6), 1810–1816

Pack, J.E., Hutchinson, C.M., Simonne, E.H. (2006) Evaluation of controlled-release fertilizers for northeast Florida chip potato production. J. Plant Nutr. **29**(7), 1301–1313. doi:10.1080/01904160600767633

Parkin, T.B., Kaspar, T.C. (2006) Nitrous oxide emissions from corn-soybean systems in the Midwest. J. Environ. Qual. **35**(4), 1496–1506. doi:10.2134/jeq2005.0183

Peoples, M.B., Boyer, E.W., Goulding, K.W.T., Heffer, P., Ochwoh, V.A., Vanlauwe, B., Woods, S., Yagi, K., van Cleemput, O. (2004) Pathways of nitrogen loss and their impacts on human health and the environment. In: Mosier, A.K., Syers, J.K., Freeney, J.R. (eds.) Agriculture and nitrogen cycle: Assessing the impacts of fertilizer use on food production and environment. pp. 53–69. Island Press, Washington, D.C.

Prasad, R., Power, J.F. (1995) Nitrification inhibitors for agriculture, health, and the environment. Adv. Agron. **54**, 233–281. doi:10.1016/s0065-2113(08)60901-3

Prather, M., Derwent, R., Ehhalt, D., Fraser, P., Sanhueza, E., Zhou, X. (1995) Other trace gases and atmospheric chemistry. In: Houghton, J.T., Filho, L.G.M., Bruce, J., Lee, H., Callander, B.A., Haites, E., Harris, N., Maskell, K. (eds.) Climate change 1994: Radiative forcing of climate change and an evaluation of the IPCC IS 92 emission scenarios. pp. 73–126. Cambridge University Press, Cambridge

Ramirez, E., Reheul, D. (2009) Statistical modelling of nitrogen use efficiency of dairy farms in Flanders. Agron. Sustain. Dev. **29**(2), 339–352. doi:10.1051/agro/2008065

Randall, G.W., Wells, K.L., Hanway, J.J. (1985) Modern techniques in fertilizer application. In: Hauck, R.D. (ed.) Fertilizer technology and use. pp. 521–560. Soil Science Society of America, Madison, WI

Raun, W.R., Johnson, G.V. (1999) Improving nitrogen use efficiency for cereal production. Agron. J. **91**(3), 357–363

Robert, P.C. (2002) Precision agriculture: a challenge for crop nutrition management. Plant Soil **247**(1), 143–149. doi:10.1023/a:1021171514148

Roberts, T.L. (2007) Right product, right rate, right time, and right place...The foundation of best management practices for fertilizer. In: Fertilizer best management practices: General principles, strategy for their adaptation and voluntary initiatives vs regulations: Proceedings of International Fertilizer Industry Association (IFA) Workshop, March 7–9, 2007, Brussels, Belgium. pp. 29–32. IFA, Paris, France

Ronald, P. (2011) Plant Genetics, Sustainable Agriculture and Global Food Security. Genetics **188**(1), 11–20. doi:10.1534/genetics.111.128553

Roy, A.H., Hammond, L.L. (2004) Challenges and opportunities for fertilizer industry. In: Mosier, A.K., Syers, J.K., Freeney, J.R. (eds.) Agriculture and the nitrogen cycle: Assessing the impacts of fertilizer use on food production and the environment. pp. 233–243. Island Press, Washington, D.C.

Sanchez, P.A. (2002) Ecology - Soil fertility and hunger in Africa. Science **295**(5562), 2019–2020. doi:10.1126/science.1065256

Schroder, J.J., Neeteson, J.J., Oenema, O., Struik, P.C. (2000) Does the crop or the soil indicate how to save nitrogen in maize production? Reviewing the state of the art. Field Crop Res. **66**(2), 151–164. doi:10.1016/s0378-4290(00)00072-1

Shaviv, A. (2001) Advances in controlled-release fertilizers. Adv. Agron. **71**, 1–49. doi:10.1016/s0065-2113(01)71011-5

Shaviv, A. (2005a) Controlled release fertilizers. In: Proceedings of the International Fertilizers Association (IFA) International workshop on enhanced efficiency fertilizers. pp. 28–30. Frankfurt, Germany

Shaviv, A. (2005b) Environmental friendly nitrogen fertilization. Science China C **48**, 937–947. doi:10.1360/062005-285

Shaviv, A., Mikkelsen, R.L. (1993) Controlled-release fertilizers to increase efficiency of nutrient use and minimize environmental degradation: A review. Fert. Res. **35**(1–2), 1–12. doi:10.1007/bf00750215

Shey, U., Kamp, T., Ruser, R. (2001) The potential of precision farming for reducing the emission of nitrous oxide from agricultural field in southern Germany. In: Midgley, P. (ed.) Transport and chemicals transformation in troposphere. Proceedings of EUROTRAC symposium. pp. 1062–1066. Springer-Verlag, Belgium

Shoji, S. (2005) Innovative use of controlled availability fertilizers with high performance for intensive agriculture and environmental conservation. Science China C **48**, 912–920. doi:10.1360/062005-282

Shoji, S., Delgado, J., Mosier, A., Miura, Y. (2001) Use of controlled release fertilizers and nitrification inhibitors to increase nitrogen use efficiency and to conserve air and water quality. Comm. Soil Sci. Plant Anal. **32**(7–8), 1051–1070. doi:10.1081/css-100104103

Shoji, S., Gandeza, A.T. (1992) Controlled release fertilizers with polyolefin resin coating. Konno Printing Co. Ltd, Sendai, Japan

Shoji, S., Kanno, H. (1994) Use of polyolefin-coated fertilizers for increasing fertilizer efficiency and reducing nitrate leaching and nitrous oxide emissions. Fert. Res. **39**(2), 147–152. doi:10.1007/bf00750913

Smaling, E.M.A., Nandwa, S.M., Janssen, B.H. (1997) 'Soil fertility is at stake!'. In: Buresh, R.J., Sanchez, P.A., Calhoun, F. (eds.) Replenishing Soil Fertility in Africa, SSSA Special Publication Number 51. pp. 47–61. Soil Science Society of America, Madison, WI

Smil, V. (2001) Enriching the earth: Fritz Haber, Carl Bosch, and the transformation of world food production. pp. The MIT Press, Cambridge, MA, USA

Smith, K.A., McTaggart, I.P., Tsuruta, H. (1997) Emissions of N_2O and NO associated with nitrogen fertilization in intensive agriculture, and the potential for mitigation. Soil Use Manage. **13**(4), 296–304. doi:10.1111/j.1475-2743.1997.tb00601.x

Smith, P., Martino, D., Cai, Z., Gwary, D., Janzen, H., Kumar, P., McCarl, B., Ogle, S., O'Mara, F., Rice, C., Scholes, B., Sirotenko, O. (2007) Agriculture. In: Metz, B., Davidson, O.R., Bosch, P.R., Dave, R., Meyer, L.A. (eds.) Climate Change 2007: Mitigation. Contribution of Working Group III to the Fourth Assessment Report of the Intergovernmental Panel on Climate Change. pp. 487–540. Cambridge University Press Cambridge, United Kingdom and New York, NY, USA

Snyder, C.S. (2008) Fertilizer nitrogen BMPs to limit losses that contribute to global warming. International Plant Nutrition Institute (IPNI).

Snyder, C.S., Bruulsema, T.W. (2007) Nutrient Use Efficiency and Effectiveness in North America: Indices of Agronomic and Environmental Benefit International Plant Nutrition Institute. On line at http://www.ipni.net/ipniweb/portal.nsf/

Snyder, C.S., Bruulsema, T.W., Jensen, T.L., Fixen, P.E. (2009) Review of greenhouse gas emissions from crop production systems and fertilizer management effects. Agric. Ecosys. Environ. **133**(3–4), 247–266. doi:10.1016/j.agee.2009.04.021

Snyder, C.S., Randall, G.W., Lamond, R.E., Hoeft, R.G. (2001) Fall nitrogen management for agronomic response and environmental protection. Fall fertilization facts 2001- Opportunities and considerations In: (IPNI), I.P.N.I. (ed.). Norcross, GA. U.S.A.

Stewart, W.M., Dibb, D.W., Johnston, A.E., Smyth, T.J. (2005) The contribution of commercial fertilizer nutrients to food production. Agron. J. **97**(1), 1–6

Subbarao, G.V., Ito, O., Sahrawat, K.L., Berry, W.L., Nakahara, K., Ishikawa, T., Watanabe, T., Suenaga, K., Rondon, M., Rao, I.M. (2006) Scope and strategies for regulation of nitrification in agricultural systems-challenges and opportunities. Crit. Rev. Plant Sci. **25**(4), 303–335. doi:10.1080/07352680600794232

Tomaszewska, M., Jarosiewicz, A. (2002) Use of polysulfone in controlled-release NPK fertilizer formulations. J. Agric. Food Chem. **50**(16), 4634–4639. doi:10.1021/jf0116808

Trenkel, M.E. (1997a) Improved Fertilizer Use Efficiency. Controlled-release and stabilized fertilizers in agriculture. pp. International Fertilizer Industry Association., Paris, France

Trenkel, M.E. (1997b) Improving fertilizer use efficiency: Controlled-release and stabilized fertilizers in agriculture. pp. International Fertilizer Industry Association (IFA), Paris, France
Trenkel, M.E. (2010) Slow- and controlled-release and stabilized fertilizers: An option for enhancing nutrient use efficiency in agriculture. pp. International Fertilizer Industry Association (IFA), Paris, France
U.S. Environmental Protection Agency (EPA) (2009) National Primary Drinking Water Regulations. US Environmental Protection Agency.
Vallejo, A., Diez, J.A., Lopez-Valdivia, L.M., Gasco, A., Jimenez, C. (2001) Nitrous oxide emission and denitrification nitrogen losses from soils treated with isobutylenediurea and urea plus dicyandiamide. Biol. Fert. Soils **34**(4), 248–257. doi:10.1007/s003740100409
van Noordwijk, M., Cadisch, G. (2002) Access and excess problems in plant nutrition. Plant Soil **247**(1), 25–40
Vanlauwe, B., Bationo, A., Chianu, J., Giller, K.E., Merckx, R., Mokwunye, U., Ohiokpehai, O., Pypers, P., Tabo, R., Shepherd, K.D., Smaling, E.M.A., Woomer, P.L., Sanginga, N. (2010) Integrated soil fertility management Operational definition and consequences for implementation and dissemination. Outlook on Agriculture **39**(1), 17–24. doi:10.5367/000000010791169998
Varel, V.H. (1997) Use of urease inhibitors to control nitrogen loss from livestock waste. Bioresour. Technol. **62**(1–2), 11–17. doi:10.1016/s0960-8524(97)00130-2
Varel, V.H., Nienaber, J.A., Freetly, H.C. (1999) Conservation of nitrogen in cattle feedlot waste with urease inhibitors. J. Anim. Sci. **77**(5), 1162–1168
Venterea, R.T., Burger, M., Spokas, K.A. (2005) Nitrogen oxide and methane emissions under varying tillage and fertilizer management. J. Environ. Qual. **34**(5), 1467–1477. doi:10.2134/jeq2005.0018
Vitousek, P.M., Naylor, R., Crews, T., David, M.B., Drinkwater, L.E., Holland, E., Johnes, P.J., Katzenberger, J., Martinelli, L.A., Matson, P.A., Nziguheba, G., Ojima, D., Palm, C.A., Robertson, G.P., Sanchez, P.A., Townsend, A.R., Zhang, F.S. (2009) Nutrient Imbalances in Agricultural Development. Science **324**(5934), 1519–1520. doi:10.1126/science.1170261
Wagner-Riddle, C., Thurtell, G.W. (1998) Nitrous oxide emissions from agricultural fields during winter and spring thaw as affected by management practices. Nutr. Cycl. Agroecosys. **52**(2–3), 151–163. doi:10.1023/a:1009788411566
Wang, F.L., Alva, A.K. (1996) Leaching of nitrogen from slow-release urea sources in sandy soils. Soil Sci. Soc. Amer. J. **60**(5), 1454–1458
Wang, W.M., Shi, Y.X., Liu, H. (2005) Effects of controlled-release nitrogen fertilizers on yield and quality of Chinese cabbage (Brassica campestris L. spp. Pekinensis). In: Zhu, Z.L., Minami, K., Xing, G.X. (eds.) Proceedings of the 3rd International Nitrogen Conference. pp. 405–410. Science Press, Beijing
Wang, X.J., Douglas, L.A. (1996) Effect of phosphoroamides on soil urease activity, and plant dry matter production and urea-N uptake by wheat plants. Agrochimica **40**(5–6), 209–215
Watson, C.J. (2000) Urease activity and inhibition - principles and practice. In: Proceeding No 454. The International Fertilizer Society. York, UK
Watson, C.J. (2005) Urease inhibitors. In: Mosier, A., Galloway, J.N. (eds.) IFA International Workshop on Enhanced-Efficiency Fertilizers. Frankfurt, Germany, 28–30 June 2005 (Available at http://www.fertilizer.org/ifa). pp. 1–8.
Watson, C.J., Miller, H., Poland, P., Kilpatrick, D.J., Allen, M.D.B., Garrett, M.K., Christianson, C.B. (1994a) Agronomic assessment and N-15 recovery of urea amended with the urease inhibitor nBTPT (N-(n-butyl) thiophosphoric triamide) for temperate grassland. Soil Biol. Biochem. **26**(9), 1165–1171. doi:10.1016/0038-0717(94)90139-2
Watson, C.J., Poland, P., Miller, H., Allen, M.B.D., Garrett, M.K., Christianson, C.B. (1994b) Soil properties and the ability of the urease inhibitor N-(n-butyl) thiophosphoric triamide (nBTPT) to reduce ammonia volatilization from surface-applied urea. Plant Soil **161**(2), 167–177. doi:10.1007/bf00046388

Weiske, A. (2006) Selection and specification of technical and management-based greenhouse gas mitigation measures in agricultural production for modeling Institute for Energy and Environment (IE), MEACAP. http://www.ieep.eu/publications/pdfs/meacap/D10a_GHG_mitigation_-measures_for_modelling.pdf.

Weisler, F., Behrens, T., Horst, W.J. (2001) The role of nitrogen-efficient cultivars in sustainable agriculture. The Scientific World J. **1 Suppl 2**, 61–69

Wen, G., Mori, T., Yamamoto, T., Chikushi, J., Inoue, M. (2001) Nitrogen recovery of coated fertilizers and influence on peanut seed quality for peanut plants grown in sandy soil. Comm. Soil Sci. Plant Anal. **32**(19–20), 3121–3140. doi:10.1081/css-120001111

Zerulla, W., Kummer, K., Wissemeier, A., Radle, M. (2000) The development and testing a new nitrification inhibitor. In: The International Fertilizer Society Proceedings No 455. pp. 5–23.

Chapter 11
Conclusions: Towards Managing Agricultural Soils for Mitigating Nitrous Oxide Emissions

Contents

Abstract Nitrogen (N) is central to living systems, and its addition to agricultural cropping systems is an essential facet of modern crop management and one of the major reasons that crop production has kept pace with human population growth. The benefits of N added to cropping systems come at significant environmental costs, including increased coastal hypoxia, atmospheric N_2O emissions, increased reactive N gases to the atmosphere, and N deposition onto forests and other natural areas as some of the consequences of inability to keep fertilizer N from leaving cropped ecosystems. The global cycling of N is complex, and mitigation of some of these consequences require a thorough understanding of both the biogeochemical pathways of N losses from ecosystems and the consequences of different management practices. A number of technologies are currently available to reduce N loss. These include adding rotational complexity to cropping systems to improve N capture by crops, providing farmers with decision support tools for better predicting crop fertilizer N requirements, improving methods for optimizing fertilizer timing and placement, and developing watershed-level strategies to recapture N lost from fields. Solutions to the problem of agricultural N loss will require a portfolio

David Ussiri and R. Lal, *Soil Emission of Nitrous Oxide and its Mitigation*,
DOI 10.1007/978-94-007-5364-8_11, © Springer Science+Business Media Dordrecht 2013

approach in which different technologies are used in different combinations to address site-specific challenges. Solutions will also require incentives that promote their adoption.

Keywords Nitrogen cycling • Policy • Reactive nitrogen • Mitigation

Abbreviations

Nr	Reactive nitrogen
NPP	Net primary production
BNF	Biological nitrogen fixation
CFCs	Chlorofluorocarbons
HCFCs	Hydrochlorofluorocarbons
GHG	Greenhouse gas
NUE	Nitrogen use efficiency
SOM	Soil organic matter
NT	No tillage

11.1 Introduction

Human activities have more than doubled the rate of transfer of the highly abundant, but biologically unavailable nitrogen gas (N_2) to biologically available forms of N (Nr) (i.e., NH_4^+, NO_3^-). Estimated global N fixation from natural and agricultural biological N fixation (BNF), NO_x combustion, and synthetic fertilizer manufacture was ~260 Tg N year^{-1} in 1990 (Galloway et al. 2004). The reactive N (Nr) creation continues to increase every year. Its creation is dominated by agricultural activities, but fossil fuel energy plays an important role also. Atmospheric concentration of N_2O is presently increasing at 0.8 ppbv year^{-1} (WMO 2011). The need to feed (125 Tg N year^{-1}) and provide energy (25 Tg N year^{-1}) for increasing population is the driving force behind this increase (Mosier 2002). The soils of the world are by far the largest overall source of N_2O emissions, accounting for ~70% of the atmospheric N_2O loading. Agricultural systems are by far the largest single source of the anthropogenic N_2O emissions, a biogenic greenhouse gas (GHG) that is also a dominant stratospheric ozone depleting substance (Ravishankara et al. 2009). Owing to the decline in chlorofluorocarbons (CFCs) and hydrochlorofluorocarbons (HCFCs) emissions, N_2O is now the most important ozone-depleting molecule in the atmosphere.

The annual global N_2O budget is currently estimated to range from 13.4 to 43.5 Tg N_2O-N year^{-1}, and microbial processes in soils and aquatic ecosystems are responsible for ~89% of the annual global N_2O emissions. Total anthropogenic emissions are estimated at 6.7 Tg N_2O-N year^{-1}, 42% of which is derived from agriculture (Denman et al. 2007). Fertilized croplands contribute 3.3 Tg

N_2O-N year^{-1}, with an additional 0.8 Tg N_2O-N year^{-1} originating from fertilized grasslands (Stehfest and Bouwman 2006). Synthetic N fertilizers or manures are the main source of increased N_2O emissions attributed to agriculture. As N fertilizer demand is expected to increase from 105.6 Tg N in 2009 to >135 Tg N in 2030 (FAO 2011), increases in N_2O are expected.

Nitrogen, an essential nutrient added to soil in synthetic and organic fertilizers to increase crop production, can be converted to N_2O by the soil microbial processes (Millar et al. 2010). Despite detrimental effects of reactive N to the environment, N fertilizer remains essential to global food production, and N availability remains the most limiting nutrient for non-leguminous plants globally. The relationship between crop productivity and N fertilizer input is non-linear and follows the diminishing return function. Therefore, combining large yield and effective fertilizer N use efficiency (NUE) is inherently difficult to achieve (Cassman et al. 2003). Highly productive agricultural systems are generally associated with large losses of reactive N to the environment, including N_2O emissions. It is estimated that for every kg of newly fixed N fertilizer, 30–50 g is lost as N_2O-N emission directly from either fertilizer N form, human waste, recycled form from manure N or indirectly after leaching as NO_3^- or volatilization as NH_3 losses (Crutzen et al. 2008).

Biologically available N is in short supply in most terrestrial ecosystems, and the availability of N limit plant growth and net primary production (NPP), fundamentally affecting the structure and function of most ecosystems (Robertson and Vitousek 2009). The N is more important in agricultural systems than in other ecosystems. The addition of N to sustain and increase crop yield is a fundamental feature of the modern crop management. Therefore, N is the most limiting nutrient in the intensive crop production system and N fertilizer is commonly applied to cereals – corn, wheat, and other non-leguminous crops. Added N enables farmers to simplify the plant community by displacing the need for N fixing plants. It also alters the microbial community structure in ways that affect decomposition, and other ecosystem processes such as nitrification, denitrification, and CH_4 oxidation. The N addition also alters the palatability of crops to pests and influencing trophic relationships (Robertson and Vitousek 2009). The net benefit of N in agricultural production systems are many, including substantially more food can be grown on a given area of land, thereby relieving some of pressure for new land clearing and abundant harvest resulting into improved human health and well-being.

However, biologically reactive N (Nr) is highly mobile and hard to contain within the agricultural systems. Even N that is efficiently conserved and harvested with crop, it eventually makes its way to the environment. For example, of the ~12 Tg N year^{-1} that are applied to agricultural system as N fertilizers in USA, only ~2 Tg N year^{-1} are consumed by humans. The remaining 10 Tg N year^{-1} of unutilized N is wasted to the environment in various forms (Robertson and Vitousek 2009). The N mobilized from agricultural ecosystems is reactive and is present in forms that are biologically active in soils and surface waters and chemically reactive in the atmosphere. Increase in Nr from agriculture is widespread and leads to losses of biological diversity, compromised air quality, threats to human health, and global warming. The main aim of agricultural N management is to provide enough N to plant to maximize growth and subsequent crop yields, while also keeping it out of other ecosystems in particular, those in which added N is harmful. Knowledge of the trade-offs between N_2O emissions, fertilizer management practices and crop yield are therefore, an essential management strategies that aims to reduce the agricultural N_2O burden without compromising productivity and economic returns. A number of management technologies have been proposed to reduce agricultural N_2O emissions (CAST 2004; Follett et al. 2005).

11.2 Mitigation Options

In general, N_2O emission from agricultural soils increases with increase in mineral N inputs (Gregorich et al. 2005). The N_2O emissions are low to moderate until the N input exceeds crop needs after which the flux nearly doubles (McSwiney and Robertson 2005), suggesting that prudent management of N input can be effective strategy to minimize N_2O emitted from croplands. Specific mitigation options can be used to reduce agriculture environmental impacts associated with N application for crop and animal production. Managing N to match crop demands can reduce N_2O emissions, while manipulating animal diet and manure management can reduce both NH_3 and N_2O emission from animal husbandry. Thus, all segments of agriculture have the management options that can reduce agriculture's GHG footprints. Opportunities for mitigating GHGs in agriculture fall into three broad categories based on the underlying principles (Smith et al. 2008): (i) Reducing emissions: the fluxes of GHG can be reduced by managing more efficiently the flows of N in the agricultural systems. The exact approaches that best reduce emissions depend on local conditions and vary from region to region. For example, practices that deliver added N more efficiently to plants often suppress emissions of N_2O. (ii) Enhancing removals: this principle is most suited for CO_2, which can be removed from atmosphere through photosynthetic input and stored in soil as soil organic carbon (SOC). (iii) Avoiding or displacing emissions: emissions of GHG can be avoided by agricultural management practices that protect cultivation of new lands currently under forest, grassland, or non-agricultural vegetation (Foley et al. 2005). The main cause of increase in anthropogenic N_2O emissions are application of N fertilizers and animal manure management and N released from fossil fuel combustion.

11.2.1 Improving Efficiency of Crop Production

Synthetic fertilizers and manure have equivalent effects on N_2O flux in most intensive cropping systems, and tillage and other cultivation techniques have little effect on flux when soil N is readily available (Robertson et al. 2000). Most cropped soils emit N_2O at a rate of 0.2–3% of the N inputs regardless of the source, therefore, decreasing N inputs in cropping systems could decrease N_2O emissions directly (Robertson et al. 2000; Delgado et al. 2010). There is a direct relationship between soil N availability and crop yields, and the main agronomic challenge for decreasing direct N_2O emission from agricultural soils is to decrease N input without decreasing crops yield. Intensively managed crops offer opportunity to impose practices that reduce net emission of GHGs. Practices such as precision farming are commonly accepted to improve NUE by adjusting N application rates to actual crop requirements or avoiding time delay between nutrient application and uptake (Smith et al. 2008; Eckard et al. 2010; Godfray et al. 2010). Such practices keep added N away from places in the field where it is less likely to be absorbed quickly by plants. Thus, N added is less available to N_2O producing microbes and less likely to be leached or volatilized. Improving crop rotation by introducing a legume in a crop rotation, which reduce the reliance on inputs of N through BNF, which produces less N_2O emissions than equivalent fertilizer N (Rochette and Janzen 2005). Also, planting a catch or cover crop between agricultural crops which can extract plant-available N unused by preceding crop can reduce N_2O emissions (Nemecek et al. 2008) and if properly planned, reduce use of synthetic N fertilizer and NO_3^- leaching per unit of crop produced (Pappa et al. 2011). Leguminous cover or catch crops have added benefits of fixing N. For example, the yield of wheat can increase by over 15% when grown after legume or rapeseed as opposed to monoculture (Cederberg et al. 2005).

Increasing crop yields may reduce net GHG emissions per unit of crop harvested (Godfray et al. 2010). Yields can be increased while reducing negative environmental impacts with the use of existing technologies (Hobbs et al. 2008). In the developing countries the gap between potential and attained crop yields is especially large and closing the yield gap has a large GHG mitigation potential. Yield gap is the difference between realized productivity and the best that can be achieved using current genetic potential, available technologies and management (Godfray et al. 2010). The yield gaps in developing countries arise from a variety of site-specific technical and socio-economic constraints ranging from high cost of farm inputs, absence of transportation and markets infrastructures variable market prices, conflicts, droughts, pest damage crop spoilage, and also side effects of globalization (Godfray et al. 2010). Insightful analysis of these localized constraints is required for assessment of true potential of increasing crop yields to reduce the contribution of developing countries to global N_2O emissions and eventually mitigate global N_2O emissions. The best yield that can be obtained locally depend on the capacity of farmers to access and use seeds, water, nutrients, pest management, soil biodiversity and knowledge. For example, it has been estimated that in Southeast Asia where irrigation is available, the average maximum climate-adjusted rice yields are 8.5 Mg ha^{-1}, yet the average actually achieved are 60% of this figure (Cassman 1999). Similar yield gaps also exist in rainfed wheat in Central Asia and rainfed cereals in Argentina and Brazil. Over the past 50 years, per capita food production in Asia has increased by approximately twofold and in Latin America

1.6-fold. In Africa, however, per capita production fell back from the mid-1970s and has only just reached the same levels as in 1961 (Evenson and Gollin 2003). Substantially more food as well as incomes to purchase food could be produced if methods are found to close the yield gap (Godfray et al. 2010).

It is estimated that improved crop NUE could decrease soil derived N_2O emissions from agriculture by as much as 35% globally with even more saving in input-intensive systems of North America, Europe, and Asia (Kroeze and Mosier 2000). Such savings could be achieved by better matching crop N needs with soil N availability. Other approaches for improving NUE of high productivity of agricultural systems can be categorized into five options, namely: (a) adjusting crop rotation to add complexity that improves the plant community's ability to take more available N, (b) providing farmers with decision support tools to allow them to better predict crop N requirements and avoid over fertilization and to better schedule irrigation to avoid driving N below the root zone, (c) using better formulation, timing and method of placement of fertilizer N management in cropping systems to ensure N is available where and when plant demand for N is greatest, (d) managing watersheds to mitigate or redirect N losses downstream from fields, and (e) managing sustainably the intensive animal production systems.

11.2.2 Cover Crop and Crop Rotation for Nitrogen Conservation

Inclusion of cover crop in a rotation, especially in temperate regions with adequate moisture (e.g., winter annual such as ryegrass), can be planted in the fall following harvest, or over-seeded into principal crop during the growing season. Fall growth captures some of the residual N that remains from the summer crop (Rasse et al. 2000; Strock et al. 2004) preventing the build-up of residual soil mineral N, and can decrease nutrient losses significantly during periods when primary crop is not growing. After winter dormancy, the cover crop is ready to grow rapidly with the onset of springtime temperatures that also stimulate N mineralization from crop residue and soil organic matter (SOM). Prior to planting the next principal summer crop, the cover crop is killed, and its subsequent decomposition releases a portion of the N captured in plant biomass since the preceding fall and make it available to the principal crop. Cover crops ensure that some plant N uptake occurs during the time of the year when decomposition and N mineralization are active, and help to conserve N within the cropping system, thereby improving NUE. Leguminous cover crops can be even more advantageous due to N accrued from N_2 fixation. Because N_2 fixation is low when adequate soil N is available, winter legumes provide the same degree of soil inorganic N scavenging as non-leguminous plants. However, they have an advantage of producing biomass with low C:N ratio, which favors rapid spring-killed biomass decomposition and make more N available earlier for the growth of the principal crop in early summer (Corak et al. 1991; Crandall et al. 2005).

The extra benefits of cover cropping includes: (a) building of SOM, and (b) drying soils earlier in spring in regions with wet springs, and can expedite spring planting. The disadvantages of cover cropping include: (i) added extra expenses of planting and killing the cover crops, and (ii) depletion of soil moisture in the absence of adequate springtime precipitation. Cover cropping is not suitable in drier regions where no-till (NT) management may be more important management intervention to minimize off-season N loss because it protects soil against losses through erosion.

Intercropping, where two crops are grown simultaneously, is another alternative of increasing NUE through rotational complexity. However, the main disadvantage of intercropping includes challenges associated with mechanized harvest of crops at different times without damaging the remaining crop. Nevertheless, intercropping can be advantageous in specialized cases such as (i) tropical agroforestry systems where trees can pull N from deeper horizons where it can be recycled when leaf litter is dropped on the surface soils (Palm et al. 2001; Mubarak et al. 2008), and (ii) strip cropping, where strip crops are planted down-slope to capture N leached from upslope position that would otherwise be lost downstream (Beaudoin et al. 2005; Laurent and Ruelland 2011).

In a multiyear crop rotation, inclusion of crops with lower N needs and more efficient at scavenging N from soil increases the overall rotation NUE. For example, winter wheat requires less N and captures more of N applied than corn, because fall-planted wheat can scavenge residual N that remains in the soil from previous crop, and when fertilizer N is applied the following spring, the wheat is well-established and actively growing leaving soil N less available for losses. Thus, corn-wheat rotation can have higher NUE than continuous corn (Robertson and Vitousek 2009). Additionally, development of crop varieties with higher efficiency of N uptake can capture more of the added N to annual cropping system and minimize loss to the environment. For example, hybrid corn containing Bt rootworm trait have higher NUE than standard corn varieties (Bruulsema 2007; Below 2007). Expanding the understanding of the genes controlling NUE and their component traits will help in developing crop varieties with enhanced N uptake efficiencies and contribute in increasing NUE and minimize the environmental impacts of excessive use of N in crop production (Moose and Below 2008).

11.2.3 Crop Nitrogen and Water Management

Soil N test prior to fertilization often improve yield goal of fertilizer N recommendation substantially, particularly when legumes or other non-fertilizer N inputs are in the rotation. For example, in the upper Midwest region of USA it has been demonstrated that pre-side dressing soil nitrate test (Magdoff et al. 1984) could reduce recommended fertilizer N rates by 92–102 kg N ha^{-1} $year^{-1}$ for systems where manure had been applied or legumes grown within the previous 3 years without any loss of the expected yield (Andraski and Bundy 2002). A reliable soil test to predict soil N mineralization is a dependable means of determining goals for yield-based fertilizer N recommendations (Sawyer et al. 2006) and thus decreases both direct and indirect N_2O fluxes. In absence of specific information about soil N availability, many crop growers tend to over-fertilize crops. In general, N_2O emission increases with increase in N input (Gregorich et al. 2005). Both fertilized and unfertilized soils emit N_2O. The N fertilizers are the main source of N_2O in fertilized soils, while mineralization of soil organic N (SON) contributes to N_2O in unfertilized soils. Fluxes of N_2O remain lower until the N input exceeds crop's need, then fluxes almost double (McSwiney and Robertson 2005), suggesting that prudent management of N inputs can be an effective strategy to minimize N_2O emitted from croplands.

In addition, water management intervention helps in retaining N within the rooting zone without leaching NO_3^- below the rooting zone. For example, applying sufficient irrigation water to satisfy crop needs is important for NUE. Irrigation at rates 25% higher than optimum water requirement exacerbates the amount of NO_3^- leached to groundwater by more than an order of magnitude (Gehl et al. 2005). Because most irrigated field crops are of high value, they tend to be over-fertilized, also making proper irrigation especially important for N retention under irrigated conditions. Drying conditions affect nitrification and favor low N_2O production, but when aerobic periods are followed by irrigation or flooding, large N_2O fluxes are observed. The N_2O emission often increases with increasing aeration (i.e., decreasing water-filled pore spaces during drainage of anaerobic flooded rice soils). Substantial N_2O emissions can also occur during freeze-thaw events (Gregorich et al. 2005).

11.2.4 Nitrogen Fertilizer Source, Timing, and Placement

Synthetic fertilizer is sold in variety of forms and formulations worldwide, but the most common form is granulated urea, which account for over 50% of world fertilizer consumption, applied to soil. If incorporated into non-alkaline soil, NH_3 hydrolyzes to NH_4^+, which is soluble and therefore retained in the soil solution. In arable lands, NH_4^+ is nitrified into NO_3^-, which is also soluble but highly mobile, and subject to loss prior to plant uptake. Other forms of N fertilizers such as anhydrous NH_3, ammonium sulfate, ammonium nitrate, solubilized urea, and ammonium nitrate are also nitrified to NO_3^- under soil conditions favorable for microbial activity. Strategies that delay nitrification and keep N in forms that are less subject to environmental loss are: (i) fertilizer formulations that delay dissolution of N in soil, and (ii) soil amendments that inhibit nitrifiers biochemically. Control-release forms of N delays enrichment of soil mineral N pool until crop growth and N demand is high (Shoji et al. 2001). The N applied as NH_4^+ or mineralized from SOM must be nitrified to NO_3^- and before it is available for denitrification. Nitrification inhibitors delay transformation of NH_4^+ to NO_3^- and help matching the timing of NO_3^- supply with peak N demand. Agronomic evidence that any nitrification inhibitor increases NUE is lacking. Combination of delivery of inhibitor to the microsites where nitrification occurs, persistent of inhibitive compound in soil and narrow range of environmental conditions (soil moisture, pH and texture) over which inhibitor effectively contributes to lack of nitrification inhibitors and to NUE. Fertilizing with NH_4^+ fertilizers like urea increases the potential for ammonia emissions (Harrison and Webb 2001), but under anaerobic flooded soils it could minimize gaseous N emissions via denitrification.

Spring plowing and fertilizing as close as possible to the period of active crop growth protects N from competing sinks and minimize N_2O emissions. Fall fertilization and fall plowing create large pool of mineral N available for winter losses through leaching and denitrification. Banding N can provide significant increase in NUE compared to broadcast application. Injecting anhydrous NH_3 into the soil near rows decreases N leaching and volatilization by as much as 35% and increases crop yields (Achon and Broder 1984).

Mismatched timing of N availability with crop need is the greatest contributor to excess N loss in cropping systems. Timing fertilization and the decomposition of crop residue can lead to improved synchrony between soil nutrient availability and crop demand. The longer N remains in soil solution before crop uptake the more likely it is to be lost as N_2O and other fates. Ideally, N should be applied in multiple small doses when plant demand is the highest. However, in most cases the timing of N application and quantity is dictated by weather, availability of equipment and labor. Commonly, for field crops such as corn, two applications, with starter rate applied at planting and side dress rate (i.e. the remaining N) several weeks later, once the crop has germinated and entered a rapid growth phase. Where the fertilizers are not side-dressed, a single application is made in the spring or even in previous fall where producers need to minimize the number of field operations in the spring. N applied in fall has 8–9 months for fertilizer N to be lost to the environment prior to crop uptake. Fall and winter-applied manure is subject to the similar fate as fall fertilizer N application. If N application occurs long before crop growth such as when crop residue starts to decompose months before the next crop, cover crops can be used to capture some of N that would otherwise be available for N_2O emissions. When killed, the cover crops will decompose and its N will be released for potential crop uptake. Any practice that tighten the coupling between soil N release and crop growth will lead to enhanced NUE and reduce need for fertilizer or manure N thereby decreasing N_2O flux.

N availability is highly variable in natural communities, with patches of more decomposable SOM leading to variation in N mineralization that can influence the plants (Robertson et al. 1997; Zhang et al. 2011), and cultivation tend to enlarge pre-existing patches rather than removing them (Hao et al. 2010). Precision agriculture technology enables the application of fertilizer to a field at variable rates selected to coincide with crop need as they vary in space and substantially increase NUE. Any practice that capture N within the system before its potential loss can conserve available N for use by the crop.

11.2.5 Strategies for Mitigating Excess Nitrogen at a Watershed Level

Management strategies can seek to retain N within agricultural landscapes and redirect losses that occur into less harmful pathways. Such strategies include: (i) plantation and management of riparian zone and other downstream vegetation to keep NO_3^- leached from croplands from entering local waterways. Perennial vegetation at stream-sides immobilizes N into growing biomass and SOM, they also trap soil particles that would otherwise erode organic N into surface waters, (ii) restoration stream channels and wetlands in agricultural watersheds to promote denitrification and other microbial processes that convert NO_3^- to inert or less mobile forms of N (Mitsch et al. 2001), and (iii) identification of targets or localized areas of high N export and planting them with forage or other perennial crops such as cellulosic biofuels or restoring native species in these areas to reduce landscape level N outputs (Robertson et al. 2011).

11.3 Management of Nitrogen Under Animal Agriculture

The N_2O emissions from animal waste can be significant. Confined animals excrete 80–95% of N in their diet, and some of this is emitted as N_2O during collection, storage, treatment, and spreading on the field, or after urine and dung deposition in grazed pastures. About 45% of agricultural N_2O emissions in Canada originates from collection, storage, and application of animal manure (Desjardins 2000), creating significant potential for mitigating N_2O emissions through better manure management practices. Manure management reduces N_2O emissions by changing livestock buildings, manure storage facilities, manure treatment, and grazing management. Adoption of slurry based instead of straw based or deep litter system reduces N_2O emission (Chadwick et al. 2011). As in synthetic fertilizers, generally, N_2O emissions increases with N content of the animal waste. In addition, the extent to which the animal waste is allowed to become aerobic allowing the initiation of nitrification-denitrification processes, and the length of storage increases N_2O emissions (Mosier et al. 1998). For N_2O to be produced, NH_4^+ from animal waste must be nitrified to NO_3^-. Nitrifying bacteria require O_2, thus, storing animal waste in liquid form or in lagoons can decrease N_2O emissions dramatically. Slurry remains predominantly anaerobic with little opportunity for nitrification and N_2O emissions. There is a possibility of increased N_2O emissions during application, however. Management options for decreasing N_2O emissions include: (i) waste storage, (ii) waste disposal, and (iii) pasture management. When animal waste is used as manure in the crop fields, same considerations as discussed in the preceding section for minimizing direct N_2O emission from soil apply. Manure applied to fallow fields in fall and winter is likely to enhance both direct and indirect N_2O

emissions. Applying manure to cover crops or just after planting will minimize N_2O emissions most effectively.

Most of N_2O emissions from grazed land occur from high-concentration of urine and dung patches. The patchiness of these excretions is difficult to control. Therefore, the main management option to minimize N_2O emissions from grazed pastures is to ensure that the grazed pastures are not over-fertilized. Emissions of N_2O are promoted when animals are grazed in previously fertilized pastures, or concentrated in small land area, since more N is added than can be absorbed by plants. Maximizing the likelihood for plant uptake of added N minimizes N_2O emissions.

Improving animal traits such as growth rate, animal milk production, fertility, efficiency of feed conversion through breeding and precision animal management may reduce net emissions because fewer animals and less feed are needed for the same amount of products (Thomassen et al. 2009; Gill et al. 2010). Improving animal productivity, therefore, reduces net GHG emissions (Steinfeld and Gerber 2010). In developed countries where animal productivity is already high, genetic selection for growth rate or milk production per animal might negatively affect animal health or fertility (Oltenacu and Broom 2010; de Vries et al. 2011) or social acceptance of animal products. In the developing countries, however, increasing animal productivity to reduce animal numbers is limited by technological and social constrains, and may interfere with multiple objectives of smallholders to maintain livestock. In addition to food, livestock in developing countries provide manure, draught power to support crop cultivation and financial security (Udo et al. 2011). Therefore, successful intensification of livestock systems in the developing countries must acknowledge the possible multifunctionality of livestock in these systems.

Optimization of N in the animal diet by precisely meeting the N requirement of animals during various stages of animal productive life reduces N excreted per unit of product produced, and therefore, reduces N_2O and NH_3 emissions. Feeding grass silage with low N content from extensively managed grassland can increase enteric CH_4 emissions, however. In contrast, reducing N content of manure increases the dependence of inorganic N fertilizers for crop production, and is related to increased NH_3 and N_2O emissions from inorganic N fertilizers. Decreasing grazing time decreases the amount of N excreted as urine and dung in grazed pastures and, therefore, N_2O emissions and nitrate leaching from grazed pastures (de Klein and Eckard 2008). Production of animal feed has a major contribution to GHG emissions, and choosing more environmental friendly livestock products in diet can mitigate environmental impacts associated with livestock production (de Vries and de Boer 2010). Emissions of N_2O from production of animal feeds ingredients can be reduced by selection of crops with higher yield or lower N demand per unit output, and also by improving efficiency of nutrient uptake by crops from applied N (i.e., NUE, De Boer et al. 2011).

In the developed countries with high share of animal protein in human diet, reducing meat products may occur without negative effects on human health (McMichael et al. 2007) and may even have positive effect (Srinivasan et al.

2006) by increasing vegetable consumption. However, the global decrease in meat products is undesirable in many developing countries since it might have negative impact on human health.

Under grazing systems, voided N is returned to soil to be recycled into new forage, and N cycling is relatively efficient under grazed pasture-based animal production system. Proper management, including avoiding overstocking and animal voids intrusion into streams minimizes N loss to the environment (Cherry et al. 2008). However, under intensive animal production systems, it is difficult to retain N within the animal-plant system because of spatial disconnect between animals in concentrated animal feeding operations (CAFOs) and the locations in which the animal feeds are grown. Strategies for improving N utilization under CAFOs includes: (i) diet adjustments to reduce the amount of N excreted per unit of production, and (ii) animal void storage and handling to avoid the loss of voided N to the environment.

Feeding digestible protein and amino acids in concentrations that meet nutritional needs without excess substantially reduce N excretion rate. For example, diets with 2–6% less protein supplemented with amino acids can decrease voided N by as much as 30–40% in poultry (Ferguson et al. 1998) and 20–50% in swine (Sutton et al. 1999). In dairy cattle, feeding lactating cows individualized rations based on of milk production decrease N excretion by 34% (Klausner et al. 1998), and adjusting metabolizable protein components to optimize microbial fermentation in rumen of dairy and beef cattle improves N efficiency significantly.

Waste storage and disposal is an important approach for containing the N fed to confined animals. Management strategies for reducing the environmental leakage of manure N include: (i) proper storage manure, especially under anaerobic conditions, which inhibits NO_3^- formation and reducing denitrification losses of N_2O, (ii) applying manure to fields only during or immediately prior to periods of active plant growth to increase the chance of N immobilization by the crop uptake rather than leaching to ground water as NO_3^- or lost to the atmosphere as N_2O, (iii) adjusting the amount of manure applied to fields to the amount that meet crop N need to prevent overfertilization, and (iv) applying liquid manure below the soil surface rather than surface application to prevent NH_3 loss (Cherry et al. 2008; Johnes 2007).

11.4 Barriers to Achieving Nitrogen Efficiency

Motivation of the implementation of technologies for lowering N loading to the ecosystems requires incentives to which the producers can respond. Three potential incentives include:

1. Creation of carbon equivalent credit market for N_2O mitigation: Providing credits for N_2O emissions could benefit N conservation in general because soil N availability controls the emission of N_2O and other reactive N forms. N_2O

emissions are disproportionately higher at N fertilization levels that exceed economical optimum N rates (McSwiney and Robertson 2005), and significant N_2O reduction could be achieved with little yield penalty (Matson et al. 1998).

2. Payments for ecosystem services valued by the society that are provided by farmers: Services such as clean water provided by better N stewardship or biocontrol provided by more diverse rotations could be promoted by direct payments or tax credits. Such payments or credits are converse of crop subsidy payments based on production, which currently promote the over-application of N fertilizers. This model is currently being used by European Union (Swinton et al. 2007).
3. Protection of farmers from perceived risks of under-fertilization: Extra N fertilizer is normally used as a protection against potential high yields in years with exceptionally favorable growing conditions. Since those conditions cannot be predicted, it becomes rational to over-fertilize, hoping that may be the extra N will pay off. Paying farmers for the yield difference between the over-fertilized and optimum N rate can keep N rate lower without any financial risks to growers. Economically optimum N rates are stable across multiple years (Vanotti and Bundy 1994), and consistent under-fertilization will be rare under this approach.

Alternative approach is a direct regulation of excess N by specifying allowable levels of N additions, surplus N balances i.e., nutrient addition relative to removals in crops or losses to aquatic systems and/or to the atmosphere. European Union has followed this approach since 1998 (i.e., nitrate protective zones), which includes stricter requirements and sanctions for noncompliance (Johnes 2007). The Netherland follows a scheme that limits applications of N (Schröder and Neeteson 2008). Similar regulations of emissions of N-containing trace gases in Europe are currently under development (Erisman et al. 2008). Overall, fundamental changes to agricultural enterprise in order to mitigate environmental damage from fertilizer N are required. Whether incentive, regulation or some combination of both is sought, it is a substantial challenge to motivate farmers to make fundamental changes to agricultural enterprise. Another challenge is that in many regions that have experienced heavy N loading, it may take decades before actions that reduce N losses from agricultural systems lead to improvements in downstream water quality (Silgram et al. 2005).

11.5 Policy Issues on Agricultural GHG Mitigation

There are significant opportunities for GHG mitigation in agriculture. However, for the potential to be realized, barriers to implementation (i.e., policy, institutional, social, educational, economic) must be overcome. The fact remains that the actual levels of GHG mitigation are far below the technical potential of the mitigation options (Smith et al. 2008). For example, the mix of agricultural GHG mitigation options that are adopted in the future may depend upon the price of CO_2

equivalents. Smith et al. (2008) estimated total agricultural GHG mitigation potential of approximately 5,500–6,000 Mg ha^{-1} CO_2 eq. $year^{-1}$, but cautioned that this potential may never be realized due to lack of appropriate policies, education, and financial incentives. It is estimated that at CO_2 equivalent prices of US $ 0–20, 0–50 and 0–100 Mg^{-1} CO_2 eq. the potentials of 1,900–2,100, 2,400–2,600, and 3,100–3,300 Mg CO_2 eq. $year^{-1}$, respectively could be achieved (Smith et al. 2007). About 70% of this potential arises from developing countries with further 10% from the countries with economies in transition (Trines et al. 2006; USEPA 2011). However, despite such significant economic potential, several barriers could prevent the implementation of these measures. Many of these barriers are particularly prevalent in developing countries and include those related to risks related to economic, political, logistical, and educational, as well as societal issues.

Various policies and approaches at national level to limit GHG emissions have been developed. These include regulations and standards, taxes and charges, tradable permits, voluntary agreements, informational instruments, subsidies and incentives, research and development and trade and development assistance (Gupta et al. 2007). These policies may be implemented through bi-lateral arrangements at national or subnational level. They can also be legally binding or voluntary. Policies can be evaluated on the basis of: (i) environmental effectiveness, (i.e., the extent to which a policy realizes positive environmental outcomes), (ii) cost effectiveness (i.e., the extent to which the policy can achieve its objectives at minimum cost to the society), (iii) distributional considerations – such as fairness and equity, and (iv) the extent to which a policy is acceptable, adaptable and implemented.

Regulatory standards specify degree of precision actions that a firm or individual must undertake to achieve environmental objectives, and can include some specific technologies or products to use or not to use. The advantage of regulatory standard is that it may be tailored to an industry or firm. In addition, the regulatory requirement and the environmental outcome are directly connected, which provide some degree of certainty about the outcome. The disadvantage of regulatory standards is that they do not provide the incentives to search for better approaches to reduce emissions, and therefore, may not perform well in inducing innovation and technological change (Sterner 2003).

Emission tax involves paying fee or charge per unit mass of GHGs released into the atmosphere, and because different GHGs have different effects on global warming, the use of CO_2 equivalents is one way of measuring relative impact of a GHG. The firm weighs the cost of emission control against the cost of emitting and paying tax, and the firm implements emission reductions that are cheaper than paying tax. The disadvantage of emission tax is lack of agreed emissions commitment.

Voluntary agreements between government authority and the private parties to implement GHG mitigation options can be used when there is strong political opposition (Thalman and Baranzini 2005). The negotiations involved to develop voluntary agreements play a significant role in raising awareness of climate change issues and potential mitigation actions. However, evaluating the effectiveness of voluntary agreements on GHG mitigation is difficult. Direct and indirect subsidies may also play significant role in mitigating GHG emissions. The advantage of subsidies is that they are politically easier to implement. Economic barriers include the cost of land, competition for land, extreme poverty, lack of existing capacity, low prices of CO_2 equivalents, population growth, transaction costs, and monitoring costs. Risk related barriers include the delay on returns from investment; issues of stability and those related to leakage and natural variation in C sink strength. Political barriers include unclear policies on land use planning and lack of clarity in GHG accounting rules and overall lack of political will. Logistical barriers include scattered nature of land holdings and conflicts of interest among landowners, accessibility to large areas and biological suitability to of the land areas for GHG farming. The educational and societal barriers include newer legislations governing the sector, stakeholders' perception, and persistence of traditional practices.

Maximizing the productivity of existing agricultural land and applying best management practices would help to reduce GHG emissions (Smith et al. 2008). Agricultural mitigation measures need to be considered within a broader framework of sustainable development. Policies to encourage sustainable development may make agricultural mitigation actions in developing countries more achievable. The barriers to implementation of mitigation actions in developing countries need to be overcome, if we are to realize even a proportion of 70% of global agricultural climate mitigation potential that is available in these countries.

11.6 Research Needs

Although GHG emissions from soils have been researched for several decades, there are still geographical regions and agricultural systems that have been neither well studied nor adequately assessed. There is an urgent need to estimate GHG emissions from a wide range of agricultural systems and geographic regions to reduce the uncertainty of the global estimates of N_2O emissions. Farmers participation are indispensable for the mitigation technology transfer of any kind, including management changes aimed at sustainable production systems. It is essential to start dialogue with farmers and other stakeholders about GHG concerns and agricultural practices that would help in mitigating the menace through various routes such as those listed below:

- Improving the farmers understanding of perception and decision making to classify different target groups for specific mitigation strategies,
- Conducting research on farmers' fields or community areas as a reality check for proposed improvements,
- Developing the management strategies in close collaboration with farmers preferably derived from indigenous knowledge on sustainable management practices,

- Focusing on farms rather than individual production system and evaluating the economic benefit to the farmers (i.e., affordability compared to profitability),
- Packaging scientific knowledge in practical and user friendly forms through easy decision support tools,
- Establishing continuous feed-back on mitigation strategies over longer time spans,
- Educating farmers and rural communities by knowledge initiatives.

11.7 Recommended Practices for Reducing N_2O Emission

Important among recommended practices of curtailing N_2O emissions include the following:

- Improving fertilizer N-use efficiency with better timing, placement and prediction of required N,
- Synchronizing N fertilizer application with plant demand throughout the growing season,
- Minimizing water movement across the soil surface and below the crop root zone,
- Accounting for the weather prediction before N application to help minimize N transport or loss,
- Scheduling irrigation for more effective nutrient utilization,
- Using slow-release inorganic and organic N fertilizers,
- Using chemical additives to inhibit N mineralization,
- Using cover-crops and deep-rooted crops for scavenging residual N,
- Testing soil and fully accounting for N in manure, residue, and residual soil N,
- Selecting plant genetics to improve plant N use efficiency,
- Managing feedlots to minimize nitrification and leaching of nitrate,
- Applying organic amendments to soil at rates and times based on agronomic principles.

References

Achon, F.P., Broder, M.F. (1984) Mechanics of applying nitrogen fertilizer. In: Hauck, R.D. (ed.) Nitrogen in Crop Production. pp. 493–506. Soil Science Society of America, Madison, WI

Andraski, T.W., Bundy, L.G. (2002) Using the presidedress soil nitrate test and organic nitrogen crediting to improve corn nitrogen recommendations. Agron. J. **94**(6), 1411–1418

Beaudoin, N., Saad, J.K., Van Laethem, C., Machet, J.M., Maucorps, J., Mary, B. (2005) Nitrate leaching in intensive agriculture in Northern France: Effect of farming practices, soils and crop rotations. Agr. Ecosyst. Environ. **111**(1–4), 292–310. doi:10.1016/j.agee.2005.06.006

Below, F.E. (2007) Triple-stacks, genetics and biotechnology in improving nitrogen use of corn. Proc. North Central Extension- Industry Soil Fertility Conference In, vol. 23:5–13. International Plant Nutrition Institute, Norcross, GA

Bruulsema, T.W. (2007) Ecologically Intensify Your corn Nutrition. In: Institute, I.P.N. (ed.). International Plant Nutrition Institute, Norcross, GA

Cassman, K.G. (1999) Ecological intensification of cereal production systems: Yield potential, soil quality, and precision agriculture. Proc. Natl. Acad. Sci. USA **96**(11), 5952–5959. doi:10.1073/pnas.96.11.5952

Cassman, K.G., Dobermann, A., Walters, D.T., Yang, H. (2003) Meeting cereal demand while protecting natural resources and improving environmental quality. Annu. Rev. Environ. Resour. **28**, 315–358. doi:10.1146/annurev.energy.28.040202.122858

CAST (2004) Climate change and greenhouse gas mitigation: Challenge and opportunities for agriculture. Council for Agricultural Science and Technology (CAST), Ames, Iowa.

Cederberg, C., Wivstad, M., Bergkvist, P., Mattsson, B., Ivarsson, K. (2005) Environmental assessment of plant protection strategies using scenarios for pia feed production. Ambio **34**(4–5), 408–413

Chadwick, D., Sommer, S.G., Thorman, R., Fangueiro, D., Cardenas, L., Amon, B., Misselbrook, T. (2011) Manure management: Implications for greenhouse gas emissions. Anim. Feed Sci. Technol. **166–67**, 514–531. doi:10.1016/j.anifeedsci.2011.04.036

Cherry, K.A., Shepherd, M., Withers, P.J.A., Mooney, S.J. (2008) Assessing the effectiveness of actions to mitigate nutrient loss from agriculture: A review of methods. Sci. Total Environ. **406**(1–2), 1–23. doi:10.1016/j.scitotenv.2008.07.015

Corak, S.J., Frye, W.W., Smith, M.S. (1991) Legume mulch and nitrogen-fertilizer effects on soil-water and corn production. Soil Sci. Soc. Am. J. **55**(5), 1395–1400

Crandall, S.M., Ruffo, M.L., Bollero, G.A. (2005) Cropping system and nitrogen dynamics under a cereal winter cover crop preceding corn. Plant Soil **268**(1–2), 209–219. doi:10.1007/s11104-004-0272-x

Crutzen, P.J., Mosier, A.R., Smith, K.A., Winiwarter, W. (2008) N_2O release from agro-biofuel production negates global warming reduction by replacing fossil fuels. Atmos. Chem. Phys. **8**(2), 389–395

De Boer, I.J.M., Cederberg, C., Eady, S., Gollnow, S., Kristensen, T., Macleod, M., Meul, M., Nemecek, T., Phong, L.T., Thoma, G., van der Werf, H.M.G., Williams, A.G., Zonderland-Thomassen, M.A. (2011) Greenhouse gas mitigation in animal production: Towards an integrated life cycle sustainability assessment. Curr. Opin. Environ. Sust. **3**(5), 423–431. doi:10.1016/j.cosust.2011.08.007

de Klein, C.A.M., Eckard, R.J. (2008) Targeted technologies for nitrous oxide abatement from animal agriculture. Austr. J. Exp. Agric. **48**(1–2), 14–20. doi:10.1071/ea07217

de Vries, M., Bokkers, E.A.M., Dijkstra, T., van Schaik, G., de Boer, I.J.M. (2011) Invited review: Associations between variables of routine herd data and dairy cattle welfare indicators. **94**(7), 3213–3228. doi:10.3168/jds.2011-4169

de Vries, M., de Boer, I.J.M. (2010) Comparing environmental impacts for livestock products: A review of life cycle assessments. **128**(1–3), 1–11. doi:10.1016/j.livsci.2009.11.007

Delgado, J.A., Del Grosso, S.J., Ogle, S.M. (2010) ^{15}N isotopic crop residue cycling studies and modeling suggest that IPCC methodologies to assess residue contributions to N_2O-N emissions should be reevaluated. **86**(3), 383–390. doi:10.1007/s10705-009-9300-9

Denman, K.L., Brasseur, G., Chidthaisong, A., Ciais, P., Cox, P.M., Dickinson, R.E., Hauglustaine, D., Heinze, C., Holland, E., D. Jacob, Lohmann, U., Ramachandran, S., Dias, P.L.d.S., Wofsy, S.C., Zhang, X. (2007) Couplings Between Changes in the Climate System and Biogeochemistry. In: Solomon, S., Qin, D., Manning, M., Chen, Z., Marquis, M., Averyt, K.B., M.Tignor, Miller, H.L. (eds.) Climate Change 2007: The Physical Science Basis. Contribution of Working Group I to the Fourth Assessment Report of the Intergovernmental Panel on Climate Change. pp. 501–587. Cambridge University Press, Cambridge, United Kingdom and New York, NY, USA

Desjardins, R.L. (2000) Agricultural greenhouse gas budget In: McRae, T., Smith, C.A.S., Gregorich, L.J. (eds.) Environmental sustainability of Canadian agriculture: Report of the Agri-environmental indicator project. pp. 133–140. Agri-Food Canada, Ottawa, Ontario

Eckard, R.J., Grainger, C., de Klein, C.A.M. (2010) Options for the abatement of methane and nitrous oxide from ruminant production: A review. **130**(1–3), 47–56. doi:10.1016/j.livsci.2010.02.010

Erisman, J.W., Bleeker, A., Hensen, A., Vermeulen, A. (2008) Agricultural air quality in Europe and the future perspectives. **42**(14), 3209–3217. doi:10.1016/j.atmosenv.2007.04.004

Evenson, R.E., Gollin, D. (2003) Assessing the impact of the Green Revolution, 1960 to 2000. Science **300**(5620), 758–762. doi:10.1126/science.1078710

Ferguson, N.S., Gates, R.S., Taraba, J.L., Cantor, A.H., Pescatore, A.J., Ford, M.J., Burnham, D.J. (1998) The effect of dietary crude protein on growth, ammonia concentration, and litter composition in broilers. Poultry Sci. **77**(10), 1481–1487

Food and Agriculture Organization of the United Nations (FAO) (2011) FAOSTAT Land resources database. Rome, Italy. On line at http://faostat.fao.org. (Accessed December 2011)

Foley, J.A., DeFries, R., Asner, G.P., Barford, C., Bonan, G., Carpenter, S.R., Chapin, F.S., Coe, M.T., Daily, G.C., Gibbs, H.K., Helkowski, J.H., Holloway, T., Howard, E.A., Kucharik, C.J., Monfreda, C., Patz, J.A., Prentice, I.C., Ramankutty, N., Snyder, P.K. (2005) Global consequences of land use. Science **309**(5734), 570–574. doi:10.1126/science.1111772

Follett, R.F., Shafer, S.R., Jawson, M.D., Franzluebbers, A. (2005) Research and implementation needs to mitigate greenhouse gas emissions from agriculture in the USA. Soil Till. Res. **83**(1), 159–166. doi:10.1016/j.still.2005.02.014

Galloway, J.N., Dentener, F.J., Capone, D.G., Boyer, E.W., Howarth, R.W., Seitzinger, S.P., Asner, G.P., Cleveland, C.C., Green, P.A., Holland, E.A., Karl, D.M., Michaels, A.F., Porter, J.H., Townsend, A.R., Vorosmarty, C.J. (2004) Nitrogen cycles: past, present, and future. Biogeochemistry **70**(2), 153–226. doi:10.1007/s10533-004-0370-0

Gehl, R.J., Schmidt, J.P., Stone, L.R., Schlegel, A.J., Clark, G.A. (2005) In situ measurements of nitrate leaching implicate poor nitrogen and irrigation management on sandy soils. J. Environ. Qual. **34**(6), 2243–2254. doi:10.2134/jeq2005.0047

Gill, M., Smith, P., Wilkinson, J.M. (2010) Mitigating climate change: the role of domestic livestock. Animal **4**(3), 323–333. doi:10.1017/s1751731109004662

Godfray, H.C.J., Beddington, J.R., Crute, I.R., Haddad, L., Lawrence, D., Muir, J.F., Pretty, J., Robinson, S., Thomas, S.M., Toulmin, C. (2010) Food Security: The Challenge of Feeding 9 Billion People. Science **327**(5967), 812–818. doi:10.1126/science.1185383

Gregorich, E.G., Rochette, P., VandenBygaart, A.J., Angers, D.A. (2005) Greenhouse gas contributions of agricultural soils and potential mitigation practices in Eastern Canada. Soil Till. Res. **83**(1), 53–72. doi:10.1016/j.still.2005.02.009

Gupta, S., Tirpak, D.A., Burger, N., Gupta, J., Höhne, N., Boncheva, A.I., Kanoan, G.M., Kolstad, C., Kruger, J.A., Michaelowa, A., Murase, S., Pershing, J., Saijo, T., Sari, A. (2007) Policies, Instruments and Co-operative Arrangements. In: Metz, B., Davidson, O.R., Bosch, P.R., Dave, R., Meyer, L.A. (eds.) Climate Change 2007: Mitigation. Contribution of Working Group III to the Fourth Assessment Report of the Intergovernmental Panel on Climate Change. pp. 745–807. Cambridge University Press, Cambridge, United Kingdom and New York, NY, USA

Hao, X.M., Thelen, K., Gao, J. (2010) Effects of Soil and Topographic Properties on Spatial Variability of Corn Grain Ethanol Yield. Agron. J. **102**(3), 998–1006. doi:10.2134/agronj2009.0481

Harrison, R., Webb, J. (2001) A review of the effect of N fertilizer type on gaseous emissions. Adv. Agron. **73**, 65–108. doi:10.1016/s0065-2113(01)73005-2

Hobbs, P.R., Sayre, K., Gupta, R. (2008) The role of conservation agriculture in sustainable agriculture. Philos. T. R. Soc. London Ser. B **363**(1491), 543–555. doi:10.1098/rstb.2007.2169

Johnes, P.J. (2007) Meeting ecological restoration targets in European waters: A challenge for animal agriculture. In: Swein, D.L., Charmley, E., Steel, J., Coffey, S. (eds.) Redesigning animal agriculture: The challenge for the 21st century. pp. 185–203. CAB International, Willingford, UK

Klausner, S.D., Fox, D.G., Rasmussen, C.N., Pitt, R.E., Tylutki, T.P., Wright, P.E., Chase, L.E., Stone, W.C. (1998) Improving dairy farm sustainability I: An approach to animal and crop nutrient management planning. J. Prod. Agric. **11**(2), 225–233

Kroeze, C., Mosier, A.R. (2000) New estimates for emissions of nitrous oxide. In: van Ham, J.E. A., Baede, A.P.M., Meyer, L.A., Ybema, R. (eds.) Non-CO_2 Greenhouse Gases: Scientific Understanding Control and Implementation. pp. 45–64. Kluwer Academic Publishers, Dordrecht, The Netherlands

Laurent, F., Ruelland, D. (2011) Assessing impacts of alternative land use and agricultural practices on nitrate pollution at the catchment scale. J. Hydrol. **409**(1–2), 440–450. doi:10.1016/j.jhydrol.2011.08.041

Magdoff, F.R., Ross, D., Amadon, J. (1984) A soil test for nitrogen availability to corn. Soil Sci. Soc. Am. J. **48**(6), 1301–1304

Matson, P.A., Naylor, R., Ortiz-Monasterio, I. (1998) Integration of environmental, agronomic, and economic aspects of fertilizer management. Science **280**(5360), 112–115. doi:10.1126/science.280.5360.112

McMichael, A.J., Powles, J.W., Butler, C.D., Uauy, R. (2007) Energy and health 5 - Food, livestock production, energy, climate change, and health. Lancet **370**(9594), 1253–1263. doi:10.1016/s0140-6736(07)61256-2

McSwiney, C.P., Robertson, G.P. (2005) Nonlinear response of N_2O flux to incremental fertilizer addition in a continuous maize (*Zea mays* L.) cropping system. Glob. Change Biol. **11**(10), 1712–1719. doi:10.1111/j.1365-2486.2005.01040.x

Millar, N., Robertson, G.P., Grace, P.R., Gehl, R.J., Hoben, J.P. (2010) Nitrogen fertilizer management for nitrous oxide (N_2O) mitigation in intensive corn (Maize) production: an emissions reduction protocol for US Midwest agriculture. Mitig. Adapt. Strateg. Glob. Change **15**(2), 185–204. doi:10.1007/s11027-010-9212-7

Mitsch, W.J., Day, J.W., Gilliam, J.W., Groffman, P.M., Hey, D.L., Randall, G.W., Wang, N.M. (2001) Reducing nitrogen loading to the Gulf of Mexico from the Mississippi River Basin: Strategies to counter a persistent ecological problem. Bioscience **51**(5), 373–388

Moose, S., Below, F.E. (2008) Biotechnology approaches to improving maize nitrogen use efficiency. In: Kritz, A.L., Larkins, B.A. (eds.) Molecular genetic approaches to maize improvement. pp. 65–67. Springer, Berlin

Mosier, A.R. (2002) Environmental challenges associated with needed increases in global nitrogen fixation. Nutr. Cycl. Agroecosyst. **63**(2–3), 101–116. doi:10.1023/a:1021101423341

Mosier, A.R., Duxbury, J.M., Freney, J.R., Heinemeyer, O., Minami, K. (1998) Assessing and mitigating N_2O emissions from agricultural soils. Climatic Change **40**(1), 7–38. doi:10.1023/a:1005386614431

Mubarak, A.R., Elbashir, A.A., Elamin, L.A., Daldoum, D.M.A., Steffens, D., Benckiser, G. (2008) Decomposition and nutrient release from litter fall in the semi-arid tropics of Sudan. Comm. Soil Sci. Plant Anal. **39**(15–16), 2359–2377. doi:10.1080/00103620802292574

Nemecek, T., von Richthofen, J.S., Dubois, G., Casta, P., Charles, R., Pahl, H. (2008) Environmental impacts of introducing grain legumes into European crop rotations. Eur. J. Agron. **28**(3), 380–393. doi:10.1016/j.eja.2007.11.004

Oltenacu, P.A., Broom, D.M. (2010) The impact of genetic selection for increased milk yield on the welfare of dairy cows. Anim. Welfare **19**, 39–49

Palm, C.A., Giller, K.E., Mafongoya, P.L., Swift, M.J. (2001) Management of organic matter in the tropics: translating theory into practice. Nutr. Cycl. Agroecosyst. **61**(1–2), 63–75. doi:10.1023/a:1013318210809

Pappa, V.A., Rees, R.M., Walker, R.L., Baddeley, J.A., Watson, C.A. (2011) Nitrous oxide emissions and nitrate leaching in an arable rotation resulting from the presence of an intercrop. Agr. Ecosyst. Environ. **141**(1–2), 153–161. doi:10.1016/j.agee.2011.02.025

Rasse, D.P., Ritchic, J.T., Peterson, W.R., Wei, J., Smucker, A.J.M. (2000) Rye cover crop and nitrogen fertilization effects on nitrate leaching in inbred maize fields. J. Environ. Qual. **29**(1), 298–304

Ravishankara, A.R., Daniel, J.S., Portmann, R.W. (2009) Nitrous Oxide (N_2O): The Dominant Ozone-Depleting Substance Emitted in the 21st Century. Science **326**(5949), 123–125. doi:10.1126/science.1176985

Robertson, G.P., Hamilton, S.K., Del Grosso, S.J., Parton, W.J. (2011) The biogeochemistry of bioenergy landscapes: carbon, nitrogen, and water considerations. Ecol. Appl. **21**(4), 1055–1067. doi:10.1890/09-0456.1

Robertson, G.P., Klingensmith, K.M., Klug, M.J., Paul, E.A., Crum, J.R., Ellis, B.G. (1997) Soil resources, microbial activity, and primary production across an agricultural ecosystem. Ecol. Appl. **7**(1), 158–170. doi:10.2307/2269414

Robertson, G.P., Paul, E.A., Harwood, R.R. (2000) Greenhouse gases in intensive agriculture: Contributions of individual gases to the radiative forcing of the atmosphere. Science **289** (5486), 1922–1925. doi:10.1126/science.289.5486.1922

Robertson, G.P., Vitousek, P.M. (2009) Nitrogen in Agriculture: Balancing the Cost of an Essential Resource. Annu. Rev. Environ. Resour. **34**, 97–125. doi:10.1146/annurev.environ.032108.105046

Rochette, P., Janzen, H.H. (2005) Towards a revised coefficient for estimating N_2O emissions from legumes. Nutr. Cycl. Agroecosyst. **73**(2–3), 171–179. doi:10.1007/s10705-005-0357-9

Sawyer, J., Nafzinger, E.D., Randall, G.W., Bundy, L.G., Rehm, G., Joern, B. (2006) Concepts and rationale for regional nitrogen rate guidelines. pp. Iowa State University Extension, Ames, Iowa

Schröder, J.J., Neeteson, J.J. (2008) Nutrient management regulations in The Netherlands. Geoderma **144**(3–4), 418–425. doi:10.1016/j.geoderma.2007.12.012

Shoji, S., Delgado, J., Mosier, A., Miura, Y. (2001) Use of controlled release fertilizers and nitrification inhibitors to increase nitrogen use efficiency and to conserve air and water quality. Comm. Soil Sci. Plant Anal. **32**(7–8), 1051–1070. doi:10.1081/css-100104103

Silgram, M., Williams, A., Waring, R., Neumann, I., Hughes, A., Mansour, M., Besien, T. (2005) Effectiveness of the nitrate sensitive areas scheme in reducing groundwater concentrations in England. Q. J. Eng. Geol. Hydrogeol. **38**, 117–127. doi:10.1144/1470-9236/04-010

Smith, P., Martino, D., Cai, Z., Gwary, D., Janzen, H., Kumar, P., McCarl, B., Ogle, S., O'Mara, F., Rice, C., Scholes, B., Sirotenko, O., Howden, M., McAllister, T., Pan, G., Romanenkov, V., Schneider, U., Towprayoon, S., Wattenbach, M., Smith, J. (2008) Greenhouse gas mitigation in agriculture. Philos. T. R. Soc. London Ser. B **363**(1492), 789–813. doi:10.1098/rstb.2007.2184

Smith, P., Martino, D., Cai, Z., Gwary, D., Janzen, H., Kumar, P., McCarl, B., Ogle, S., O'Mara, F., Rice, C., Scholes, B., Sirotenko, O., Howden, M., McAllister, T., Pan, G., Romanenkov, V., Schneider, U., Towprayoon, S. (2007) Policy and technological constraints to implementation of greenhouse gas mitigation options in agriculture. Agric. Ecosyst. Environ. **118**(1–4), 6–28. doi:10.1016/j.agee.2006.06.006

Srinivasan, C.S., Irz, X., Shankar, B. (2006) An assessment of the potential consumption impacts of WHO dietary norms in OECD countries. Food Policy **31**(1), 53–77. doi:10.1016/j.foodpol.2005.08.002

Stehfest, E., Bouwman, L. (2006) N_2O and NO emission from agricultural fields and soils under natural vegetation: summarizing available measurement data and modeling of global annual emissions. Nutr. Cycl. Agroecosyst. **74**(3), 207–228. doi:10.1007/s10705-006-9000-7

Steinfeld, H., Gerber, P. (2010) Livestock production and the global environment: Consume less or produce better? Proc. Natl. Acad. Sci. USA **107**(43), 18237–18238. doi:10.1073/pnas.1012541107

Sterner, T. (2003) Policy instruments for Environmental and Natural Resource Management. 504 pp. Resources of the Future Press, Washington, DC

Strock, J.S., Porter, P.M., Russelle, M.P. (2004) Cover cropping to reduce nitrate loss through subsurface drainage in the northern US Corn Belt. J. Environ. Qual. **33**(3), 1010–1016

Sutton, A.L., Kephart, K.B., Verstegen, M.W.A., Canh, T.T., Hobbs, P.J. (1999) Potential for reduction of odorous compounds in swine manure through diet modification. J. Anim. Sci. **77** (2), 430–439

Swinton, S.M., Lupi, F., Robertson, G.P., Hamilton, S.K. (2007) Ecosystem services and agriculture: Cultivating agricultural ecosystems for diverse benefits. Ecol. Econ. **64**(2), 245–252. doi:10.1016/j.ecolecon.2007.09.020

Thalman, P., Baranzini, A. (2005) An overview of the economics of voluntary approaches in climate policies. 304 pp. Edward Elgar Publishing Ltd, Northampton, MA, USA

Thomassen, M.A., Dolman, M.A., van Calker, K.J., de Boer, I.J.M. (2009) Relating life cycle assessment indicators to gross value added for Dutch dairy farms. Ecol. Econ. **68**(8–9), 2278–2284. doi:10.1016/j.ecolecon.2009.02.011

Trines, E., Home, N., Jung, M., Skutsch, M., Petsonk, A., Silva-Chavez, G., Smith, P., Nabuurs, G. J., Verweij, P., Schlamadinger, B. (2006) Integrating agriculture, forestry and other land use in future climate regimes. Methodological issues and policy options. A Report for the Netherlands Research Programme on Climate Change (NRP-CC). Netherlands Programme on Scientific Assessment and Policy Analysis (WAB).

Udo, H.M.J., Aklilu, H.A., Phong, L.T., Bosma, R.H., Budisatria, I.G.S., Patil, B.R., Samdup, T., Bebe, B.O. (2011) Impact of intensification of different types of livestock production in smallholder crop-livestock systems. Livest. Sci. **139**(1–2), 22–29. doi:10.1016/j.livsci.2011.03.020

U.S.Eenvironmental Protection Agency (USEPA) (2011) Global Anthropogenic Non-CO_2 Greenhouse Gas Emissions: 1990–2030. U.S. Environmental Protection Agency.

Vanotti, M.B., Bundy, L.G. (1994) Corn nitrogen recommendations based on yield response data. J. Prod. Agric. **7**(2), 249–256

World Meteorological Organization (WMO) (2011) The state of greenhouse gases in the atmosphere based on the observations through 2010. World Meteorological Organization Greenhouse Gas Bulletin **7** 1–3

Zhang, W.T., Weindorf, D.C., Zhu, Y.D. (2011) Soil Organic Carbon Variability in Croplands: Implications for Sampling Design. Soil Sci. **176**(7), 367–371. doi:10.1097/SS.0b013e31821eb7d2

Index

David Ussiri and R. Lal, *Soil Emission of Nitrous Oxide and its Mitigation*,
DOI 10.1007/978-94-007-5364-8, © Springer Science+Business Media Dordrecht 2013

F

O

P

MIX
Papier aus verantwortungsvollen Quellen
Paper from responsible sources
FSC® C105338

If you have any concerns about our products,
you can contact us on
ProductSafety@springernature.com

In case Publisher is established outside the EU,
the EU authorized representative is:
Springer Nature Customer Service Center GmbH
Europaplatz 3, 69115 Heidelberg, Germany

Printed by Libri Plureos GmbH
in Hamburg, Germany